SCIENCE OF
CERAMIC CHEMICAL PROCESSING

Front row left to right: Dr. Ralph Iler, Keynote Awardee for inorganic chemistry; Prof. Paul Flory, Keynote Awardee for organic and polymer chemistry; Prof. Per-Olov Löwdin, Keynote Awardee for quantum chemistry. Second row left to right are presenters of Keynote Awards: Dr. George Parshall, Director of Chemistry Research, E. I. duPont de Nemours presenter of Dr. Iler's Keynote Award; Mr. Marshall Criser, President of the University of Florida, presenter of Prof. Löwdin's Keynote Award; Dr. Leo Young, Director, Research and Laboratory Management, Office of the Under Secretary of Defense, presenter of Prof. Flory's Keynote Award.

S C I E N C E O F
CERAMIC CHEMICAL PROCESSING

EDITED BY

LARRY L. HENCH
University of Florida
Gainesville, Florida

DONALD R. ULRICH
Air Force Office of Scientific Research
Washington, D.C.

A WILEY-INTERSCIENCE PUBLICATION
JOHN WILEY & SONS
New York   Chichester   Brisbane   Toronto   Singapore

Library of Congress Cataloging in Publication Data

Main entry under title:

Science of ceramic chemical processing.

   "A Wiley-Interscience publication."
   Proceedings of the Second International Conference
on Ultrastructure Processing of Ceramics, Glasses, and
Composites, held February 25–March 1, 1985 in Palm
Coast, Florida . . . sponsored by the Department of
Materials Science and College of Engineering,
University of Florida"—Pref.
   Includes index.
   1. Ceramics—Congresses.   2. Glass—Congresses.
3. Composite materials—Congresses.   4. Colloids—
Congresses.   I. Hench, L. L.   II. Ulrich, Donald R.
III. University of Florida.   Department of Materials
Science and Engineering.   IV. University of Florida.
College of Engineering.   V. International Conference
on Ultrastructure Processing of Ceramics, Glasses, and
Composites (2nd: 1985: Palm Coast, Fla.)

TP785.S35 1986   666   85-22490
ISBN 0-471-82645-6

CONTRIBUTORS

W. WADE ADAMS
AFML
Department of the Air Force
Wright-Patterson AFB, Ohio

ILHAN A. AKSAY
Department of Materials Science
   and Engineering
University of Washington
Seattle, Washington

I. ARTAKI
Department of Chemistry
School of Chemical Sciences
University of Illinois
Urbana, Illinois

JAMES H. AUBERT
Sandia National Laboratories
Albuquerque, New Mexico

P. H. BARBOUX
Spectrochimie due Solide,
Université Paris
Paris, France

E. A. BARRINGER
Ceramics Process System
   Corporation
Lexington, Massachusetts

KRIS A. BERGLUND
Departments of Agricultural and
   Chemical Engineering
Michigan State University
East Lansing, Michigan

ELIZABETH K. BONDERSON
Department of Chemistry
North Dakota State University
Fargo, North Dakota

PHILIP BOUDJOUK
Department of Chemistry
North Dakota State University
Fargo, North Dakota

H. K. BOWEN
Ceramic Processing Research
   Laboratory
Materials Processing Center
Massachusetts Institute of
   Technology
Cambridge, Massachusetts

M. BRADLEY
Department of Chemistry
School of Chemical Sciences
University of Illinois
Urbana, Illinois

C. J. BRINKER
Sandia National Laboratories
Albuquerque, New Mexico

H. A. BUMP
Rockwell International
Thousand Oaks, California

Larry W. Burggraf
Department of Chemistry
Air Force Academy
Colorado Springs, Colorado

LEE A. CARMAN
Department of Materials Science
and Engineering
Pennsylvania State University
University Park, Pennsylvania

YEU-CHYI CHENG
Department of Materials Science
and Engineering
University of Florida
Gainesville, Florida

D. E. CLARK
Department of Materials Science
and Engineering
University of Florida
Gainesville, Florida

M. J. CRIMP
Department of Metallurgy and
Materials Science
Case Western Reserve University
Cleveland, Ohio

LARRY P. DAVIS
Department of Chemistry
Air Force Academy
Colorado Springs, Colorado

I. DJUROVICH
Department of Chemistry
University of Wisconsin
Madison, Wisconsin

R. G. DOSCH
Sandia National Laboratories
Albuquerque, New Mexico

W. F. DOYLE
Department of Metallurgy and
Materials Science
Massachusetts Institute of
Technology
Cambridge, Massachusetts

B. D. FABES
Department of Metallurgy and
Materials Science
Massachusetts Institute of
Technology
Cambridge, Massachusetts

D. L. FEKE
Department of Chemical
Engineering
Case Western Reserve University
Cleveland, Ohio

Paul J. Flory
Department of Chemistry
Stanford University
Stanford, California

SETH FRADEN
Martin Fisher School of Physics
Brandeis University
Waltham, Massachusetts

K. J. FRANKLIN
Atomic Energy of Canada Ltd.
Chalk River Laboratories
Chalk River, Ontario, Canada

K. G. FRASE
National Bureau of Standards
Gaithersburg, Maryland

DAVID R. GAGNON
Department of Polymer Science
and Engineering
University of Massachusetts
Amherst, Massachusetts

MARK S. GORDON
Department of Chemistry
North Dakota State University
Fargo, North Dakota

THIERRY GRANIER
Department of Polymer Science
   and Engineering
University of Massachusetts
Amherst, Massachusetts

J. W. HALLORAN
Department of Metallurgy and
   Materials Science
Case Western Reserve University
Cleveland, Ohio

R. H. HEISTAND II
Dow Chemical USA
New England Laboratory
Wayland, Massachusetts

THADDEUS E. HELMINIAK
Polymer Branch
Materials Laboratory
Air Force Wright Aeronautical
   Labs
Wright Patterson AFB, Ohio

LARRY L. HENCH
Department of Materials Science
   and Engineering
University of Florida
Gainesville, Florida

GEORGE K. HENRY
Loker Hydrocarbon Research
   Institute
Department of Chemistry
University of Southern California
Los Angeles, California

ARLON J. HUNT
Applied Science Division
Lawrence Berkeley Laboratory
University of California
Berkeley, California

ALAN J. HURD
Department of Chemistry
University of New Mexico
Albuquerque, New Mexico

RALPH K. ILER
811 Haines Avenue
Wilmington, Delaware

N. A. IVES
Chemistry and Physics Laboratory
The Aerospace Corporation
El Segundo, California

D. W. JOHNSON, JR.
AT&T Bell Laboratories
Murray Hill, New Jersey

R. E. JOHNSON, JR.
E.I. DuPont de Nemours & Co.
Central Research & Development
   Department
Experimental Station
Wilmington, Delaware

J. P. JOLIVET
Chimie des Polymeres Inorganiques
Université Pierre et Marie Curie
Paris, France

JIRI JONAS
Department of Chemistry
School of Chemical Sciences
University of Illinois
Urbana, Illinois

Alfred Kaiser
Fraunhofer-Institute für
   Silicatforschung
Wurzburg, West Germany

FRANK E. KARASZ
Materials Research Laboratory
University of Massachusetts
Amherst, Massachusetts

K. D. KEEFER
Sandia National Laboratories
Albuquerque, New Mexico

BRUCE KELLETT
Department of Materials Science
   and Engineering
University of California
Los Angeles, California

R. KIKUCHI
Department of Materials Science
    and Engineering
University of Washington
Seattle, Washington

SUNUK KIM
Institute of Glass Science
    and Engineering
New York State College of Ceramic
    Ceramics
Alfred University
Alfred, New York

L. C. KLEIN
Department of Ceramics
Rutgers University
The State University of
    New Jersey
Piscataway, New Jersey

SRIDHAR KOMARNENI
Materials Research Laboratory
Pennsylvania State University
University Park, Pennsylvania

G. KORDAS
Mechanical and Materials
    Engineering
Vanderbilt University
Nashville, Tennessee

MASATO KUMAGAI
Department of Materials Science
    and Engineering
Pennsylvania State University
University Park, Pennsylvania

WILLIAM C. LaCOURSE
Institute of Glass Science
    and Engineering
Alfred University
Alfred, New York

J. J. LAGOWSKI
Department of Chemistry
University of Texas
Austin, Texas

F. F. LANGE
Rockwell International Science
    Center
Thousand Oaks, California

BURT I. LEE
Department of Materials Science
    and Engineering
University of Florida
Gainesville, Florida

AXEL LENTZ
Abteilung für Anorganische
    Chemie
Universität Ulm
Ulm, West Germany

ROBERT W. LENZ
Department of Polymer Science
    and Engineering
University of Massachusetts
Amherst, Massachusetts

M. S. LEUNG
Chemistry and Physics Laboratory
The Aerospace Corporation
El Segundo, California

H. C. LING
AT&T Bell Laboratories
Murray Hill, New Jersey

R. A. Lipeles
The Aerospace Corporation
Los Angeles, California

J. LIVAGE
Spectrochimie due Solide,
Université Paris
Paris, France

PER-OLOV LÖWDIN
Department of Chemistry
University of Florida
Gainesville, Florida

KEVIN D. LOFFTUS
Applied Science Division
Lawrence Berkeley Laboratory
University of California
Berkeley, California

J. B. MACCHESNEY
AT&T Bell Laboratories
Murray Hill, New Jersey

J. D. MACKENZIE
Department of Materials Science
   and Engineering
University of California
Los Angeles, California

J. E. MARK
Department of Chemistry
University of Cincinnati
Cincinnati, Ohio

EGON MATIJEVIĆ
Department of Chemistry
Clarkson University
Postdam, New York

JAMES L. MCARDLE
Department of Materials Science
   and Engineering
Pennsylvania State University
University Park, Pennsylvania

GARY L. MESSING
Department of Materials Science
   and Engineering
Pennsylvania State University
University Park, Pennsylvania

ROBERT B. MEYER
Martin Fisher School of Physics
Brandeis University
Waltham, Massachusetts

W. C. MOFFATT
Ceramic Processing Research
   Laboratory
Materials Processing Center
Massachusetts Institute of
   Technology
Cambridge, Massachusetts

P. E. D. MORGAN
Rockwell International
Thousand Oaks, California

BRIJ M. MOUDGIL
Department of Materials Science
   and Engineering
University of Florida
Gainesville, Florida

B. NOVICH
Ceramic Processing Research
   Laboratory
Materials Processing Center
Massachusetts Institute of
   Technology
Cambridge, Massachusetts

Y. OGURI
Massachusetts Institute of
   Technology
Cambridge, Massachusetts

H. OKAMURA
Nippon Soda Company
New York, New York

GEORGE Y. ONODA
IBM
Thomas J. Watson Research Center
Yorktown Heights, New York

GERARD ORCEL
Department of Materials Science
   and Engineering
University of Florida
Gainesville, Florida

CARLO G. PANTANO
Materials Science and Engineering
Pennsylvania State University
University Park, Pennsylvania

S. C. PARK
Department of Materials Science
   and Engineering
University of Florida
Gainesville, Florida

I. PETER
Department of Chemistry
University of Wisconsin
Madison, Wisconsin

T. M. PETTIJOHN
Department of Chemistry
University of Texas
Austin, Texas

J. PHALIPPOU
CNRS Glass Laboratory and
    Materials Science Laboratory
Montpellier, France

PARAS N. PRASAD
Department of Chemistry
State University of New York
    at Buffalo
Buffalo, New York

M. PRASSAS
Corning Europe Inc.
Centre Europeen de Recherche
Avon, France

E. A. PUGAR
Rockwell International
Thousand Oaks, California

E. M. RABINOVICH
AT&T Bell Laboratories
Murray Hill, New Jersey

PETER B. RAND
Sandia National Laboratories
Albuquerque, New Mexico

J. J. RATTO
Rockwell International
Thousand Oaks, California

W. W. RHODES
AT&T Bell Laboratories
Murray Hill, New Jersey

J. J. RITTER
National Bureau of Standards
Gaithersburg, Maryland

E. P. ROTH
Sandia National Laboratories
Albuquerque, New Mexico

RUSTUM ROY
Materials Research Laboratory
Pennsylvania State University
University Park, Pennsylvania

MICHAEL RUDOLPH
Abteilung für Anorganische Chemie
Universität Ulm
Ulm, West Germany

MICHAEL D. SACKS
Department of Materials Science
    and Engineering
University of Florida
Gainesville, Florida

DALE W. SCHAEFER
Sandia National Laboratories
Albuquerque, New Mexico

G. W. SCHERER
Corning Glass Works
R&D Division
Corning, New York

HELMUT SCHMIDT
Fraunhofer-Institut für
    Silicatforschung
Würzburg, West Germany

DIETMAR SEYFERTH
Department of Chemistry
Massachusetts Institute of
    Technology
Cambridge, Massachusetts

RICHARD A. SHELLEMAN
Department of Materials Science
    and Engineering
Pennsylvania State University
University Park, Pennsylvania

RONG-SHENG SHEU
Department of Materials Science
  and Engineering
University of Florida
Gainesville, Florida

L. SILVERMAN
Department of Materials Science
  and Engineering
Massachusetts Institute of
  Technology
Cambridge, Massachusetts

YOSHIKO SOWA
Materials Research Laboratory
Pennsylvania State University
University Park, Pennsylvania

R. F. STEWART
ICI Corporate Colloid Science
  Group
The Heath Runcorn
Cheshire, England

HAROLD STÜGER
Department of Chemistry
  University of Wisconsin
Madison, Wisconsin

D. SUTTON
ICI PLC
Corporate Colloid Science Center
The Heath Runcorn
Cheshire, England

D. R. TALLANT
Sandia National Laboratories
Albuquerque, New Mexico

PARAM H. TEWARI
Applied Science Division
Lawrence Berkeley Laboratory
University of California
Berkeley, California

E. TRONC
Spectrochimie due Solide,
Université Paris
Paris, France

THANH N. TRUNOG
Department of Chemistry
North Dakota State University
Fargo, North Dakota

C. W. TURNER
Atomic Energy of Canada Limited
Chalk River Nuclear Laboratories
Chalk River, Ontario,
Canada

D. R. UHLMANN
Department of Metallurgy and
  Materials Science
Massachusetts Institute of
  Technology
Cambridge, Massachusetts

DONALD R. ULRICH
Department of the Air Force
Air Force Office of Scientific
  Research
Bolling AFB, Washington, D.C.

S. Wallace
Department of Materials Science
  and Engineering
University of Florida
Gainesville, Florida

S. H. WANG
Department of Materials Science
  and Engineering
University of Florida
Gainesville, Florida

S. B. WARNER
Kimberly-Clark
Roswell, Georgia

WILLIAM P. WEBER
Department of Chemistry
University of Southern California
Los Angeles, California

ROBERT WEST
Department of Chemistry
University of Wisconsin
Madison, Wisconsin

GARY H. WISEMAN
Department of Chemistry
Massachusetts Institute of
  Technology
Cambridge, Massachusetts

K. W. WISTROM
Department of Materials Science
  and Engineering
University of Florida
Gainesville, Florida

M. F. YAN
AT&T Bell Laboratories
Murray Hill, New Jersey

J. ZARZYCKI
Materials Science Laboratory
University of Montpellier
Montpellier, France

B. J. ZELINSKI
Department of Materials Science
  and Engineering
Massachusetts Institute of
  Technology
Cambridge, Massachusetts

T. W. ZERDA
Department of Chemistry
School of Chemical Sciences
University of Illinois
Urbana, Illinois

XING-HUA ZHANG
Department of Chemistry
University of Wisconsin
Madison, Wisconsin

In Memoriam
PAUL J. FLORY
Nobel Laureate
June 19, 1910–Sept. 8, 1985

RALPH K. ILER
July 12, 1909—November 9, 1985

PREFACE

This book contains the proceedings of the "Second International Conference on Ultrastructure Processing of Ceramics, Glasses, and Composites, held February 25–March 1, 1985 in Palm Coast, Florida. The conference was sponsored by the Department of Materials Science and College of Engineering, University of Florida and supported by the Directorate of Chemical and Atmospheric Sciences of the Air Force Office of Scientific Research. More than 250 scientists and engineers from university, industry, and government laboratories attended the conference, including researchers from the United States, Canada, England, France, Italy, Japan, and West Germany.

Three Keynote Award lectures were presented. Dr. Ralph Iler's Keynote Award was given for a lifetime's contribution to inorganic chemistry. His pioneering work in the study of silica polymerization and colloidal chemistry is the foundation of many of the concepts explored in Parts 1, 2, and 5 of this book.

Professor Paul Flory's Keynote Award was given for a lifetime's contribution to polymer chemistry. His developments in the theory of organic networks and structures provide the basis for interpreting many of the new materials developments discussed in Parts 1, 3, and 4 of this book.

Professor Per-Olov Löwdin's Keynote Award was given for a lifetime's contribution to quantum chemistry. His organization and chairmanship of the Sanibel Conference for 25 years have been a major influence on quantum calculations being directed toward practical applications such as the silicon-based systems discussed in Parts 3 and 4.

The concept of chemically based ultrastructure processing involves a synthesis of the fields of inorganic chemistry, organic chemistry, polymer chemistry, surface chemistry, and quantum chemistry, all oriented toward producing a new generation of high-performance materials. The three keynote awardees have provided much of the foundation for that synthesis.

In addition to the three keynote lectures, 34 of 36 oral presentations and 23 of 42 poster papers are included, selected after peer review by the conference review board. The resulting 60 chapters were organized into six parts: sol–gel science, applications of sol–gel processing, materials from organometallic precursors, ultrastructure in macromolecular materials, micromorphology (fine particulate) science, and quantum chemistry (a review).

Consequently, this book provides a comprehensive treatment of the broad scientific basis of producing ceramic, glass, and composite materials using chemistry-based processing methods.

It is the goal of ultrastructure processing to control the structure, surfaces, and interfaces of materials and devices at the molecular level in the earliest stages of production. The scientific understanding of molecular structure control of complex materials is beginning to emerge, as is evident in this volume. However, there is still much to be learned that will require multiple investigator efforts. The beginnings of such interdisciplinary efforts are evident herein. The long-term consequences of this new approach to creating complex materials from a molecular viewpoint are just beginning to emerge. The potential rewards are enormous.

LARRY L. HENCH
DONALD R. ULRICH

Gainesville, Florida
Washington, D.C.
January 1986

CONTENTS

PART 2 APPLICATIONS OF SOL–GEL PROCESSING

PART 3   MATERIALS FROM ORGANOMETALLIC
PRECURSORS

PART 4   ULTRASTRUCTURE IN
MACROMOLECULAR MATERIALS

PART 5   MICROMORPHOLOGY SCIENCE

PART 6 QUANTUM CHEMISTRY

SCIENCE OF

CERAMIC CHEMICAL PROCESSING

PART 1

Sol−Gel Science

1

INORGANIC COLLOIDS FOR FORMING ULTRASTRUCTURES

RALPH K. ILER
811 Haines Avenue
Wilmington, Delaware

INTRODUCTION

The first tools used by humans were made of wood, a fiber-reinforced lignin polymer, and from bone, a composite of inorganic microcrystals and protein polymer. Then it was found that of all the kinds of stones, flint, and jadeite are much the strongest. Both have unique micrograin structures. Jade consists of a tangled mass of fibrous crystals.[1]

According to Sanders,[2] flint consists of microcrystalline quartz, some crystals smaller than 100 nm, bonded with some amorphous silica. He found that the closely related lace agate to be microcrystalline quartz containing rodlike crystals of cristobalite (Fig. 1.1). Flint has a transverse rupture strength of 207 MPa (30,000 psi) while that of nephrite (mutton fat) jade is 275 MPa (40,000 psi).[3] These materials were ancient relatives of modern composites and reinforced plastics.

Metals and ceramics were the earliest synthetic materials. Metals were continuously improved for weapons of war but ceramics reached a high plateau of development in ancient China. Only in this century have ceramics been developed with new properties. Today we have the remarkably strong partially stabilized zirconia and zirconia-modified alumina.[4] New processes and structures now in development are discussed in this volume.

The First International Conference on "Ultrastructure Processing" held in Gainesville, Florida in 1983[5] defined the subject as the "reproducible production

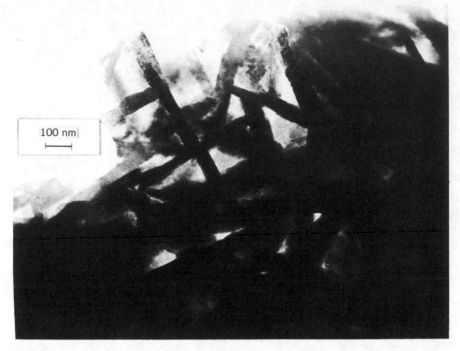

Figure 1.1   Electron micrograph of lace agate composed of microcrystalline quartz containing rod-like crystals of cristobalite.

of uniquely homogeneous structures, extremely fine scale (10 nm) second phases, controlled surface gradients and unique combinations of physical properties." Some portions of this field are also referred to as "high-tech ceramics."[4]

The main deficiency of conventional ceramics is not a lack of inherent strength or stiffness but rather the undependable brittleness due to the ease with which cracks start and propagate. It is now known that cracks start at defects or pores which in present products cannot be avoided. However cracks are stopped when energy is consumed as they encounter grain boundaries, especially if an energy-absorbing phase transformation occurs as in transformation toughening.

THE TWO APPROACHES TO ULTRASTRUCTURE PROCESSING

Conventional ceramics are generally molded from plastic masses of water-wet powders of mixed particle sizes well above 1 μm in diameter. However, when the particles are of submicrometer size, as is required for ultrastructure processing, special techniques are required to form the molded green body.

In the proceedings of the first conference in 1984[5] it appears that there are two quite different approaches to the fabrication of pore-free micrograin ceramics.

The first part of the proceedings was devoted to "sol–gel" technology, which largely involves amorphous silica, but is also applicable to other materials.

In the second part, attention was turned to the use of discrete uniform particles and no gels were involved. Barringer et al.[6] described the processing of monosized particles. Matijevic[7] reviewed how monosized particles of inorganic materials may be made. Aksay and Schilling[8] showed how discrete uniform particles can be packed into a uniform microstructure by filtration. It is proposed to call this the "densely packed colloid" process.

It was pointed out by Barringer et al.[6] that the ideal powder is monosized between 0.1 and 1.0 μm. I would expect that with improved techniques, particles as small as 0.05 μm may be processed. This is still 10 or 20 times the size of the typical ultimate particles in the sol–gel process.

The Sol —Gel Process

In sol–gel processing (fig. 1.2a) very small colloidal particles are first formed in solution, usually by hydrolysis of organic compounds of metals to the hydrous oxides. These particles are usually only 3 to 4 nm in size. Rijnten[9] has pointed out the surprising fact that when particles of hydrated oxides of iron, thorium, titanium, zirconium, or aluminum are first formed in water they are all in the above size range. In sufficient concentration, these very small particles link together in chains and then 3-D networks that fill the liquid phase as a gel.

One advantage of this procedure is that extremely intimate and uniform mixtures of different colloidal oxides can form a gel that is molecularly homogeneous. Another is that the gel can be molded in the shape of the final desired object, with dimensions enlarged to allow for shrinkage during drying and sintering.

Disadvantages have been that generally the hydrolizable organic derivatives of the metals are expensive. Also the molded body tends to crack during drying and sintering due to the considerable shrinkage.

The Densely Packed Uniform Colloid Process

This alternative process involves the preparation and compaction of colloidal particles of uniform size (Fig. 1.2b) so as to make a green body with minimum pore volume and uniform pore diameter. Thus there is minimum shrinkage during sintering. However, to achieve a high packing density the particles must be in the 0.05–1.0 μm range. Although uniform oxide particles in this range can be made by hydrolysis of alkoxides, for example, ethyl silicate, they can also be made by purely inorganic reactions which will no doubt be the method of future low-cost, large-scale production.

A body of hexagonally packed uniform particles has a relative pore volume

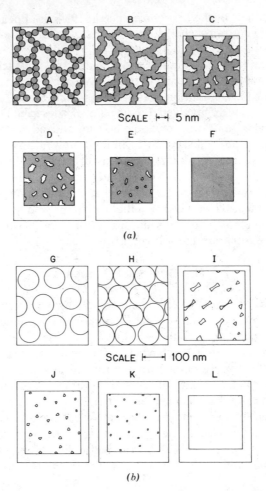

Figure 1.2 (*a*) Two-dimensional representation of the sol–gel process: A—colloidal particles 3–4 nm in diameter form gel network, B—particles coalesce into chains, hardening the body; C—gel shrinks as it is being dried; D and E—further shrinkage and closure of pores while being sintered; F—final pore-free body. (*b*) Two-dimensional representation of the "densely packed colloid" process: G—a sol of uniform colloidal particles, average diameter in the range 50 to 500 nanometers; H—particles densely packed in the molded green body; I, J, and K—progressive sintering and shrinkage of body with decreasing pore diameter but no pore closures; L—final pore-free body.

of about 25% and shrinks 10% linearly when sintered to a pore-free state. A more practical, randomly close-packed body has a porosity of 30 to 45% by volume and shrinks linearly about 15%.

The ideal process requires:[6] (1) an unagglomerated powder of monosized particles, 1 μm or less in diameter, preferably between 0.1 and 1.0 μm, (2) a green microstructure with a uniform distribution of void space (pores) and as many particle-to-particle contacts as possible, approaching that for close packing, and (3) selection of the thermochemical variables of temperature, pressure, and composition such that densification rather than particle coarsening occurs and so that pores are pinned to grain boundaries.

Most of the problems involved in this process have been described by Sacks and Tseung[10] who studied the effects of ordered versus random packing of silica spheres of uniform size in the diameter range of 0.1–1.0 μm. The degree of uniformity of packing controlled the degree of uniformity of the pore diameter which in turn controlled the temperature required for sintering to 100% relative desntiy. Aggregation or flocculation of the particles precludes dense packing and uniform pore size.

Before discussing the practical difficulties, let us consider what probably happens during sintering.

SINTERING A UNIFORM-PORED GREEN BODY

It is assumed that uniform silica spheres in the diameter range of 0.1–1.0 μm are packed in such a way as to form a body with a uniform 3-D network of uniform pores. As the body is sintered it shrinks uniformly and the diameter of the pores decreases, but if the pores are all of the same size the network still remains throughout until the pores are closed.

Whether the sintering mechanism involves lattice diffusion, grain boundary diffusion, or viscous flow, the changes in surface geometry and decrease in surface area are analogous to the changes that occur in the surface of amorphous silica as silica is dissolved and deposited from water solution.[11] For this reason it is interesting to review this behavior of silica in some detail.

Solids tend to decrease in area so as to reduce surface energy to a minimum. This also applies to the surfaces within pores. In 1900 Ostwald and Freundlich developed an equation relating the solubility of a curved solid surface to the radius of curvature,

$$\log S_r - \log S_i = KET^{-1}r^{-1} \tag{1}$$

where S_r is the solubility of a particle or a surface having a radius of curvature r, S_i is the solubility of a flat surface with a radius of curvature of infinity, E is the surface energy of the solid in ergs/cm^2, T is the temperature °K, and K is a constant.

Figure 1.3 represents a cross section of the surface of silica. At the upper right the otherwise flat surface has a series of rounded projections with increasing radii. Just above, there are spherical particles having the same radius of curvature as the projections below. All of these rounded surfaces have positive radii of curvature.

On the upper left, the surface has a series of depressions or crevices of increasing radii of curvature but in this case the curvature is negative. Above these are two spherical particles in contact where there is an annular crevice or negative radius of curvature.

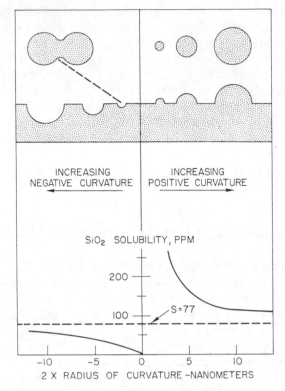

INCREASING
NEGATIVE CURVATURE

INCREASING
POSITIVE CURVATURE

SiO$_2$ SOLUBILITY, PPM

200

100   S=77

-10   -5   0   5   10
2 X RADIUS OF CURVATURE –NANOMETERS

Figure 1.3   Solubility of silica varies with the radius of curvature of the surface according to the Ostwald–Freundlich equation. (From *The Chemistry of Silica* by Ralph K. Iler, 1979, p. 80, by permission of the publisher, John Wiley & Sons, Inc., New York.)

In the lower portion of Fig. 1.3 on the right, is shown the relation between the positive radius of curvature and the solubility of that surface. As the radius becomes smaller in the range of a few nm, the solubility increases sharply. This applies both to the surface of the projection as well as to the colloidal particles. It is seen at once that silica will dissolve from smaller particles and will be deposited on larger particles or flat surfaces.

On the lower left, the equation shows that a surface with a negative radius of curvature is less soluble than a flat surface. Thus if the negative radius is small, as at the bottom of a crevice or at the contact of two spheres, silica will be deposited.

Thus edges and corners and small projections dissolve and cracks and depressions become filled, all resulting in a smoother surface.

There is an analogy to the changes in the pore geometry in close packed spheres as sintering progresses. Here there is no solubility involved and no transport of silica. Yet pores become smaller and the rounded surfaces of the spherical particles in the pore walls become flatter. Thus the pores, while

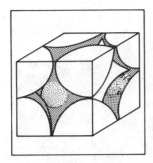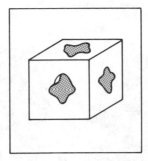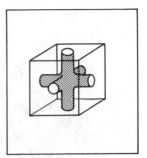

Figure 1.4 Pore cavity, between eight cubic-packed uniform silica spheres with a diameter in the range 50–500 nm, changes shape and shrinks during sintering to become a 3-D network of cylindrical pores.

decreasing in diameter, also change in shape and become converted to a 3-D network of intersecting cylindrical channels (Fig. 1.4).

This phenomenon has been described by Brinker and Scherer.[12] The pore diameter continues to decrease even as theoretical density is approached. It is probable that instead of transport of silica there is instead a transport of vacancies along the channels to the outside of the body. This can continue as long as the 3-D system of channels is maintained, regardless of their small diameter.

If the spherical particles are of uniform size and relatively large (100 nm), some decrease in diameter does not greatly increase the rate at which they shrink. (The situation is different in a silica gel with pores only 2 or 3 nm in diameter where a small decrease in size results in a great increase in the rate of closure.)

It can be assumed from the geometry of the system that when particles of diameter D (nm), initially in cubic packing, are sintered, there remains at one stage a cubic network of cylindrical channels with a diameter of d, where d is much smaller than D. If the fractional pore volume is V, then $d = 0.5DV^{0.5}$. Thus when particles 425 nm in diameter are compacted and sintered to 96.5% of theoretical density, d is calculated to be about 40 nm. This is reasonably close to the observed diameter of around 50 nm.

The actual presence of a continuous pore network even at this high density was reported by Bergna and Simko,[13] Fig. 1.5. They started with silica powder consisting of relatively uniform particles 425 nm in average diameter. The particles were compacted in a mold at 20 psi, which was insufficient to close the network of pores between the particles. The particles consisted of a dense silica gel of 15-nm silica particles. These gel particles sintered to density before the larger pores between the 425-nm particles began to close. The body was sintered for 5 hr at 1100°C. The resulting opaque silica body had a density of 96.5% of theoretical. Surprisingly the transverse rupture strength was twice that of optical grade fused silica cut and tested under the same conditions.

Electron micrographs showed that there was an internal network of pores

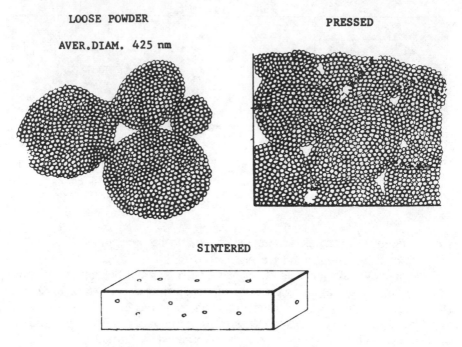

Figure 1.5 Pressed silica gel particles averaging 425 nm in diameter sintered to 96.5% relative density, still contains a 3-D network of pores about 50 nm in diameter.

about 50 nm in diameter. When ink was applied to one side of a sliced specimen, it appeared at numerous pore openings on the opposite side. When the specimen was further sintered to close the pores to obtain a translucent body, the strength dropped to the same value as that of fused silica.

It was theorized that the 3-D network of cylindrical channels stopped the propagation of cracks.

The above sintering behavior of silica is characteristic of particles of relatively uniform large particles of the order of 0.5 nm in diameter. A quite different behavior is noted in the case of conventional silica gels with pore diameters of 1 or 2 nm.

When such a fine-pore gel is sintered in air or vacuum at increasing temperatures, the specimen begins to shrink and the pore volume and the specific surface area decreases, but the *remaining pores remain unchanged in diameter*.[14]

When pores are only a few nm diameter, they follow the behavior predicted by the Ostwald–Freundlich equation. Once they begin to decrease in size the rate of decrease becomes ever faster and they quickly close and disappear. This has an exact analog in the case of small colloidal particles of silica in a solvent such as molybdic acid. Once a particular particle begins to dissolve and the diameter decreases to a few nm the solubility increases rapidly and the particle disappears.

Thus when the fine-pore gel begins to sinter there are regions where there is

TABLE 1.1   **When Silica Gel is Sintered in Air or Vacuum, Pore Diameter Remains Unchanged**

Temperature, °C	Surface, m^2/g	Pore Volume, cm^2/g	Pore Diameter, nm
300[a]	1004	1.48	1.48
500	800	1.26	1.58
700	477	0.9	1.89
900	162	0.3	1.85
100[b]	618	0.57	2.2
1000	162	0.3	2.2
200[c]	700	—	0.6
800	450	—	0.6

[a]S. J. Teichner and G. A. Nicolaon, *Adv. Colloid Interface Sci.* **5**, 245 (1976).
[b]H. W. Kohlschütter and G. Kampf, *Z. Anorg. Chem.* **292**, 298 (1957).
[c]J. F. Goodman and S. J. Gregg, *J. Chem. Soc.* **1959**, 694.

an above-average concentration of the finest pores so that the region shrinks and becomes dense. This applies a compression on the immediately adjacent pores which then also close. The whole body thus shrinks but the intervening regions of the gel between the dense regions remain unchanged and retain the same pore diameter.

Examples of this general behavior are shown in Table 1.1. It was shown that no pores were isolated or closed since the density of the body measured in He remained constant.

The above behavior occurs in air or vacuum but not in steam or under hydrothermal conditions. When water is present the silica becomes mobile. As small pores disappear, the larger ones grow in size and there is no shrinkage of the body which becomes ever more coarsely porous.[15] In this case the silica is transported either through the water or as vapor since it is known to be somewhat volatile in steam. In dry steam the silica probably migrates as $Si(OH)_4$ along the surface. In the absence of water, the changes must be due to migration of vacancies which travel to the exterior along the surface of the small pore network but do not remain to enlarge the pores.

EFFECT OF PARTICLE SIZE

There is an optimum range of size of particles used in the densely packed uniform particle process, as emphasized by Barringer et al.[6] When particles are too small they adhere to each other upon contact and the forces between them are very large. For example, it is found that when a sol of 200-nm silica particles is concentrated and dried, random close packing is attained. On the other hand, under similar conditions, particles smaller than about 50 nm tend to bridge, leaving pores almost as large as the particles.[16] Thus aggregation must be avoided.

If the particles are too large, for example, over a μm in diameter, then the pores, even in a closely packed body, are so large that a higher sintering temperature is required to begin pore closure. If all other factors are constant, on a theoretical basis, doubling the particle size increases the sintering time tenfold. However, this effect is strongly offset by an increase in temperature. Experiments with silica have shown that a gel with pores only 1 nm in diameter can be sintered to zero surface area (presumably zero porosity) at 900–1000°C. In comparison a body of closely packed particles 200 nm in diameter, with pores estimated to be about 40 nm in diameter, can be sintered to a nonporous, clear silica glass at 1000 to 1200°C, or only 200°C higher.[14] Such examples may not be strictly comparable unless the alkali content is very low and constant. In this case the 200-nm particles were prepared from ethyl silicate and contained no sodium that could have alowered the sintering temperature. Thus an increase in sintering temperature of 200°C was sufficient to compensate for a $40\times$ increase in pore diameter. The particle diameter and pore diameter are not as important as having pores of uniform size between uniform particles, providing the latter are less than about 1 μm or preferably less than 500 nm in diameter.

PROBLEMS IN DENSE-PACKING COLLOIDAL PARTICLES

Even with colloidal particles as large as 100 nm in diameter, the main problem is that as a sol is concentrated or dry particles compressed, interparticle bonding occurs, forming irregular agglomerates that cannot be further compacted. Spherical particles are more easily close packed than those of irregular shape. In either case for dense packing, spheres must be separate and mobile enough to move about near the surface of the growing body so as to reach a location where contact can be made with the greatest number of particles already on the surface.

Examples of close-packed particles in regular array are the virus particles which aggregate into crystals, uniform latex particles which form iridescent uniform close-packed masses, and 200–300 nm silica spheres which aggregate uniformly to form brilliant colored precious opal. For the latter to form in the laboratory by sedimentation concentration, the pH must be near neutrality so that there is only a slight ionic charge on the particles. This prevents one particle from adhering to another in suspension, but when the particle, through Brownian motion, finds a spot on the growing opal surface where contact is made with two or three other particles simultaneously then the multiple attraction overcomes the slight repulsion forces and the particle thus becomes part of the uniform compact opal structure.

If the pH is too high the more highly charged particles mutually repel each other and close packing cannot occur. If the pH is below about 6, the particles in suspension collide and aggregate through hydrogen bonding and inter-particle siloxane bonds.

Formation of closely packed structures of the above type is of course imprac-tical for commercial processing. Other means to prevent interparticle bonding before compaction is complete must be found. One may speculate that an extremely thin coating of an organic nature might promote dense packing. Certain polar organic molecules may be adsorbed as a monolayer on the surface of the particles.

This is purely speculative, but, for example, low molecular weight poly-ethylene oxides are adsorbed on the silica surface. Numerous lower alcohols are hydrogen bonded as a monolayer on the surface.[17] Polyvinyl alcohol can lie flat along the surface, making it organophilic so that colloidal silica separates from water as a concentrated oily liquid phase.[18] These surface films may pro-mote particle packing during filtration at the wall of the mold in slipcasting or may provide a lubricant effect during dry pressing.

In the case of other oxides such as alumina, the surface is known to adsorb oriented monolayers of carboxylic acids that may perform the same function. These are chemisorbed films in contrast to the above physically adsorbed hydrogen-bonded films on silica. For chemisorption on silica a basic or cationic molecule is needed. Short-chain cationic polymers should lie flat along the silica surface as does polyvinyl alcohol, but may give a problem with flocculation by interparticle bridging unless the right proportions are used. A monolayer of a cationic surfactant would give a much thicker film that might be a better lubri-cant for compacting molding of dry powder.

If a suitable coating and compacting method for silica is found, it should be recalled that colloidal particles of other metal oxides can be made to act like silica by applying a coating of silica about 1 nm in thickness by known methods.

It is possible that forming molded bodies of some kinds of particles may be done more successfully from an organic liquid than from water. Colloidal silica can be transferred from water to organic liquid media by several procedures. For some organosols an organic molecular coating may be required for good dispersion.

There remains the probable difficulty that upon removal of the organic component the resulting voids will make the body impractically fragile.

MOLDING

A major problem is how to achieve the most dense packing of colloidal particles in a molded shape. Different procedures may be used: (a) slip casting, (b) mech-anical compacting, (c) centrifuging. Tape casting or injection molding involve blending the conventional ceramic powder with a polymeric organic carrier to form a mass that can be extruded. The organic material is later removed during sintering. Much finer submicrometer powders will probably require a higher proportion of organic component.

Slipcasting conventional ceramic slips offers no problem because the particles are larger than the submicrometer pores in the plaster mold so that only water penetrates the wall. When the slip consists of a suspension of submicrometer particles, the process may be referred to as colloidal filtration or ultrafiltration.

This method was discussed at the first conference by Aksay and Schilling[8] who emphasized the importance of avoiding any flocculation or aggregation of the particles as the filter cake is formed. They found that even in a one-component colloidal system there were formed densely packed regions that were the building blocks of the primary structure. Apparently the flow liquid through the filter cake developed some channels that resulted in variations in density of the body and caused inhomogeneous sintering.

Possibly this inhomogeneity was caused by aggregation when the concentration of the colloid reached a high level just before it solidified. Some colloids show an indication of aggregation when a very high solids concentration is reached.

Another possibility is that the filter pores may vary slightly over microscopic regions. This too can cause variations in the density of the filter cake.

Ideally the filter wall of the mold should have pores only just small enough to retain the colloidal particles but pass the liquid as well as small colloidal particles without becoming blinded. The absence of particles smaller than 50 nm is important. Such particles enter the pores of the filter and reduce the filtration rate drastically. In a 50% suspension of the large colloidal particles that form the body, as little as 1% of fine colloid can completely stop the flow.

An ultrafilter membrane on the inner wall of the mold may be made by depositing a film of uniform particles 25–40 nm in diameter which have been already slightly aggregated. A fibrous colloid with a fiber diameter of 25 to 50 nm forms an even smoother and thinner coating. Colloidal cellulose, well-dispersed attapulgite clay,[19] or colloidal chrysotile asbestos[20] are practical possibilities.

Pressure may not be applied to the mold if sufficient time is allowed for drying. The suction due to capillary forces as water evaporates from the outside may be adequate. For example Zarzycki[21] showed that pores 120 nm in diameter in the outer wall of the mold have a capillary pressure of 0.5 MPa or around 100 psi.

Mechanical compaction and extrusion methods applicable to conventional ceramic powders will probably not give a very uniform packing of colloidal particles. However, if the particles are first coated with a molecular film of an organic lubricant, possibly heavy compaction in a die might result in a close-packed density. Conventional ceramic compositions are extruded with the aid of a polymer binder but with colloidal particles an organophilizing film may have to be applied to maintain flow. The problem may be that with uniform spherical colloids with no intervening smaller particles to fill the volume, at least 50% by volume of polymer may be required to attain flow. In this case some of the organic material must be retained as a binder up to the temperature at which sinter necks are formed between the particles or the body will not remain coherent.

Centrifugal packing of the colloidal particles from a sol into a mold is presumably a possibility for small objects. I have noted that the cake formed from a 200-nm silica sol by centrifuging at a speed sufficient for slow settling is still very weak after it is dried. High-speed centrifuging may well give a density approaching random close packing. Presumably the body would be dried in the mold.

IMPROVING THE STRENGTH OF MOLDED BODIES

Practical methods for dense-packing uniform particles in the size range of 50–1000 nm seem to give dried green bodies that are very fragile or even microcracked. The self-adhesion of particles of this size is very slight. Organic binders involve problems of burnout and residual porosity.

An answer may be to use an inorganic colloid of very small particle size. Thus, for alumina particles, it is likely that basic aluminum chloride, which is actually a colloid with a particle size of the order of 1 nm, has been used. Similarly basic zirconium chloride can probably be used with colloidal zirconia. The idea is that the gel of small particles will fill in the necks between the large particles, thus providing green strength and reducing total porosity before sintering (Fig. 1.6).

As for colloidal silica, I have added 3% by weight of a 3-nm silica sol to a 30% silica col of 200-nm particles with the idea that when the mixture is ultrafiltered, the resultant cake of the larger particles would be filled with the small particle sol which would gel upon drying, giving strength to the green body. At pH 8 the dried body was weak and not very dense because of the negative charges on the particles. At pH 2, where the 3-nm sol gels most slowly, the dried body was more dense but also very weak, no doubt because microgel had formed even before filtration was complete as evidenced by a final very slow filtration rate. Possibly a silica sol of particles about 10 nm in size would work at pH 2 where there is no surface charge, yet where the 10-nm particles would be much slower to aggregate.

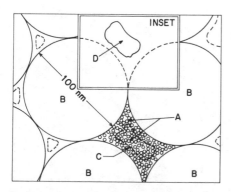

Figure 1.6 In the "densely packed colloid" process, the green body may be strengthened with a colloidal inorganic colloidal binder in the pores: A—small colloidal inorganic particles gelled in pores between larger particles: B—small particles should sinter at lower temperature leaving the larger pores to be sintered at higher temperature.

REINFORCED SILICA GLASS

There is a possibility that colloidal particles of a refractory oxide uniformly dispersed throughout silica glass (so-called "fused silica") might improve the physical properties. Oxides such as alumina and zirconia have higher coefficients of expansion than amorphous silica. Thus if these were incorporated uniformly in a silica glass body they would shrink upon cooling and put the glass under compression. Such an expansion misfit may cause interfacial microcracks if they were $>1\ \mu m$ but if the particles were of colloidal dimensions the linear shrinkage may be insufficient to break the strong bonding with silica at the particle interface.

Such colloidal particles may be incorporated into silica glass by either of the ultrastructure processes that have been discussed. Thus in the sol–gel process the refractory colloid particles 50 or 100 nm in size could be dispersed in the silica sol from ethyl silicate as employed in making clear silica glass (see Chapter 23). Up to 7% by volume of the oxide could be added, based on the final volume of the silica body. This would give particles dispersed in the glass with an average separation equal to the particle diameter. This should not interfere with gel formation or drying or sintering, providing the particles remain dispersed and do not form large clusters (Fig. 1.7).

The other approach may be to make silica particles of uniform size, for example, 122 nm in diameter, each with a core of a 50-nm particle of refractory oxide. If alumina particles can be obtained, the deposition of silica to grow the size to 122 nm can be done by a known process. Such particles may be sinterable to the same glass composition described above containing 7% by volume of alumina based on silica (Fig. 1.8). There should be no extensive reaction between alumina and silica phases at the relatively low sintering temperature of silica.

Still another approach would produce a 3-D network of zircon or mullite throughout silica glass. The network may be so fine that no light will be scattered and the glass will remain clear. In this case 100-nm particles of silica of uniform size would be coated with a very thin (1–5 nm) layer of a refractory oxide by known means. I have observed that when chromia is used the particles can be

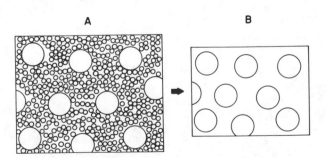

Figure 1.7 Hybrid process of including larger refractory colloidal particles in the sol–gel body: A—before sintering; B—after sintering.

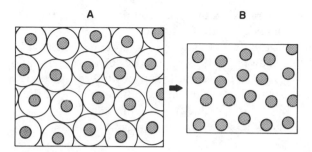

Figure 1.8 Uniform refractory colloidal particles are coated with silica to obtain uniform particles in the range 50–500 nm diameter which sinter to a particle-reinforced silica glass.

heated to over 900°C with no particle adhesion. The powder remains very soft and can be rubbed out to a monolayer on glass. However, if this powder were compacted and sintered to a sufficient temperature to react the chromia with the silica the film would be broken and the body would sinter to full density. The chromia would remain as a residue along the path of the closed pores. In the case of alumina the residue would be either alpha alumina or mullite. With zirconia the network would consist of either zirconia, or, more likely, zircon. The question remains whether the refractory particles will permit the silica to sinter to 100% relative density.

MIXED COLLOIDS

Sols of 100-nm particles of different oxides can be blended to obtain non-aggregated sols if the particle charges are alike. If the particle charge cannot be made equivalent by adjusting the pH then it may be possible to first coat the one sol with a molecular layer of the other. For example, colloidal silica cannot be mixed with colloidal alumina at any pH without flocculation. However, if the silica particles are coated with a 1-nm film of alumina, then a homogeneous sol can be obtained. Alternately alumina particles can be coated with any desired thickness of silica before blending with the silica particles. Since the surface chemistry of the particles is the same, the particles may be packed as densely as silica alone.

However, the behavior during sintering cannot be predicted. Presumably this might prove to be a route to very fine-grained mullite. It would be interesting to see the result of blending two materials that form a eutectic.

THE NEED FOR SUITABLE COLLOIDS

In addition to the fabrication problems that must be solved there is undoubtedly a need for adequate research quantities of refractory oxides in the form of particles of suitable size and uniformity.

There are two general ways in which colloidal refractory materials can be made: (1) nucleation and growth of particles in a liquid phase such as water or molten salts, or (2) formation of the particles in the gas phase at high temperature by chemical reaction or by condensation of the vapor. In all cases the problem is to form a certain concentration of uniform nuclei and then grow these with no further nucleation under conditions such that the growing particles do not become aggregated. Another approach is to create a high concentration of very small particles and then permit these to aggregate into uniform spheres or crystals.

For example, colloidal silica is formed in aqueous solution by forming nuclei from supersaturated solution. Silica continues to deposit from solution on these nuclei while the degree of supersaturation is kept sufficiently low to avoid forming more nuclei. Commercial sols are thus made with uniform particles from 4 to 100 nm in diameter. On the other hand, the Stöber process of hydrolyzing ethyl silicate in alcoholic aqueous ammonia creates a sudden high concentration of colloidal particles less than 5 nm in diameter of incompletely hydrolized ester which aggregates into a coacervate or droplets of an immiscible silica-rich liquid phase which then further hydrolize and form silica particles from 50 to 500 nm in diameter of very uniform size.

In a somewhat similar manner colloidal zirconia is first formed as colloidal particles which are small crystals less than 10 nm in diameter which then aggregate into spherical clusters in which the crystals self-orient themselves and the sphere becomes a single cubic crystal.

This is mentioned only to show that there are at least two mechanisms by which colloidal particles in the size range of 50–500 nm can be formed. These mechanisms can operate in water but which one is applicable to which oxide depends on the solubility. Silica is relatively soluble and thus particle growth by deposition of silica from solution is the preferred procedure. In the case of the much less soluble zirconia, and probably also other refractory oxides, the initially formed particles are exceedingly small and formation of large colloidal particles will involve a uniform aggregation process with formation of initially spherical particles.

For the less soluble oxides which can be formed in hydrothermal conditions, it does not appear that the technique of nucleation followed by growth by slow addition of reagents has yet been applied under pressure. This may well lead to a process operating at much higher concentrations for growing particles to a controlled size as is done with the silica system at 100°C at ordinary pressure.

It is hoped that the desired type of colloidal particles will become available in one way or another. The encouraging results of making spherical particles of alumina–titania reported by Gani and McPherson[21] suggest that such techniques using plasma methods may prove practical.

CONCLUSIONS

In developing ultrastructure processing there are two separate approaches which involve colloid chemistry in different ways. The sol–gel process starts with soluble precursors, usually organic–metal derivatives, which react to form extremely small colloidal oxide particles less than 5 nm in diameter which link together into a strong gel network. This is formed in a mold in the shape of the final body, with allowance for shrinkage during later drying and sintering to a pore-free product.

The other approach, which may be described as the "densely packed colloid" process, starts with much larger, uniform, preferably spherical, particles of the component oxide or oxides, in the size range of 50–500 nm. These are uniformly packed into a mold in such a way as to obtain a body with pores as uniform as possible. The body is then sintered to 100% density.

The success of this approach will depend to a large extent upon the development of sources for the colloidal components at minimum cost. It is assumed that research will improve the integrity of the bodies during shrinkage while being processed.

Eventual success will lead to a standardized technology for making a variety of compositions into new fine-grained, high-strength, impact-resistant ceramics and composites.

REFERENCES

1. D. J. Rowcliffe and V. Fruhauf, The Fracture of Jade, *J. Mater. Sci.* **12**, 35–42 (1977).

2. J. V. Sanders, Division of Materials Science, CSIRO, Melbourne, Australia, personal communication.

3. R. K. Iler, Strength and Structure of Flint, *Nature* **199** No. 4900, 1278–1279 (1963).

4. H. J. Sanders, High-tech Ceramics, *Chemical and Engineering News*, July 9, 1984, pp. 26–40.

5. *Ultrastructure Processing of Ceramics, Glasses and Composites*, L. L. Hench and D. R. Ulrich, Eds, Wiley, 1984.
    (a) E. Barringer, N. Jubb, B. Fegley, R. L. Pober, and H. K. Bowen, Chapter 26, Processing Monosized Powders, pp. 315–333.
    (b) E. Matijević, Chapter 27, Monodispersed Colloidal Metal Oxides, Sulfides and Phosphates, pp. 334–352.
    (c) I. A. Aksay and C. H. Schilling, Chapter 34, Colloidal Filtration Route to Uniform Microstructures, pp. 439–447.
    (d) C. J. Brinker and G. W. Scherer, Chapter 5, Relations Between Sol-to-Gel and Gel-to-Glass Conversions, pp. 43–59.
    (e) J. Zarzycki, Chapter 4, Monolithic Xero and Aerogels of Gel-Glass Processes, pp. 27–42.

6. H. K. Bowen et al., see Ref. 5a.

7. E. Matijevic, see Ref 5b.

8. I. A. Aksay and C. H. Schilling, see Ref. 5c.

9. H. Th. Rijnten, *Physical and Chemical Aspects of Adsorbents and Catalysts*, B. G. Linsen, Ed., Academic Press, London, New York, 1970, p. 370.

10. Michael D. Sacks and Tseung-Yuen Tseng, Properties of SiO_2 Glass from Model Powder Compacts, *J. Am. Ceramic Soc.* **67**(8), 526–537 (1984).

11. R. K. Iler, *The Chemistry of Silica*, Wiley, New York, 1979, p. 50.

12. C. J. Brinker and G. W. Scherer, see Ref. 5d.

13. H. E. Bergna and F. A. Simko, Jr., *Molded Amorphous Silica Bodies and Powders for Manufacture*, U.S. Patent 3,301,635 (DuPont), 1967.

14. R. K. Iler, see Ref. 11, pp. 544–547.

15. R. K. Iler, see Ref. 11, pp. 539–543.

16. R. K. Iler, see Ref. 11, pp. 480–483.

17. R. K. Iler, see Ref. 11, pp. 652–695.

18. R. K. Iler, see Ref. 11, p. 295.

19. R. K. Iler, *The Colloid Chemistry of Silica and Silicates*, Cornell Univ. Press, Ithaca, New York, 1955, pp. 211–214.

20. G. D. Barbaras, *Colloidal Dispersions of Chrysotile Asbestos*, U.S. Patent 2,661,287 (DuPont), 1953.

21. J. Zarzycki, see Ref. 5e.

22. M. S. J. Gani and R. McPherson, The Structure of Plasma-Prepared Al_2O_3–TiO_2 Particles, *J. Mater. Sci.* **15**, 1915–1925 (1980).

2

PHYSICAL–CHEMICAL FACTORS IN SOL–GEL PROCESSES

J. ZARZYCKI

Materials Science Laboratory
University of Montpellier
MONTPELLIER, FRANCE

INTRODUCTION

Numerous physical–chemical factors influence the various stages of the gel–glass process which ends with the formation of monolithic glass. All subsequent stages, however, closely depend on the initial structure of the "wet" gel formed in the reaction bath. It is therefore essential to have a better understanding of the mechanisms involved in order to be able to control the structure and texture finally obtained.

The existing theories of gel formation should in principle provide a convenient framework for discussing this process. The classical Flory theory[1] in general use in polymer chemistry and the recent percolation theories[2] more in favor with physicists give a description of the sol–gel transition.

GELATION THEORIES

Classical Theory

In Flory's theory the clustering of molecules of functionality (f) consists in establishing for each of them zero to f bonds with other molecules which leads to a progressive formation of larger units (Fig. 2.1a). This collection of finite

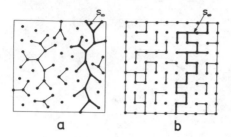

Figure 2.1 Gelation theories (schematic). (*a*) Flory's model ($f = 3$). (*b*) Percolation on a square lattice. In each case a spanning s_∞ gel molecule is indicated embedded in the sol.

clusters is termed a "sol." As the reaction progresses a stage is reached where an "infinite" cluster appears which spans the whole volume of the sample. This infinite cluster is called a "gel" and the system is said to have reached gel point. This concept is evidently valid only in the thermodynamic limit of large samples, otherwise no sharp transition is to be expected. The gel then coexists with the sol, the finite clusters being trapped in the mesh of the gel. Beyond the gel point the clusters join progressively the infinite network; the gel fraction increases at the expense of the sol fraction.

In the case of silica, the clustering sol fraction before the gel point is reached was termed "microgel" by Iler.[3]

Percolation Theories

These are in reality only generalizations of the classical gelation theory using lattices other than Bethe's lattice (or Cayley's tree) implicit in Flory's treatment. They are probabilistic treatments of random bond formation between entities on a fixed lattice ending with the appearance of a spanning (infinite) network at the "percolation point" (Fig. 2.1*b*). These theories can be used to predict the variation of numerous properties of the system during gelation. Their interest resides in the fact that the "scaling relations" so obtained are of a general character for phenomena belonging to the same "universality class."

Table 2.1 assembles these relations together with the characteristic exponents both for classical and percolation theories. Three main kinds of relations are of interest:

1. Relations describing the asymptotic evolution of the various statistical parameters of the system as the *conversion factor p* approaches the critical (percolation) value p_c. These are: average molecular weight M_w, gel fraction G, spatial extent of the cluster R.
2. Relations giving the statistical distribution of the clusters in the reaction bath at the critical point ($p = p_c$), the infinite (gel) molecule excluded. These are the fraction n_s and the spatial extent R_s of clusters containing s units.
3. Relations for rheological (viscosity) and mechanical (moduli) properties in the vicinity of the critical point.

TABLE 2.1   **Scaling Relations in the Sol–Gel Transition Theories**

Property	Scaling Relation	Classical (Flory) Theory	Percolation (3-D)
Gel fraction	$G(p) \propto (p - p_c)^\beta$	$\beta = 1$	$\beta = 0.40$
Average molecular weight	$M_w(p) \propto (p_c - p)^{-\gamma}$	$\gamma = 1$	$\gamma = 1.7$
Spatial extent of the cluster	$R(p) \propto (p_c - p)^{-\upsilon}$	$\upsilon = \frac{1}{2}$	$\upsilon = 0.85$
Fraction of s clusters at $p = p_c$	$n_s \propto s^{-\tau}$	$\tau = 2.5$	$\tau = 2.2$
Spatial extent of the s cluster at $p = p_c$	$R_s \propto s^\rho$	$\rho = \frac{1}{4}$	$\rho = \frac{1}{2.6} = 0.38$
Viscosity	$\eta \propto (p_c - p)^{-k}$	$k = 0$	$k = 0.7 - 0.8$
Elastic modulus	$E \propto (p - p_c)^t$	$t = 3$	$t = 1.65 - 2.6$

Conditions of Application

Several preliminary remarks have to be made concerning the above theories:

1. In Flory's theory the "percolation threshold" is fixed

$$p_c = \frac{1}{1 - f} \tag{1}$$

where f is the functionality of the molecule. On the contrary p_c is not universal in the percolation theories and its value depends on the type and dimensionality of the lattice on which the percolation takes place.

2. *Random bond percolation* supposes a lattice; however, *continuous percolation* simulations without an underlying lattice structure lead to similar (or identical?) exponents.

3. In the real situations besides the f-functional monomers solvent molecules are also present. This supposes an extension of the theory to *site and bond percolation* described by two probabilities. There is, however, an indication that the whole percolation line is also described by random percolation exponents.

4. In silica gels the bonds formed are irreversible (permanent). SiO_2 gel belongs to a class of "strong" gels as opposed to "weak" (reversible) gels.

5. The structure of the monomer is in reality more complicated than a simple mathematical point. Gelation does not occur randomly but correlations between various bonds exist. This in turn supposes a *correlated* site–bond percolation model.

6. Dynamic effects (role of molecular mobility) have been ignored. Coagulation effects will influence the cluster size distribution—if two molecules

are joined no other coagulation would be expected around that site for some time.

7. The extent of the critical region (how small the $p - p_c$ factor should be in order to get the theoretical asymptotic behavior) is still under debate.

EXPERIMENTAL TESTS

Three fundamental difficulties appear when one wishes to compare the gelling theories with experiment.

1. Definition of the conversion factor scale. Most of the experiments are indeed performed simply as a function of *time*, implicitly assuming a proportionality between the time (t) and the conversion factor (p).
2. Inadequate experimental definitions of the gelling point ($p - p_c$). This is generally linked with a more or less arbitrary viscosity level ($\eta = 0.5$ to 3000 P) which is equivalent to a phenomenological criterion: absence of flow when the container is tilted, some resistance opposed to a penetrometer, and so on. In more refined experiments the arrest of Brownian motion of a suspended tracer particle has been used to characterize the viscosity.

   All these criteria are arbitrary and furthermore linked to theoretically least-well-defined scaling exponents (k) and (t).
3. Lack of experiments close enough to the critical threshold. This prevents the verification of asymptotic behavior in most cases.

Elimination of the p Scale

Some of the above difficulties can in principle be avoided if *several* physical–chemical properties are measured simultaneously on the system leading to the elimination of the $|p - p_c|$ factor. The following scaling relations can be derived:

$$R_s \propto G^{-v/\beta} \tag{2}$$

$$M_w \propto G^{-\gamma/\beta} \tag{3}$$

which lead to the ratios v/β and γ/β, respectively.

Light scattering (LS) or small-angle x-ray or neutron scattering (SAXS or SANS) give the average molecular weight M_w

$$M_w = \frac{\displaystyle\int_1^{s_{max}} s^2 n_s \, ds}{\displaystyle\int_1^{s_{max}} s n_s \, ds} \tag{4}$$

and the so-called z average of the gyration radius R_s of the cluster population

$$\langle R_s^2 \rangle_z = \frac{\int_1^{s_{max}} s^2 n_s R_s^2 \, ds}{\int_1^{s_{max}} s^2 n_s \, ds} \tag{5}$$

The gel fraction (G) can be obtained by weighing after separation by ultracentrifuging the sol and gel fractions of the reaction bath. This may, however, prove difficult. To our knowledge no joint experiments of this type have yet been performed on silica gels. Existing experiments rather involve the determination of *viscosity* which is more straightforward. Here, however, other difficulties arise if we wish to check the gelling theories.

Viscoelastic Properties

The intrinsic viscosity

$$[\eta] = \lim_{c \to 0} \left[\frac{\eta - \eta_0}{\eta_0 c} \right] \tag{6}$$

where η_0 is the solvent viscosity and c the concentration is difficult to evaluate theoretically; it can in principle be approximated by

$$[\eta] = \frac{\int_1^{s_{max}} s n_s \frac{R_s^Z}{s} \, ds}{\int_1^{s_{max}} s n_s \, ds} \tag{7}$$

The exponent Z depends on the type of hydrodynamic interaction inside the clusters. The choice lies between

Total interaction $Z = 3$ (Zimm clusters)

No interaction $Z = 2 + 1/\rho$ (Rouse clusters).

In Zimm's hypothesis

$$[\eta] \propto \log M_w \tag{8}$$

and for Rouse's model,

$$[\eta] \propto (M_w)^{\frac{2v - \beta}{\gamma}} \tag{9}$$

The $[\eta]$ versus M_w graph (Staundinger plot) is widely used in polymer chemistry where it can give indications on the type of polymerization. Acker has used this approach in his study of acid-set silica hydrosols;[4] he found the relation

$$[\eta] \propto M_w^{0.65} \tag{10}$$

up to gel point. Using his results, we have attempted to calculate the separate variations of M_w and $[\eta]$ on the approach to gelation time t_G. The $\log_{10} M_w$ versus $\log_{10}(t_G - t)$ plot (Fig. 2.2) shows a reasonably linear relationship leading to $\gamma = 1.66$, close to the 3-D percolation value $\gamma = 1.7$. This leads to $2\upsilon - \beta = 1.07$ closer to the expected 1.3 in percolation theories than to the zero required in Flory's approach.

The corresponding $\log_{10}[\eta]$ versus $\log_{10}(t_G - t)$ plot (Fig. 2.2) shows, however, an important deviation near the gel point.

This is to be expected as the experimental viscosity remains finite and does not diverge as would require the percolation theories.

In spite of this an average value $k = 0.9$ may be adopted to represent the data, not far from the $k = 0.8$ value of the scaling exponent deduced from superconductor electrical analogy for viscosity.

The scaling exponent (t) relative to the *elastic modulus* is also difficult to obtain both theoretically and experimentally. The recent experiments of Baza[5] using a microbalance penetrometer device led to $t \sim 0.5$. It is possible that the measured property reflected rather the variation of the *gel fraction* G for which $\beta = 0.4$. In fact it has been shown[6] that for the gelatine gel the attenuation $\Delta\alpha$ of the ultrasonic waves at 790 MHz increased linearly with $(t - t_G)$.

As $\Delta\alpha \propto G^2/f$ where G is the gel fraction and f a friction coefficient between the solvent and the network, if f remains finite G should vary as $(t - t_G)^{0.5}$ as was found experimentally for the SiO_2 gels.

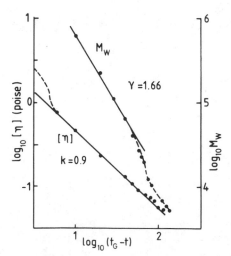

Figure 2.2 Molecular weight M_w and reduced viscosity $[\eta]$ of silica sols as a function of time to gel $(t_G - t)$ calculated from Acker's data.[4]

Determination of the p Scale

In all the preceding experiments the $t_G - t$ scale was used which is not necessarily proportional to the $p_c - p$ scale of gelation theories.

On rare occasions the gelling reaction was monitored in detail using gas-chromatographic analytic techniques.[7,8] We used the recent Yamane's et al. results[7] on the hydrolysis of silicon methoxide (TMS) in an attempt to deduce a possible conversion factor describing the advancement of the gelling reaction.

Assuming that the reactions can be approximated by (1) simultaneous hydrolysis, (2) condensation, and (3) esterification reactions,

$$m\mathrm{Si(OR)_4} + 4m\mathrm{H_2O} + q\mathrm{ROH} \rightarrow m\mathrm{Si(OH)_4} + (4m + q)\mathrm{ROH} \qquad (11)$$

$$m\mathrm{Si(OH)_4} + (4m + q)\mathrm{ROH} \rightarrow m\mathrm{Si(OH)}_x\mathrm{(OR)}_{4-x} + (q + xm)\mathrm{ROH}$$
$$+ (4 - x)m\mathrm{H_2O} \qquad (12)$$

$$m\mathrm{Si(OH)}_x\mathrm{(OR)}_{4-x} \rightarrow m\mathrm{SiO}_{(x-y)/2}\mathrm{(OH)}_y\mathrm{(OR)}_{4-x} + m\left(\frac{x-y}{2}\right)\mathrm{H_2O} \quad (13)$$

the following global reaction may be written

$$m\mathrm{Si(OR)_4} m\mathrm{Si(O)}_{(x-y)/2}\mathrm{(OH)}_y\mathrm{(OR)}_{4-x} + mx\mathrm{ROH} - \frac{p(x+y)}{2}\mathrm{H_2O} \qquad (14)$$

From the gas-chromatographic analyses of the $\mathrm{Si(OR)_4}$, ROH, and $\mathrm{H_2O}$ content as a function of time, the fraction m of decomposed TMS and the composition of the resulting gel

$$\mathrm{Si(O)}_a\mathrm{(OH)}_b\mathrm{(OR)}_c$$

were obtained.

The fraction of bridging oxygens per Si atom is equal to $2a$ and thus the fraction of possible Si–O bridging bonds found in the system is

$$\psi = \frac{ma}{2} \qquad (15)$$

The quantity ψ could be used in principle as the conversion factor p for percolation theories. The stoechiometric coefficients b and c provide additional information on OH groups and organic fraction content.

The three cases studied by Yamane corresponded respectively to

$$\mathrm{pH} = 7, \qquad \mathrm{pH} = 9.5, \qquad \text{and } \mathrm{pH} = 3 \qquad (\text{Figs. 2.3--2.5})$$

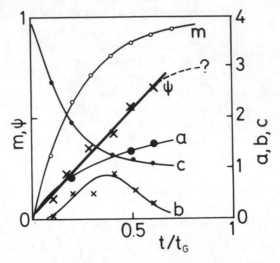

Figure 2.3   Analysis of reaction parameters during gelling of a TMOS solution for pH = 7. (See text for definitions.) Calculated from Yamane's et al. data.[7]

Figure 2.3 relative to neutral condition (pH = 7) shows a progressive increase of m and a accompanied by a transient evolution of b and a steady decrease of c.

ψ shows a remarkably linear increase with reduced gelling time t/t_G in the interval studied. Initially, $t_G - t$ can be taken as $\psi_c - \psi$ here but it is probable that, because of incomplete hydrolysis, ψ would tend towards a value of the

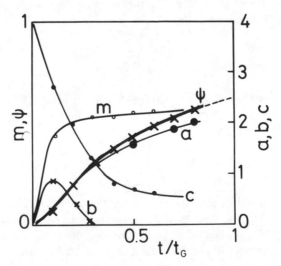

Figure 2.4   Analysis of reaction parameters during gelling for pH = 9.5. (See text for definitions.) Calculated from Yamane's et al. data.[7]

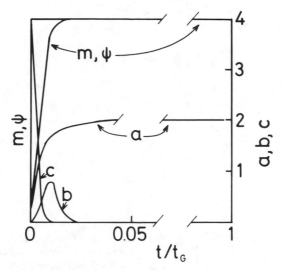

Figure 2.5 Analysis of reaction parameters during gelling for pH = 3. (See text for definitions.) Calculated from Yamane's et al. data.[7]

order of

$$\psi \sim 0.84$$

Figure 2.4 shows analogous results for the same system studied at pH = 9.5. The decomposed fraction m quickly saturates close to 0.55, b shows a rapid transient behavior, and ψ seems to tend toward $\psi \sim 0.6$.

Base-catalysed decomposition leads to quicker initial condensation but the organic fraction remains high.

Figure 2.5, relative to an acid catalysed system (pH = 3), shows an extremely fast and total decomposition of TMS, m reaching unity for $t_G/t < \frac{1}{100}$. The very brief transient for b and rapid decrease for c do not enable a significant parameter ψ to be defined; it reaches the maximum value close to unity almost instantaneously. The gel formation is not accompanied by a measurable variation of OH or OR content.

Analysis of the transformed fraction m points to an autocatalytic reaction (Fig. 2.6):

$$\ln\left(\frac{m}{1-m}\right) = k\frac{t}{t_G} + Cte \tag{16}$$

with a rate constant $k/t_G = 0.20 \, \text{hr}^{-1}$ for pH = 7 and equal to $0.12 \, \text{hr}^{-1}$ for pH = 9.5, respectively. From these plots limiting values m_∞ may be extrapolated for m.

These results show that it should indeed be possible in some cases to define a useful p scale. The precision of the analysis, however, should be increased and

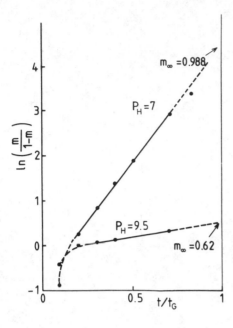

Figure 2.6 Transformed fraction m of TMOS as a function of reduced gelling time for solutions at pH $= 7$ and pH $= 9.5$ corresponding to Figs. 2.3 and 2.4. Extrapolated limiting fractions m_∞ are indicated. Calculated from Yamane's et al. data.[7]

the monitoring carried out closer to the gel point if useful results on scaling coefficients are to be obtained.

Gelling Mechanism

The most frequent measurements are relative to viscosity versus time behavior before gelling. It is possible to use these results to evaluate the *total fraction* ϕ of polymerized sol before the gel point is reached (or "microgel" fraction in Iler's sense). ϕ is related to the viscosity's increase by relations of the type

$$\frac{\eta}{\eta_0} = 1 + \frac{5}{2}\phi + \cdots \tag{17}$$

where η_0 is the viscosity of the starting solution, or, more closely by Mooney's formula

$$\ln\left(\frac{\eta}{\eta_0}\right) = \frac{2.5}{1 - k_1\phi} \tag{18}$$

With $k_1 = 1.43$. Iler has demonstrated that this relation can be used to evaluate ϕ for silica sols made at pH $= 2$.[3]

Using Acker's results,[4] he showed[3] that log ϕ presents a linear approach versus time when the gel point is approached.

We have similarly analysed recent results of Mizuno[9,10] who made a syste-

matic study of viscosity variations of TMS/methyl alcohol solutions during
gelling. Viscosity was measured by a falling sphere method.

Figure 2.7 shows typical examples corresponding to *neutral* conditions for
various water contents. Similar results were also obtained for acidic conditions.
In each case a linear relationship was reached for log ϕ:

$$\log \phi = Kt + Cte \tag{19}$$

Is it possible to use this type of determination to check gelling theories? The
fraction ϕ is a polydisperse suspension of s-clusters with a volume R_s^3 which
fill the fraction $\phi_s \propto n_s R_s^3$ of the sample. Neglecting cluster–cluster interaction,

$$\phi \propto \int_1^{s_{max}} n_s R_s^3 \tag{20}$$

$n_s R_s^3$ varies as

$$s^{3\rho-\tau} \bar{f}[(p_c - p)s^\sigma] \tag{21}$$

with a scaling function $\bar{f}$ decaying rapidly for $s \to \infty$. Classically ($\rho = \frac{1}{4}$,
$\tau = \frac{5}{2}$) $s^{3\rho-\tau} \to s^{-7/4}$ and ϕ remains finite. In percolation theories $3\rho = \tau - 1$,
the leading term varies as s^{-1}, and ϕ diverges logarithmically at p_c. Mooney's
formula[6] leads to $\eta \to \infty$ when $\phi \to 1/k_1 \sim 0.7$. This roughly corresponds to a
volume fraction $\phi \sim 0.63$ of disordered assembly of identical spheres.

To progress further more information is needed on the behavior of n_s and R_s
before the gel point is reached. To have an idea of the behavior of R_s, X-ray small-
angle scattering results may be used.

We analysed Brinker et al.'s published data obtained on a high-precision
SAXS goniometer.[8] Unfortunately the evolution could only be expressed as a

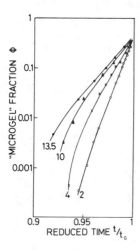

Figure 2.7 Polymerised sol fraction Φ as a function of reduced
gelling time t/t_G for 50:50 TMOS/methyl alcohol solutions.
Numbers indicate the ratio mole H_2O/mole TMOS. Calculated
from Mizuno's data.[9]

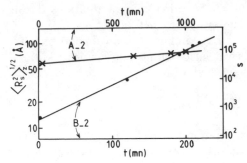

Figure 2.8 Variation of gyration radius $\langle R_s^2 \rangle^{\frac{1}{2}}$ obtained from SAXS measurements as a function of time for TEOS solution A-2 (acid catalyzed) and B-2 (base catalysed) An estimate of s-values is indicated. Calculated from Brinker's data.[8]

function of time and not on a more appropriate p scale. Fig. 2.8 shows the results for acid- (A-2) and base-catalysed (B-2) tetraethoxysilane (TEOS) solutions. The behavior observed in both cases during the whole gelation time is expressed by

$$\ln \langle R_s^2 \rangle_z^{1/2} = K_0 t + Cte \qquad (22)$$

with $K_0 = 0.43 \ hr^{-1}$ for base and $K_0 = 0.017 \ hr^{-1}$ for acid-catalyzed cases, respectively.

This clearly indicates an *agglomeration mechanism*; the results can be explained if the increase of the "microgel" fraction $d\phi$ is proportional to the total surface S of the fraction

$$d\phi = dV = kS \ dt \qquad (23)$$

Then

$$\frac{dV}{V} = k\left(\frac{S}{V}\right) dt \qquad (24)$$

The ratio $x = S/V$ can be estimated from the number of OH groups present on the s clusters.

Figure 2.9 shows the ratio x = number of surface OH per number of silicon atoms calculated for various cluster models together with the results of measurements of Hoebbel and Wieker.[11] This ratio shows a dramatic decrease in the first stages of polymerization and then stabilizes toward a value $x \sim 0.36$. As $x \sim$ constant, the equation

$$\frac{d\phi}{\phi} = kx \ dt \qquad (25)$$

can be integrated to

$$\ln \phi = kx(t - t_0) + \ln \phi_0 \qquad (26)$$

which is precisely the behavior observed.

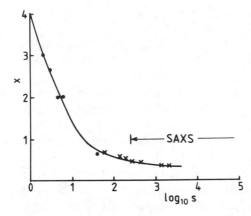

Figure 2.9 Ratio x = (number of surface OH groups)/number of Si atoms. ● calculated from cluster models; × measurements of Hoebbel and Wieker.[11]

In the region accessible to SAXS the variations of the factor x are small: 0.5 to 0.36, and a similar behavior is observed for the gyration radius:

$$\ln\langle R^2\rangle_z^{1/2} = \frac{kx}{3}(t - t_0) + C^{te} \tag{27}$$

The agglomeration leads successively to the formation of primary clusters then to their linking to larger units. This is in accord with usual observations where micro- and macroporosity of different orders is present in the dried gel.

Fractal Aspects of the Problem of Gelation

The behavior of the factor x should be linked with the properties of percolating clusters at $p = p_c$. It can be shown that the ratio x then tends to a fixed value close to unity rather than decreasing to zero:

$$\lim(x)_{s\to\infty} = \frac{1 - p_c}{p_c} \tag{28}$$

Close to the percolation point

$$R_s \propto s^{1/D} \tag{29}$$

where $D = 1/\rho$ is the *fractal dimension* of the cluster.

Fractal geometry, pioneered by Mandelbrot,[12] provides the necessary framework to deal with highly irregular surfaces such as in particular result from the percolation mechanisms.

For an object which is (statistically) invariant over a certain range of scale transformations, that is, self-similar upon change of resolution power, a fractional dimension D can be defined by a relation

$$D = -\frac{\ln N(r)}{\ln(1/r)} \tag{30}$$

where $N(r)$ is the number of parts obtainable from the whole object by a similarity of ratio $1/r$.

Figures 2.10(a), (b), and (c) illustrate this concept of self-similarity for an arc of Koch's curve, the 2-D Sierpinski's "carpet," and a 3-D Mengers's "sponge."

In each case D is a fractional number $1 < D < 3$ and the departure $d - D$ from the Euclidean dimension d represents the Hausdorff co-dimension of the fractal object.

The gelling mechanisms and agglomeration processes described above are likely to lead to fractal surfaces in the case of silica gels.

Avnir et al.,[13] using adsorption studies, were able to demonstrate that at the molecular range the surfaces of most materials are indeed fractals. The whole range of fractal dimensions $2 < D < 3$ was found. In particular it was shown that a highly porous silica gel has a fractal dimension $D = 2.94 \pm 0.04$[14] and a nonporous "aerosil" fumed silica $D = 2.02 \pm 0.06$.[15]

Using luminescence methods a porous glass (VYCOR 7930) was also shown to have a fractal dimension $D = 1.74 \pm 0.12$.[16]

It is certain that the fractal dimension which is an *intensive* property of the sample provides an extremely interesting parameter for characterizing the particularly intricate surfaces of silica gels for which most usual models (spherical or cylindrical voids) are oversimplifications, especially as the structures are probed at molecular level where the concept of smooth geometrical surfaces is meaningless.

It is possible to propose a *fractal model of a cluster* resulting from *a multistep particle agglomeration.*

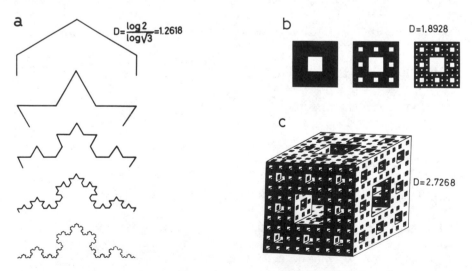

Figure 2.10   Illustration of fractal concepts. (a) Progressive generation of an arc of Koch's curve; $D = 1.2618$. (b) Generation of a Sierpinski's "carpet"; $D = 1.8928$. (c) Menger's "sponge"; $D = 2.7268$.

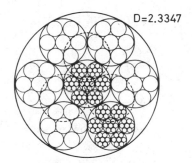

D=2.3347

Figure 2.11 Fractal model of multiple cluster $D =$ 2.3347 . . . . Only a limited number of spheres of fourth order are indicated for clarity.

Consider (Fig. 2.11) a sphere of radius unity and inscribe in this sphere 13 identical spheres of radius 1/3 in close contact with each other; then repeat the process indefinitely by subdividing each time by 3 (increase resolution by 3).

For this set,

$$N = 13; \qquad r = 3$$

and the fractal dimension is

$$D = -\frac{\ln 13}{\ln(1/3)} = 2.3347 \cdots \tag{31}$$

This represents a highly regular idealized model of multiple clusters. It is also possible to imagine other algorithms for subdivision of spheres with random packing $N \sim 9 + 1$; in this case,

$$D' \simeq \frac{\ln 10}{\ln(1/3)} \sim 2 \tag{32}$$

which would correspond to a "smoother" surface.

Fractal dimension D is linked to other scaling coefficients by relations

$$D = \frac{d + \gamma/v}{2} \tag{33}$$

and

$$D = d - \frac{\beta}{v} \tag{34}$$

where d is Euclidean dimensionality.

Independent determination of D would permit the evaluation of the ratios γ/v and β/v and the comparison of them with values obtained as indicated above from parallel measurements of M_w, G, and R_s in the reaction bath.

This would allow a close check of the validity of percolation theories and would shed light on microscopic evolution of the surface of wet gels during drying.

CONCLUSION

Existing classical and percolation theories provide a convenient framework for discussing the sol–gel transition. However, in the case of silica gels, the present experimental evidence is too scarce and, in particular, parallel determinations of various physico–chemical factors are necessary, together with an adequate monitoring of the chemical evolution of the system to provide a conversion factor scale as close as possible to the gel point.

Multiple agglomeration seems to be the general mechanism of gelation. It is shown that fractal concepts will help in elucidating the intricate structure of gels. A model of a fractal cluster applicable to silica gels is proposed.

REFERENCES

1. P. J. Flory, *Principles of Polymer Chemistry*, Cornell University Press, Ithaca, NY, 1953.
2. D. Stauffer, A. Coniglio and M. Adam, Gelation and Critical Phenomena, *Adv. Polymer Sci.* **44**, 103–158 (1982).
3. R. K. Iler, *The Chemistry of Silica*, Wiley, New York, 1979.
4. E. G. Acker, *J. Colloid. Interf. Sci.* **32**, 41–54 (1970)
5. S. Baza, Propriétés mécaniques de gel de silice précurseurs de verres, Thesis, University of Lyon, France, 1984.
6. J. Dumas and J. C. Bacri, *J. Phys. (Paris)* **41**, L279 (1980).
7. M. Yamane, S. Inoue, and A. Yasumori, *J. Non Cryst. Solids* **63**, 13–22 (1984).
8. C. J. Brinker, K. D. Keefer, D. W. Schaefer, R. A. Assink, B. D. Kay, and C. S. Ashley, *J. Non Cryst. Solids* **63**, 45–60 (1984).
9. T. Mizuno, Procédé sol-gel et génie des matériaux Thesis Ing. Dr., Montpellier, France, 1984.
10. T. Mizuno, J. Phalippou, and J. Zarzycki, Evolution of the Viscosity of Solutions Containing Metal Alkoxides, Meeting of Rheological Properties of Amorphous Materials, London, 25 Jan. 1984.
11. D. Hoebbel, W. Wieker, et al., see Ref. 3, p. 266.
12. B. B. Mandelbrot, *The Fractal Geometry of Nature*, Freeman, New York, (1977).
13. D. Avnir, D. Farin, and P. Pfeifer, Nature **308**, 261 (1984).
14. D. Avnir and P. Pfeifer, *Nouv. J. Chim.* **7**, 71 (1983).
15. D. Avnir, D. Farin, and P. Pfeifer, *J. Chem. Phys.* **79**, 3566 (1983).
16. U. Even, K. Rademan, J. Jortner, N. Manor, and R. Reisfeld, *Phys. Rev. Lett.* **52**, 2164 (1984).

3

RELATIONSHIPS BETWEEN SOL TO GEL AND GEL TO GLASS CONVERSIONS: STRUCTURE OF GELS DURING DENSIFICATION

C. J. BRINKER, E. P. ROTH, AND D. R. TALLANT
SANDIA NATIONAL LABORATORIES,
Albuquerque, New Mexico

G. W. SCHERER
Corning Glass Works
R & D Division
Corning, New York

INTRODUCTION

Although most structural studies of inorganic gels are based on concepts developed by Iler[1] for aqueous silicates, it is now well established that the shrinkage behavior of such colloidal gels differs markedly from that of metal alkoxide gels synthesized in alcoholic solutions.[2] From these differences (illustrated in Fig. 3.1 for three silica gels), we infer that for many synthesis schemes employed, the structures of alkoxide-derived gels prepared in nonaqueous solvents (or mixed solvents) are unlike the structures of aqueous silicate gels described by Iler.

In aqueous silicate systems fully hydrolyzed monomers condense in such a manner to maximize the number of siloxane bonds and minimize the number of terminal silanols.[1] In addition, since silica exhibits partial solubility in water

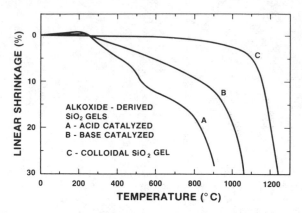

Figure 3.1   Linear shrinkage at 2°C/min for: A—an acid-catalyzed silica gel (A2, H_2O:TEOS ~ 4); B—a base-catalyzed silica gel (B2, H_2O:TEOS ≈ 4); and C—a colloidal gel prepared using flame-pyrolyzed $SiCl_4$.

(increasing at elevated temperature and pH) Ostwald ripening occurs, further biasing growth toward the formation of fully crosslinked (anhydrous) silica colloids. In contrast to aqueous systems, for most alkoxide syntheses, condensation begins before hydrolysis is complete (as evidenced by ^{29}Si NMR[3] and ^1H NMR[4]). Thus in all but the initial stage, condensation reactions occur between oligomeric species which cannot easily coalesce to a fully crosslinked state. Furthermore, the possible condensation sites can be limited by reducing the extent of hydrolysis. Computer simulations show that this results in highly ramified, fractal species (i.e., wispy structures whose density decreases with distance from the center of mass) rather than dense colloids.[5] Finally, because the solubility of silica in alcohol is much less than in water, there can be comparatively little rearrangement or ripening to more highly compacted states. Recent small-angle X-ray scattering (SAXS) experiments confirm the fractal nature of alkoxide-derived polymers in a wide range of gel compositions (silicates,[6] borates,[7] and borosilicates[7]) whereas Ludox,* an aqueous silica sol, is proven to consist of dense colloids.[6]

Because few bridging bonds are broken and reformed during solvent removal, the structure of the resulting porous, xerogel (the standard glass preform) bears some relation to the structure at the gel point. Thus, the structural starting point for subsequent heat treatments to form a dense glass may vary significantly depending on the original gelation conditions. Colloidal gels (composed of a fully crosslinked skeleton) are analogous to porous glass. Their densification behavior is accurately predicted by application of a viscous sintering model.[2,8,9] The densification of alkoxide-derived gels cannot be described on the basis of viscous sintering alone. This paper reviews recent work concerning the densification of alkoxide-derived gels from which both structural and thermodynamic

*Registered trademark, E. I. DuPont de Nemours, Inc. Wilmington, Delaware.

information have been derived. We show that the metastable nature of alkoxide-derived gels dramatically influences both the densification kinetics and thermodynamics. Thus, it is not appropriate to view the densification of colloidal gels (and porous glasses) as representative of the densification of alkoxide-derived gels.

GEL DENSIFICATION

Gel densification in glass-forming, alkoxide-derived systems occurs by: (1) polycondensation reactions which serve to crosslink the network and expel water; (2) structural relaxation; and (3) viscous sintering. Mechanisms 1 and 2 cause the density of the solid phase (skeleton) to increase toward the density of the corresponding melt-prepared glass. Mechanism 3 eliminates the continuous, porous phase so that the *bulk* density approaches that of the melted glass. In certain temperature regimes all three mechanisms are operative. The following discussion attempts to eluciate the complex interdependency of these densification mechanisms.

Skeletal Densification

When metal alkoxide-derived gels are heated continuous shrinkage occurs. Its degree depends on the synthesis method employed.[10] This is not the case for colloidal gels (Fig. 3.1) which shrink only at elevated temperatures by viscous sintering. Although it has been suggested (e.g., Ref. 11) that shrinkage at intermediate temperatures, for example, 150–550°C, in metal alkoxide-derived gels results from the rearrangement of primary particles or secondary agglomerates to higher coordination sites (Fig. 3.2), it has recently been demonstrated[12] that this shrinkage can result exclusively from skeletal densification, that is, the densification of the solid phase comprising the porous gel toward that of the corresponding melt-prepared glass (Fig. 3.3).

Reduced skeletal density is a consequence of starting with a rather weakly crosslinked, hydroxylated network at the gel point. The amorphous skeleton

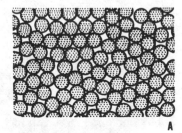
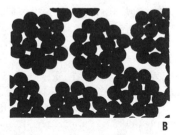

A   B

Figure 3.2 Schematic representations of two low density xerogels $\rho_{\mathrm{bulk}}/\rho_{v-\mathrm{SiO_2}} \cong 0.5$. A—random close packing of low-density ($\rho_{\mathrm{particle}}/\rho_{v-\mathrm{SiO_2}} = 0.75$) particles; and B—hierarchical packing of fully dense spheres. A is representative of most alkoxide-derived gels.

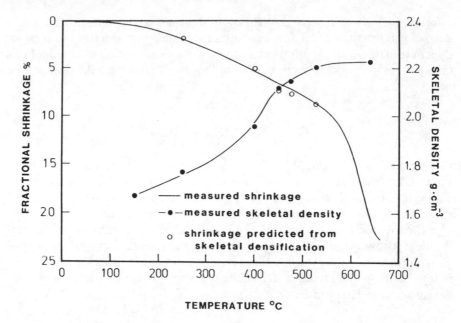

Figure 3.3 Shrinkage and calculated skeletal density for a borosilicate gel heated at 2°C/min. Open circles represent the shrinkage accounted for by the increasing skeletal density between 150 and 525°C (after Ref. 15).

resulting after solvent evaporation exhibits a lower density than that of the corresponding melted glass.[12-15] For example, whereas Iler reports the skeletal density of commercial silica aquasols and gels to be nearly identical to that of vitreous silica,[1] the relative skeletal density ($\rho_{skeleton}/\rho_{glass}$) of alkoxide-derived silica gels has been reported as low as 0.78[12] and the relative skeletal density of borosilicate gels as low as 0.61.[14]

The reduced skeletal density of alkoxide-derived gels (as measured by He[13] and N$_2$[12,15] adsorption studies and by a modified Archimedes method[14]) results from: (1) increased hydroxyl and alkoxy contents (decreased crosslinking) and (2) greater excess free volume.

Experimental observations show that the average number of nonbridging oxygens (OH + OR) remaining in desiccated silica gels can be as high as 1.48/Si.[16] Reduced crosslinking results in reduced density as observed by Bartholomew in hydrated $Na_2O \cdot SiO_2$ glass where density decreases linearly with water content (OH + H$_2$O) up to about 60 mole% H$_2$O.[17] Excess free volume in alkoxide-derived gels has not been measured directly but experimental results support its existence as described in the following discussion.

As a consequence of increased hydroxyl content and excess free volume skeletal densification occurs by both continued crosslinking and structural relaxation (i.e., the approach of the structure toward the configuration characteristic of the metastable liquid). In silica gels, skeletal dehydration occurs below 500°C (at higher temperatures hydroxyls are located principally on surfaces in which case they do not affect skeletal density).[18] By analogy to melted glass, structural relaxation is expected to occur in the vicinity of the T_g of the skeleton (i.e., at sufficiently low viscosity to allow diffusive motions of the network). In this latter case, the skeleton densifies exothermically with no associated weight loss.

The T_g of the skeleton depends on the hydroxyl content and the excess free volume. Bartholomew shows that the glass transition temperatures and viscosities of hydrated alkali silicate glasses decrease in a linear fashion with water additions up to 50 mole%.[19] For melted glasses it is well established that viscosity is also related to the excess free volume according to

$$\eta = \eta_0 \exp\left(\frac{xQ}{RT} + \frac{(1-x)Q}{RT_f}\right) \tag{1}$$

where T_f is the fictive temperature, Q is the activation energy for viscous flow, and x is a constant normally equal to ~ 0.5.[20] We assume that, although the gel network is not formed during cooling from the melt, its structure can be represented by a high fictive temperature, because it is rapidly solidified from a liquidlike state during gelation. High T_f (corresponding to a large excess free volume) results in reduced viscosity, lowering the temperature at which structural relaxation can occur.

In melted glass, structural relaxation is observed during heating as the glass transition temperature T_g is approached. Of course, the glass transition occurs over a range of temperature; the higher the initial fictive temperature, the broader that range. Since the low skeletal density of a gel is equivalent to a very high T_f, the relaxation is observed at relatively low temperatures during heating. As the temperature rises, crosslinking and structural relaxation tend to increase the viscosity and skeletal density, or (equivalently) to reduce the fictive temperature. The concurrence of rising T and falling T_f can cause the viscosity to remain almost constant, so that T_g seems to increase with the temperature. That is, one has the impression of being continually at T_g as the temperature increases. The same situation occurs upon heating of conventional glasses, but the effect is enormously exaggerated in alkoxide-derived gels.

In only one of the systems we have investigated did structural relaxation occur in such a dramatic fashion to be experimentally observable. This gel was prepared from TEOS using a two-step hydrolysis procedure at pH = 1 with a final water to Si molar ratio of 5. At the gel point the network was weakly crosslinked as quantified by its fractal dimension, 1.9.[6] Desiccation resulted in a microporous xerogel (average pore radius ~ 1 nm). Whereas complete bulk densification ($\rho_{final} = 2.21$ g/cm^3) occurred by viscous sintering between 800 and 900°C as shown in Fig. 3.1, curve A, skeletal densification occurred at much lower temperatures (between 450 and 600°C). Figures 3.4 and 3.5 show that the skeleton densified exothermically with little associated weight loss. The repeat DSC scan shows that this densification process is irreversible. The surface area decrease accompanying skeletal densification can be accounted for by the

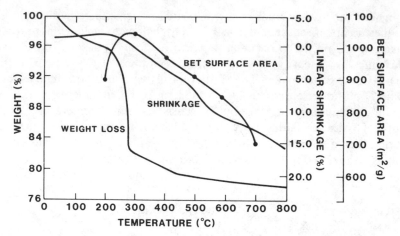

Figure 3.4   Weight loss, shrinkage, and surface area for an acid-catalyzed silica gel (A2) heated at 2°C/min in air. H_2O:TEOS ~ 5 (after Ref. 12).

contraction of the skeleton according to

$$\left(\frac{\rho_{\text{skeleton(initial)}}}{\rho_{\text{skeleton(final)}}} \right)^{2/3} = \frac{S_{\text{(final)}}}{S_{\text{(initial)}}} \qquad (2)$$

and, thus, does not result from sintering of open pores. The net energy of this process (-4.7 cal/g from DSC-1) was determined to be more negative than the sum of the energies associated with the reduction in surface energy, the heat of dehydration, and the skeletal heat capacity.[12] Thus, structural relaxation is completely consistent with the experimental observations. Sintering of isolated

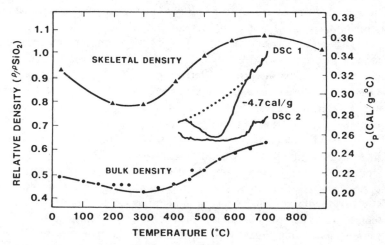

Figure 3.5   Calculated bulk and skeletal densities and first and second DSC scans, DSC-1 and DSC-2, for A2 (after Ref. 12).

pores residing within the skeletal framework could also account for these observations; however, since 1.0-nm radii pores shrink at or above 800°C, the size of isolated "pores" which could sinter at lower temperatures would be much smaller and essentially indistinguishable from the dimensions of excess free volume.

These observations suggest that the gel network prior to skeletal densification may be described by a high fictive temperature. As such, molecular dynamics calculations which simulate high fictive temperatures may be useful for understanding gel structure during the gel-to-glass conversion. By comparison colloidal gels show very limited shrinkage at intermediate temperatures (e.g., Fig. 3.1).

Densification Kinetics

In virtually all glass-forming systems, final densification is achieved by viscous sintering. The densification rate is proportional to the surface energy divided by the product of viscosity and pore size and total shrinkage is proportional to the integral of time divided by viscosity. As described earlier, viscosity is, in turn, dependent on the hydroxyl concentration and excess free volume. Therefore, the densification rate and total shrinkage depend in a complex fashion on prior thermal history.

Having studied the densification of numerous silicate and borosilicate systems, we can generally represent the temperature and heating rate dependence of viscosity as shown schematically in Figs. 3.6(a) and 3.6(b) for the elevated temperature regime in which viscous sintering is the predominant shrinkage mechanism. Figure 3.6a shows that, for constant rates of heating, increased heating rates reduce the viscosity. Larger reductions occur at lower temperatures. This phenomenon results because increased heating rates maintain the metastable low-temperature structure (i.e., high hydroxyl contents and excess free volume) to higher temperatures, reducing the viscosity. As the temperature rises, the rates of structural relaxation and condensation are increased, reducing the heating rate dependence of η. During isothermal conditions, the viscosity rises with time toward an equilibrium value equivalent to the viscosity of the corresponding melt-prepared glass (Fig. 3.6b). The viscosity increase is greater at lower temperatures and after higher initial heating rates. The initial rate of viscosity increase is greater for higher temperatures and higher heating rates. It is worthwhile to note that no isothermal changes in viscosity are observed for colloidal silica gels prepared from aqueous silicates[9] or flame-pyrolyzed SiCl$_4$.[8]

Isothermal increases in viscosity appear related to [OH] as shown in Fig. 3.7 for a borosilicate gel. However, as evidenced in this figure at 595°C, an increase in viscosity occurs with no change in [OH]. This increase is attributed entirely to structural relaxation.[21]

The practical consequences of the temperature, time, and heating rate dependences of viscosity are illustrated in Fig. 3.8 for a silica gel. Since shrinkage is

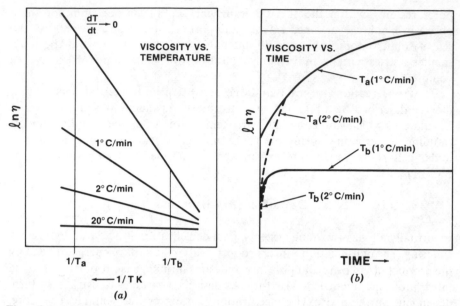

Figure 3.6 (a) Schematic representation of the relationship between viscosity (η) and temperature for gels heated at different rates. $DT/dt \rightarrow 0$ represents the isothermal limit where true Arrhenius behavior is observed. (b) Viscosity versus time for two isothermal temperatures, T_a and T_b ($T_a < T_b$), and two initial heating rates.

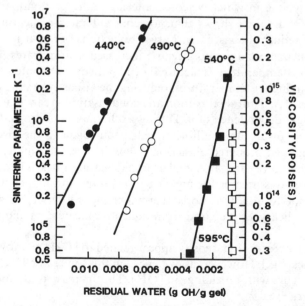

Figure 3.7 Sintering parameter (K^{-1}) and viscosity versus hydroxyl content during isothermal treatments for a borosilicate gel (after Ref. 21).

44

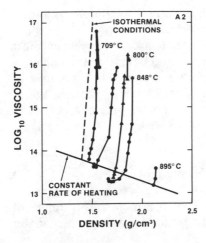

Figure 3.8 Viscosity and density of A2 during isothermal treatments (1000 min) after initial heating at 2°C/min and 20°C/min (▲). Isothermal treatments cause increases in viscosity, whereas constant rate of heating treatments maintain reduced viscosity leading to complete densification.

proportional to the integral of time divided by viscosity, heat treatments that maximize the time spent at the lowest viscosities are required to achieve complete densification at the lowest temperatures. For alkoxide-derived gels these conditions are often met by treatments in which the temperature is continually increased. Figure 3.8 shows viscosity versus density for a silica gel heated at 2°C/min (and 20°C/min) to temperatures ranging from 709–855°C followed by isothermal anneals of about 1000 min. Although isothermal conditions obviously maximize the time allowed for viscous flow, the dramatic increases in viscosity (three orders of magnitude at 709°C) which occur isothermally reduce the shrinkage rate practically to zero. By comparison, constant heating at 2°C/min maintains reduced viscosities (as evidenced by the initial value of η at each isothermal temperature) and complete densification is achieved at 895°C. More rapid heating (e.g., 20°C/min) permits even greater densification (13%) by the start of the 800°C isotherm because the reduced viscosity is maintained to higher temperature. Thus, although the amount of time in which viscous flow can occur is decreased, the viscosity is sufficiently reduced so that the amount of shrinkage is actually increased. For bulk specimens, the maximum heating rate is often limited by entrapment of organic residues and water; however, thin films can be heated at very high rates, permitting their densification at the lowest possible temperatures.

ENERGY ASSOCIATED WITH DENSIFICATION

The metastable nature of gels requires that their conversion to dense glass be a net exothermic process. Since we have already addressed the exothermic contribution of structural relaxation, this discussion will be limited to the energy changes associated with consolidation of the skeletal network. Although Iler does not analyze the energy of gel densification, he does define the energy change attributable to increasing particle size in aqueous solution.[1] Two contributions are pertinent to the present discussion: (1) the heat of dehydration (to form a pure siloxane surface) and (2) the reduction in surface free energy.

We assume that after heating to about 150°C (the temperature by which most physically adsorbed water is removed) the silica gel surface is completely hydroxylated ($\sim$4.5 OH/nm^2). Therefore at higher temperatures, the heat of dehydration represents the sum of the heat of formation of siloxane bonds resulting from condensation reactions plus the heat of vaporization of the by-product, water. The heat of formation of bulk vitreous silica has been calculated for the overall equilibrium[1]

$$SiO_2(s) + mH_2O(l) \leftrightharpoons H_4SiO_4(aq) \tag{3}$$

and ranges from $\Delta H_{f298°K} = -2.65$ to -3.35 kcal/mole.[1] This range of values is similar to the value Gibbs obtains for the following dimerization reaction[22]

$$\tag{4}$$

when the Si—O—Si bond angle, ϕ, is close to the most probable Si—O—Si angle in bulk vitreous silica, viz. 150°.

Galeener proposes that this minimum energy structure consists of puckered n-fold rings where $n \geqslant 4$.[23] Siloxane bonds formed by dehydroxylation of the gel surface, however, are constrained by the underlying skeletal structure so that ϕ may differ from 150° causing H_f to become less negative. For example, when ϕ is reduced (corresponding to $n < 4$), ΔH_f becomes positive. Twofold rings (edge-sharing tetrahedra) represent the highest energy structures. Molecular orbital calculations predict $\Delta H_{f298°K} = 50$ kcal/mole[22]

$$\tag{5}$$

Because of the sensitivity of ΔH_f to ϕ the surface structure can be deduced by measuring the heat of dehydration. Fig. 3.9 shows the change in surface area, weight loss, and heat capacity for silica gels heated at 20°C/min in Ar between 350 and 650°C after an initial 10-hr heat treatment at 350°C in air. The integrated heat capacity represents the energy change resulting from the reduction in surface area plus the heat of dehydration. In Table 3.1, the average heat of formation of siloxane bonds over the temperature interval 350–650°C is calculated to be very positive, greater than 25 kcal/mole.

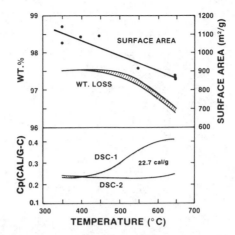

Figure 3.9 Change in surface area, weight, and heat capacity (DSC-1) for B2 heated between 350 and 650°C. DSC-2 is a repeat scan and represents the heat capacity of the skeleton.

CHARACTERIZATION BY RAMAN SPECTROSCOPY

Raman spectra of silica gels heated between 300 and 700°C (see Figs. 3.10 and 3.11) contain both features that are significantly enhanced in intensity when compared to spectra of vitreous silica and features that are absent in spectra obtained from vitreous silica. These features include relatively narrow Raman bands near 490, 610, 980, and 3740 cm^{-1} which have been reported by a number of investigators.[12,18,23] The bands near 980 and 3740 cm^{-1} have been assigned, respectively, to Si-OH and SiO-H stretching modes of "surface vicinal"[18] or "isolated" silanols. Isolated silanols are here defined as those silica groups that contain three oxygens bridged to other silica groups and one hydroxyl group which is negligibly hydrogen-bonded. Hydrogen-bonded vicinal silanol groups have been associated with Raman bands at 3615 and 3680 cm^{-1}.[18] The Raman bands near 490 and 610 cm^{-1} have been attributed to silica ring structures by Galeener,[24] who designated the bands D_1 and D_2. The relative intensities of the isolated silanol bands decrease over the temperature interval 350–650°C as the relative intensities of D_1 and D_2 increase (see Fig. 3.10 and references 12,

TABLE 3.1 Heat of Formation of $\equiv$Si—O—Si$\equiv$ (350–650°C) Assuming 2 $\equiv$SiOH → $\equiv$Si—O—Si$\equiv$ + H$_2$O

Change in surface area (m^2/g)	263
Change in surface energy, assuming $\gamma = 250$ ergs/cm^2	−16 cal/g
Moles of water evolved	$(6-11) \times 10^{-4}$/g
Heat of vaporization of water, assuming $\Delta H_v = 30$ cal/mole	5.5–10.4 cal/g
Range of energy change measured by DSC	12–23 cal/g
Average heat of formation of $\equiv$Si—O—Si$\equiv$	26–56 kcal/mole

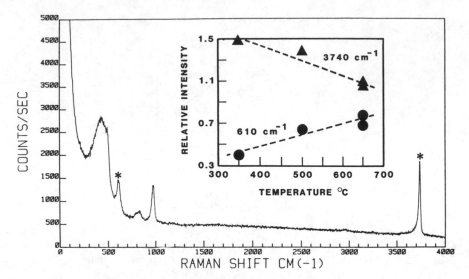

Figure 3.10 Raman spectrum of B2 after dehydration at 350°C and (inset) the change in D_2 and SiO—H during heating between 350 and 650°C at 20°C/min.

18, and 23). This behavior suggests that the structures giving rise to D_1 and D_2 are condensation products of isolated silanols. This hypothesis for the source of D_1 and D_2 recently was confirmed[25] by experiments involving isotopic substitution of ^{18}O for some of the hydroxyl oxygen atoms in the isolated silanols of silica gels. The D_1 and D_2 bands resulting from heating the isotopically substituted gels to 600°C appear at lower frequency than the D_1 and D_2 bands from gels with natural isotopic contents of oxygen. The observed frequency shifts are interpreted as a consequence of the incorporation of ^{18}O from the silanol into the structures giving rise to the D_1 and D_2 bands.

Heat treatments which cause complete sintering of the gel structure, eliminating all internal surfaces, reduce the relative intensities of D_1 and D_2 to approximately the intensities found in vitreous silica (Fig. 3.11). The structures giving rise to D_1 and D_2 therefore appear to be metastable and relax to a more thermodynamically favored structure when the viscosity of the gel is reduced sufficiently. Although the heat associated with this process cannot be measured for silica, which sinters above 727°C (the upper temperature limit of DSC), combined Raman and DSC measurements on two similar alkali borosilicate gels whose Raman spectra also exhibit a strong D_2 band show the removal of D_2 and an associated exotherm between 660 and 725°C.[15,27]

Michalske and Bunker[27] addressed the question of what constitutes an isolated silanol on a silica surface. Their modeling suggests that the closest OH distance between neighboring silanols [sharing a common siloxane bond; see Eq. (6)] is 0.32 nm, which is approximately the van der Waals contact distance for nonbonded oxygen atoms and is ~0.05 nm longer than typical strong hydrogen bonds involving O—H—O.[28] Therefore, even neighboring

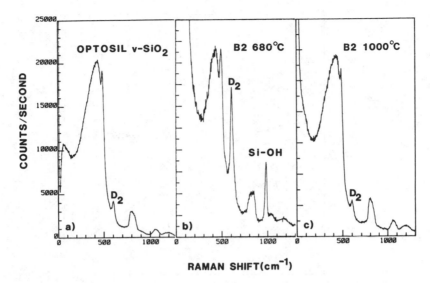

Figure 3.11 Raman spectra of (a) vitreous-SiO$_2$ type III, (b) porous silica gel (B2) dehydrated at 680°C, and (c) B2 completely densified at 1000°C.

$$\overset{OH}{\underset{O}{\overset{|}{Si}}}\overset{\cdot}{\underset{\longleftarrow>3\text{Å}\longrightarrow}{}}\overset{OH}{\underset{O}{\overset{|}{Si}}} \longrightarrow \quad Si \underset{O}{\overset{O}{\diamond}} Si \quad + \quad H_2O \tag{6}$$

silanols should appear spectroscopically as isolated. Silanols which are not neighbors would not hydrogen bond unless puckering of the surface topography brings the hydroxyl groups into proximity.[27] Such hydrogen-bonded silanol groups are the most easily removed from a silica surface via dehydroxylation and disappear at relatively low temperatures.[27] Condensation of two neighboring silanols via Eq. (6) would yield a twofold ring structure (edge-sharing tetrahedra). Condensation of nonneighboring isolated silanols would lead to higher order rings such as the threefold ring proposed by Galeener to account for D$_2$.[24] The heat of formation of threefold rings (Si–O–Si angle = 130°) is reported[22] to be + 19 kcal/mole, slightly below our lower estimate of the heat of formation of Si–O–Si from silanol species (Table 3.1). Thus the available data suggest that the structures giving rise to the D$_1$ and D$_2$ Raman bands are small, strained ring structures formed by condensation of isolated silanols but do not uniquely link either Raman band with a specific structural species.

CONCLUSION

Our results show that, for metal alkoxide-derived gels, the network structure is established as a product of condensation reactions which occur in solution prior

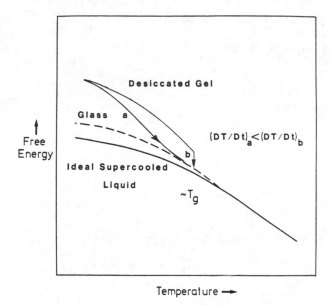

Figure 3.12 Schematic representation of energy change during gel densification. Rapid heating causes the metastable structure to persist to higher temperatures.

to gelation and during the gel-to-glass conversion. In solution these reactions lead to metastable, hydroxylated, open networks which may be retained (at least in part) to elevated temperatures where they dramatically influence viscosity and hence densification kinetics. Condensation reactions at elevated temperatures lead to further changes in the intermediate range order of the network. Depending on the heating rate, there are many pathways by which gels can be converted to glass. The energetics associated with this conversion are presented schematically in Fig. 3.12.

Higher heating rates retain the metastable structure to higher temperatures, ultimately providing a very large driving force for densification and, in some cases, reducing the densification temperature. Surface area, excess free volume, and the formation of metastable surface species all contribute to this high-energy configuration. By comparison colloidal gels have generally lower surface areas and reduced hydroxyl contents and excess free volume reducing their associated energy. This distinction (between colloidal and alkoxide derived gels) may be exploited by intelligent processing to produce dense glasses (especially films) at the lowest possible temperatures.

REFERENCES

1. R. K. Iler, *The Chemistry of Silica*, Wiley, New York, 1979.
2. C. J. Brinker, W. D. Drotning, and G. W. Scherer, in *Better Ceramics Through Chemistry*, C. J. Brinker, D. E. Clark, and D. R. Ulrich, Eds., Elsevier-North Holland, New York, 1984.

3. S. Melpolder, to be published.

4. R. A. Assink and B. D. Kay, in *Better Ceramics Through Chemistry*, C. J. Brinker, D. E. Clark, and D. R. Ulrich, Eds., Elsevier-North Holland, New York, 1984.

5. K. D. Keefer, these proceedings.

6. D. W. Schaefer and K. D. Keefer, in *Better Ceramics Through Chemistry*, C. J. Brinker, D. E. Clark, and D. R. Ulrich, Eds., Elsevier-North Holland, New York, 1984.

7. K. D. Keefer and C. J. Brinker, unpublished results.

8. G. W. Scherer and J. C. Luong, *J. Non-cryst. Solids* **63**, 163–172 (1984).

9. M. D. Sacks and T. Y. Tseng, *J. Am. Ceram. Soc.* **67**, 532–537 (1984).

10. C. J. Brinker and G. W. Scherer, in *Ultrastructure Processing of Ceramics, Glasses, and Composites*, L. L. Hench and D. R. Ulrich, Eds., Wiley, New York, 1984.

11. J. Zarzycki, M. Prassas, and J. Phalippou, *J. Mater. Sci.* **17**, 3371–3379 (1982).

12. C. J. Brinker, E. P. Roth, G. W. Scherer, and D. R. Tallant, accepted for publication, *J. Non-Cryst. Solids*.

13. T. A. Gallo and L. C. Klein, to be published.

14. N. Tohge, G. S. Moore, and J. D. Mackinzie, *J. Non-Cryst. Solids* **63**, 95 (1984).

15. C. J. Brinker, G. W. Scherer, and E. P. Roth, accepted for publication in *J. Non-Cryst. Solids*.

16. M. Yamane, S. Aso, S. Okano and T. Sakaino, *J. Mater. Sci.* **14**, 607 (1979).

17. R. F. Bartholomew, in *Treatise on Materials Science and Technology*, 22, Glass Vol. III, M. Tomozawa and R. H. Doremus, Eds., Academic, New York, 1982.

18. D. M. Krol and J. G. vanLierop, *J. Non-Cryst. Solids* **3**, 131 (1984).

19. R. F. Bartholomew, *J. Non-Cryst. Solids* **56**, 331 (1983).

20. O. S. Narayanaswamy, *J. Am. Ceram. Soc.* **54**, 491 (1971).

21. T. A. Gallo, C. J. Brinker, G. W. Scherer, and L. C. Klein, in *Better Ceramics Through Chemistry*, C. J. Brinker, D. E. Clark, and D. R. Ulrich, Eds., Elsevier-North Holland, New York, 1984.

22. G. V. Gibbs, private communication.

23. F. L. Galeener, *J. Non-Cryst. Solids* **49**, 53 (1982).

24. C. J. Brinker and D. R. Tallant, to be published.

25. T. A. Michalske and B. C. Bunker, *J. Appl. Phys.* **19**, 2686 (1984).

26. V. Gottardi, M. Gugliemi, A. Bertoluzza, C. Fagnano, and M. A. Morelli, *J. Non-Cryst. Solids* **63**, 71 (1984).

27. S. Melpolder, unpublished results.

4

USE OF DRYING CONTROL CHEMICAL ADDITIVES (DCCAs) IN CONTROLLING SOL–GEL PROCESSING

LARRY L. HENCH

Department of Materials Science and Engineering
University of Florida
Gainesville, Florida

INTRODUCTION

Rapid, reliable production of gel monoliths is necessary to realize many of the potential advantages of this type of ultrastructure processing.[1] Zarzycki has reviewed the many factors that lead to drying stresses and cracking of gels.[2] He shows that the drying stress is a function of pore size and rate of evaporation of the pore liquor, which depends on the liquor vapor pressure. In a recent series of papers we have demonstrated the use of organic additions to alkoxide sols, termed drying control chemical additives (DCCA), to control the rate of hydrolysis and condensation, pore size distribution, pore liquor vapor pressure, and drying stresses.[3–10] These studies followed an initial inquiry by M. Prassas in 1982 while at the University of Florida which showed the potential of using formamide as an additive in alkoxide sol–gel processes. Shoup[11] had previously shown that formamide could be added to colloidal silica gels to improve processibility. However, the silica products so produced are opaque. By use of the DCCAs, which includes formamide (NH_2CHO), glycerol ($C_3H_8O_3$), and several organic acids, such as oxalic acid ($C_2O_4H_2$), and alkoxide precursors it is

52

possible to produce with 100% reliability a wide range of sizes and shapes of optically transparent dried gel monoliths of SiO$_2$, Li$_2$O—SiO$_2$, Na$_2$O—SiO$_2$, and Na$_2$O—B$_2$O$_3$—SiO$_2$ within a several-day processing schedule in ambient atmospheres.

SiO$_2$ GEL PROCESSING

Figure 4.1 shows an optimized flow diagram for producing pure SiO$_2$ gels using formamide in the tetramethylorthosilicate (TMOS) system. The DCCA reduces gelation, aging and drying times, the drying stress, and increases the size of gel monoliths that can be made up to >100 cm^3 after drying for 2 days. This is facilitated by using along with the formamide DCCA, acid catalysis, an optimum solvent volume, and optimum DCCA/solvent ratio.[7] The acid catalysis of the formamide–TMOS system increases the hydrolysis rate and a stronger gel forms. FTIR liquid-cell spectroscopy[7] confirms the extremely rapid formation of siloxane bonds in the acid-catalyzed formamide DCCA–TMOS system, Eqs. (1) and (2):

$$Si(OCH_3)_4 + 4H_2O \rightarrow Si(OH)_4 + 4ROH \qquad (1)$$

and

$$\equiv Si-OH + OH-Si\equiv \rightarrow \equiv Si-O-Si + H_2O \qquad (2)$$

Flow Diagram of Optimized 100S Gel Manufacture

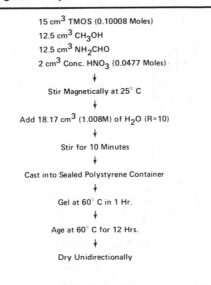

Figure 4.1   Sol–gel processing of pure SiO$_2$ (100S).

Recent studies[12–14] using ^{29}Si NMR have shown that formamide slows down the hydrolysis reaction by five times. However, use of acid catalysis together with formamide results in a 3.5 times faster rate of hydrolysis. The subsequent condensation reaction [Eq. (2)] of the acid-catalyzed formamide system has a much smaller rate constant than the formamide-containing sols that are neutral or basic. Use of the formamide DCCA, compared with methanol alone, leads to a larger gel network. Consequently, a larger pore size distribution is developed in the gel but still with a narrow distribution of pores (see Fig. 6.6 of Chapter 6 and Fig. 4.2). The gel network has substantially larger necks and greater strength. Therefore, large silica gel monoliths made with formamide DCCAs can be dried much more rapidly without cracking (Fig. 4.3). Use of oxalic acid as a DCCA also controls the size and shape of the pore distribution curve (Fig. 4.2 and Ref. 15). Either formamide or the organic acid DCCA greatly decreases the breadth of the pore distribution which decreases the magnitude of capillary stresses induced during drying.

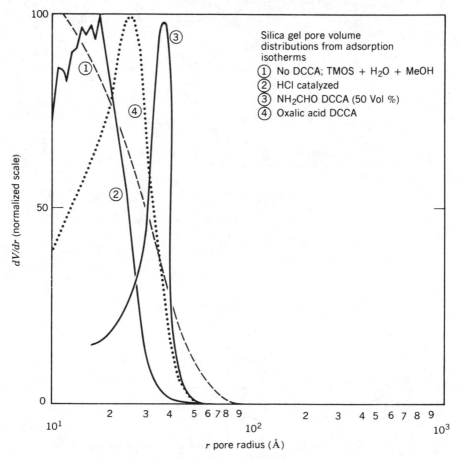

Figure 4.2   Pore distributions of dried SiO$_2$ gel monoliths.

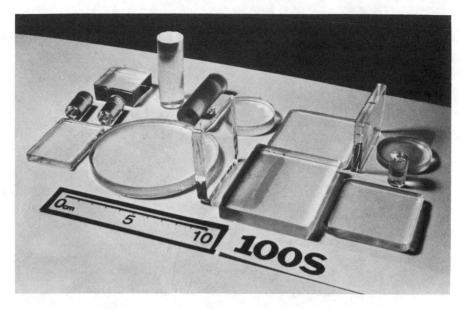

Figure 4.3 As dried pure SiO₂ (100S) transparent optical components made by sol–gel casting.

DCCA Control Mechanisms

The above results suggest the mechanisms for DCCA control of silica sol–gel processing as depicted schematically in Figs. 4.4–4.7. Figure 4.4 illustrates the sequence of structural changes that must be controlled in order to produce large-scale monoliths with a range of densities. Our conclusion is that addition of DCCAs in the sol stage, Fig. 4.5, Step 1 affects each of the succeeding stages, Figs. 4.6–4.7, Steps 2–5.

Addition of a basic DCCA such as formamide produces a large sol–gel network with uniformly larger pores. An acidic DCCA, such as oxalic acid, in contrast results in a somewhat smaller scale network after gelation but also with a narrow distribution of pores. Thus, either basic or acidic DCCAs can minimize differential drying stresses by minimizing differential rates of evaporation and ensuring a uniform thickness of the solid network that must resist the drying stress. Achieving a uniform scale of structure at gelation also results in uniform growth of the network during aging which thereby increases the strength of the gel and its ability to resist drying stresses (see Ref. 6). Recent work[12-14] indicates that this ultrastructural control is due to the DCCA's effect on the rates of both hydrolysis and polycondensation, as discussed above.

Without a DCCA a wide range of pore sizes (Fig. 4.2) and diameter of solids network are produced (Figs. 4.5–4.7) when gelation occurs. Differential growth of the silica network will thereby occur during aging (see Iler, Chapter 1) due to local variations in solution-precipitation rates. The net effect is an aged gel

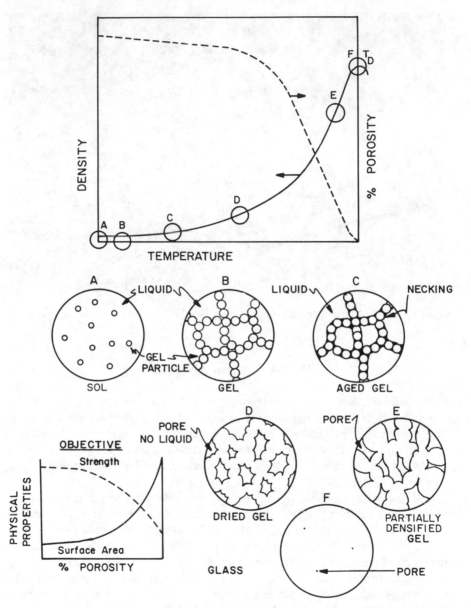

Figure 4.4   Schematic of sol–gel processing steps and resulting structures.

structure such as depicted in Figs. 4.6–4.7, with many regions susceptible to cracking during drying.

An effect of DCCA, however, must also be capable of being removed during densification before pore closure (Step 5, Fig. 4.7). The DCCA must also be capable of removal during drying without producing a residue which is sensitive to moisture. This is a problem with formamide, as described previously.[16]

Control of Sol--Gel Processing with Organic Acid DCCA's

STEP 1: Sol FORMATION

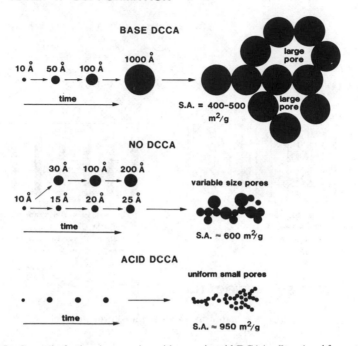

BASE DCCA

10 Å   50 Å   100 Å   1000 Å

time

large pore

S.A. = 400–500 m²/g

large pore

NO DCCA

30 Å   100 Å   200 Å

10 Å   15 Å   20 Å   25 Å

time

variable size pores

S.A. ≈ 600 m²/g

ACID DCCA

uniform small pores

time

S.A. ≈ 950 m²/g

Figure 4.5   Control of sol–gel processing with organic acid DCAAs. Step 1: sol formation.

STEP 2: GELATION

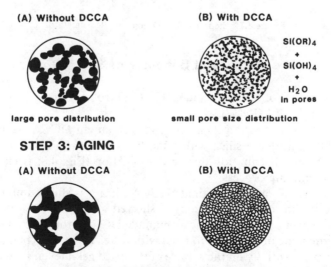

(A) Without DCCA

large pore distribution

(B) With DCCA

$SI(OR)_4$
+
$SI(OH)_4$
+
H_2O
in pores

small pore size distribution

STEP 3: AGING

(A) Without DCCA

(B) With DCCA

Figure 4.6   Steps 2 and 3: Gelation and aging.

STEP 4: DRYING (Minimize Drying Stress Due to

$$p_i = \frac{2\gamma \cos\theta}{d_i} \text{ by}$$

Minimizing Δd_i; $\Delta p_i = 0$ when

$$\Delta d_i = 0)$$

(A) Without DCCA   (B) With DCCA

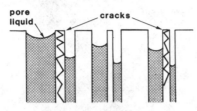 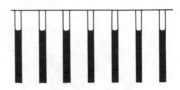

pore
liquid   cracks

large differential evaporation   little differential evaporation
large stresses, large σ uniform σ
distribution   uniform stress distribution
→ CRACKING   → NO CRACKS

STEP 5: DENSIFICATION

DCCA Eliminated as Vapor before Pore Closure

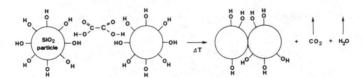

Figure 4.7   Steps 4 and 5: Drying and densification.

Properties and Behavior of Silica Gels

Physical properties of the formamide DCCA controlled silica gels dried at 60°C are listed in Table 4.1 (Ref. 7).

Viscous sintering of the gels made with formamide DCCA starts at about 800°C resulting in dense silica with a hardness of 400 DPH at 1000°C.[7] The evolution of pore distribution during densification (Fig. 4.8) is characteristic of a viscous sintering process.[7]

X-ray diffraction of the formamide DCCA silica gels showed no evidence of devitrification; however, FTIR analysis showed a 926 cm^{-1} SiOH peak still present.[7] The major problem in the formamide DCCA process is a tendency for residual formamide in the pores to react with water vapor.[16] When the adsorption occurs preferentially on the surface of the dried gel uneven stresses develop between the surface and the bulk and cracks develop. Therefore, it is essential

TABLE 4.1 Physical Properties of DCCA-Modified Silica Gel Dried at 60°C

Density	1.42 g/cm^3
Hardness	4 DPH (100-g load)
Tensile strength	200 psi
Index of refraction	1.442
Surface area	750 m^2/g
Pore volume	1.0 cm^3/g
Average pore size	40 Å

to eliminate residual formamide without exposure to water vapor if full densification of monoliths is to be achieved.

In order to avoid the moisture related densification problems associated with the formamide DCCA, organic acid DCCAs have also been used with TMOS and H$_2$O to form large monolithic silica gels.[9,10] The specific surface area of the fully dried silica gels made in this manner[10] using oxalic acid DCCA is 690 m^2/g prior to densification with an average pore size of only 20 Å. The

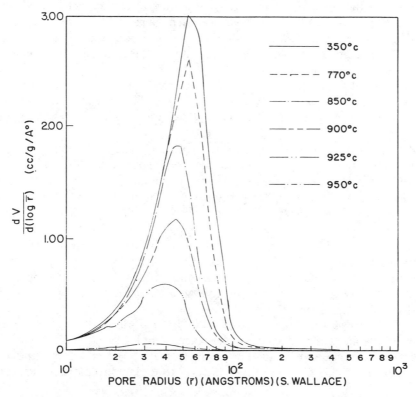

Figure 4.8 Change in 100S gel pore distribution with sintering temperature.

tail of the pore distribution does not exceed 70 Å which accounts for the excellent optical transparency of these silica monoliths.

There is little change in the pore distribution of the DCCA silica gels until densification occurs at $T > 600°C$, Fig. 4.8. The most interesting feature of densification of these gels is that the mesopores are eliminated first and subsequent densification occurs with the mean pore size remaining small. Densification primarily is due to only the number of pores decreasing.

The silica gel monoliths made with oxalic acid DCCA show a broad range of physical properties depending upon densification temperature (Fig. 4.9), while optical transparency is maintained throughout.

By varying the densification temperature it is possible to obtain an index of refraction of the silica matrix between $n = 1.397$ (at $\lambda = 0.6328 \mu m$) and that of dense vitreous silica, $n = 1.457$. This makes it possible to produce silica lenses of very low index of refraction when the microporosity is taken into account. If we assume the same dispersion reported by Malitson for vitreous silica,[17] these gel-derived silicas provide a family of optical components with properties not previously available from melt-derived processes. With the exception of the lowest temperature samples (150°C) this wide range of n is achieved without sacrifice of the IR absorption edge.[9]

For certain applications, an important feature of the sol–gel derived silica lenses is their low density. A sol–gel silica lens densified at 750°C is environmentally stable and half the strength of vitreous silica and only 61% of the density. Thus, large lenses will have substantially lower weight requiring less support structure, and so on.

For optical applications such as large mirrors the sol–gel silica monoliths offer the further weight saving advantage of being able to be cast with an array of indentations avoiding the assembly of individual pieces to form an egg-crate configuration.

Another feature of the gel silica system is that the linear mechanical property–density relationships discussed in Ref. 10 and based on the data shown in Fig. 4.9 make it possible for a designer to optimize a component for a given application. Furthermore, the processing temperature–property relationships shown herein make it possible to realize the design objectives within the device by simply altering the densification temperature or time.[15] This can be done remotely by using on-line measurement of any density sensitive property such as IR reflectivity with feedback to the furnace controls. Thus, final properties can be carefully tailored to meet a broad range of design specifications. This type of on-line process control is generally not possible with either traditional processing of ceramics or melt casting of glasses.

The 10 × variation in microhardness (DPN)* of gel-derived silica components (Fig. 4.9) is another major advantage when complex grinding and polishing operations are required. Very fast removal of material is possible for components densified up to 750°C. A final densification to design specifications can then

---

*DPN = Diamond Pyramid Number.

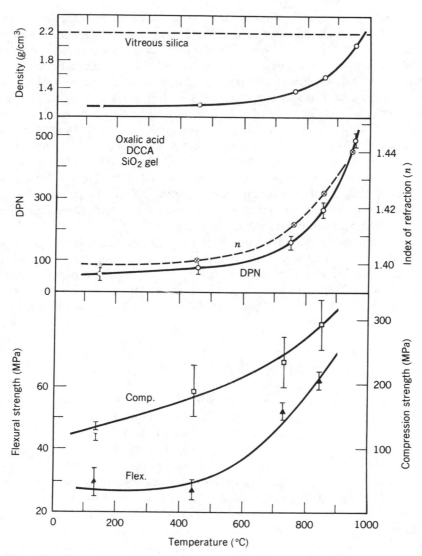

Figure 4.9 Change in physical properties of 100S silica gel monoliths with temperature.

follow the rapid machining, grinding, and polishing. This feature coupled with the ease of casting near net shapes in molds makes it possible to fabricate complex optical components very rapidly. Because of the very low thermal expansion coefficient of silica and the superb thermal shock resistance of the microporous monoliths it is possible to heat components to the upper densification temperatures very rapidly, densify, and quickly cool. This simplifies sequential machining and densification operations. It also means that complex shapes do not crack when rapidly heated which is important for some applications.

The data shown herein also indicate that permanent gradients in physical properties, such as index of refraction, can be achieved by imposing thermal gradients during densification. Since the density varies by as much as 0.005 $(g/cm^3)/°C$ over the range from 750–950°C it takes only a small positive or negative thermal gradient to make a density and index gradient. Focused laser beams, or microwave heaters can produce precise local variations in properties. Use of cryogenic sinks at a critical densification stage can induce negative property gradients. Because of the nonlinear dependence of $\Delta n/\Delta T$ during thermal treatments, Fig. 4.9, it is possible to induce very sharp spatial index gradients, $\Delta n/\Delta X$, in precise configurations such as required for optical networks or digital storage. As far as the authors know, this is the first method of achieving a wide range of positive *and* negative index gradients without resort to chemical compositional profiles. Consequently, it may be possible to avoid potential environmental instabilities associated with compositional gradients by using thermally induced densification gradients.

The nonlinear $\Delta n/\Delta T$ dependence of Fig. 4.9 also suggests the possibility for producing spatial arrays or configurations of a wide range of differing indices of refraction or dispersion with very small separation between lines or dots. Complex waveguide configurations heretofore impossible can thereby be produced in pure silica without use of chemical diffusion and its associated complications and restrictions.

$Na_2O - SiO_2$ GELS

A similar procedure to that outlined above was developed for 20 mole% Na_2O–80 mole% SiO_2 (20N) and 33 mole% Na_2O–67 mole% SiO_2 (33N) gels.[3,4,6,8] It was found that by using formamide and glycerol as DCCAs the critical shrinkage rate of the 20N gels was substantially increased. Consequently, the evaporation rate of the pore liquor was increased and the drying time shortened severalfold. In addition the DCCAs accelerate the aging process such that the gel strength is nearly doubled in a 12-hr period.[6] A substantially higher temperature can be used if required to remove the formamide compared with methanol which enables interparticle necks in the gel to develop sufficiently to resist appreciable drying stresses. A combination of aging times and drying temperatures which can lead to large monolithic dried 20N gels are above the processing curves described in Ref. 6. The net effect is to reduce total processing time of $20N$ monoliths to 1–2 days with 100% reliability. Similar results have been obtained for the 33N and the 42S–30B–28N systems.

$Li_2O - SiO_2$ GELS

With control of processing variables such as R, pH, solvent concentration, and lithia and metal organic precursors, it is possible to produce 20L gel monoliths in 2 days or less.[5] $LiOCH_3$ is the preferred alkoxide precursor for lithia, $LiNO_3$

is the preferred inorganic lithia precursor, and TMOS is the preferred silica precursor for 20L gel monoliths.

$SiO_2-Al_2O_3-TiO_2-Li_2O$ GELS

70 mole% $SiO_2-19Al_2O_3-6TiO_2-5Li_2O$ monolithic gels can be prepared using formamide as a DCCA.[18] The pore size, pore volume, and surface area can be controlled by varying the aging conditions. It is possible to follow the crystallization of the gels using hot stage FTIR and x-ray diffraction. The stable crystalline phase is beta spodumene which appears between 375 and 400°C.

CONCLUSION

Basic or acidic additions to an alkoxide sol can greatly alter the rates of hydrolysis and condensation reactions and thereby control the distribution of pore sizes and solid network of the gel. The highly uniform network increases its strength during aging which permits an increase in drying rates without cracking of monolithic components. Use of drying control chemical additives such as formamide or oxalic acid in TMOS sol–gel systems result in optically transparent solids which can be densified in ambient atmospheres with a wide range of physical properties depending upon the densification temperature and time at temperature.

ACKNOWLEDGMENT

The author gratefully acknowledges the U.S. AFOSR Contract #F49620-83-C-0072 for support of this work and the enthusiastic help of S. Wallace (Fig. 4.8), G. Orcel, S. H. Wang, and S. Park (Fig. 4.9).

REFERENCES

1. J. D. Mackenzie, Applications of Sol-Gel Methods for Glass and Ceramics Processing, in *Ultrastructure Processing of Ceramics, Glasses and Composites*, L. L. Hench and D. R. Ulrich, Eds., Wiley, New York, 1984.

2. J. Zarzycki, Monolithic Xero and Aerogels for Gel-Glass Process, in *Ultrastructure Processing of Ceramics, Glasses and Composites*, L. L. Hench and D. R. Ulrich, Eds., Wiley, New York, 1984.

3. S. H. Wang and L. L. Hench, Processing Variables of Sol-Gel Derived (20N) Soda Silicates, Proceedings of the 8th Annual Conference on Composites and Advanced Ceramic Materials, Cocoa Beach, FL, January 15–18, 1984.

4. G. Orcel and L. L. Hench, Effect of the Use of a Drying-Control Chemical Additive (DCCA) on the Crystallization and Thermal Behavior of Soda Silicate and Soda Borosilicate, Proceedings

of the 8th Annual Conference on Composites and Advanced Ceramic Materials, Cocoa Beach, FL, January 15–18, 1984.

5. S. Wallace and L. L. Hench, Metal Organic Derived 20L Gel Monoliths, Proceedings of the 8th Annual Conference on Composites and Advanced Ceramic Materials, Cocoa Beach, FL, January, 15–18, 1984.

6. S. H. Wang and L. L. Hench, Processing and Properties of Sol-Gel Derived 20 Mol %–80 Mol % SiO_2 (20N) Materials, in *Better Ceramics Through Chemistry*, C. Jeffrey Brinker, David E. Clark, and Donald R. Ulrich, Eds., Elsevier, New York, 1984, pp. 71–78.

7. S. Wallace and L. L. Hench, The Processing and Characterization of DCCA Modified Gel-Derived Silica, in *Better Ceramics Through Chemistry*, C. Jeffrey Brinker, David E. Clark, and Donald R. Ulrich, Eds., Elsevier, New York, 1984, pp. 47–52.

8. G. Orcel and L. L. Hench, Physical-Chemical Variables in Processing $Na_2O–B_2O_3–SiO_2$ Gel Monoliths, in *Better Ceramics Through Chemistry*, C. Jeffrey Brinker, David E. Clark, and Donald R. Ulrich, Eds., Elsevier, New York, 1984, pp. 79–83.

9. L. L. Hench, S. H. Wang, and S. C. Park, SiO_2 Gel Glasses, Proceedings of SPIE's 28th Annual International Technical Symposium on Optics and Electro-Optics, San Diego, California, August 19–24, 1984.

10. L. L. Hench and S. C. Park, Microporous Silica Monoliths, submitted to Bulletin American Ceramic Society.

11. R. D. Shoup and W. J. Wein, U.S. Patent #4,059,658, Nov. 22, 1977.

12. G. Orcel and L. L. Hench, The Chemistry of Silica Sol-Gels: Part 1: NMR of Silica Hydrolysis, submitted to *J. Non-Crystalline Solids*.

13. I. Artaki, M. Bradley, T. W. Zerda, J. Jonas, G. Orcel, and L. L. Hench, Chapter 6 this volume.

14. J. Jonas, Chapter 5 this volume.

15. S. C. Park and L. L. Hench, Chapter 18 this volume.

16. L. L. Hench, Environmental Effects in Gel Derived Silicates, in *Better Ceramics Through Chemistry*, C. Jeffrey Brinker, David E. Clark, and Donald R. Ulrich, Eds., Elsevier, New York, 1984, pp. 101–110.

17. I. H. Malitson, *J. Opt. Soc.* **5**, 10 (1965).

18. G. Orcel and L. L. Hench, Chapter 24 this volume.

5

KINETICS AND MECHANISM OF SOL–GEL POLYMERIZATION

JIRI JONAS
Department of Chemistry
School of Chemical Sciences
University of Illinois
Urbana, Illinois

INTRODUCTION

In spite of the remarkable activity in the area of sol–gel processes,[1-4] an atomic and molecular level understanding of the sol–gel process is lacking. In this chapter several examples of specific NMR[5,6,8] and Raman[7] studies will illustrate the potential of these techniques to study the sol–gel process.

Englehardt and collaborators[9] have clearly demonstrated that ^{29}Si NMR can be used to study the details of the condensation reactions of the monosilicic acid prepared by hydrolysis of $Si(OCH_3)_4$ in diluted HCl. The kinetics of condensation reactions was investigated as a function of SiO_2 concentration and pH. There have also been studies by Harris and collaborators [10-12] who have carried out detailed investigations of aqueous solutions of silicates by ^{29}Si NMR and also reported results on the hydrolysis of $Si(OCH_3)_4$. Assink and Kay[13] used proton NMR to study these reactions as well.

As an example of the high information content of the ^{29}Si NMR technique, Fig. 5.1 shows a typical ^{29}Si NMR spectrum obtained 9 hr after hydrolysis under 5-kbar pressure at $-30°$C. The four resonances depicted here are identified by the arbitrary letter assignment previously used in the literature.[10,14] Peaks B, C, and E lie at -9.4, -19.2, and -29 ppm with respect to peak A. In accordance with the line assignments made in the literature,[10,14] peak A

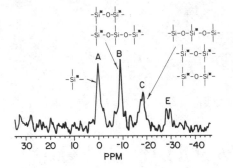

Figure 5.1 ^{29}Si NMR spectrum obtained 9 hr after hydrolysis at $-30°C$ and 5-kbar pressure. Peaks B, C, and E lie at -9.4 ppm, -19.2 ppm, and -29 ppm with respect to peak A.

can be unambiguously attributed to monosilicic acid. Peak *B* is assigned to both the end groups of disilicic and linear trisilicic acid. Peak *C* corresponds to the middle group of linear trisilicic acid and to the four groups of cyclotetra-silicic acid. Peak *E* is believed to arise from the branched groups of the higher polycyclic acids.

In contrast to Fig. 5.1 it is interesting to compare the dimer region of the ^{29}Si NMR spectrum at conditions[8] under which the hydrolysis is not complete. Figure 5.2 shows the dimer frequency region of the ^{29}Si spectrum for pH $\cong 6$ and time $0.3t/t_g$ at 25°C. The asterisk denotes the ^{29}Si being recorded. Since replacement of an alkyl group by a hydroxyl group causes a downfield chemical shift, one observes four main peaks. Again, each peak is split due to the shielding effect of alkyl replacement at the other distant Si atom. Obviously, there is a wealth of information about the mechanism of hydrolysis but quantitative evaluation of the mechanism is quite complex. Assink and Kay[13] have commented on this problem when they introduced so-called sol–gel matrix.

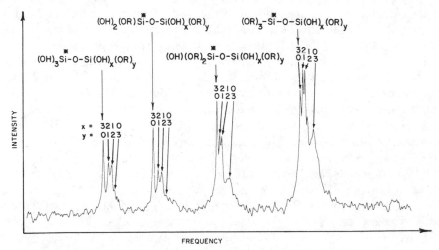

Figure 5.2   Dimer frequency region of the ^{29}Si NMR spectrum obtained at time $0.3t/t_g$ and 25°C; pH $\cong 6$.

TABLE 5.1   Sol–Gel Systems Investigated (Mole%)

Study	H_2O	CH_3OH	$Si(OCH_3)_4$	Additive	T (°C)	pH
HP NMR	54	40.6	5.4	—	−30	3.5
HP Raman	54	40.6	5.4	—	25	3.5
Mo reaction	~54	~20.3	~5.4	~20.3	25	~6
Raman	~54	~20.3	~5.4	~20.3	25	~6

In addition to ²⁹Si one has to realize that modern NMR spectrometers[15] allow us to study nuclei such as ¹¹B, ²³Na, ²⁷Al, ³¹P and that the high-resolution solid-state NMR techniques have not yet been explored to characterize the wet or dry gels.

EXPERIMENTAL

The NMR and Raman experiments have been described in detail in our earlier publications.[5,6,8] The composition of the sol–gel systems studied is given in Table 5.1. The molybdic acid reagent was prepared according to Iler's recommended procedure.[16]

HIGH-PRESSURE²⁹ Si NMR STUDY

Extensive research[17,18] of both scientific and technological nature has been devoted to the optimization of the parameters of the gelation process which produce the highest purity and most homogeneous monolithic glasses, glass–ceramics, and ceramics. However, the optimum solution parameters which generate the most favorable sol to gel transition almost invariably require excessively long gelation times, often in the order of days, even months.

The object of our study[5,6] was to investigate the role of pressure on the polymerization kinetics of sol–gel processes. ²⁹Si NMR was used in this study to monitor the time evolution of all condensed species as a function of pressure in order to explore the possibility of reducing gelation times without affecting the desired characteristics of the resulting wet gels. To the best of our knowledge, this was the first instance of the monitoring of the NMR signal of ²⁹Si nucleus under high pressure to investigate the kinetics of the sol–gel process.

The NMR spectra were analyzed to monitor quantitatively the time evolution of all condensed species in solution during the initial stages of the polymerization reaction. Excessive line broadening and weakening of the NMR signal intensity precluded measurements during the final stages of the polymerization just prior to gelation. The NMR data show that the effect of pressure is to accelerate the condensation process without altering either the path of the

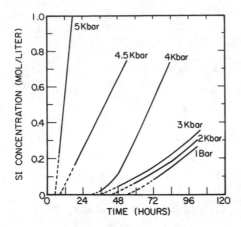

Figure 5.3 Time evolution of the calculated concentration of the higher polymer species as a function of pressure (solid lines). The dashed lines represent extrapolations to zero concentration.

polymerization reaction or the structure of any of the condensed species. At both ambient and higher pressures, the disappearance of the monomer, $Si(OH)_4$ is accompanied by the formation of the dimer and subsequently that of the linear trimer and cyclic tetramer, which in turn leads to the gradual disappearance of the dimer. The formation of the higher polymeric species, not detectable under our experimental conditions, can be indicated by the difference between the total Si concentration ($1.6M$) and the sum of the concentrations of the lower polymeric species (monomer, dimer, trimer, and cyclotetramer). At 1 bar, their presence is detected approximately 50 hr after hydrolysis, whereas at 4.5 kbar they are detected during the initial 5 hr.

The formation of the high polymer species was quantified by subtracting from the total Si concentration, the sum of the monomer, dimer, trimer, and cyclotetramer concentrations. Figure 5.3 shows the time dependence of the concentration of these high polymer species as a function of pressure. Best fit lines were drawn through the points and extrapolated to zero concentration. A detailed analysis of the pressure effects can be found in our earlier paper.[6]

HIGH-PRESSURE RAMAN STUDY

The results of the high-pressure NMR study of the sol–gel process provided motivation for a series of exploratory high-pressure experiments[7] using laser Raman scattering. It is interesting to note that no investigation of the sol to gel transition using Raman spectroscopy has appeared in spite of the well-accepted use of this technique to study dried gels and the gel-to-glass transition.[19,20]

As discussed in more detail in our earlier study[7] we used the Raman band at $830\,cm^{-1}$ to follow the formation of SiO_2 polymers. This band has been assigned to the SiO_2 network stretching vibration.[19,20]

From basic light scattering theory[21] it is known that the Raman intensity I_j of the normal mode j is proportional to the number of oscillating molecules N

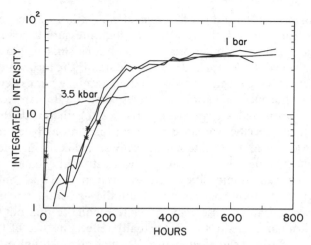

Figure 5.4 Time dependence of normalized intensities of the 830 cm^{-1} band of the SiO$_2$ gels at different pressures. Asterisk denotes the gelation time.

in the scattering volume

$$I_j \propto I_0 v_j^4 N \left(\frac{\partial \alpha}{\partial q}\right)^2 \qquad (1)$$

where v_j is the frequency of the oscillations. Since the intensity of the incident light, I_0, is difficult to determine accurately, and also since the polarizability derivatives for the O—Si—O modes are unknown it is impossible to evaluate from Eq. (1) the number of the Si—O bands in the sample. Also the time evolution study of the Raman spectrum is very difficult. In general, the intensities of two Raman spectra cannot be directly compared, and therefore the only available solution is the use of an internal standard which in this case is methanol.[7]

Our high-pressure Raman experiment[7] showed that pressure can speed up appreciably the condensation process. As seen in Fig. 5.4 the gelation time decreases from about 168 hr (average value) at normal pressure down to 6 hr at 3.5 kbar. As discussed earlier in the NMR study made in our laboratory[5,6] pressure was shown to influence significantly the initial stage of polymerization when dimers, tetramers are formed. The Raman result indicates that pressure also enhances the later stages of the condensation process (the rates increase about eight times). From the final intensity of the Raman band it appears, however, that although the particles grow faster when pressure is applied, the sizes of polymer particles are smaller.

EFFECT OF ADDITIVES ON PARTICLE SIZE

Preliminary results of several promising exploratory experiments[8] are now discussed. The result of the high-pressure Raman study[7] and the investigation of the effect of formamide additive[22] provided motivation for these experiments. As it follows from Fig. 5.4, the result of the high-pressure Raman experiment would indicate that under high-pressure conditions smaller and more dense particles are formed. In contrast, the Raman data for the formamide[22] showed that larger particles are formed. In view of these results we considered it worthwhile to follow the effect of various additives using the Raman technique. The compounds investigated are given in Table 5.2 together with their physical properties. In the specific Raman experiment (conditions listed in Table 5.1) the wet gels were aged at 70°C in hermetically sealed containers for 12 hr and then the intensity of the 830 cm^{-1} SiO$_2$ network stretching band was recorded. The normalized Raman intensities are also given in Table 5.2. Since from the Raman experiments above one cannot obtain the size of the polymer particles formed, it was desirable to use some other technique to find out whether the normalized Raman intensities [$I(R)$] actually reflect the size of the particles. We selected the classical chemical method[16] which gives a rough estimate of the particle size.

In the excellent monograph Iler[16] has critically reviewed various studies of polymerization of silicic acid. There were numerous studies using the well-known reaction of silicic acid with molybdic acid. It is important to emphasize that only monosilicic acid Si(OH)$_4$ (monomer) can react directly with the molybdic acid. Each higher polymeric species depolymerizes at a characteristic rate k.

It was suggested that this reaction rate could be used to estimate the size of particles. Therefore, we carried out the molybdic acid experiments for the Si(OH)$_4$ system with the various additives listed in Table 5.2. The conditions for the reactions are given in Table 5.1. It is appropriate to mention that the depolymerization rate constant k was determined at time $0.9t/t_g$, because at gelation this method fails. From the literature[16] it follows that the depolymerization rate k is inversely proportional to the size of particles ($k \propto 1/\text{size}$), and if

TABLE 5.2 Physical Properties of Additives[a]

Molecule	BP (°C)	Viscosity (cP)	Dipole Moment (debye)	H bond[b]	Gel Time (hr)	$I(R)^c$
Methanol (M)	65	0.62	1.7	A	39	220
Formamide (F)	200	4.32	3.7	D, A	20	343
Dimethyl Formamide (DF)	150	1.1	3.8	D, A	71	300
Acetonitrile (A)	82	0.37	3.9	A	55	450
Dioxane (D)	110	1.44	0	A	75	475

[a]For composition and experimental conditions see Table 5.1.
[b]D denotes donor properties, A denotes acceptor properties.
[c]I(R) denotes normalized Raman intensity (for details, see the text).

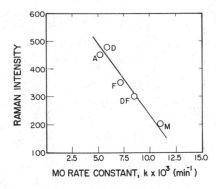

Figure 5.5 Relationship between normalized Raman intensity of the 830 cm^{-1} band versus depolymerization rate constant k obtained from the molybdic acid reaction in systems containing various additives. The Raman data are for a gel kept at 25°C for $10t_g$ and then aged for 12 hr at 70°C. The Mo rate constants are reported for wet gel at 25°C and time $t/t_g = 0.9$. (For details and symbols, see the text.)

our argument about the normalized Raman intensity is correct [$I(R) \propto$ size]. then one would expect that the experimental k's for various additives should be inversely proportional to the normalized Raman intensity $k \propto 1/I(R)$.

It was reassuring to find in Fig. 5.5 that this relation holds quite well. Since we are well aware of the crude approximative nature of our arguments, we have to postpone a detailed analysis till quantitative data about the size and nature of the polymer particles are available from other experimental techniques. However, it is worth pointing out that a simple photometric method and Raman experiment seem to be useful for rapid characterization of the sol–gel process.

ACKNOWLEDGMENT

This work was partially supported by the Air Force Office of Scientific Research under the Grant USAFOSR 81-0010.

REFERENCES

1. H. Dislich and P. Hinz, *J. Non-Cryst. Solids* **48**, 11, (1982).

2. J. D. Mackenzie, *J. Non-Cryst. Solids* **48**, 1 (1982).

3. *Ultrastructure Processing of Ceramics, Glasses and Composites*, L. L. Hench and D. R. Ulrich, Eds., Wiley, New York, 1984.

4. *Better Ceramics Through Chemistry*, C. J. Brinker, D. E. Clark, and D. R. Ulrich, Eds., Elsevier, North-Holland, New York, 1984.

5. I. Artaki, S. Sinha, and J. Jonas, *Mater. Lett.* **88**, 5425 (1984).

6. I. Artaki, S. Sinha, A. D. Irwin, and J. Jonas, *J. Non-Cryst. Solids*, **72**, 391 (1985).

7. T. W. Zerda, M. Bradley, and J. Jonas, *Mater. Lett.* **3**, 124 (1985).

8. I. Artaki, M. Bradley, T. W. Zerda, and J. Jonas, *J. Phys. Chem.* **89**, 4399 (1985).

9. G. Englehardt, W. Altenburg, D. Hoebbel, and W. Wieker, *Z. Anorg. Allg. Chem.* **428**, 43 (1977).

10. R. K. Harris, C. T. G. Knight, and D. N. Smith, *J. Chem. Soc. Chem. Comm.*, 726 (1980).

11. R. K. Harris and C. T. G. Knight, *J. Chem. Soc. Chem. Comm.*, 726 (1980).

12. R. K. Harris, C. T. G. Knight, and W. E. Hull, *ACS Sym. Ser.* **194**, 79 (1982).

13.   R. A. Assink and B. D. Kay, in *Better Ceramics Through Chemistry*, C. J. Brinker, D. E. Clark, and D. R. Ulrich, Eds., Elsevier, North-Holland, New York, 1984, p. 301.

14.   D. Hoebbel, G. Garzo, G. Engelhardt, and A. Till, *Z. Anorg. Allg. Chem.* **450**, 5 (1977).

15.   B. Coleman, in *NMR of Newly Accessible Nuclei*, Vol. 2, P. Laszlo, Ed., Academic Press, New York, 1983, p. 197.

16.   R. K. Iler, *The Chemistry of Silica*, Wiley, New York, 1979, p. 97.

17.   C. J. Brinker, K. D. Keefer, D. W. Schaefer, and C. S. Ashley, *J. Non-Cryst. Solids* **48**, 47 (1982).

18.   B. E. Yoldas, *J. Non-Cryst. Solids* **63**, 145 (1984).

19.   A. Bertoluzza, C. Fagnano, M. A. Morelli, V. Gottardi, and M. Guglielmi, *J. Non-Cryst. Solids* **48**, 117 (1982).

20.   D. M. Krol and J. G. van Lierop, *J. Non-Cryst. Solids* **63**, 131 (1984).

21.   D. A. Long, *Raman Spectroscopy*, McGraw-Hill, London, 1977.

22.   I. Artaki, M. Bradley, T. W. Zerda, J. Jonas, G. Orcel, and L. L. Hench, Chapter 6 this volume.

6

NMR, RAMAN STUDY OF THE EFFECT OF FORMAMIDE ON THE SOL–GEL PROCESS

I. ARTAKI, M. BRADLEY, T. W. ZERDA, AND JIRI JONAS
Department of Chemistry
School of Chemical Sciences
University of Illinois
Urbana, Illinois

G. ORCEL AND LARRY L. HENCH
Ceramics Division
Department of Materials Science and Engineering
University of Florida
Gainesville, Florida

INTRODUCTION

During the past decade, many studies have demonstrated the possibility of using sol–gel processing to prepare glasses and ceramics with potential commercial applications.[1-3] However, until recently the rapid production of monolithic samples on a large scale had not been satisfactorily demonstrated. A new approach[4] using organic drying control additives (DCCAs) such as formamide, greatly facilitates the drying of large monolithic gels. The role of formamide (NH_2CHO) DCCA on hydrolysis and condensation mechanisms has been studied using high-resolution ^{29}Si NMR, Raman spectroscopy, and N_2 adsorption–desorption. Some structural features of the resulting silica gel monoliths are reported as well.

73

RESULTS AND DISCUSSION

Sample Preparation

Sols were prepared by the dropwise addition of an H_2O/methanol mixture to the initiating Si alkoxide reagent, $Si(OCH_3)_4$, (TMOS), which had been previously mixed with the DCCA, formamide. For comparative purposes, additional sols in the absence of the DCCA were prepared by substituting an equivalent amount of methanol for the formamide. The solutions prepared in the presence and in the absence of formamide are denoted by I and II respectively. The silicon concentration was adjusted to $1.6M$ and a 10:1 mole ratio of H_2O to Si was used. The sols were allowed to gel either at 14°C (NMR experiment) or at 20°C (Raman experiment) and were later stored at room temperature in sealed glass tubes.

NMR

^{29}Si NMR measurements were performed on a Nicolet-250 pulsed spectrometer operating at 49.69 MHz, equipped with a variable temperature capability and a spin-decoupler accessory. Spectra were recorded every hour using 1000 pulses of 10 μs duration at 2-s intervals. Figure 6.1 shows a typical natural abundance

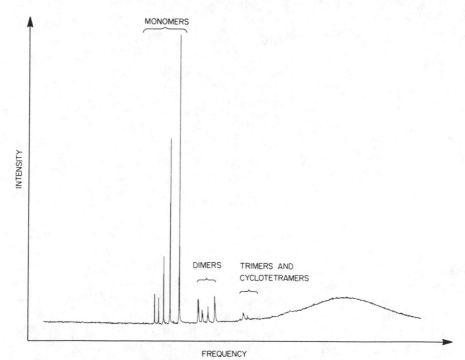

Figure 6.1 Typical ^{29}Si NMR spectrum of solution I, 12 hr after preparation. The individual peaks have been assigned to specific chemical species as marked in the figure and described in the text.

^{29}Si NMR spectrum for solution I, approximately 12 hr after preparation. The characteristic features of the spectra, such as the nature of the NMR resonances along with their corresponding chemical shifts, are identical for solutions I and II. However, time evolution of the intensity of the individual peaks is markedly different in the two solutions. The NMR evidence does not support the possibility of formamide being chemically bonded to the Si network through the scissioning of stressed bridging oxygen bonds.[4]

In accordance with previous line assignments[5] the spectra may be divided into groups of resonances which correspond to the monomeric, dimeric and trimeric, or cyclotetrameric silicon species, as labeled in Fig. 6.1. Since replacement of an alkyl (OCH$_3$) group by a hydroxyl group causes a downfield chemical shift, the resonances belonging to the monomeric group are assigned from left to right, to Si(OH)$_4$, Si(OH)$_3$(OCH$_3$), Si(OH)$_2$(OCH$_3$)$_2$, Si(OH)-(OCH$_3$)$_3$, and TMOS respectively. The same principle applies to the dimeric group. The varied chemical environment in trimers and higher order oligomers produces overlapping resonances which result in broad peaks with little information content.

Figure 6.2 depicts the rate of disappearance of TMOS, along with the time evolution of one of the partially hydrolyzed monomers, Si(OH)(OCH$_3$)$_3$, in solutions I and II. The significantly slower rate of disappearance of TMOS in solution I indicates that the role of formamide is to inhibit the hydrolysis reaction. Assuming that immediately after the preparation of the sol, TMOS only participates in the hydrolysis reaction according to

$$H_2O + Si(OCH_3)_4 \xrightarrow{\ k_H\ } Si(OH)(OCH_3)_3 + CH_3OH, \qquad (1)$$

and since an excess of H$_2$O is present, the disappearance of TMOS should follow first order kinetics, such that the initial slope of ln[Si(OCH$_3$)$_4$] versus time should predict the magnitude of k_H. The ratio of the initial slopes in solutions I

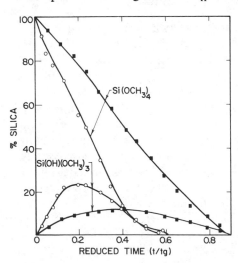

Figure 6.2 Time evolution of starting reagent, Si(OCH$_3$)$_4$ and partially hydrolyzed monomer Si(OH)(OCH$_3$)$_3$ in solutions I ($\square$) and II ($\bigcirc$).

and II was determined to be 1 : 5 which suggests that the presence of formamide may lead to a fivefold reduction of the hydrolysis rate.

The following explanation can be provided for this observed behavior. The hydrolysis reaction proceeds by a nucleophilic substitution mechanism[6] whereby the Si atom is attacked by a negatively charged hydroxyl group. To achieve a maximum charge separation, the attacking and departing groups must be situated on opposite sides of the Si atom.[6] The bulk viscosity of solution I was measured and found to be approximately 15% greater than the viscosity of solution II. As a result, the mobility of the molecules in the formamide solution may be restricted.

A second reason for the retardation of the hydrolysis rate may be the presence of bulky substituents directly attached to the Si atom.[6] Generally, condensation between incompletely hydrolyzed species leaves bulky and highly basic alkoxy groups attached to Si. It is likely that this process is more prominent in solution I rather than in solution II because the silanol groups are known to form stronger hydrogen bonds with formamide than with methanol.[7] Hydrogen-bonded formamide provides a more effective shield for the silanol group hindering its reaction with $\equiv$Si—O$^-$ in the process of condensation and causing instead a preferential attack of a nonhydrolyzed $\equiv$Si—OCH$_3$ group to displace an OCH$_3^-$ rather than an OH$^-$ ion. This would promote condensation between incompletely hydrolyzed species, leaving more bulky substituents which results in a gradually decreasing rate of hydrolysis.[8]

Raman Spectroscopy

The Raman spectra were measured approximately every 12 hr with an Ar ion laser operating at 488 nm and 0.6-W output power. A slit width of 1.2 cm^{-1} was used, and the data were scanned at 4 cm^{-1} interval. Only the nonpolarized spectra were recorded, as it appeared that there is no depolarized component for the SiO$_2$ network vibrations.

A typical spectrum for a sample after complete hydrolysis is shown in Fig. 6.3. The sample is composed of different species (H$_2$O, CH$_3$OH, NH$_2$CHO, TMOS, and various SiO$_2$ polymers) and all of the characteristic bands for each species are present in the spectrum. The bands due to SiO$_2$ network vibrations are easily identifiable as their intensity grows continuously with time, reflecting the formation of SiO$_2$ polymers. From light scattering theory it is known that the Raman intensity is proportional to the number of bonds in the scattering volume. Unfortunately, quantitative analysis is impossible, and any qualitative comparison of two spectra registered at various times or of different samples is possibly only using an internal standard technique.[9] For formamide solutions any of the formamide bands may be used, but in order to compare the growth of the SiO$_2$ polymers in various solvents it is better to choose the C—O band of methanol as the standard. This technique, however, has one disadvantage: the number of methanol molecules in solution increases in time due to the ongoing hydrolysis reaction. During the sol-to-gel transformation the TMOS

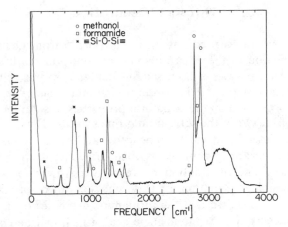

Figure 6.3 Raman spectrum of solution I, 13 hr after preparation. The marked bands are assigned to vibrations of different components.

bands decreased steadily, and after approximately 20 hr they disappear, indicating completion of the hydrolysis process. This observation is confirmed by the NMR results; Fig. 6.2 indicates that only 4% of TMOS is still present in the solutions after about $0.7T_g$, where T_g is the gelation time for a sol. This allows us to assume that after gelation the amount of methanol remains constant.

Figure 6.4 presents the growth of the normalized intensities of the 830 cm^{-1} band originating from the SiO_2 network stretching vibrations.[6] In both solvents, the peak increases rapidly in time, continues to grow after the gelation point, and after approximately $2T_g$ it remains constant. The main difference lies in the total intensities. The peak in the formamide solution is 40% more intense than its counterpart in methanol which indicates that large or more condensed particles exist in the formamide system.

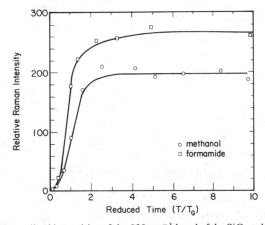

Figure 6.4 Normalized intensities of the 830 cm^{-1} band of the SiO_2 gels versus time.

Nitrogen Desorption

The pore analysis was carried out on a Autosorb 6 from Quantachrome. A technical description of the apparatus can be found in the literature,[10] and the method of data analysis is discussed in Chapter 16 of this volume.

Three samples were prepared in order to determine the influence of the concentration of formamide on the structural properties: 2 solution I gels with 25 and 50 vol% of DCCA in the solvent, and one solution II sample. The results are reported in Figs. 6.5 and 6.6. It can be seen that the pore radius, pore volume, and the specific surface area increase when the concentration of formamide DCCA increases. The distribution curves are narrower for higher DCCA contents. A possible explanation for this behavior can be provided using the NMR results. It was shown that the formamide decreases the hydrolysis rate. In fact the concentration of monosilicic acid, which can be considered as a

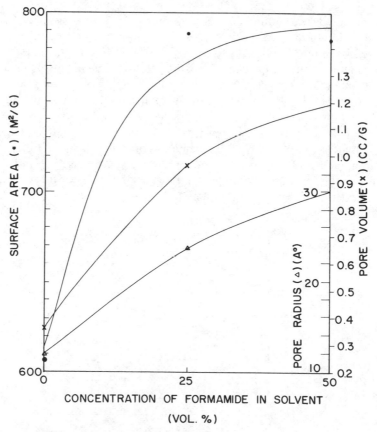

Figure 6.5 Variation of the surface area, pore radius, and pore volume with the concentration of formamide.

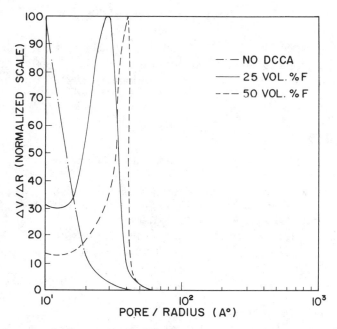

Figure 6.6 Pore size distribution of gels prepared with different amounts of formamide, DCCA.

potential nucleus for a gel particle decreases when the amount of formamide increases. This leads to larger particle sizes at higher DCCA contents, and thus larger pores are obtained. Due to the large amount of water used, and the low temperatures involved, a high density of silanol groups is expected on the surface of the gel particles. By hydrogen bonding to these silanol groups, the formamide can form a layer which may decrease the density of gel particles in the sol. This increases the pore volume and the mean pore diameter by increasing the distance between the particles which are not in contact.

CONCLUSIONS

The formamide DCCA has been shown to hinder the hydrolysis reaction and possibly favor the condensation of incompletely hydrolyzed species, thereby entrapping a larger percentage of alkyl groups in the Si polymers of the sol. As a result, larger and more highly branched but less dense SiO_2 networks are formed as confirmed by the Raman experiments and pore size determinations. Lower drying stresses will be generated since there are larger pores. Also larger necks will make the gel stronger and more capable of resisting drying stresses. Therefore, the net effect of the formamide DCCA is to enhance the rate of drying without inducing cracks in the gel monolith.

ACKNOWLEDGMENTS

The authors gratefully acknowledge partial financial support of the AFOSR (GO and LLH) #F49620-83-C-0072 and (IA and JJ) #81-0010, and the encouragement of D. R. Ulrich for interinstitutional research.

REFERENCES

1. R. K. Iler, *The Colloid Chemistry of Silica and Silicates* Cornell University Press, Ithaca, NY, 1955.
2. N. J. Arfsten, R. Kaufmann, and H. Dislich, in *Ultrastructure Processing of Ceramics, Glasses and Composites*, L. L. Hench and D. R. Ulrich, Eds., Wiley, New York, 1984.
3. J. Livage, *Mater. Res. Soc. Symp. Proc.* **32**, 125 (1984).
4. G. Orcel and L. L. Hench, *Mater. Res. Soc. Symp. Proc.* **32**, 79 (1984).
5. D. Hoebbel, G. Garzo, G. Engelhardt, and A. Till, *Z. Anorg. Allg. Chem.* **450**, 5 (1977).
6, K. D. Keefer, *Mater. Res. Soc. Symp. Proc.* **32**, 15 (1984).
7. P. Schuster, G. Zundel, and C. Sandorfy, *The Hydrogen Bond*, Vol. III, North Holland Publishing, Amsterdam, 1976.
8. I. Artaki, M. Bradley, T. W. Zerda and J. Jonas, *J. Phys. Chem.* **89**, 4399 (1985).
9. A. Bertoluzza, C. Fagnano, M. A. Morelli, V. Gottardi, and M. Guglielmi, *J. Non-Cryst. Solids* **48**, 117 (1982).
10. J. E. Shields and S. Lowell, *Am. Lab.*, November 1984, p. 17.

7

A MULTINUCLEAR (^{1}H, ^{29}Si, ^{17}O) NMR STUDY OF THE HYDROLYSIS AND CONDENSATION OF TETRAETHYLORTHOSILICATE (TEOS)

C. W. TURNER AND K. J. FRANKLIN
Atomic Energy of Canada Limited,
Chalk River Nuclear Laboratories
Chalk River, Ontario, Canada

INTRODUCTION

There has been increasing interest over the past five years in the sol–gel synthesis of glasses and ceramics based on the hydrolysis and condensation of metal alkoxides. TEOS [$Si(OCH_2CH_3)_4$] is the usual starting material for silica glasses and the hydrolysis and condensation reactions are perceived to occur by reactions such as

hydrolysis $Si(OCH_2CH_3)_4 + H_2O \rightleftharpoons Si(OCH_2CH_3)_3OH + CH_2CH_3OH \cdots$

$$(1)$$

condensation

$$2Si(OCH_2CH_3)_3OH \rightleftharpoons Si(OCH_2CH_3)_3-O-Si(OCH_2CH_3)_3 + H_2O \cdots$$

$$(2)$$

Water is consumed by hydrolysis and released by condensation such that the overall reaction is given by

$$Si(OCH_2CH_3)_4 + 2H_2O \rightleftharpoons SiO_2 + 4CH_3CH_2OH \qquad (3)$$

where 2 moles of water are required to convert 1 mole of TEOS to an amorphous silica gel.

The purpose of this investigation is to characterize the various reaction intermediates and to provide information about the rates and mechanisms of the hydrolysis and condensation reactions using multinuclear NMR spectroscopy. The initial work reported here has focused on the early stages of the reaction by using substoichiometric concentrations of water.

EXPERIMENTS

The reaction mixtures studied and NMR experiments performed are displayed in Table 7.1. HCl was added to determine its effect as a catalyst. Proton spectra were obtained on a Bruker HX90 at 90 MHz, ^{29}Si spectra were obtained on a Bruker AM400 at 79.4 MHz, and ^{17}O spectra were obtained on a Bruker XP200 at 27.1 MHz. Cr(III) acetylacetonate was added as a relaxation agent for the Si experiments.

TABLE 7.1   NMR Experiments

Experiment	Mole Ratio[a] TEOS:water:acid	^1H[b]	^{29}Si[b]	^{17}O[c]
A-1	1:0.81:0.008	X		
A-2	1:0.81:0.00008	X		
A-3	1:0.81:0.0008	X		
A-4	1:0.16:0.0008	X		
B	1:1.16:0.0007		X	
C	1:1.16:0.0007			X

[a]Sufficient ethanol was added to make the water and TEOS miscible.
[b]Tetramethylsilane was added as an internal reference.
[c]Water enriched to 20 at% in ^{17}O was used.

RESULTS AND DISCUSSION

Figure 7.1 shows the methylene region of the ^1H NMR spectrum after two hours of reaction for each of the experiments A-1, A-3, and A-4. The quartets centered at 3.59 and 3.80 ppm with respect to tetramethysilane are due to ethanol and TEOS, respectively. Two additional quartets are evident at +0.028 and +0.056 ppm with respect to TEOS in the A-1 and A-3 experiments. In

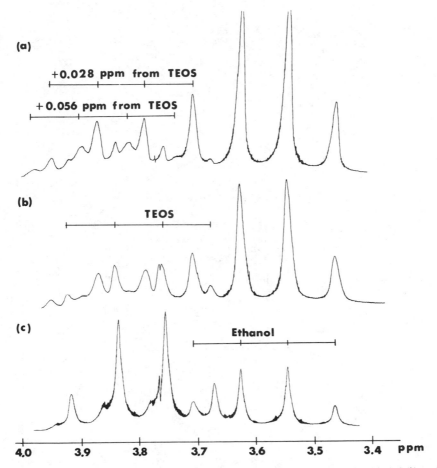

Figure 7.1 Methylene region of the ¹H NMR spectrum after 2-hr reaction for A-1 (*a*), A-3 (*b*), and A-4 (*c*). Chemical shifts shown on the horizontal axis at the bottom of the figure are with respect to tetramethylsilane which was used as an internal reference.

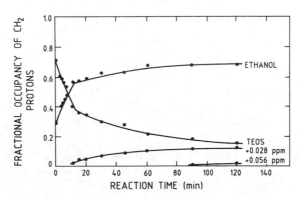

Figure 7.2 Time dependence of the fractional site occupancy of the CH_2 protons for A-3.

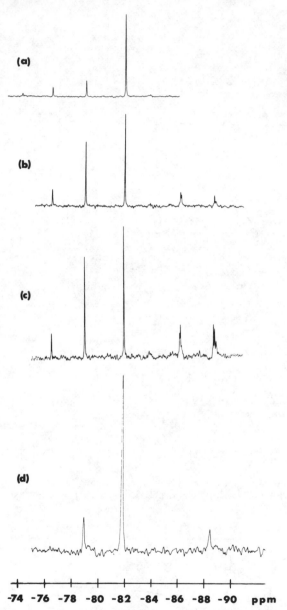

Figure 7.3 ^{29}Si NMR spectra after (*a*) 20 min, (*b*) 40 min, and (*c*) 80 min reaction for experiment B. For comparison a spectrum is shown (*d*) after 80-min reaction that was taken without the use of a relaxation agent. Chemical shifts are measured in ppm with respect to tetramethylsilane.

A-1, where the concentration of acid is higher, the intensities of both of these new quartets are greater than in A-3, indicating a higher rate of reaction in A-1. At a lower acid concentration (e.g., A-2) the rate of reaction is correspondingly lower. In the absence of acid no change in the NMR spectrum was observed after several hours. In all cases the breadth of the lines indicates that more than one chemical species is giving rise to a particular quartet. In experiment A-4, with a fivefold reduction in the initial concentration of water, only the peak at +0.028 ppm appears after 2 hr although the one at +0.056 ppm is present after several days.

Figure 7.2 shows the fractional site occupancy of the methylene protons for A-3 during the first 2 hr of reaction. The dramatic change in the relative intensities of the TEOS and ethanol quartets reflects the initial hydrolysis of TEOS, and since this change occurs before the new quartets appear we conclude that these quartets are not hydrolyzed monomers but products of the condensation reaction. The rates of hydrolysis and condensation were both observed to increase with increasing acid concentration. Lowering the initial concentration of water (e.g., A-4) lowered the rate of condensation, presumably through a reduction in the concentration of hydrolyzed monomers, but did not appear to greatly affect the rate of the initial hydrolysis.

Figure 7.3 shows ^{29}Si spectra after 20, 40, and 80 min of reaction for experiment B and, for comparison, a spectrum obtained after 80 min in the absence of a relaxation agent. The importance of using a relaxation agent is clearly demon-

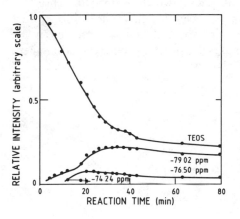

Figure 7.4 The time dependence of the relative concentrations of TEOS and the hydrolyzed monomers at -76.02, -76.50, and -74.24 ppm for experiment B as measured by ^{29}Si NMR spectroscopy. Chemical shifts are measured with respect to tetramethylsilane which was present as an internal reference.

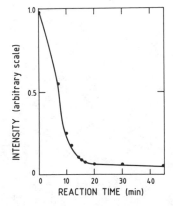

Figure 7.5 Time dependence of the concentration of free water for experiment C as measured by ^{17}O NMR spectroscopy.

strated. The peak at -81.94 ppm is TEOS and the peaks at -79.02, -76.50, and -74.24 ppm have been attributed to hydrolyzed monomers, $Si(OCH_2CH_3)_{4-n}(OH)_n$, $n = 1$, 2, and 3, as each clearly represents a unique Si environment. The peaks in the regions -86.2 and -88.8 ppm are attributed to condensation products. Figure 7.4 shows the time dependence of TEOS and the hydrolyzed monomers for series B. The curve corresponding to the singly hydrolyzed monomer shows a discontinuity at approximately 15 min corresponding to initiation of condensation reactions.

^{17}O NMR was used to monitor the free-water concentration during the reaction for experiment C and the results are plotted in Fig. 7.5. The initial inventory of water is largely consumed within the first 15 min after which a steady state is reached where the rate of removal of water by hydrolysis equals the rate of production by condensation.

CONCLUSION

We have found that TEOS reacts rapidly with substoichiometric amounts of water in the presence of acid to form monomers which in turn condense at a slower rate to form various polymeric species. New insights concerning product speciation have been obtained by correlating spectral changes observed in ^{1}H, ^{17}O, and ^{29}Si NMR spectroscopy. The use of a paramagnetic relaxation agent is necessary in order to observe all ^{29}Si signals.

8

CONTRIBUTION TO THE KINETICS OF GLASS FORMATION FROM SOLUTIONS

HELMUT SCHMIDT AND ALFRED KAISER
Fraunhofer-Institut für Silicatforschung,
Würzburg, West Germany

MICHAEL RUDOLPH AND AXEL LENTZ
Abteilung für Anorganische Chemie,
Universität Ulm, West Germany

INTRODUCTION

The kinetics of hydrolysis of tetraalkoxysilanes have been investigated by different authors.[1-5] Depending on reaction conditions different orders of the overall reaction were found and different mechanisms of the transition states were postulated. Generally, these papers do not take into account the possibility of different rates in the single reaction steps of the hydrolysis of the different species $(RO)_n Si(OH)_{4-n}$ ($n = 4$ to 1) for the reaction

$$(RO)_n Si(OH)_{4-n} + H_2O \rightarrow (RO)_{n-1} Si(OH)_{4-(n-1)} + ROH \qquad (1)$$

If different rate constants for the reaction with different values of n are assumed, it is unlikely that the reaction can be simply described. The results described elsewhere[6,7] show that for special reaction conditions one can describe at least the beginning of hydrolysis as first order overall kinetics. In this paper the experimental conditions of hydrolysis were varied with respect to higher and lower catalyst concentrations. Furthermore, mechanistic aspects of the hydrolysis reaction are discussed.

87

EXPERIMENTAL

The reaction of $Si(OR)_4$ with water was monitored by measuring the H_2O concentration via IR spectroscopy. The concentration of H_2O represents the number of unhydrolyzed OR groups exactly, as long as no condensation takes place.[6] It could be shown that in a $1:1$ mixture of silane and ethanol the absorbance of water at 1650 cm^{-1} follows Lambert–Beer's law up to 8% by volume and with minor corrections up to about 10% by volume. This was valid for HCl concentrations up to 1 mmole/L and NH_3 concentrations up to 100 mmole/L. The IR data were transmitted to a data processing system, where baseline corrections and kinetics plots could be made.

Tetramethoxysilane (TMOS) and tetraethoxysilane (TEOS) were chosen to be studied. The water to silane ratios were $1:2$ to $2:1$ (by moles), yielding a ratio of hydrolysable groups to water ($SiOR:H_2O$) from $8:1$ to $2:1$. Ethanol was used as the solvent for silane and catalyst. The volume ratio of silane: ethanol was kept at $1:1$. HCl was used as catalyst in concentrations from 0.1 to 1.0 mmole/L and NH_3 in concentrations from 10 to 100 mmole/L. The high NH_3 concentrations were necessary for a reasonable reaction time. The reaction temperature was kept at $20°C$ (in all experiments) by use of a thermostated cell.

RESULTS AND DISCUSSION

General Considerations

Figure 8.1 shows a comparison between the hydrolysis of TMOS and TEOS with identical HCl concentrations and the NH_3 catalyzed hydrolysis of TMOS.

Generally the reaction rate of hydrolysis of TMOS is remarkably faster than that of TEOS if similar reaction conditions are used. In the case of HCl catalysis, the half life (with respect to water consumption) is about 10 to 20 times higher with TEOS than with TMOS. Further systematic experiments confirm that these results are independent of HCl concentration and of the starting amount of water which is consistent with former data. With NH_3 as a catalyst the differences between the reaction rates seem to be even higher than with HCl. However, with base catalysis condensation takes place much earlier than with acid catalysis and therefore interpretation of long-term experiments is more difficult. Another general difference between HCl and NH_3 catalysis is that NH_3-catalyzed reactions require higher catalyst concentrations for similar reaction rates. This might be due to the fact that HCl is a strong and nearly completely dissociated acid, whereas NH_3 is a weak base which remains mainly undissociated as NH_4OH in the reacting system.

Though these experiments were carried out with an understochiometric amount of water with respect to hydrolysis of all SiOR groups, hydrolysis was not complete and stopped at a residual amount of about 5–20% water. There was a tendency for residual concentrations to increase with increasing starting

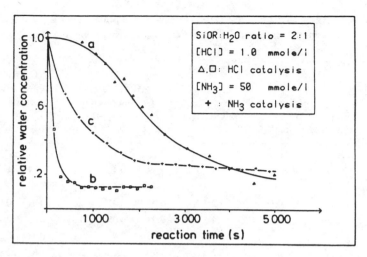

Figure 8.1   Hydrolysis of TEOS with HCl catalysis (curve *a*) and of TMOS with HCl (curve *b*) and NH₃ catalysis (curve *c*).

concentrations. With NH_3 catalysis sometimes an intermediate increase of water concentration could be observed, indicating that the condensation reaction overcomes the effect of water consumed by the hydrolysis reaction.

Kinetic Results

Hydrolysis experiments were carried out with TMOS and TEOS using HCl and NH_3 as catalysts in concentrations of 0.1 to 1.0 mmole/L (HCl) and 10 to 100 mmole/L (NH_3), which confirmed the tendencies described above. The curves obtained from reaction conditions with half lifes of less than about 500 s generally can be fitted approximately with first order kinetics according to the restrictions mentioned above. A typical example of this type of curve is shown in Fig. 8.1, curve *b*. Curves with half lifes of significantly more than 500 s show something like "inhibition phases" (curve *a*, Fig. 8.1). The hydrolysis curve of the latter type cannot be explained by a single reaction using only one reaction order and one rate constant *k*. As a consequence, *k* values computed under the assumption of a definite order of reaction should vary with turnover [k_I, Eq. (2) $\triangleq$ first order; k_{II}, Eq. (3) $\triangleq$ second order].

$$k_I t = \ln \frac{1}{x} \tag{2}$$

$$k_{II} t = \left(\frac{1}{a-b} \right) \ln \left(\frac{b}{a} \cdot \frac{a-x}{b-x} \right) \tag{3}$$

$$a = c_{o(H_2O)}; \qquad b = c_{o(SiOR)}; \qquad x = c_{t(SiOH)} = c_{o(H_2O)} - c_{t(H_2O)}$$

TABLE 8.1   Rate Constants k_I (First Order Assumption) and k_{II} (Second Order Assumption) Calculated After Different Consumption of Water

Silane	Catalyst	Catalyst Concentration (mmole/L)	SiOR:H_2O Ratio	$k_I \times 10^3$ (s^{-1}) After Water Consumption of			$k_{II} \times 10^4$ (L·mole^{-1}·s^{-1}) After Water Consumption of		
				20%	30%	50%	20%	30%	50%
TEOS	HCl	0.5	2:1	0.11	0.14	0.19	1.14	1.51	2.27
			4:1	0.39	0.42	0.49	3.97	4.35	5.29
			8:1	0.49	0.57	0.72	5.00	5.82	7.44
		0.1	2:1	0.01	0.01	0.01	0.09	0.12	0.17
			4:1	0.04	0.05	0.07	0.36	0.48	0.71
			8:1	0.06	0.08	0.11	0.62	0.80	1.10
TMOS	HCl	0.5	2:1	1.70	2.10	2.98	18.0	22.8	36.7
			4:1	2.32	3.21	4.95	23.9	33.5	53.3
			8:1	3.28	4.20	5.59	33.2	42.8	57.9
		0.1	2:1	0.27	0.30	0.39	2.9	3.2	4.5
			4:1	0.29	0.42	0.68	3.0	4.4	7.3
			8:1	0.66	0.77	0.99	6.6	7.9	10.2
TMOS	NH$_3$	50	2:1	1.16	1.06	0.94	12.24	11.59	10.96
			4:1	0.74	0.84	0.88	7.64	8.74	9.48
			8:1	0.47	0.62	0.92	4.71	6.31	9.58
		10	2:1	2.98	0.13	0.11	1.41	1.44	1.25
			4:1	0.10	0.13	0.17	1.04	1.32	1.81
			8:1	0.04	0.05	0.09	0.45	0.56	0.88

Table 8.1 compares turnover dependence of k values calculated according to assumptions of first or second order reactions.

The most important consequence of these results is that a first order assumption as suggested from earlier results cannot be maintained, if the reaction conditions are varied over a wider range. Since the type b curve cannot be fitted with any other reaction order, and since it is unlikely that the chemical mechanisms of the reaction change remarkably with the relatively modest change of pH and H_2O concentrations, the influence of different rates of the single reaction steps has to be considered.

Effect of Water Concentration

Regardless of the formal treatment and the rate of water turnover, Table 8.1 shows a remarkable effect of the SiOR:H_2O ratio on the reaction rate. In the case of HCl catalysis the reaction rates generally increase as the starting concentration of water decreases. Figure 8.2 shows the dependence of the relative hydrolysis rate constants of TEOS on HCl concentration and on the SiOR:H_2O

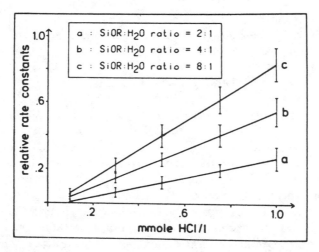

Figure 8.2 Dependence of the relative rate constants $k_1 \cdot f$ of the hydrolysis of TEOS with HCl as catalyst from the HCl concentration and from the starting $SiOR : H_2O$ ratio. (Normalization factor $f = 500$.)

ratio. The constants are calculated from Eq. (2) after 50% consumption of the initial water. Similar results are obtained by using Eq. (3) and also for the TMOS system.

With NH_3 as catalyst the effect of the starting concentration of water on the reaction rate is less clear. If one calculates relative k values from an early state of the reaction, there often is an interference from the "inhibition phase", but later in the reaction the influence of the condensation reaction becomes stronger and can lead to errors. Therefore Fig. 8.3 shows the relative rate constants for

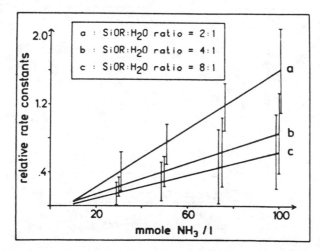

Figure 8.3 Dependence of the relative rate constants $k_1 \cdot f$ of the hydrolysis of TMOS with NH_3 catalysis from the NH_3 concentration and from the $SiOR:H_2O$ starting ratio. (Normalization factor $f = 500$.)

SiOR:H_2O starting ratios of 2:1, 4:1, and 8:1 using TMOS. Despite the scattered data the trend toward higher reaction rate constants with higher starting amounts of water becomes clear.

Mechanistic Aspects

The results of the measurements of hydrolysis reaction rates in the systems TEOS and TMOS with HCl and NH_3 as catalysts show that considering only an overall hydrolysis reaction can be misleading. Therefore the single reaction steps of Eq. (1) have to be taken into consideration. From theoretical considerations it becomes probable that the hydrolysis rate increases with an increasing number of hydrolyzed OR groups of the reacting silane due to a decrease in the stabilizing effect of alkyl groups in the transition state. In order to find out the effect on the hydrolysis reaction of this assumption, a computer simulation of the single hydrolysis steps was performed. For computing the actual concentrations of the $(RO)_nSi(OH)_{4-n}$ species first order kinetics and a ratio of 1:2:4:8 for the k values of the single steps of reaction (1) was assumed. Figure 8.4 shows the relative concentrations of each $(RO)_nSi(OH)_{4-n}$ species versus H_2O content. The resulting curve (dashed line) represents the total "activity" of hydrolysable species derived from the sum of the concentrations of each alkoxy- or hydroxy-silane multiplied with its relative rate constant. This curve indicates that a maximum of "activity" occurs somewhat after the beginning of the reaction. This behavior shows an "inhibition phase" followed by a region of maximum activity where water is consumed very rapidly due to an "acceleration phase" in the H_2O consumption curve. These results however do not indicate whether the reactions of the single species follow first or second order kinetics.

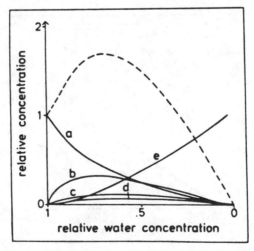

Figure 8.4 Computer simulation of the relative concentrations of the different hydrolysis intermediates [$a = (RO)_4Si$; $b = (RO)_3SiOH$; $c = (RO)_2Si(OH)_2$; $d = ROSi(OH)_3$; $e = Si(OH)_4$]. Dashed line: resulting summarized "activity" (explanation see text).

The different effects of water in acid and base catalyzed systems may be due to the fact that HCl is a strong acid which is completely dissociated. Increasing amounts of water reduce the activity of protons by hydration with respect to

$$H_2O + H^+ \rightleftharpoons H_3O^+ \tag{4}$$

$$SiOR + H^+ \rightleftharpoons SiO^+R \tag{5}$$
$$\underline{\quad} H$$
$$(I)$$

the formation of (I) which is considered as an important intermediate in the hydrolysis path. In the case of NH_3 as catalyst water is needed to build up the catalytically active OH^- species; so increasing amounts of water lead to higher OH^- activity:

$$NH_3 + H_2O \rightleftharpoons NH_4OH \rightleftharpoons NH_4^+ + OH^- \tag{6}$$

These mechanistic ideas have not yet been proved, but they can explain the results obtained from the experiments reported above. In order to complete the analysis more concentrations of the reaction components, especially the $SiOR:H_2O$ ratio need to be studied. More information is also expected from extended computer calculations with systematic variations of the single rate constants. Additional data also can be expected by quantitative chemical analysis of $(RO)_nSi(OH)_{4-n}$ species as a function of reaction time and by comparing the effects of different catalysts (weak acids or strong bases).

ACKNOWLEDGMENTS

The authors want to thank Prof. Dr. H. Scholze for his helpful discussions and the Bundesminister für Forschung und Technologie for financial support.

REFERENCES

1. R. Aelion, A. Loebel, and F. Eirich, *J. Am. Chem. Soc.* **72**, 5705 (1950).
2. E. Åkerman, *Acta Chem. Scand.* **10**, 298 (1956); **11**, 298 (1957).
3. C. J. Brinker, K. D. Keefer, D. W. Schaefer, R. A. Assink, B. D. Kay, and C. S. Ashley, *J. Non-Cryst. Solids* **63**, 45 (1984).
4. M. G. Voronkov, V. P. Mileshkevich, and Y. A. Yushelevskii, *The Siloxane Bond*, Plenum, New York, 1978.
5. R. E. Timms, *J. Chem. Soc. A* 1969 (1971).
6. H. Schmidt, H. Scholze, and A. Kaiser, *J. Non-Cryst. Solids* **48**, 65 (1981).
7. H. Schmidt and A. Kaiser, *Glastechn. Ber.* **54**, 338 (1981).

9

TIME-RESOLVED RAMAN SPECTROSCOPY OF TITANIUM ISOPROPOXIDE HYDROLYSIS KINETICS

KRIS A. BERGLUND
Departments of Agricultural and Chemical Engineering
Michigan State University
East Lansing, Michigan

D. R. TALLANT AND R. G. DOSCH
Sandia National Laboratories
Albuquerque, New Mexico

INTRODUCTION

The hydrolysis of metal alkoxides is a commonly used method of preparing ceramic precursors.[1] In the case of multicomponent ceramics the hydrolysis rates of the alkoxides can be very important in determining the degree of homogeneity which can be achieved in the precursors. Published information regarding the hydrolysis kinetics of metal alkoxides is very sparse, partially due to the difficulty in obtaining such data. In this work a technique utilizing Raman spectroscopy is presented for the study of such kinetics.

MATERIALS AND METHODS

Titanium isopropoxide (TiPT) from DuPont was redistilled and used for all experiments. Electronic grade isopropanol with a water content of less than

94

0.03% was used to prepare all solutions. Raman spectra were taken using a standard 90° scattering configuration with a SPEX model 1404 spectrometer equipped with an Ar$^+$ ion laser (5145-Å line used). The spectrometer slits were set at 500 μm and the laser power used was 200 mW. All spectra were taken at room temperature.

For the Raman spectra of the reactants and products of the hydrolysis reaction, samples were loaded into melting point capillaries and sealed. The experimental apparatus used for the hydrolysis kinetics studies was a standard rapid mixing device consisting of a dual channel syringe pump, a mixing tee, and a quartz capillary tube with 0.8 mm ID. The reactants were loaded into the syringes in the form of TiPT/isopropanol and H$_2$O/isopropanol solutions. The syringe pump provided a steady flow of each reactant through the tee where they were mixed. During continuous flow the Raman spectrum was recorded at a known distance along the quartz capillary attached to the mixing tee. By varying the flow rate the spectrum could be taken at different times after mixing, thus allowing the study of kinetics.

RESULTS AND DISCUSSION

Figure 9.1 presents the Raman spectra of the reactants TiPT and isopropanol. These results indicate that while both compounds have rather complicated spectra there are "windows" in the isopropanol spectrum which allow study of hydrolysis reactions. This can be seen more easily in Figs. 9.2 and 9.3 where the concentration dependence of TiPT/isopropanol solutions (W/W) on their Raman spectra are shown. Figure 9.2 shows that in the higher wave number

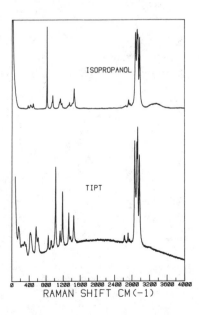

Figure 9.1   Raman spectra of pure isopropanol and pure titanium isopropoxide (TiPT).

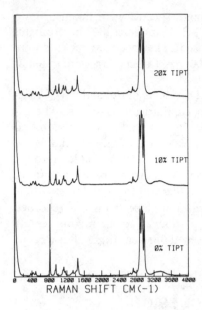

Figure 9.2 Concentration dependence (on a weight basis) of the Raman spectrum of titanium isopropoxide (TiPT) in isopropanol solution.

region (>1600 cm^{-1}) no drastic changes occur with concentration. However, as can be seen in Fig. 9.3, several concentration dependent features occur in the lower wave number region (<1600 cm^{-1}). The most important bands related to the present work are: (1) 159 cm^{-1} and 329 cm^{-1}, which are probably due to large concerted vibrations of TiPT, (2) 564 cm^{-1} and 612 cm^{-1}, due to the Ti—O symmetric stretch,[2] and (3) 1025 cm^{-1}, due to the (C—O)Ti stretch.[2] The bands

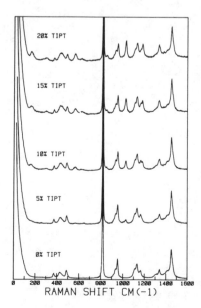

Figure 9.3 Concentration dependence (on a weight basis) of the Raman spectrum of titanium isopropoxide (TiPT) in isopropanol solution.

listed under (2) and (3) were found to have intensities directly proportional to concentration over the range studied.

Figure 9.4 shows the results of hydrolyzing a 10% (W/W) TiPT/isopropanol solution with a $2.8H_2O/Ti$ containing H_2O/isopropanol solution. It can clearly be seen that all of the bands related to high symmetry about the titanium are absent in the earliest time observation made (0.5 s after mixing). Conversely, the band corresponding to the (C—O)Ti stretch is present even at long times. These same general results were obtained when using ratios of $2.0H_2O/Ti$ and $1.0H_2O/Ti$.

In attempting to determine the reaction kinetics, a plot of the intensity of the (C—O)Ti band was made as a function of time. The intensity decreased greatly at first, but increased slightly at a later time. Figure 9.5 shows a greatly enlarged view of this band as a function of time. Initially it is symmetric, indicating a single species, but as the reaction proceeds a large amount of asymmetry is present, indicating a reaction intermediate. Since the band does not remain symmetric as it decreases in intensity, the intensity alone should not be used to determine the concentration of (C—O)Ti bonds. This presents the possibility of studying the reaction intermediates attributable to the asymmetry through deconvolution of the spectrum.

The product formed by the mixing device was collected and held for several hours. The powder was dried and its Raman spectrum is shown prior to and after calcination in Fig. 9.6. Clearly, the powder contains significant amounts of the isopropyl group still attached to the titanium. After calcination at 750°C the powder was analyzed as 65% rutile and 37% anatase by X-ray diffraction.

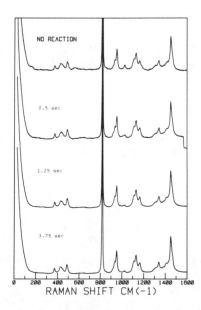

Figure 9.4 The Raman spectrum of titanium isoproxide (TiPT) undergoing hydrolysis in isopropanol. Each spectrum is taken at a different time during the reaction.

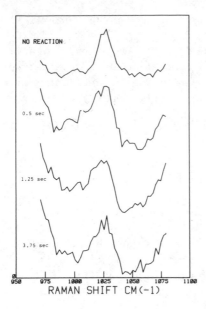

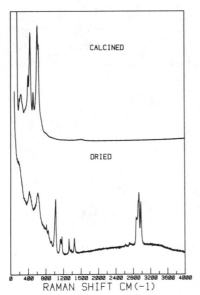

Figure 9.5 The band contour of the (C—O)Ti stretch vibration during the hydrolysis of titanium isopropoxide (TiPT) in isopropanol. Each spectrum is taken at a different time during the reaction.

Figure 9.6 The Raman spectrum of the powder formed by titanium isopropoxide hydrolysis before and after calcination.

SUMMARY

Raman bands sensitive to the symmetry about the titanium atom were found to disappear very quickly, indicating that the initial steps of the hydrolysis reaction are very fast. The subsequent steps are apparently much slower as is evidenced by the presence of the (C—O)Ti band at later times. Raman spectra of the powder product indicate incomplete hydrolysis. The results of this work demonstrate that Raman spectroscopy can be a valuable tool for studying titanium alkoxide hydrolysis reactions and kinetics.

ACKNOWLEDGMENTS

This work was performed at Sandia National Laboratories and supported by the U.S. Department of Energy under contract number DE-AC04-76DP00789.

REFERENCES

1. K. S. Mazdiyasni, *Cer. Int.* **8**(2), 42 (1982).
2. D. C. Bradley, R. C. Mehrotra, and D. P. Gaur, *Metal Alkoxides*, Academic Press, London, 1978.

10

RHEOLOGICAL CHARACTERIZATION DURING THE SOL–GEL TRANSITION

MICHAEL D. SACKS AND RONG-SHENQ SHEU
Department of Materials Science and Engineering
University of Florida
Gainesville, Florida

INTRODUCTION

The processing and physical properties of sol–gel materials have been studied extensively in recent years.[1–3] However, investigations of the rheological behavior of sol–gel materials are relatively limited. Much of the recent work in this area has been carried out by Sakka and coworkers.[4–7] They used intrinsic viscosity measurements to provide information about the structure of polysiloxane species formed from tetraethylorthosilicate (TEOS) under various hydrolysis/condensation conditions.[6,7] Viscosity measurements have also been used to identify points during the sol–gel transition when sols are suitable for various processing operations, for example, fiber drawing, substrate coating, and so on.[4–10] Rheological considerations are also used to define the point of gelation. Gel formation is associated with rapid increases in viscosity and/or observations of a maximum viscosity.[5–7,11,12]

One limitation to previous studies of sol–gel rheology is that most viscosity measurements were made using poorly defined shear rate conditions. Unless the flow behavior of the sol is Newtonian, viscosity values will vary with shear rate. This is important because a wide range of shear rates are typically encountered in various processing operations, that is, drawing, coating, extrusion, and so on. The objective of the present study was to provide more complete rheological characterization during the sol–gel transition. Studies were carried out with acid-catalyzed, TEOS-derived sols containing different concentrations of water.

EXPERIMENTAL PROCEDURE

Two acid-catalyzed "silica" sols were prepared using tetraethylorthosilicate (TEOS),* ethanol,* deionized water, and $1M$ nitric acid.* Sol 1 had $H_2O/$ TEOS and acid/TEOS mole ratios of 20 and 0.01, respectively, while sol 2 had $H_2O/$TEOS and acid/TEOS mole ratios of 2.0 and 0.1, respectively. Under conditions of continuous mixing, the TEOS and half of the ethanol were heated at 80°C for 20 min in a flask equipped with a reflux condenser. The solution was subsequently cooled to room temperature. A second solution was prepared by mixing the water, nitric acid, and the other half of the ethanol. This was added slowly to the TEOS/ethanol solution under conditions of continuous mixing. Samples were aged at 25°C until gelation occurred.

Rheological flow characteristics were determined using a concentric cylinder viscometer.† Steady rotational flow curves (i.e., shear stress vs. shear rate) were generated by increasing the shear rate from zero to the maximum desired value in 2 min, immediately followed by decreasing the shear rate back to zero in another 2 min. Viscoelastic properties were measured in oscillatory rotational flow using a frequency of 0.5 s^{-1} and a displacement amplitude of 10°.

RESULTS AND DISCUSSION

Figure 10.1 is a plot of viscosity (at shear rates of 3,300, and 2700 s^{-1}) versus aging time for sol 1. Up to 240-hr aging, the viscosity is independent of shear rate. For longer aging times, the viscosity becomes increasingly dependent on shear rate. Figure 10.2A shows shear stress versus shear rate curves for sol 1 at three aging times. Corresponding plots of viscosity versus shear rate are shown in Fig. 10.2B. The 7-hr aged sol shows Newtonian flow behavior. After 262 hr, shear thinning flow behavior is observed, that is, the viscosity decreases with increasing rate. With further aging, hysteresis is observed in the flow curve, that is, the sol becomes thixotropic.

The transitions in flow behavior in Fig. 10.2 are similar to changes observed in flocculated particle/liquid suspensions as the solids loading (i.e., particle concentration) is increased. In the case of "silica" sols, the changes in flow behavior reflect the condensation growth and agglomeration of polysiloxane

*Fisher Scientific Co., Fair Lawn, NJ.
†Model RV-100/CV-100, Haake, Inc., Saddle Brook, NJ.

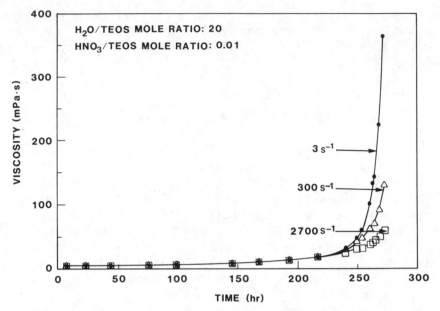

Figure 10.1 Plots of viscosity (at indicated shear rates) versus aging time for sol 1.

"particles." The hydrolysis/condensation conditions used to prepare sol 1 (i.e., acid-catalyzed and high water content) produce highly branched, polymeric clusters.[13-15] During the Newtonian period of aging, the concentration of these polysiloxane species is relatively low. The sol flow behavior is qualitatively similar to a suspension of colloidal particles with a low solids loading. The presence of particles (or polysiloxane species) in a flowing liquid produces perturbations in the liquid streamlines and, thus, increases the rate of energy dissipation. For a given shear rate, a higher shear stress (and a higher viscosity) is measured relative to the pure liquid. When the particle concentration is relatively low, the flow behavior is not influenced much by particle–particle interactions.

  As aging proceeds, there is continued formation and condensation growth of polysiloxane clusters, along with agglomeration of the clusters. This results in a rapid increase in the "effective" solids loading since the void space (within the cluster agglomerates) contains liquid that is unavailable for flow. Shear thinning behavior is observed because the cluster agglomerates will break down at high shear rates. This releases entrapped liquid and reduces the "effective" solids loading in the sol. With continued aging, condensation growth and agglomeration of polysiloxane clusters results in the formation of a continuous, 3-D network (gel formation). The elastic character of the polysiloxane network is indicated by the observation of yield behavior in the shear stress versus shear rate curve at 276-hr aging (Fig. 10.2). Breakdown of the network occurs after the yield stress is exceeded. (It may be useful to define the gel point in terms of a yield stress value determined under specific measuring conditions. It is evident

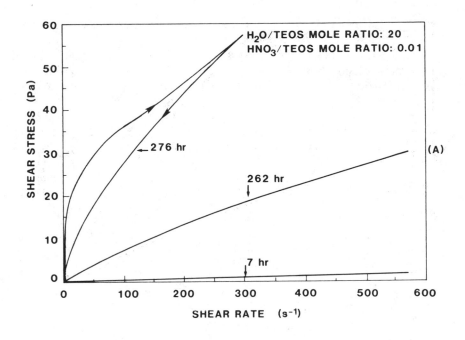

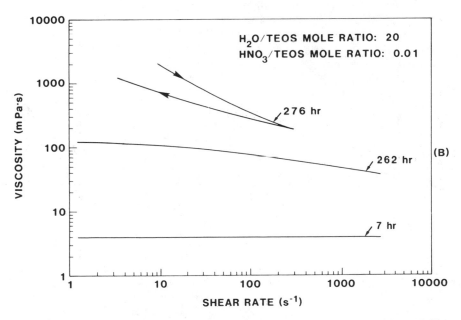

Figure 10.2 (A) Plots of shear stress versus shear rate for sol 1 at indicated aging time. (B) Plots of viscosity versus shear rate for sol 1 at indicated aging time.

from Fig. 10.1 that some ambiguity is associated with defining the gel point in terms of a region of rapidly increasing viscosity.)

The structural changes that occur due to the condensation growth and agglomeration of polysiloxane species are also reflected in measurements of the viscoelastic properties during the sol–gel aging period. Plots of storage

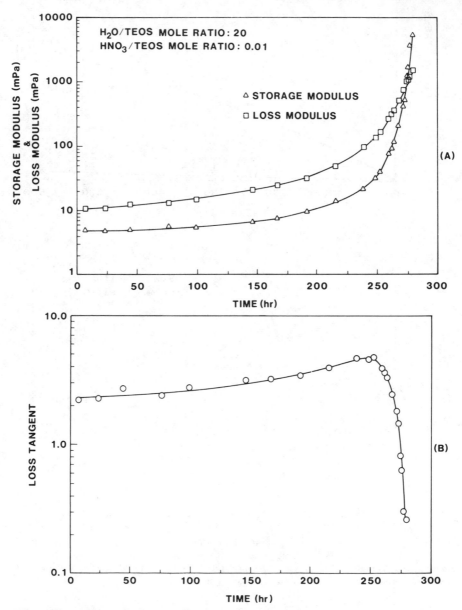

Figure 10.3 (A) Plots of storage modulus and loss modulus versus aging time for sol 1. (B) Plot of loss tangent versus aging time for sol 1.

modulus, G', and loss modulus, G'', versus aging time are shown for sol 1 in Fig. 10.3A. The loss tangent, tan δ, is plotted versus aging time in Fig. 10.3B (Note that tan $\delta = G''/G'$.) During the period of Newtonian flow behavior (i.e., through 240 hr) both moduli increase slowly. However, it is evident from Fig. 10.3B that the loss modulus increases more quickly than the storage modulus. This behavior is associated with the increase in the "effective" solids loading during

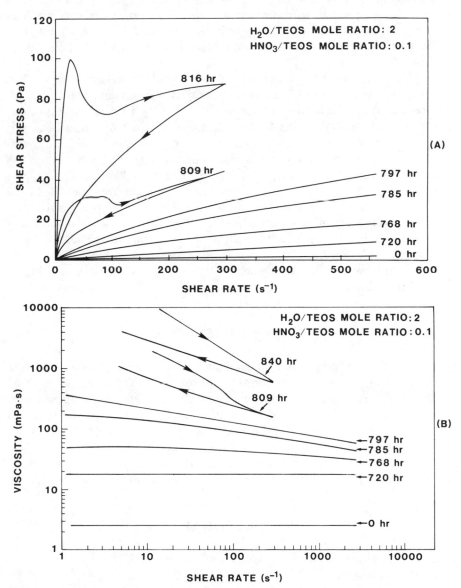

Figure 10.4   (A) Plots of shear stress versus shear rate for sol 2 at indicated aging times. (B) Plots of viscosity versus shear rate for sol 2 at indicated aging times.

the Newtonian period (i.e., due to condensation growth and agglomeration of polysiloxane species). The increase in loading will result in an increased rate of energy dissipation during flow (which is reflected in the increasing loss modulus). However, since the *overall* "effective" solids loading remains relatively low during the Newtonian period, polysiloxane cluster–cluster interactions have relatively little influence on flow behavior. This is indicated by the slower increase in the storage modulus. With further aging, that is, during the periods of shear thinning and thixotropic flow behavior, the storage modulus increases very rapidly compared to the increase in the loss modulus (Fig. 10.3B). This reflects the importance of cluster–cluster interactions and the development of an extensive 3-D network structure.

Figure 10.4A shows the shear stress versus shear rate behavior for sol 2 at various aging times. Plots of viscosity versus shear rate are given in Fig. 10.4B. The results are similar to those obtained with sol 1 (Fig. 10.2) in that transitions are observed from Newtonian to shear thinning to thixotropic flow behavior.

The hydrolysis/condensation conditions used to prepare sol 2 (i.e., acid-catalyzed and low water content) produce polysiloxane species which are less branched and more chainlike.[6,7,13,14] According to Sakka,[4-7] this type of polysiloxane species is necessary in order to draw fibers from the sol. In the present study, aged samples were tested for "spinnability" by dipping a glass rod in the sol and attempting to draw a fiber. "Spinnability" was not observed until 785-hr aging, that is, *after* the sol became shear thinning. The best "spinnability" was obtained at 797 hr, that is, *prior* to the transition to thixotropic behavior. Sols were no longer "spinnable" when large yield stresses and highly thixotropic behavior occurred.

CONCLUSION

The rheological properties of acid-catalyzed, TEOS-derived "silica" sols were investigated. During aging, transitions were observed from Newtonian to shear thinning to thixotropic flow behavior. These changes reflect the condensation growth and agglomeration of polysiloxane species. Dissipative (viscous) processes dominate the flow properties during the Newtonian flow period. Elastic interactions become increasingly important during the shear thinning and thixotropic flow periods. In low water content sols, optimum "spinnability" was observed when flow behavior was highly shear thinning, but not thixotropic.

ACKNOWLEDGMENTS

The authors gratefully acknowledge E. Lunghofer and Dresser Industries for supporting this work.

REFERENCES

1. *Ultrastructure Processing of Ceramics, Glasses, and Composites*, L. L. Hench and D. R. Ulrich, eds., Wiley, New York, 1984.

2. *Better Ceramics Through Chemistry, Materials Research Society Symposia Proceedings*, Vol. 32, C. J. Brinker, D. E. Clark, and D. R. Ulrich, Eds., North-Holland, New York, 1984.

3. *J. Non-Cryst. Solids* **48** (1), 1982.

4. S. Sakka, Gel Method for Making Glass, in *Treatise on Materials Science and Technology*, Vol. 22, *Glass III*, M. Tomozawa and R. Doremus, Eds., Academic Press, New York, 1982, pp. 129–167.

5. S. Sakka and K. Kamiya, Preparation of Shaped Glasses Through the Sol-Gel Method, in *Materials Science Research*, Vol. 17, *Emergent Methods for High-Technology Ceramics*, R. F. Davis, H. Palmour III, and R. L. Porter, Eds., Plenum Press, New York, 1984, pp. 83–94.

6. S. Sakka and K. Kamiya, The Sol–Gel Transition in the Hydrolysis of Metal Alkoxides in Relation to the Formation of Glass Fibers and Films, in Ref. 3, pp. 31–46.

7. S. Sakka, Formation of Glass and Amorphous Oxide Fibers from Solution, in Ref. 2, pp. 91–99.

8. W. C. LaCourse, Strength of Gel-Derived SiO_2 Fibers, in Ref. 2, pp. 53–58.

9. E. Leroy, C. Robin-Brosse, and J. P. Torre, Fabrication of Zirconia Fibers from Sol–Gels, in Ref. 1, pp. 219–231.

10. S. P. Mukherjee, Deposition of Transparent Noncrystalline Metal Oxide Coatings by the Sol–Gel Process, in Ref. 1, pp. 178–188.

11. M. M. Akhtar, Factors Controlling the Sol-Gel Conversion in TEOS, Ph.D. Thesis, Alfred University, 1983.

12. M. Nogami and Y. Moriya, Glass Formation Through Hydrolysis of $Si(OC_2H_5)_4$ with NH_4OH and HCl Solution, *J. Non-Cryst. Solids* **37**, 191 201 (1980).

13. C. J. Brinker, K. D. Keefer, D. W. Schaefer, and C. S. Ashby, Sol-Gel Transition in Simple Silicates, in Ref. 3, pp. 47–64.

14. K. D. Keefer, The Effect of Hydrolysis Conditions on the Structure and Growth of Silicate Polymers, in Ref. 2, pp. 15–24.

15. D. W. Schaefer and K. R. Keefer, Structure of Soluble Silicates, in Ref. 2, pp. 1–14.

11

EFFECTS OF WATER CONTENT ON GEL STRUCTURE AND SOL-TO-GEL TRANSFORMATIONS

G. KORDAS
Mechanical and Materials Engineering
Vanderbilt University
Nashville, Tennesse

L. C. KLEIN
Department of Ceramics
Rutgers University
The State University of New Jersey
Piscataway, New Jersey

INTRODUCTION

During the past 10 years the synthesis of glasses with the sol–gel method has found considerable interest, primarily because homogenous new glass compositions not achievable by oxide melting techniques can be prepared.[1] Several investigators measured the density, viscosity, hardness, linear thermal expansion, Young's modulus, and IR spectra of SiO_2–sol–gel glasses.[2] These physical properties of the SiO_2 sol–gel glasses are within the error of the measurements the same as those reported for commercial available silica glasses.[2] Based on these results Mackenzie[2] concluded that the "overall structure" of the sol–gel glasses cannot be different from those produced by melting the oxides. This hypothesis was contradicted by Brinker et al.[3] using the results of other experimental techniques.

It is the aim of this work to detect the ESR spectra of vacuum-dried SiO_2 gels and gels heat treated at temperatures, T_H, between 400 and 1000°C produced by $n = 2, 4, 8, 16$. Using these results, we attempt to establish a relationship between the SiO_2 glass structures produced by sol–gel technique and oxide melting techniques.

EXPERIMENTAL PROCEDURE

Gels with $n = 2, 4, 8, 16$ moles water/mole TEOS were produced by the technique described in previous publications.[4a,b] These gels were vacuum dried at 80°C and subsequently heat treated up to 400, 500, 600, 700, 800, 900, 1000°C with constant heating rate (3°C/s). These samples were exposed to γ-ray irradiation in order to induce paramagnetic defect centers. The ESR spectra were recorded using a VARIAN V4500 spectrometer operating at 9.5 GHz. The microwave frequency was measured with a Systron Donner counter (model 1017/1255A) and the magnetic field with a NMR magnetometer (model ANAC-SENTEC 1001).

RESULTS AND DISCUSSION

Figure 11.1 shows the ESR spectrum of the γ-ray irradiated vacuum-dried $n = 2$ gels. The same asymmetric broad resonance was also detected in the irradiated vacuum-dried $n = 4, 8, 16$ gels.

Figure 11.2 displays the ESR spectra of the γ-ray irradiated gels produced with 4 and 16 mole water/mole TEOS as a function of the temperature of heat treatments, T_H. One can perceive from this figure that the ESR spectra of the densified gels depend both on n and T_H.

The asymmetric line detected in the irradiated vacuum-dried gels dominated the ESR spectra of the $n = 16$ gels heat treated up to 500°C (Fig. 11.2a,b). At $T_H = 500$°C a new signal becomes evident (Fig. 11.2b) indicating that major changes occur at this temperature of heat treatments. These changes of the structure are apparent in the vacuum-dried $n = 16$ gels heat treated at T_H 900, and 1000°C (Fig. 11.2c,d). The ESR spectrum (Fig. 11.2c) of the $n = 16$ gels with $T_H = 900$°C consists of the resonance of the E' center[5] and of the resonance generated by nonbridging oxygen.[5] At $T_H = 1000$°C three different para-

Figure 11.1   ESR spectrum of γ-ray irradiated $n = 4$-gel.

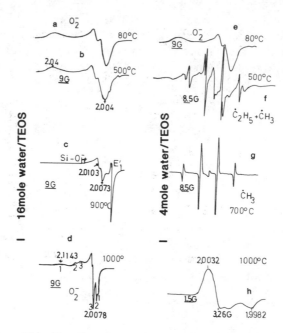

Figure 11.2   The gel-to-glass transformation for $n = 4$ and $n = 16$.

magnetic centers were detected due to O_2^- ions probably in three different interstitial positions (Fig. 11.2d).[5]

Although the O_2^- ions were detected in the irradiated $n = 4$ gels (Fig. 11.2e), these ions were not obtained in the densified $n = 4$ gels heat treated up to $T_H = 1000°C$ (Fig. 11.2f, g, h). The spectra of the $n = 4$ gels with $T_H = 500, 700°C$ consists of two ($T_H = 500°C$, Fig. 11.2f) and of one ($T_H = 700°C$, Fig. 11.2g) set of hfs splittings. Hydrogen atoms are the only one element in appreciable concentration having nonzero nuclear spin which can cause these hfs splittings. The separation and intensity ratio of these lines are about the same as those reported for $\cdot C_2H_5$ and $\cdot CH_3$ radicals.[6] Therefore, we attributed these signals to $\cdot C_2H_5$ and $\cdot CH_3$ radicals. The $\cdot CH_3$ radical is the dominant defect center in the $n = 4$ gels with $T_H = 700°C$. In the $n = 4 T_H = 1000°C$ glass only the central component of the $n = 4$ gels with $T_H = 700°C$ was detected (Fig. 11.2h). Figure 11.3 displays this signal recorded at various microwave power levels. This signal depends on the microwave power level which may indicate that this resonance is formed by superposition of several individual lines.[7] The defect centers generating these signals have not been identified yet. We believe that at 0.004 mW the E' center is present.

The detection of the ESR spectra of the gels and heat-treated gels at various T_H produced with various n was the first purpose of this study. The second purpose of this study was the comparison of the SiO_2 structures prepared by the sol–gel technique and oxide melting techniques as observed with the ESR method. This can be made by comparing the defect centers detected in these

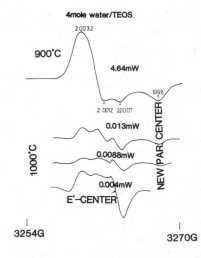

4mole water/TEOS

2.0032

900°C

4.64mW

1998

2.0012 2.0007

0.013mW

NEW(PAR) CENTER

0.0088mW

1000°C

0.004mW

E'–CENTER

3254G   3270G

Figure 11.3 The microwave power dependence of the ESR spectrum of the $n = 4$, $T_H = 1000°C$ sol–gel glass.

glasses. In SiO_2 glasses produced by melting the oxides E'_1 centers, nonbridging oxygens, and peroxy radicals were reported.[8] In the SiO_2 sol–gel glasses a variety of new centers and O_2^- ions probably in interstitial positions were found which have not been observed in the silica glasses obtained from commercial sources. Based on this comparison, we believe that the structure of the $n = 16$-SiO_2 sol–gel glasses is not identical to the structure of the silica glasses produced by melting the oxides.

CONCLUSIONS

The most important results of this study can be summarized as follows:

1. The paramagnetic centers detected in vacuum-dried gels were attributed to O_2^- in interstitial positions.
2. In 16 mole water/TEOS gels:
    (a) The O_2^- ions were preserved in the gels heat treated up to 500°C.
    (b) Major changes occur at T_H from 500 to 900°C. These changes favor the formation of E'_1 centers and nonbridging oxygen.
    (c) Further increase of T_H induces three different O_2^- radicals present in different interstitial positions.
3. In 4 mole water/TEOS gels:
    (a) The $\cdot C_2H_5$ and $\cdot CH_3$ radicals were detected in the gels with $T_H = 500°C$.
    (b) The $\cdot CH_3$ radical is the predominant center at T_H above 700°C.
    (c) At temperatures higher than 700°C, a variety of new centers (probably E'_1 centers) were induced.
4. In the sol–gel glasses a variety of new centers and O_2^- ions, not reported

in commercially available silica glasses, were detected in the gels. This comparison may indicate that the "overall structure" of these glasses is not identical.

ACKNOWLEDGMENTS

The authors wish to express their appreciation for financial support of this work under ARO/D Contract #DAAG-29-81-K-0118.

REFERENCES

1. S. P. Mukherjee and S. K. Sharma, *J. Non-Cryst. Solids*, **71**, 317 (1985).
2. J. D. Mackenzie, *J. Non-Cryst. Solids* **48**, 1 (1982).
3. C. J. Brinker, E. P. Roth, G. W. Scherer, and D. R. Tallant, *J. Non-Cryst. Solids*, **71**, 171 (1985).
4. (a) L. C. Klein and G. J. Garvey, *J. Non-Cryst. Solids* **48**, 97 (1982). (b) L. C. Klein and G. J. Garvey, *J. Non-Cryst. Solids* **38** and **39**, 45 (1980).
5. G. Kordas, R. A. Weeks, and L. C. Klein, *J. Non-Cryst. Solids*, **71**, 327 (1985).
6. A. A. Wolf, E. J. Friebele, and D. C. Tran, *J. Non-Cryst. Solids*, **71**, 345 (1985).
7. G. Kordas, B. Camara, and H. J. Oel, *J. Non-Cryst. Solids* **50**, 79 (1982).
8. E. J. Friebele, D. L. Griscom, M. Stapelbroek, and R. A. Weeks, *Phys. Rev. Lett.* **42**, 1346 (1979).

12

APPLICATIONS OF THE SOL–GEL METHOD: SOME ASPECTS OF INITIAL PROCESSING

J. D. MACKENZIE
Department of Materials Science and Engineering
University of California
Los Angeles, California

INTRODUCTION

By the end of 1984, the sol–gel technique has been used to prepare crystalline and noncrystalline oxides from over 50 chemical systems.[1] It has been estimated that in the next two decades some 200 oxide systems will have been studied by the sol–gel method.[1] (For instance, SiO_2 and ZrO_2–SiO_2 are considered as systems.) Opportunities for new applications and new products are encouraging. Examples such as glass and ceramic fibers, coatings and films, porous solids and microballoons, nuclear fuels, radioactive waste encapsulation, and electronically conductive gels were cited in the First International Conference on Ultrastructure Processing of Ceramics, Glasses and Composites.[2] Applications to the fabrication of contact lenses,[3] filters,[4] and catalyst supports[5] were described in recent publications. In order to exploit the uniqueness of the sol–gel method and the unusual noncrystalline solids derived from gels, such that marketable products can actually be produced by industry from "laboratory curiosities," the science of sol–gel processing must be understood. At present, such a science is still in its infancy. This report is concerned with some of the interdependent variables of great importance to sol–gel processing.

The various types of products which can be directly and indirectly made by

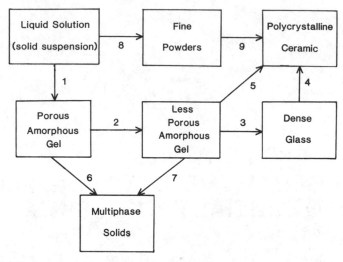

Figure 12.1 Various types of products prepared directly and indirectly from the sol–gel method.

the sol–gel technique are shown in Fig. 12.1. Although this report is specifically concerned with Steps 1 and 2 of that figure, the results presented are of obvious importance to some of the other steps. For instance, if different solvents used in Step 1 to produce a porous gel have significant effects on pore size and porosity, and the porous gel is subsequently impregnated with a polymer according to Step 6, then the nature of the solvent must be of importance to Step 6 as well.

The interdependent variables which can play an important role in Step 1 and Step 2 are:

1. Solvents.
2. Catalysts.
3. Compounds of the same metal.
4. Temperature.

Systematic studies of the effects of these variables on any oxide sol–gel system have not been reported in the open literature. Results of some on-going research on SiO_2 sol–gels in the author's laboratory are presented below.

ROLE OF THE SOLVENT

In a typical laboratory process, a gel is made from a solution consisting of metal-containing compounds (tetraethyoxysilane, for example), a solvent, water, and a catalyst. Although ultrasonic agitation can be applied in place of a solvent to obtain a homogeneous liquid,[6] it is presently assumed that for most gel systems, the use of a solvent is desirable. The most common solvents used are the alcohols and most alcohols have relatively high vapor pressures, even at room temperature. For Step 1, reported gelation times can last for many hours and even days. The loss of solvents through evaporation can be very appreciable. One has the choice of a closed system, an opened system, or partially opened system during gelation. Frequently, this aspect of the process is not quantitatively controlled. A cursory consideration would suggest that the amount of solvent lost during gelation must have a significant influence on the bulk density of the gel and hence its porosity and probably the average size of the pores. Because the pores are related to the cracking of the gel,[7] solvent losses must be one of the most important factors for Step 1 and Step 2.

Recently, a systematic study of the effects of solvent on the formation of SiO_2 gel has been initiated in the author's laboratory. Table 12.1 shows the solvents used in two series of experiments. The *constant mole* series involved the mixing of 4 moles of solvent, 4 moles of water, 1 mole of tetramethoxysilane and 0.05 mole of HCl as catalyst. The *constant volume* series involved 162 cm³ of each solvent (equivalent to 4 moles of methanol). The gelation point was arbitrarily taken as the time when the solution viscosity reaches 10^4 P. Figure 12.2 shows the weight loss of the starting solutions as a function of the vapor pressure of the solvents for the constant volume series for a partially opened system. When the gels were fired at 150 and 600°C, dramatic differences in the bulk density as well as the specific surface areas were observed. Figure 12.3 shows the specific surface area differences. Preliminary results have indicated that for the constant mole series, significant differences could arise from the size and shape of the solvent molecule.

There are other less pronounced effects due to solvent differences. For instance, the viscosity and surface tensions of the solution are influenced by the

TABLE 12.1   Solvents Used in Gelation of SiO_2

		Vapor Pressure at 25°C (torr)	Molar Volume (mL/mole)
1. Methanol	CH_3OH	122.0	40.47
2. Ethanol	C_2H_5OH	56.7	58.37
3. n-Propanol	C_3H_7OH	19.7	74.79
4. i-Propanol	$CH_3CH(OH)CH_3$	43.1	76.50
5. n-Butanol	C_4H_9OH	6.44	91.51
6. s-Butanol	$CH_3CH(OH)CH_2CH_3$	17.1	91.73
7. t-Butanol	$(CH_3)_3COH$	42.0	94.30
8. Ethylene glycol	$HOCH_2CH_2OH$	0.118	55.77
9. Glycerol	$HOCH(CH_2OH)_2$	0.0395	73.03
10. Acetone	CH_3COCH3	229.2	73.54
11. Formamide	$HCONH_2$	0.0945	39.72

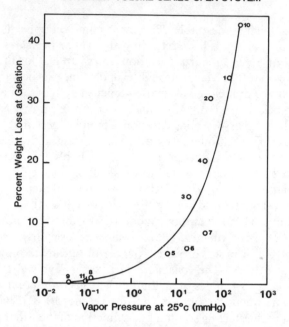

Figure 12.2 Weight loss of the original solution at gelation as a function of solvent vapor pressure.

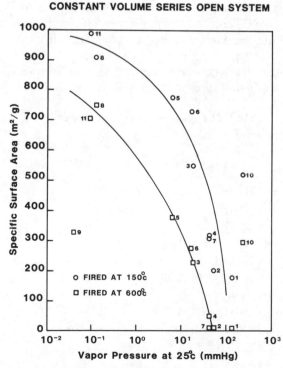

Figure 12.3 Difference in specific surface area of fired gels as a function of solvent vapor pressure.

solvent used and this can be of importance in the preparation of fibers. Secondly, the possibility of ligand exchange between solvent and alkoxide exists[8] as shown by Eq. (1):

$$C_2H_5OH + CH_3OM(OCH_3)_{x-1} \rightarrow C_2H_5OM(OCH_3)_{x-1} + CH_3OH \qquad (1)$$

Thirdly, the polymerization reaction itself can generate alcohols as shown in Eq. (2):

$$(RO)_{x-1}MOH + ROM(OR)_{x-1} \rightarrow (RO)_{x-1}MOM(OR)_{x-1} + ROH \qquad (2)$$

During gelation, any one or all of the above changes would lead to an overall variation of the composition of the starting solution. A homogeneous solution can become phase separated and give rise to further complexity. Lastly, the amount of solvent used can obviously influence the gelation time as well as the microstructure of the gel since the concentration of the reactants in the solution is altered.

ROLE OF THE CATALYST

It is well known that the catalyst used in a gelation reaction can have large effects on the microstructure of the gels formed as well as on the rapidity of the gelation process.[9,10] Some recent results from a constant mole series experiments involving a partially opened system (approximately 1% of surface area of solution exposed to ambient atmosphere) held at 20°C are shown in Table 12.2. TEOS was dissolved in EtOH in the same ratio as discussed above. The gelation

TABLE 12.2 Gelation Times for Different Catalysts

Catalyst	Concentration (mole: TEOS)	Apparent Initial pH of Solution	Gelation Time (hr)
HF	0.05	1.90	12
HCl	0.05	0.05[a]	92
HBr	0.05	0.20	285
HI	0.05	0.30	400
HNO$_3$	0.05	0.05[a]	100
H$_2$SO$_4$	0.05	0.05[a]	106
HOAc	0.05	3.70	72
NH$_4$OH	0.05	9.95	107
No catalyst		5.00	1000

[a]Between 0.00 and 0.05.

TABLE 12.3   Properties of Gels Dried at 25°C for Different Catalysts

Catalyst	Volume Shrinkage (%)	25°C Bulk Density (g/cm³)	Apparent Density (g/cm³)	Volume Shrinkage (%)	600°C Bulk Density (g/cm³)	Apparent Density (g/cm³)	Vickers Hardness
HOAc	84.0	1.32	1.33		2.08	2.12	666.5
HCl	81.3			85.2	2.06	2.12	429
HNO₃	79.9	1.14	1.16	85.2	1.82	2.02	470
H₂SO₄	71.6			80.0	1.46	2.12	224
HF	78.4	0.54	1.24	82.7	0.71	2.13	75
NH₄OH	67.8	0.49	1.13	71.7	0.70	2.21	28
No catalyst	87.5	0.95	2.09		1.25	2.21	

times differ widely according to the catalyst used. The pH of the solutions varied only slightly with time. The differences between HF, HCl, HBr, and HI are due to the difference in the anions of these catalysts. For the acids shown in Table 12.2, the differences in gelation times are not explainable from pH differences. It is expected that the concentration of any one catalyst, within reasonable limits, will have an effect on the gelation times, as observed by Sakka and Kamiya.[11]

Besides gelation times, the mirostructure and hence physical properties of the gels formed under Steps 1 and 2 are also drastically influenced by the catalyst used. Examples of some such differences are shown in Table 12.3. Column 2 refers to the shrinkage of the original liquid solutions.

The overall gelation process for alkoxides can be conveniently treated as a two-step process involving first hydrolysis and then polymerization. To what extent each catalyst can affect each of these two steps is a highly complex problem which is dependent on the catalyst. The early studies of Aelion[12] and Iler[13] are useful in that they offer at least partial explanations of the role of catalysts. Detailed understanding of the mechanisms involved, especially the interesting behavior of HF, must await the results of further studies.

ROLE OF THE METAL-CONTAINING COMPOUNDS

Most of the studies on SiO_2 gel were based on the use of TEOS as the starting reactant. We will conveniently refer to TEOS as the "metal-containing" compound. For other gel systems, especially polycomponent ones, other compounds such as nitrates have been used. The solubility and costs of these raw materials are of obvious importance. The structural difference between an alkoxide and a nitrate of the same metal is also obvious. The mechanisms of gelation and the structures of the gel must also be different. Again, systematic studies of the role played by the starting metal-containing compounds are lacking.

Some preliminary results at 25°C are presented to illustrate the interesting effects of raw materials on the preparation of SiO_2 gels. The silicon compound to water to alcohol to catalyst molar ratios were 1:4:4:0.07 with HCl as the catalyst. The surfaces of the reacting solutions were partially exposed as for the experiments described above.

Table 12.4 shows three series of experimental results. In Series (A), EtOH was the solvent. Four different silicon compounds were reacted. The most obvious difference between these compounds is the very short gelation time with silicon tetracetate. This can be attributed to the reaction between the acetate and EtOH as shown by Friedel and Landenburg:[14]

$$Si(OOCCH_3)_4 + 4C_2H_5OH \rightarrow 4CH_3COOC_2H_5 + SiO_2 + 2H_2O \quad (3)$$

Alternately, this short gelation time can be attributed to the easier hydrolysis of an acetate group as compared to an alkoxy group.[15] The absence of shrinkage for the acetate sample is due to the minimal evaporative losses in a very short time. In Series (B), the gelation times are not a simple function of the molecular weight of the silicon compounds when MeOH is the solvent. A comparison of the results for samples (1) and (8) indicates that ligand exchange is not a dominant reaction. In Series (C) when n-propanol is the solvent, the gelation times are directly proportional to the chain length of the alkoxy groups. In addition to the possible difference in rates of hydrolysis, the generation of different alcohols according to Eq. (2) can also influence the gelation times. For instance, in the polymerization phase for sample (12) MeOH is generated. However, for sample (14), BuOH is the corresponding product. Table 12.1 shows that the vapor pressures of these two alcohols are very different.

TABLE 12.4   Gelation Characteristics of SiO_2 Gels

			Gelation Characteristics			
	Starting		Gelation Time	Initial Volume	Volume Gel pt.	$\dfrac{V_{Gel}}{V_{Initial}}$
Sample	Compounds	Solvent	(hr)	(cm^3)	(cm^3)	(vol%)
(A) (1)	Si(OCH$_3$)$_4$	C$_2$H$_5$OH	152	24.0	16.7	70
(2)	Si(OC$_2$H$_5$)$_4$	C$_2$H$_5$OH	242	27.7	17.3	63
(3)	Si(OC$_4$H$_9$)$_4$	C$_2$H$_5$OH	243	36.0	30.6	85
(5)	Si(OOCCH$_3$)$_4$	C$_2$H$_5$OH	1	25.0	25.0	100
(B) (7)	Si(OCH$_3$)$_4$	CH$_3$OH	44	16.5	12.0	70
(8)	SI(OCH$_2$H$_5$)$_4$	CH$_3$OH	108	21	14.7	70
(9)	Si(OC$_4$H$_9$)$_4$	CH$_3$OH	64	29.0	27.7	96
(C) (12)	Si(OCH$_3$)$_4$	n-C$_3$H$_7$OH	131	24.2	21.5	89
(13)	Si(OC$_2$H$_5$)$_4$	n-C$_3$H$_7$OH	246	28.3	23.5	83
(14)	Si(OC$_4$H$_9$)$_4$	n-C$_3$H$_7$OH	550	37.5	29.5	79

TABLE 12.5 Physical Properties of Xerogels Prepared by Different Starting Materials and Solvents after 600°C, 24 hr Heat Treatment

Sample	Starting	Solvents	Density Bulk (g/cm³) ρ_B	Density Apparent (g/cm³) ρ_A	Porosity $P(\%)$	Specific Surface Area (m₂/g) A
(A) (1)	Si(OCH₃)₄	C₂H₅OH	1.59	2.18	27	169
(2)	Si(OC₂H₅)₄	C₂H₅OH	1.59	2.21	28	180
(3)	Si(OC₄H₉)₄	C₂H₅OH	1.30	2.00	35	773
(5)	Si(OOCCH₃)₄	C₂H₅OH	1.17	2.11	45	605
(B) (7)	Si(OCH₃)₄	CH₃OH	1.56	2.04	24	385
(8)	Si(OC₂H₅)₄	CH₃OH	1.64	2.19	25	160
(9)	Si(OC₄H₉)₄	CH₃OH	1.24	2.01	38	673
(C) (12)	Si(OCH₃)₄	n-C₃H₇OH	1.38	2.09	34	305
(13)	Si(OC₄H₉)₄	n-C₃H₇OH	1.64	2.12	23	285
(14)	Si(OC₄H₉)₄	n-C₃H₇OH	—	—	—	—

The gels described in Table 12.4 were subsequently heat treated at 600°C for 24 hr. Some results are shown in Table 12.5. Different silicon compounds are seen to give gels with very different densities, porosities, and specific surface areas. Because of the interdependent effects of solvents, catalysts, and raw materials and because of the difficulties associated with the separation of the hydrolysis phase and polymerization phase of the gelation process, it is not possible to offer quantitative explanations for the above results.

EFFECTS OF TEMPERATURE

For a chemical reaction such as the sol gel process of step 1 it is obvious that temperature must have an effect on the gelation rate. Surprisingly, there is practically no published information on the effects of temperature on the gelation process. This is perhaps because most known studies have involved EtOH and MeOH, and in an open or partially open system, temperatures in excess of 25°C would result in rapid losses of the solvents. Some preliminary results are shown in Table 12.6. Gelation was carried out in a closed system to avoid evaporative losses. The effects of temperature are obviously very pronounced. Yamane and Okana[16] reported on gelation experiments for SiO_2 gels at 54 to 70°C in an open system. No rates were reported but the bulk densities of the gels were found to decrease from 1.46 g/cm³ to 0.98 g/cm³ at these two temperatures, respectively. There is no explanation for these apparently anomalous results.

TABLE 12.6   Effect of Temperature on Gelation
Time for TEOS in a Closed System (TEOS/
H_2O/EtOH/Catalyst = 1/4/4/0.05 moles)

	Gelation Time (hr)		
Temperature, °C	HF	HCl	None
4	48.5	1440	
25	9.2	380	2520
70	0.3	20	27

WATER CONCENTRATION

Not all gelation processes require the presence of water. In alkoxide reactions, water is of course necessary for it is a reactant. The amount of water in a sol–gel solution of alkoxides and alcohol can have a very large effect on the structure and properties of the gel as shown by Yoldas[17] and Klein.[18]

SUMMARY AND CONCLUSIONS

The gelation time as well as the physical properties and microstructure of silica gels are significantly governed by the solvent, the catalyst, the chemical compound containing the silicon, and by the temperature of the gelation process. The interdependence of these various factors and the lack of systematic studies contribute to the difficulty of designing optimum processes for industrial exploitation of sol–gel technology. The science of sol–gel processing is still at its infancy at present.

ACKNOWLEDGMENT

The support of the Directorate of Chemical and Atmospheric Sciences, AFOSR, and particularly that of Drs. Donald Ball and Donald Ulrich of the sol–gel research program at UCLA are gratefully acknowledged. This report would have been impossible without the contributions of H. Nasu, A. Osaka, K. C. Chen, Mary Colby, Edward Pope, and Greg Moore of my laboratory.

REFERENCES

1. J. D. Mackenzie, *J. Non-Cryst. Solids*, in print (1985).
2. J. D. Mackenzie, *Ultrasonic Processing of Ceramics, Glasses and Composites*, Chapter 3, L. L. Hench and D. R. Ulrich, Eds., Wiley, New York, 1984.

3. G. Philipp and H. Schmidt, *J. Non-Cryst. Solids* **63**, 283 (1984).

4. A. Kaiser and H. Schmidt, *J. Non-Cryst. Solids* **63**, 261 (1984).

5. G. Carturan et al., *J. Non-Cryst. Solids* **63**, 273 (1984).

6. M. Tarasevich, presented at American Ceramic Society Annual meeting, May 1, 1984.

7. J. Zarzycki, *Ultrastructure Processing of Ceramics, Glasses and Composites*, Chapter 4, L. L. Hench and D. R. Ulrich, Eds., Wiley, New York, 1984.

8. D. C. Bradley, R. C. Mehrotra, and D. P. Gaur, *Metal Alkoxides*, Academic Press, New York, 1978.

9. C. J. Brinker et al., *J. Non-Cryst. Solids* **48**, 47 (1982).

10. M. Nogami and Y. Moriya, *J. Non-Cryst. Solids* **37**, 191 (1980).

11. S. Sakka and K. Kamiya, *J. Non-Cryst. Solids* **42**, 403 (1950).

12. R. Aelion et al., *J. Am. Chem. Soc.* **72**, 5705 (1950).

13. R. K. Iler, *The Chemistry of Silica*, Wiley, New York, 1979.

14. C. Friedel and A. Landenburg, *The Chemistry of Aliphatic Orthoester*, H. W. Post, Eds., American Chemical Society Monograph Series No. 92, p. 145, Reinhold, New York, 1943.

15. K. A. Andrianov, *Thin Film Technology for Microelectronics*, Chapter 2, A. Abraham, Ed., FPH/OP, Israel, 1975.

16. M. Yamane and S. Okana, *Yogyo-Kyokai-Shi* **87**, 434 (1979).

17. B. E. Yoldas, *J. Non-Cryst. Solids* **63**, 145 (1984).

18. L. C. Kelin and G. J. Garvey, *J. Non-Cryst. Solid* **48**, 97 (1982).

13

STRUCTURE AND CHEMISTRY OF SOL−GEL DERIVED TRANSPARENT SILICA AEROGELS

PARAM H. TEWARI, KEVIN D. LOFFTUS,
AND ARLON J. HUNT
Applied Science Division
Lawrence Berkeley Laboratory
University of California
Berkeley, California

INTRODUCTION

Aerogel is a microporous silica material containing a high fraction of voids (up to 97% by volume). It is transparent rather than translucent because the pore and particle sizes are smaller than the wavelength of light; therefore, it transmits rather than scatters light.

Aerogels were first made by Kistler[1−2] in 1931, but their application as transparent insulating materials has attracted attention only recently.[3−7] Silica aerogels, prepared by base-catalyzed hydrolysis and condensation of $Si(OCH_3)_4$, have been used in Chevenkov radiation detectors.[8−11] Here, aerogels with index of refraction between 1.02 and 1.10 replaced the pressurized cryostatic gas systems with low indices of refraction.

Since $Si(OCH_3)_4$, the starting material for transparent silica is extremely toxic, it is desirable to utilize alternative materials to ensure commercial viability. Tetraethylorthosilicate, TEOS, $Si(OC_2H_5)_4$, seems to be the logical choice. Schmitt[3−4] prepared aerogels based on $Si(OC_2H_5)_4$ hydrolysis and condensation using acid catalysis. However, the aerogels are not as clear as obtained by base catalysis of $Si(OCH_3)_4$. Also, the alcogels shrink during removal of the alcohol which further reduces the optical transparency. A base-catalyzed hydrolysis and gelation of $Si(OC_2H_5)_4$ combines the desirable properties of base catalysis with a low-toxicity starting material. However Schmitt[4] reported that aerogel synthesis by base-catalyzed $Si(OC_2H_5)_4$ was not possible because base-catalyzed $Si(OC_2H_5)_4$ gave white powdery material instead of a transparent aerogel. Russo and Hunt reported the base catalysis of $Si(OC_2H_5)_4$.[12] In this paper we report the results of our studies of the base-catalyzed hydrolysis and condensation of $Si(OC_2H_5)_4$ to optimize the desired transparency, strength, and stability of silica aerogels.

The overall goals of the research program are to improve the optical and thermal properties of aerogels, develop methods to protect them from environment, discover less expensive synthesis methods, and develop a technology base for production of transparent aerogels. Using a factorial design set of experiments, process parameters have been varied over a wide range of conditions to achieve the desired properties of the aerogels.

EXPERIMENTAL

Methods and Materials

Commercial $Si(OC_2H_5)_4$ was used without distillation. C_2H_5OH was 200 proof and NH_4OH and NH_4F were analytical grade. Viscosity was measured by a Brookfield viscometer with small sample adapter and thermostatic control. Light scattering intensity measurements as a function of angle of scattering were performed during gelation and after drying of aerogels as described earlier.[6-7] Optical transmission was obtained using Perkin Elmer Lamda 9 UV/VIS/NIR spectrophotometer.

The aerogel production has two segments: (1) hydrolysis and condensation of alkoxides giving alcogels, and (2) removal of alcohol from the alcogels to achieve aerogels. These are discussed below.

Hydrolysis and Gelation from $Si(OC_2H_5)_4$

Both acid- and base-catalyzed hydrolysis and condensation reactions give alcogels from alcohol solutions of alkoxides according to reactions

$$Si(OC_2H_5)_4 + 4H_2O \xrightarrow{\text{catalyst}} Si(OH)_4 + 4C_2H_5OH \qquad (1)$$

$$nSi(OH)_4 \rightarrow n(SiO_2) + 2nH_2O \qquad (2)$$

We used ammonia and ammonium fluoride as the catalyst for the gelation in the hydrolysis reaction.

Supercritical Drying of Alcogels

The alcogel structure contains more than 90% by volume fine pores containing alcohol. This alcohol must be removed to obtain aerogels.

Because the radius of the pores in the alcogels is extremely small, the surface tension at the interface between liquid and gas is extremely high. To prevent damage to the gel structure due to these high interfacial forces, drying was done under supercritical conditions, where interfacial forces are minimum. For ethyl alcohol, supercritical conditions are 270°C and 1700 psi pressure. Therefore, the process requires a high-temperature and high-pressure system. The autoclave used for supercritical drying was computer controlled with a data acquisition capability up to 300°C and 3000 psi. Typical drying conditions were 270°C and 1700 psi and the heating rate was 0.2 to 0.5°C/min. The pressure release was at high temperature and the typical drying time was 48 hr. These requirements made the process expensive and slow (total drying time was 2 to 3 days for each batch).

Near Ambient Temperature Drying

Performing the supercritical drying at near ambient temperature not only simplifies the requirements from those of high-temperature drying but also reduces the time for drying and makes the process much more economical. Based on the critical constants of some common fluids,[13] CO_2 seems to be the most practical choice. The cost of liquid CO_2 bottles is low ($6.00 per 50-lb bottle). Liquid CO_2 has been successfully used in critical point drying of biological samples for scanning electron microscopy. N_2O, on the other hand, has not been used successfully for biological samples.[13] Other fluids which seem attractive are Freons 13, 23, and 116. However, they are relatively more expensive. Water and alcohols are obviously poor choices. Therefore we have used CO_2 substitution followed by supercritical drying and have compared the properties of aerogels dried by both methods; CO_2 drying at 40°C and 1200 psi and high-temperature drying at 270°C and 1700 psi. The results are discussed below.

RESULTS AND DISCUSSION

Light Scattering of Alcosol, Alcogel, and Aerogels

Light scattering intensities for the gelling solutions were obtained during and after gelation. During gelation, the scattering intensity increases with time and reaches a limiting value (Fig. 13.1). This observation is consistent with light scattering intensity from the nucleation and growth of silica particles with subsequent gel formation. The light scattering intensity reaches a limiting value when the microstructural changes cease. However, on drying the alcogel, light

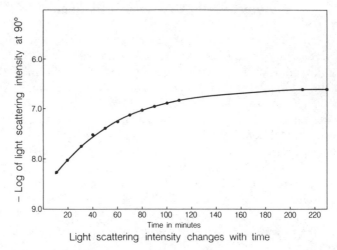

Figure 13.1 Light scattering intensity from a base-catalyzed alcosol during gelation as a function of time.

scattering intensity increases from the limiting value by a factor of 10 to 20 depending on the shrinkage, heterogeneity, and so on, in the aerogel. A theoretical estimate based on the change of index of refraction between air and alcohol for the two materials[15] gives roughly a factor of 10 change in scattering intensity. Therefore, the change in scattering intensity arises from both intrinsic changes and from structural changes during drying. The changes in light scattering intensity during drying are similar for both the CO_2 dried and high-temperature dried aerogels.

Process Optimization using Factorial Design

Using a factorial design set of experiments process parameters were varied over a wide range of conditions to explore the properties of aerogels. Concentration of the alkoxide, water content, amounts of alcohol, ammonia, and ammonium fluoride in the alcosol mixture, were simultaneously varied. Light scattering, optical transmission spectra, rheology, pH, shrinkage, surface area, and transmission electron microscopy of the final aerogels were studied for the optimization. The results were represented by using a polynomial equation containing the parameters and displayed using contour plots (Fig. 13.2). These calculations gave an evaluation of the significance and importance of the parameters and the direction of change to optimize the process. For example, it is concluded from Fig. 13.2 that to improve the light scattering (i.e., increasing the negative log of the scattered intensity) the temperature of gelation and the concentration of NH_4OH has to be lower than what was used. In addition, at gelation temperatures higher than 35°C, the light scattering can not be improved by the variation of NH_4OH (as indicated by the dotted line at the top in Fig. 13.2).

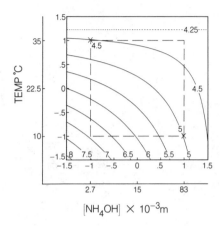

TEOS AEROGEL
–LOG(L.S.)

H_2O:TEOS = 8:1
$[NH_4F] = 1.6 \times 10^{-3} m$

Figure 13.2 Contour plots of the dependence of intensity of light scattering (LS) on temperature and ammonia content at H_2O:TEOS ratio of 8:1; NH_4F, 1.6×10^{-3} m and $V_{TEOS}/V_{total} = 0.15$. The inner scale gives the scaled change of the variable.

Contour plots with other parameters helped to optimize the process conditions. We have achieved a formulation which produces aerogels comparable to that produced using $Si(OCH_3)_4$. A final optimization will be done to determine the exact process conditions for the desired properties of the aerogels.

Optical Spectra

Transmission spectra [Figs. 13.3(a) and 13.3(b)] show that in the visible region, the transparency of the aerogels dried by the two methods is very similar. The difference in the NIR region is probably due to water absorption bonds. Since in CO_2 drying, the aerogels are subjected to temperatures ⩽40°C compared to ⩾270°C in the high-temperature supercritical method, water peaks are more pronounced in the CO_2-dried sample [Fig. 13.2(b)]. Water peaks have been reported[14] between 700 and 900, 900 and 1150, 1350 and 1800, and 1800 and 2200 nm. After heating the CO_2-dried aerogels in air at 250°C for 2–3 hr, certain water peaks disappear between 1700 and 1800 nm [Figs. 13.4(a) and 13.4(b)], and other peaks are sharpened and reduced in overall intensity. The fine structure in some of the water peaks disappears, suggesting desorption of water from certain sites. The spectra of aerogels dried by the CO_2 method, after further heating becomes similar to that for the high-temperature supercritically dried aerogels [compare Figs. 13.3(a) and 13.4(c)], confirming our other observations that the two drying methods produce similar quality aerogels. Figure 13.5 shows the comparison of a high-temperature supercritically dried base-catalyzed TMOS aerogel with a CO_2 dried base-catalyzed TEOS aerogel (4 mm thick).

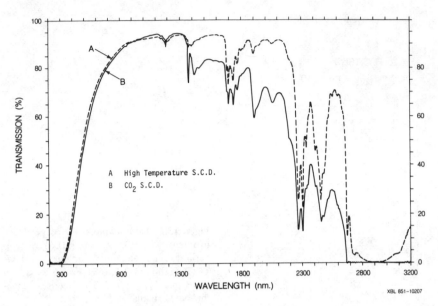

Figure 13.3 Transmission spectra of (A) base-catalyzed $Si(OC_2H_5)_4$ aerogel prepared by high-temperature supercritical drying; (B) base-catalyzed $Si(OC_2H_5)_4$ aerogel prepared by CO_2 super-critical drying (3 mm thick).

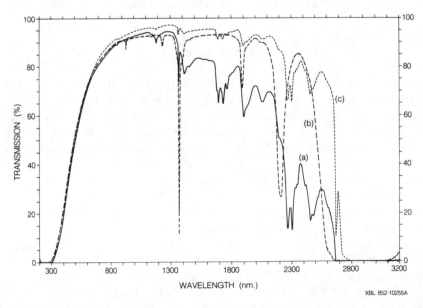

Figure 13.4 Transmission spectra of (a) base-catalyzed $Si(OC_2H_5)_4$ aerogel dried by CO_2 super-critical drying; (b) spectra of the same aerogel after heating for 3 hr in air at 250°C; (c) spectra of the same aerogel after heating for 4 hr in air at 450°C (samples 3 mm thick).

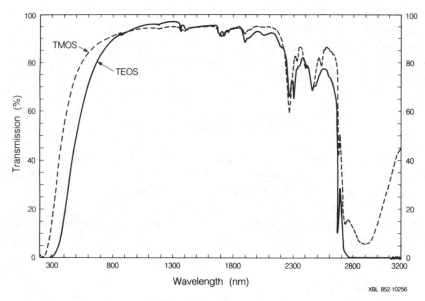

Figure 13.5 Transmission spectra of base-catalyzed $Si(OCH_3)_4$ (TMOS) aerogel dried by the high-temperature supercritical method (dashed line) and base-catalyzed $Si(OC_2H_5)_4$ (TEOS) aerogel dried by the CO_2 method and heated to 450°C for 4–5 hr; both samples are 4 mm thick.

Reproducibility and Quality of Aerogels in CO_2 Drying

The ease of operation, reproducibility, and repeatability of supercritical CO_2 drying of aerogels is far better than that of high-temperature supercritical drying. The incidence of cracking or fracture of aerogels is significantly lower using the CO_2 method. The batch-to-batch reproducibility in drying is also improved. Above all, there is a marked time saving from 2 to 3 days for the high-temperature process, to 8–10 hr for the CO_2 method. In addition, the equipment for the CO_2 method is simpler and less expensive than that required for the high-temperature supercritical drying.

SUMMARY AND CONCLUSIONS

Optical transmission, light scattering, and other data show that CO_2 supercritical drying produces aerogels similar in quality to those produced by high-temperature supercritical drying. The reproducibility of the CO_2 drying method is far better than that of the high temperature supercritical drying.

   CO_2 supercritical drying is simpler and less expensive because it is done at $\leqslant 40°C$ and 1200 psi instead of at $\geqslant 270°C$ and $\geqslant 1700$ psi. The CO_2 method also saves significant time in the drying process.

   By exploring a wide range of process variables by a factorial design set of experiments, we have achieved a base-catalyzed $Si(OC_2H_5)_4$ aerogel similar in quality to the one prepared by $Si(OCH_3)_4$.

Supercritical drying using CO_2 may offer a unique opportunity in ceramic processing for controlling pore and particle sizes in gels, thereby providing a better control of the microstructure.

ACKNOWLEDGMENTS

The authors are grateful to Drs. J. S. Bastacky and M. D. Rubin for allowing the use of the critical point drying apparatus and UV/VIS/NIR spectrophotometer, respectively, and to Drs. R. E. Russo and P. Berdahl for helpful discussions.

This work was supported by the Assistant Secretary for Conservation and Renewable Energy, Office of Solar Heat Technologies, Passive and Hybrid Solar Energy Division of the U.S. Department of Energy under Contract No. DE-AC03-76SF00098.

REFERENCES

1. S. S. Kistler, *Nature* **127**, 741 (1931).
2. S. S. Kistler, *J. Phys. Chem.* **34**, 52 (1932).
3. W. J. Schmitt, M. S. Thesis, Dept. of Chemical Engineering, University of Wisconsin, 1982.
4. W. J. Schmitt, Annual Meeting of the AIChE, New Orleans, 1981.
5. M. Rubin and C. Lampert, *Solar Energy Mater.* **7**, 393 (1983).
6. A. J. Hunt, in *Ultrastructure Processing of Ceramics, Glasses and Composites*, L. L. Hench and D. R. Ulrich, Eds., Wiley, New York, 1984, p. 549.
7. A. J. Hunt and P. Berdahl, *Material Res. Soc. Symp. Proc.* **32**, 275 (1984).
8. S. S. Henning and L. Svensson, *Phys. Scripta* **23**, 697 (1981).
9. G. A. Nicholaon and S. J. Teichner, *Bull. Soc. Chim* **5**, 1900, 1909 (1968).
10. M. Cantin, M. Casse, L. Coch, R. Jouan, P. Mestreau, and D. Roussel, *Nucl. Instrum. Methods* **118**, 177 (1974).
11. M. Bourndinard, J. B. Cheze, and J. C. Thevenin, *Nucl. Instrum. Methods* **136**, 99 (1976).
12. R. E. Russo and A. J. Hunt, Comparison of Ethyl vs. Methyl Sol–Gels for Silica Aerogels using Polar Nepholometry, to be communicated.
13. A. L. Cohen, Critical Point Drying Principles and Procedures in Scanning Electron Microscopy, 1979/I SEM. Inc. AMF O'Hare, IL 60066, USA.
14. J. A. Curcio and C. C. Petty, *J. Opt. Soc. Am.* **41**, 302 (1951).
15. P. Berdahl, private communication.

14

A MODEL FOR THE GROWTH OF FRACTAL SILICA POLYMERS

K. D. KEEFER
Sandia National Laboratories
Albuquerque, New Mexico

INTRODUCTION

It is not at all obvious that objects which grow in a random, indeterministic fashion can have characteristic structures. Yet one of the most interesting phenomena in random growth is that systems with only short-range forces and no long-range order can form particles whose large-scale structures are statistically well defined and distinctive. The structure of silicate species which grow during the formation of silica sols and gels are good examples. By varying the chemical conditions under which silica is polymerized, structures can form which range from randomly branched polymers to colloidal particles.[1] Although numerous hypothetical growth processes have been studied by computer simulation, only a few appear to describe any objects in nature. This paper will present a new model of random growth, whose rules are based on the local chemistry of the hydrolysis and polymerization of tetraethoxysilane, $Si(OC_2H_5)_4$, and which mimics some of the different random structures that can form under different chemical conditions.

FRACTAL GEOMETRY

Interest in materials resulting from random growth processes has been spurred by the development of fractal geometry, which provides a means of quantitatively describing the average structure of certain random objects.[2] The fractal dimension, d_f, of an object may be defined by the relation

$$M \propto R^{d_f} \tag{1}$$

where M is the mass of the object and R is some measure of the radius. For Euclidian (nonfractal) objects, d_f equals the dimension of space, d (i.e., in three dimensions the mass of a sphere scales as the radius cubed). Fractals are objects for which d_f is less than the dimension of space. For example, the root mean square radius of a random walk polymer in three dimensions is proportional to the square root of the mass of the polymer, hence it is fractal with $d_f = 2$. Because d_f has many of the properties of a dimension but is often fractional, it is called the fractal dimension.

The fractal dimension of objects generated in computer simulations may be measured in several ways. One way is simply to measure the radius of gyration of the object as a function of mass as it grows. Another is to calculate scattering curves for the object, the interpretation of which is discussed in the next section. A third uses an algorithm called the "sandbox," in which each site in the object is used as the center of a box and the mass inside the box is counted as the length of its side is increased. The result is a power law like Eq. (1), with R equal to the length of the side of the box. All three methods were used; however the values quoted here were obtained by the latter method which gives the most consistent results for the model described.

For surface fractals, the surface area is related to the radius by a fractional power

$$S \propto R^{d_s} \tag{2}$$

For nonsurface fractals, the dimension of the surface, d_s, is one less than the dimension of space, that is, in three dimensions the surface area of a sphere varies as the square of the radius, $3 - 1 = 2$. Objects for which $d_f = d$ and $d > d_s > d - 1$ have very rough surfaces and will be referred to as surface fractals, as distinct from those which obey Eq. (1) with $d_f < d$, which will be referred to as mass fractals. If $d_f < d$, then $d_s = d_f$, that is, mass fractals are all surface. However, if $d_f = d$, the condition $d_s > d - 1$ may still hold. Such objects have homogeneous cores and rough surfaces and appear to result from the hydrolysis of $Si(OC_2H_5)_4$ and the polymerization of the products.

SCATTERING EXPERIMENTS

The fractal dimension of objects with radii on the order of 10^2–10^4 Å may be measured in a small-angle x-ray scattering or static light scattering experiment. Although the details are outside the scope of this paper, it can be shown that if an object is a surface or mass fractal, the scattered intensity will obey a power law over a certain range of scattering angle.[3] Specifically, for mass fractals

$$I(h) \propto h^{-d_f}, \qquad R_g \gg h^{-1} \gg a \tag{3}$$

where $I(h)$ is the scattered intensity, $h = 4\pi \sin \theta/\lambda$, 2θ is the scattering angle, R_g is the electronic radius of gyration, and a is the size of a monomer. Thus, mass fractals yield power law exponents in the range $d > d_f \geqslant 1$. In the real world of three dimensions, the slope lies between -1 and -3.

For surface fractals, the scattered intensity obeys the relation[4]

$$I(h) \propto h^{d_s - 2d} \tag{4}$$

For nonfractal surfaces in three dimensions, this gives the well-known Porod law.[5] Since d_s must lie in the range $d_f = d \geqslant d_s > d - 1$, the slope of a plot of $\log I$ versus $\log h$ will lie in the range -3 to -4.

The above expressions were derived assuming the scattering curves are measured with point collimation geometry. A complication arises if the experimental X-ray scattering curves are measured with a slit collimation geometry as were those reported here. The scattering is "smeared" over a range of angles. However, if a power law is observed in slit smeared curves, then the exponent of the power law is one less than that which would have been observed with point collimation. This relation, rather than numerical "desmearing," has been used in the interpretation of the observed scattering curves reported here. Thus, slit smeared slopes in the range of -2 to -3 will be interpreted as arising from surface fractals and those less than -2 from mass fractals.

GROWTH AND STRUCTURE OF SILICA POLYMERS

When $Si(OC_2H_5)_4$ is mixed with water in an alkaline ($0.01 M$ NH_4OH) ethanol solution the alkoxide groups on silicon are hydrolyzed to form silanol groups ($\equiv SiOH$) which subsequently undergo condensation polymerization to form $\equiv Si—O—Si\equiv$ linkages according to the reactions

$$(RO)_3SiOC_2H_5 + H_2O \xrightarrow{OH^-,H^+} (RO)_3SiOH + C_2H_5OH \tag{5}$$

$$2(RO)_3SiOH \xrightarrow{OH^+} (RO)_3SiOSi(OR)_3 + H_2O \tag{6}$$

where $R = -C_2H_5, -H$, or $-Si(OR)_3$. The small-angle x-ray scattering curves from solutions hydrolyzed with different amounts of water [molar ratios of $H_2O:Si(OC_2H_5)_4$ ranging from 1 to 4] are shown in Fig. 14.1. The details of the experiment have been presented elsewhere.[1] The curves from the solution with the smallest amount of water may be interpreted as arising from particles which are mass fractals with $d_f = 2.84$ and those from the solutions with higher water

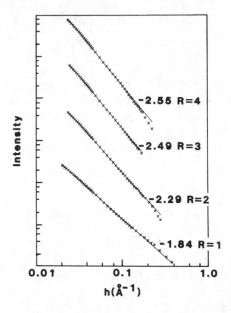

Figure 14.1 Small angle X-ray scattering curves (slit smeared) from $1M$ solutions of $Si(OC_2H_5)_4$ hydrolyzed with varying amounts of H_2O. R is the molarity of H_2O. 0.01 NH_4OH was used as a catalyst.

contents as arising from particles which are not mass fractals, but which have fractally rough surfaces with d_s ranging from 2.45 to 2.71.

No existing random growth models produce structures with homogeneous cores and fractal surfaces, so a new class of growth model based on the chemistry of silica was developed to account for these results. The aggregates which grow from fully hydrolyzed orthosilicic acid $(Si(OH)_4)$ under alkaline conditions have been shown to be fully dense and relatively uniform in size.[6] The reason is that polymerization occurs by a nucleophilic substitution mechanism which favors reaction between weakly acidic and therefore neutral SiOH groups and strongly acidic, negatively charged $\equiv SiO^-$ groups.[6] The silanols on monomers of $(Si(OH)_4)$ are less acidic than those on polymers, hence in the pH range of the experiment growth occurs primarily by monomers attaching to polymers. The essential features of this system to be accounted for in a growth model are that growth occurs primarily between a monomer and a polymer and not between two monomers or two polymers, and that each monomer is fully tetrafunctional.

A random growth model which contains these essential features was first proposed by Eden to describe the growth of cell colonies.[7] In a 2-D computer simulation of the growth of an Eden cluster, one starts with a square lattice which has at its center a nucleus or "seed." Each of the four sites adjacent to the starting seed site is designated a "growth" site. To grow the cluster, one of the growth sites is selected at random and a monomer is placed on that site. All empty sites (i.e., neither occupied, nor existing growth sites) adjacent to the new monomer then become growth sites. This process is continued until the cluster grows to some desired size. A cluster of 3200 sites is shown in Fig. 14.2 in which monomers are represented by filled squares and growth sites by "+" signs.

Figure 14.2   Computer simulation of growth from fully hydrolyzed monomers (Eden model).

The salient features of the particles produced in the Eden growth model are that they are equant, dense, and their surfaces, though rough, are not fractally rough. Thus the Eden model, based only on the short-range chemistry, has yielded the larger structure of the colloidal particles which are, in fact, produced under the chemical conditions described above. An important aspect of the Eden model is that the monomers do not "diffuse" on the lattice. This is of practical importance in a computer simulation because the clusters grow quickly and to obtain good statistics a large number of clusters must be measured. However, it does imply that the growth is reaction limited rather than diffusion limited since all potential growth sites are accessible to a monomer with equal probability. Although this is a reasonable assumption for growth sites which are on the surface of the particle, it also means that growth can occur in "lakes" which do not connect to the surface, which is unreasonable. It is believed that "lakes" occur much more frequently in 2-D models than in 3-D, so this artifact should not detract from conclusions drawn about real systems.

Although the Eden model appears to describe growth of colloidal silica particles from orthosilicic acid, it does not account for the small-angle scattering curves in Fig. 14.1 since it generates neither fractal particles nor fractal surfaces. The reason the Eden model generates nonfractal objects is clear; sooner or later all potential growth sites will be occupied, so that "old" growth sites are rare. Since the cluster grows more or less equally in all directions the oldest sites are at the center; thus there are no unoccupied sites and the center is completely dense.

INCOMPLETE HYDROLYSIS

There are several significant differences between the chemistry of the solutions studied and those which generate colloidal particles. The solutions studied contain relatively low water and catalyst concentrations, compared to those which generate colloidal silica, so hydrolysis is slow and incomplete. It is possible, then, that the silicate monomers are not fully hydrolyzed, that is, they still have alkoxy groups bonded to them. Due to steric hindrance, alkoxy groups on polymers tend not to hydrolyze as rapidly as those on monomers, hence if a polymer is formed from partially hydrolyzed monomers, it will have sites which are unavailable to further polymerization or become active only at a relatively slow rate.

A random growth model which includes these features was developed to see if incomplete hydrolysis might perturb the normal growth pattern sufficiently to produce fractal structures such as those observed in the scattering experiments. In this model, growth occurs from a fixed seed, as in the Eden model, but not every monomer has a full complement of growth sites. Some of the sites on each monomer are "poisoned" to imitate the effect of unhydrolyzed alkoxy groups. When both the average ratio of growth sites to poisoned sites on a monomer and the distribution of those sites among the monomers are varied, different structures with different fractal dimensions are formed.

The rules for this model of growth in partially hydrolyzed systems are as follows. Fixed ratios of singly, doubly, triply, and fully hydrolyzed monomers are selected, these being the only parameters in the model. A seed with four (the exact number is unimportant) growth sites is placed on a lattice. One of the four types of monomer is then chosen with the specified probability, a growth site on the particle is chosen at random, and the monomer is then placed on the growth site in a random orientation. Lattice sites adjacent to sides of the monomer which are hydrolyzed are added to the list of growth site and sites adjacent to unhydrolyzed sides are "poisoned" and can never be occupied.

Some of the different random structures which can form in this model are shown in Figs. 14.3, 14.4, and 14.5. These structures are grown with an average of one unhydrolyzed site per monomer. The only difference is the distribution of doubly, triply, and fully hydrolyzed monomers from which they were grown. The cluster in Fig. 14.3 was grown entirely from triply hydrolyzed monomers. Compared to the Eden cluster (Fig. 14.2), its periphery is considerably rougher and it is porous, with a distribution of pore sizes. However, the various measurements described above show that it is not fractal; therefore, its structure is essentially the same as that of the Eden cluster. For an object to be fractal, it must have holes or surface roughness on all length scales, up to the length of the object itself. Apparently, having an even distribution of one unhydrolyzed site per monomer prevents the density fluctuations that characterize a fractal.

When the number of unhydrolyzed sites on a monomer is allowed to fluctuate, fractals may result. Clusters grown from equal numbers of doubly, triply, and fully hydrolyzed monomers (Fig. 14.4) have a fractal dimension of 1.8 and those made from equal number of doubly and fully hydrolyzed species (Fig. 14.5) have a fractal dimension of 1.67. The growth sites ("+" sign in Figs. 14.2–14.5) on each are localized in a few areas and are not evenly distributed as they are in the nonfractals in Fig. 14.2 and 14.3. This causes the periphery or "hull"

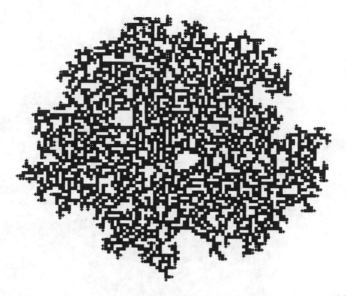

Figure 14.3 Simulation of random growth from triply hydrolyzed monomers which generates porous but nonfractal objects.

Figure 14.4 Simulation of random growth from equal numbers of doubly, triply, and fully hydrolyzed monomers which generates a fractal object of fractal dimension 1.8.

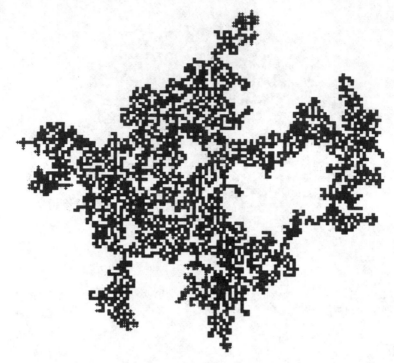

Figure 14.5 Simulation of random growth from equal numbers of doubly and fully hydrolyzed monomers which generates an object of fractal dimension 1.67.

of these objects to be fractal as well. Close inspection of these clusters reveals many of the apparent "lakes" to be connected to the exterior and hence represent a part of the "hull," which contributes to the very high hull fractal dimensions of 1.67 and 1.75 of Figs. 14.4 and 14.5, respectively.

An important feature of this model is that the apparent fractal dimension varies continuously with certain parameters. For example, the fractal dimension of models composed only of doubly and fully hydrolyzed monomers varies continuously from 2, when only fully hydrolyzed species are used (the Eden model), to 1.67, when the ratio is 1:1. It is possible that this continuous variation is due to finite size effects and that only a few discrete values would be observed in the limit of infinite molecular weight. However, real polymers are also of finite size so that the asymptotic behavior of the model described here is not critical to the interpretation of the experiment.

In its current state of development, the model described here does not generate the homogeneous core, fractal surface objects which would account for the experimental scattering curves measured at the higher water concentrations. Certain chemically realistic variations may change this. For example, if hydrolysis of the polymer can occur, it will turn into a nonfractal Eden cluster since growth can occur at all sites. If the rate of polymer hydrolysis is slow, however, the core may become dense but the surface may remain fractally rough. Such an effect could account for the experimentally observed transition from mass fractals to surface fractals as the amount of water in the solution is increased. In any event, the purpose of this model is not to predict the structure of silica polymers under different hydrolysis conditions but to demonstrate how particular aspects of the hydrolysis can alter the structure in definable ways.

CONCLUSION

A new model of random growth in solutions of partially hydrolyzed $Si(OC_2H_5)_4$ gives rise to polymers with fractal structures, while an analogous model for growth from fully hydrolyzed orthosilicic acid results in dense particles. These models qualitatively account for difference in the small-angle X-ray scattering curves obtained from alkaline solutions with different degrees of hydrolysis. Further development of random growth models may help to explain the great diversity of silicate polymers.

REFERENCES

1. K. D. Keefer, The Effect of Hydrolysis Conditions on the Structure and Growth of Silicate Polymers, in Proceedings of the Materials Research Society Symposium, *Better Ceramics Through Chemistry*, C. J. Brinker, D. E. Clark, and D. R. Ulrich, Eds., Elsevier, New York 1984, pp. 15–24.

2. D. W. Schaefer and K. D. Keefer, Fractal Geometry of Silica Condensation Polymers, *Phys. Rev. Lett.* **53**(14), 1383–1386 (1984).

3. D. W. Schaefer and K. D. Keefer, Structure of Soluble Silicates, in Proceedings of the Materials Research Society Symposium, *Better Ceramics Through Chemistry*, C. J. Brinker, D. E. Clark, D. R. Ulrich, Eds., Elsevier, New York, 1984, pp. 1–14.

4. H. D. Bale and P. W. Schmidt, Small-Angle Scattering Investigation of Submicroscopic Porosity with Fractal Properties, *Phys. Rev. Lett.* **53**(6), 596 (1984).

5. A. Guinier, G. Fournet, C. Walker, and K. Yudowitch, *Small Angle Scattering of X-rays*, Wiley, New York, 1955.

6. R. K. Iler, *The Chemistry of Silica, Solubility, Polymerization Colloid and Surface Properties and Biochemistry*, Wiley, New York, 1979.

7. H. E. Stanley, F. Family, and H. Gould, Kinetics of Aggregation and Gelation, to be published in Proceedings of the Institute for Theoretical Physics Workshop on Polymer Dynamics, *Journal of Polymer Science; Polymer Physics* (1982).

15

STRUCTURE OF POROUS MATERIALS: SILICA AEROGELS AND ORGANIC FOAMS

DALE W. SCHAEFER, K. D. KEEFER JAMES H. AUBERT,
AND PETER B. RAND
Sandia National Laboratories
Albuquerque, New Mexico

INTRODUCTION

In spite of the numerous applications of porous materials, few structural studies exist.[1] In this paper we report a study of random porous materials by small-angle X-ray scattering (SAXS). We outline the principles used to establish structure from SAXS curves and trace the observed structures to the physics and chemistry of the precursor solutions.

SCATTERING FROM RANDOM MATERIALS

Figure 15.1 shows a generic small-angle scattering curve[2] from an amorphous material. The scattered intensity, I, is plotted as a function of the magnitude of the scattering vector, $K = (4\pi/\lambda) \sin(\theta/2)$, where λ is the incident wavelength, and θ is the scattering angle.

The variable K in Fig. 15.1 is the analogue of the Bragg period extracted from wide-angle scattering from crystalline materials. In amorphous materials, of course, there are no regular planes of atoms so Bragg peaks are not observed. There are, nevertheless, spatial variations of electron density which give rise to

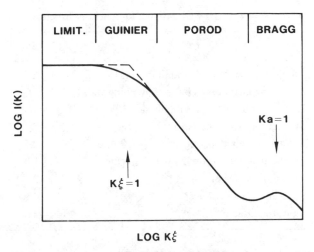

Figure 15.1   Schematic scattering curve for a random porous material.

small-angle scattering. By Fourier analysis of these random variations one identifies scattering at a particular scattering angle with fluctuations having a spatial Fourier frequency K or an equivalent Bragg spacing of $2\pi K^{-1}$.

In the limiting regime shown in Fig. 15.1, one probes very long wavelength fluctuations. Typically, density fluctuations are uncorrelated for dimensions exceeding 1000 Å so $I(K \to 0)$ is just proportional to the mean square fluctuation in electron density.

In the "Guinier" regime one probes dimensions over which fluctuations are correlated. In a porous material the correlation range, ξ, is roughly the mean pore size and the position of the break in $I(K)$ in the Guinier regime provides a measure of ξ. If the pores are very large then the Guinier and limiting regimes may not be observable using typical x-ray apparatus ($K > 0.001$ Å^{-1}).

In the "Porod" regime, one probes dimensions which are small compared to the pore size ($K\xi \gg 1$) but large compared to typical chemical bond distances, $a(Ka \ll 1)$. In the Porod regime power law $[I(K) \sim K^{-x}]$ scattering curves are common,[2] indicating that scattering results from structures which have no characteristic length. These structures often have the properties of random fractals[3-6] and the concept of fractal geometry is useful in characterizing their essential geometric properties.

Table 15.1 lists the Porod exponent (x) expected for a variety of structures. Exponents more positive than -3 are expected for so-called volume fractals, of which polymers are the prototype. These objects have the property that the mass, M, varies as a power of length, $M \sim$ (length)^D, D being the fractal dimension. For volume fractals the Porod slope, x, is related to D as $x = -D$. The concept of fractal geometry is useful because measured D's can be compared with model predictions to ascertain the origin of random structures.

Porod slopes more negative than -3 are found when the scattering arises from an interface. If the interface is smooth,[7] the slope is -4. If the surface is

TABLE 15.1   Porod Slopes for Various Structures

	Slope
Linear ideal polymer[13] (random walk)	-2
Linear swollen polymer[13] (self-avoiding walk)	$-5/3$
Randomly branched ideal polymer[13]	$-16/7$
Swollen branched polymer[13]	-2
Diffusion limited aggregate[14]	-2.5
Multiparticle diffusion limited aggregate[15]	-1.8
Percolation cluster[16]	-2.5
Fractally rough surface[8]	-3 to -4
Smooth surface[7]	-4

fractally rough, slopes of between 3 and 4 are expected.[8,9] Fractal surfaces have the property that the surface area, S, varies as a noninteger power of length, $S \sim (\text{length})^{D_s}$, D_s being the surface fractal dimension. For smooth surfaces $D_s = 2$, whereas for rough surfaces $2 < D_s < 3$. Fractal surfaces are characterized by a Porod slope[8] $x = D_s - 6$.

The "Bragg" region of Fig. 15.1 contains local configurational information and is generally not of concern in small-angle scattering.

In what follows we concentrate primarily on the Porod regime and exploit power law analysis to infer the structure of porous materials.

SILICA AEROGEL

Two aerogels prepared from different solution precursors were studied. Commercial aerogel (Fig. 15.2) was obtained from Airglass, Inc., Byggaregrand 1, S-27500, Sjobo, Sweden. This material is prepared[10] by base-catalyzed hydrolysis and condensation of tetramethoxysilane in alcohol. Critical point drying after gelation yields a porous material with a density of 0.088 g/cm^3. The second material (Fig. 15.3) was prepared under similar conditions but with lower H_2O/Si ratio using tetraethoxysilane. The resulting solution (ungelled) was air dried.

The data in Fig. 15.2 show the general structure of Fig. 15.1. At small K the curve is flat, indicating that the material is uniform for dimensions above $\xi = 85$ Å. The high-angle portion of the curve is power law with a slope of -4. From Table 15.1, this slope is consistent with sharp interfaces. At intermediate angles, a second power law region is found with a slope of -2. From Table 15.1, this value is consistent with structures whose backbone geometry is basically chainlike such as a branched polymer. The break in the curve occurs at $K^{-1} = 12$ Å, suggesting a smooth surface on length scales below 12 Å and a branched linear structure above 12 Å.

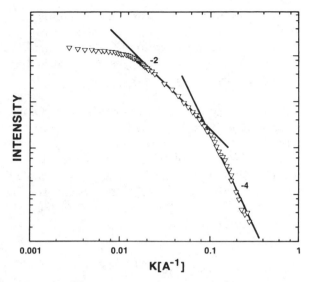

Figure 15.2 SAXS for Airglass™ Aerogel. Density = 0.088 g/cm³. The data are desmeared results from a Kartky SAXS system. H₂O/Si = 4.1 in the solution precursor.

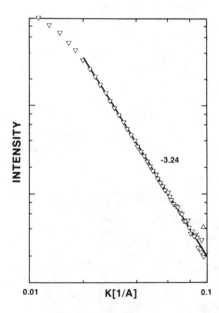

Figure 15.3 SAXS curves of a dried solution of fractally rough particles. This material is prepared by drying a solution similar to that used in the R = 2 curve in Fig. 1 of Ref. 6. Data measured in point collimation. The density is 0.8 g/cm³.

The above observations indicate that the aerogel is best represented as a random jungle gym structure resulting from a colloid aggregation process.[9] In the solution precursor, colloidal particles grow to a hard sphere diameter of 40 Å (consistent with a 12-Å radius of gyration) and then aggregate and ultimately gel. Critical point solvent removal leaves a structure identical to the aggregate.

Support for the above idea comes from structural data on solution grown colloidal aggregates.[4] Solution aggregates also show two power law regimes with slopes of -2.1 and -4, very close to the values found above. The random jungle gym model is consistent with the ideas proposed by Iler[11] for aqueous silica gels. We find, however, that a variety of other structures can be produced within the silica system.[5,6]

Figure 15.3 shows the scattering pattern for gels produced under less aggressive hydrolysis conditions. A single power law regime is seen with an exponent of -3.24 consistent with fractal surfaces of dimension $D_s = 2.76 \pm 0.2$. Once again this structure is traceable to the solution precursor in which SAXS results and model simulations indicate the existence of fractally rough colloidal particles.[6] In fact, the solution precursor of this system gives $D_s = 2.71$ (taken from the $R = 2$ curve of Fig. 15.1, Ref. 6). Thus the fractally rough pores arise from fractally rough precursors.

ORGANIC FOAMS

Whereas polymerization and aggregation are the dominant growth processes for aerogels, phase separation is the dominant process which controls the structure of microporous foams. These foams are prepared[12] by rapid quench and freeze drying of polymer solutions. We have studied both aqueous and non-aqueous systems prepared by subspinodal quenches and solvent freezing. Although the long-range structure ($>1 \mu$m) of these foams can be radically different (Fig. 15.4), we find the structure below 100 Å is similar.

Figure 15.5 shows the scattering curves for a series of polystyrene (PS)/cyclohexane (CH) foams prepared by spinodal quench along the critical isochore. This system phase separates at $\sim 30°$C. The temperature shown is the final temperature to which the foams were quenched prior to freeze drying. In all cases the slopes are near -4, indicating sharp interfaces.

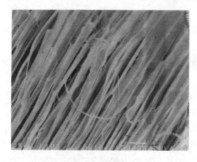 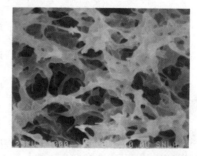

SOLVENT FREEZING   SPINODAL DECOMPOSITION

PS/BENZENE   PS/CYCLOHEXANE

Figure 15.4 Electron micrograph of PS foams prepared in CH (spinodal decomposition) and benzene (solvent freezing).

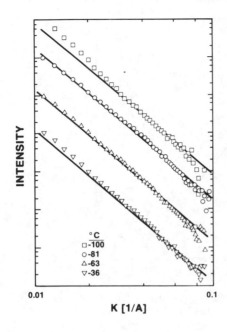

Figure 15.5 SAXS curves for PS/CH foam as a function of quench temperature. Data measured in point collimation. Lines have a slope of − 4.0. Foams have density of 0.04 g/cm^3. $M = 1.8 \times 10^6$.

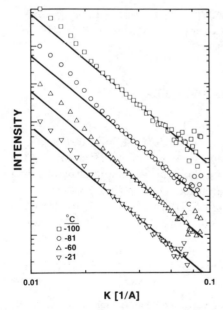

Figure 15.6 SAXS curves for PS/BZ foam as a function of quench depth. The scattering vector is perpendicular to the fiber axis. No scattering is observed in the parallel geometry. Data were measured in point collimation. Lines have a slope of − 4.0. The foam density is 0.04 g/cm^3. $M = 1.8 \times 10^6$.

In the PS/benzene (BZ) system, the solvent freezes giving rise to the lamellar structure of Fig. 15.4. The scattering curves, nevertheless, are very similar to the PS/CH system. The PS/BZ results as a function of quench depth are shown in Fig. 15.6.

Finally, a single nonaqueous system (polyacrylic acid/water) was briefly

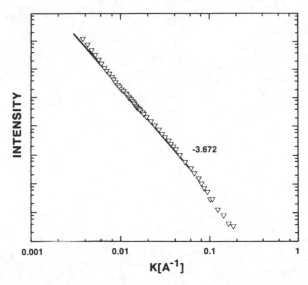

Figure 15.7   SAXS curve for polyacrylic acid/water. Data were measured in line collimation with "analytic" desmearing. $I_{point} = I_{line} \div K$. The foam density is 0.02 g/cm³.

studied. The data in Fig. 15.7 indicate a fractally rough surface with $D_s = 2.33 \pm$ 0.3. Like Fig. 15.2, the data appear to cross over to Porod's law at $K \simeq 0.06 \,\text{Å}^{-1}$ indicating smooth interfaces at short length scales.

ACKNOWLEDGMENTS

This work was performed at Sandia National Laboratories supported by the U.S. Department of Energy under Contract No. DE-AC04-76DP00789.

We thank David Evans for technical assistance in construction of the Kratky SAXS instrument and for collecting the data. We thank Bruce Bunker for bringing the Airglass material to our attention. Data in Figs. 15.4–15.6 were taken at the National Center for Small Angle Scattering at Oak Ridge National Laboratory with the supervision and assistance of J. S. Lin and C. Spooner.

REFERENCES

1. P. Debye, H. R. Anderson, and H. Brumberger, Scattering by an Inhomogeneous Solid II, *J. Appl. Phys.* **28**, 679 (1957).
2. D. W. Schaefer and K. D. Keefer, Structure of Soluble Silicates, in *Better Ceramics Through Chemistry, Mater. Res. Soc. Symp. Proc.* **32**, 1 (1984).
3. B. B. Mandebrot, *Fractals, Form and Chance*, Freeman, San Francisco, 1977.
4. D. W. Schaefer, J. E. Martin, D. S. Cannell, and P. Wiltzius, Fractal Geometry of Colloidal Aggregates, *Phys. Rev. Lett.* **52**, 2371 (1984).

5. D. W. Schaefer and K. D. Keefer, Fractal Structure of Silica Condensation Polymers, *Phys. Rev. Lett.* **53**, 1383 (1984).

6. K. D. Keefer, Chapter 14, this volume.

7. A. Guinier and G. Fournet, *Small Angle Scattering of X-Rays*, Wiley, New York, 1955.

8. H. D. Bale and P. W. Schmidt, Small Angle X-Ray Scattering Investigation of Submicroscopic Porosity with Fractal Properties, *Phys. Rev. Lett.* **53**, 596 (1984).

9. D. W. Schaefer and K. D. Keefer, Structure of Random Porous Materials: Silica Aerogel, to be published.

10. S. Henning and L. Svensson, Production of Silica Aerogel, *Phys. Scripta* **23**, 697 (1981).

11. R. K. Iler, *The Chemistry of Silica*, Wiley, New York, 1979.

12. J. H. Aubert and R. L. Clough, Low Density Microcellular Polystyrene Foams, to be published.

13. M. Daoud and J. F. Joanny, Conformation of Branched Polymers, *J. Phys. (Paris)* **42**, 1359 (1981).

14. T. A. Witten and L. M. Sander, Diffusion Limited Aggregation, a Kinetic Critical Phenomena, *Phys. Rev. Lett.* **47**, 1400 (1981).

15. P. Meakin, The Formation of Fractal Clusters and Networks by Irreversible Diffusion Limited Aggregation, *Phys. Rev. Lett.* **51**, 1119 (1983).

16. H. E. Stanley, *Connectivity: A Primer in Phase Transitions and Critical Phenomena for Students of Particle Physics in Structural Elements in Particle Physics and Statistical Mechanics*, K. Fredenhagen, and J. Honerkamp, Eds. Plenum, New York, 1982.

16

COMPARATIVE PORE DISTRIBUTION ANALYSIS OF A SOL–GEL DERIVED SILICA

S. WALLACE AND LARRY L. HENCH
Department of Materials Science and Engineering
University of Florida
Gainesville, Florida

INTRODUCTION

If the mechanism of densification of silica gels is to be fully understood, a knowledge of shape, size, and distribution of pores and how pore size distributions change during processing is required. The difficulty is obtaining information about pores in acid-catalysed silica gels, which are less than 40 Å across. Mercury porisimetry is only useful above 50 Å.[1] TEM quantitative analysis is very difficult and time consuming as the gels are amorphous and brittle and the scale of porosity is so small. Gas adsorption, using nitrogen as the adsorbate at 77°K, is a common technique, but very slow when done manually. The advent of automatic sorption isotherm instruments has opened the way for routine use of gas adsorption analyses in the measurement of the ultrastructural characteristics of gels.

The purpose of this paper is a comparison of the three automatic instruments available on the market in the United States. Three silica gels, made in identical fashion, were analysed, one on each instrument, and the results compared. The influence of degassing variables and sample inhomogeneity on sample analysis were also examined.

REVIEW OF GAS ADSORPTION FOR PORE ANALYSIS

Gas adsorption data are generated by measuring the equilibrium volume of nitrogen adsorbed (V_{ads}) by a gram of sample as the relative pressure (P/P_0) is raised by increments from 0 to 1, after the sample has first been degassed to empty the pores. The desorption isotherm measures the volume of nitrogen desorbed (V_{des}) as the P/P_0 is lowered by increments from 1 to 0.

The shape of the plot of V_{ads} and V_{des} against P/P_0, called the sorption isotherm, depends on the shape and size of the pores in the gel[1] (this reference covers gas adsorption theory in great detail). Correct interpretation of an isotherm is required for pore analysis.

The gel used for comparing the three instruments is a silica gel made with an organic acid drying control chemical additive. The gel contains a mixture of micropores (radius $r < 15\,\text{Å}$) and mesopores ($r < 500\,\text{Å}$). Consequently the surface area (S) is calculated using the BET[2] theory for $P/P_0 < 0.2$. The pore volume (V) is calculated from V_{ads} converted to liquid at $P/P_0 = 0.999$. The hydraulic, or average, pore radius $r_H = 2V/S$, assumes cylindrical pores. The pore volume and surface area distributions are calculated using the BJH method,[3] assuming open ended cylindrical pores, which gives values for V and S of the pores within a range of pore radii r_i to $r_i + dr$. Cumulative addition of S and V plotted against the average radius ($\bar{r}$) of each pore range give the cumulative V_{TOT} and S_{TOT} curves. The derivative of these curves gives the normalized distributions. A plot of V_{ads} against the statistical thickness of a monolayer of N_2 adsorbed on a nonporous solid, which is related to P/P_0 by the Halsey equation,[4] is called a t plot. Analysis of this curve gives the micropore volume V_t and mesopore surface area S_t.[5]

COMPARISON OF THE THREE AUTOMATIC SORPTION INSTRUMENTS

The Digisorb 2600* automatic sorption system degasses a sample, then measures V_{ads} and V_{des} at preset P/P_0 values, analyzing five samples sequentially. The Autosorb 6† works in the same way, analyzing six samples simultaneously. The Omnisorb 360‡ works on a different principle using a mass flow meter to measure V_{ads} and V_{des} as P/P_0 is continuously varied, analyzing four samples sequentially. All three instruments differ in their peripheral hardware. Table 16.1 shows the data reduction techniques available and how they are calculated. The software on the Omnisorb 360 appears to be for the analysis of zeolites

*Micromeretic Instrument Corp., 5680 Goshen Springs Rd., Norcross, GA 30093.
†Quantachrome Corp., 6 Aerial Way, Syosset, NY 11791.
‡Omicron Technology Corp., 160 Sherman Ave, Berkeley Heights, NJ 07922.

TABLE 16.1   The Data Reduction Calculations Available for Each Instrument

Type of data	Digisorb 2600	Autosorb 6	Omnisorb 360
Surface area	Yes	Yes	Yes
Pore volume	Cumulative V_{TOT} from BJH calculation	C from V_{ads} (liquid) at $P/P_0 = 0.999$	Cumulative V_{TOT} from BJH calculation
Hydraulic radius	No	Yes	No
t-Plot analysis	No	Yes, from $t = 3$ to 10Å	Yes, from $t = 2$ to $4.5\,\text{Å}$
Sorption isotherm	Tabulated and graphical forms from $P/P_0 = 0$ to 1	Tabulated and graphical form from $P/P_0 = 0$ to 1	For V_{ads} from $P/P_0 = 0 \to 0.2$ For V_{des} from $P/P_0 = 1 \to 0.2$
Pore size distributions available	$\Delta V_{TOT} - \bar{r}, \Delta S_{TOT} - \bar{r}$ for V_{ads} and V_{des} on log-normal graph	$V_{TOT} - \bar{r}, S_{TOT} - \bar{r},$ $dV_{TOT} - \bar{r}$ and $dV_{TOT} - \bar{r}$ for V_{ads} and V_{des} on log-normal Graph	$V_{TOT} - \bar{r}, S_{TOT} - \bar{r},$ $dV_{TOT} - \bar{r}$, and $dV_{TOT} - \bar{r}$ for V_{des} on linear graph

and other microporous samples, while the Digisorb 2600 and Autosorb 6 will analyze more general samples.

EXPERIMENTAL PROCEDURE FOR GEL PREPARATION

A drying control chemical additive (DCCA) was used, in this case an organic acid, to make the silica gel (G1). Three grams of the acid was dissolved in 150 cm^3 of deionized (D.I) H$_2$O. Seventy-five cubic centimeters of tetramethoxysilane was added while stirring. The sol was cast into four 25-cm^3 polystyrene vials (G1A, B, C and D), allowed to gelate, aged for 28 hr at 80°C, and dried unidirectionally at 60°C. Before analysis each sample was outgassed for 24 hr at 200°C.

EXPERIMENTAL RESULTS

Table 16.2 shows the numerical results obtained from each instrument. Figure 16.1 shows the sorption isotherm of the gels with the organic acid DCCA gels (G1 A, B, and C). It is not possible to construct a complete sorption isotherm from the data produced by the Omnisorb 360. Figure 16.2 shows the corresponding cumulative pore volume plots, Fig. 16.3 the pore volume distribution, and Fig. 16.4 the t plots for each instrument.

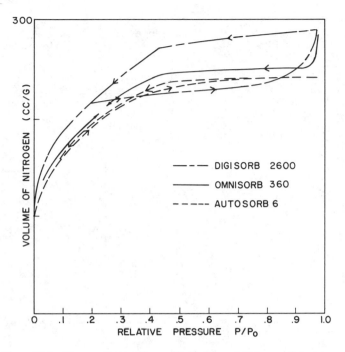

Figure 16.1 The sorption isotherm obtained for gel G-1 on each instrument.

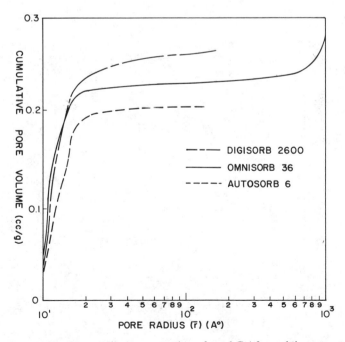

Figure 16.2 The cumulative pore volume for gel G-1 for each instrument.

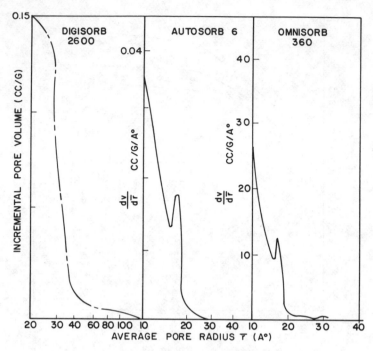

Figure 16.3 The pore volume distribution for gel G-1 from each instrument.

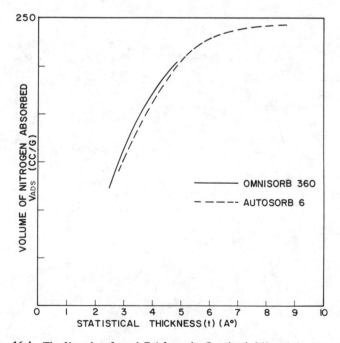

Figure 16.4 The $V-t$ plots for gel G-1 from the Omnisorb 360 and the Autosorb 6.

152

TABLE 16.2   The Physical Properties of Organic Acid Catalysed Gel G1
Analysed by the Instruments

Type of Data	Digisorb 2600 (G1A)	Autosorb 6 (G1B)	Omnisorb 360 (G1C)
Surface area S	796 m^2/g	701 m^2/g	723 m^2/g
Pore volume	$V_{TOT} = 0.414$ cm^3/g	$V = 0.380$ cm^3/g	$V_{TOT} + V_t = 0.307$ cm^3/g
Hydraulic pore radius r_H	—	10.85 Å	—
Micropore volume V_t	—	0.362 cm^3/g	0.069 cm^3/g
Mesopore area S_t	—	16 m^2/g	542 m^2/g

DISCUSSION OF RESULTS

The silica gel used for comparison (G1) is microporous so it would be expected to have a type 1 isotherm with type E hysteresis due to bottleneck pores.[1] Only the autosorb 6 isotherm in Fig. 16.1 is this type. The Digisorb 2600 isotherm is type 1 but with a much larger hysteresis loop. The Omnisorb 360 only shows V_{ads} up to 0.2 and V_{des} down to 0.2, and as these are not related a complete isotherm cannot be drawn. Nevertheless, it is obvious that the Omnisorb 360 data is not a type 1 isotherm due to the large desorption occurring at 0.96. The differences in the pore volume, the micropore volume, and mesopore surface areas are due to different methods of calculation and are not directly comparable. But, whatever caused the variation in the isotherms would also cause the variation seen in S and in Figs. 16.2, 16.3, and 16.4 as they are all calculated from the isotherm.

These variations could be due to: (1) sample to sample variation or internal sample inhomogeneity, (2) degassing variables such as particle size or temperature, or (3) variations in the instruments. The gels used for comparison should have been identical as they were all from the same processing batch. To examine possibility (1) a new batch of 10 organic acid DCCA silica gels (G2) were made and analyzed in an Autosorb 6 for sample to sample variation. Also, one of these gel samples (G2A) was divided into five portions and tested for variability due to internal inhomogeneity. Table 16.3 shows the results of these homogeneity studies along with the standard deviation. There is a larger sample to sample variation (G2) than internal sample variation (G2A); that is, $S_{10}(G2) > S_5$ (G2A) for all three characteristics. To examine possibility (2), an original gel (G1D) was divided into 15 portions and degassed for 24 hr under a 10-mtorr vacuum using different particle sizes and temperatures before analysing in an Autosorb 6. A Student t-distribution calculation shows that within a 95% confidence level, the true mean lies within one standard deviation of the

**TABLE 16.3   Silica Gel Physical Property Variation Due to Sample-to-Sample (G2),
Internal Sample (G2A, G1D), Particle size (G1D), and Degassing Temperature (G1D)
Variation**

	Degassing Variables				
Gel No.	Particle Size (mm)	Temperature (°C)	BET Surface Area (m²/g)	Pore Volume V_{TOT} (cm³/g)	Hydraulic Pore Radius r_H (Å)
G2	2	200	768; Standard Deviation of 10 samples $S_{10}=18$	0.478; $S_{10}=0.028$	12.47; $S_{10}=0.81$
G2A	2	200	766; $S_5=12$	0.445; $S_5=0.10$	11.63; $S_5 = 0.35$
G1D	2	200	768; $S_6=26$	0.444; $S_6 = 0.028$	11.55; $S_6 = 0.43$
G1D	2	350	739; $S_4=17$	0.415; $S_4=0.16$	11.23; $S_4=0.21$
G1D	0.1	200	732; $S_5=19$	0.434; $S_5 = 0.026$	11.85; $S_5 = 0.48$

measured mean values, so the variation that results (Table 16.3) is no larger than that seen in G2, and using a smaller particle size or increasing the temperature slightly decreases the surface area. The largest error actually comes from weighing, as for a 760 m²/g sample about 0.1 g is used and unless extreme care is taken an error of ± 0.005 g is possible.

A total of 30 samples of the same silica gel was run on the Autosorb 6 and all had isotherms identical to that of G1B, that is, type 1 with type E hysteresis. Gel GIC was run twice on the Omnisorb 360 and each time the isotherm in Fig. 16.1 was obtained. The large hysteresis loop produced by the Digisorb 2600 was also seen in a nitric acid catalysed gel which gave an isotherm with type E hysteresis on the Autosorb 6, so the differences in isotherms in Fig. 16.1 are reproducible. Since the variations cannot be explained by degassing variables or sample variability they seem to be characteristic differences of the instruments.

CONCLUSIONS

The Digisorb 2600, Autosorb 6, and Omnisorb 360 will all run sorption isotherms quickly and reproducibly. Identical samples run on each instrument produce different results which do not appear to be due to sample to sample variation, internal inhomogeneity, or degassing variables but instead are due to the instruments. Increasing the degassing temperature or decreasing the particle size decreases the surface area of a silica gel. Within a confidence level of 95%, the internal inhomogeneity of a sample is smaller than the sample to sample variation.

ACKNOWLEDGMENT

The authors are grateful to the financial support from the U.S. Air Force Office of Scientific Research under contract AFOSR #F49620-83-C-0072 for this work.

REFERENCES

1. S. Lowell and J. E. Shields, *Powder Surface Area and Porosity*, 2nd ed., Wiley, New York, 1984.
2. S. Brunauer, P. H. Emmett, and E. J. Teller, *Am. Chem. Soc.* **60**, 309 (1938).
3. E. P. Barrett, L. G. Joyner, and P. P. Halenda, *J. Am. Chem. Soc.* **73**, 373 (1951).
4. G. D. Halsey, *J. Chem. Phys.* **16**, 931 (1948).
5. J. H. de Boer, B. C. Liffens, B. G. Linsen, J. C. P. Brockhoff, A. van den Heuvel, and Th. V. Osinga, *J. Colloid Interface Sci.* **21**, 405 (1966).
6. L. L. Hench, S. H. Wang, and S. C. Park, "SiO₂ Gel Glasses", SPIE's 28th Annual International Technical Symposium on Optics and Electro-Optics, 19–24 August 1984, San Diego, California.

17

SINTERING OF MONOLITHIC SILICA AEROGELS

M. PRASSAS
Corning Europe
Centre Europeen de Recherche
Avon, France

J. PHALIPPOU AND J. ZARZYCKI
Materials Science Laboratory and C.N.R.S Glass Laboratory,
University of Montpellier
Montpellier, France

INTRODUCTION

The main requirement to obtain a monolithic glass by the sol–gel process is to have initially a completely crack-free dried gel. So far, monolithic gels made by hydrolysis and polycondensation of metal alkoxides have been achieved by slow drying procedures.[1] Depending on the size and the chemical composition of the alcogel, 2 to 4 weeks are needed to completely remove the solvent.

Capillary stresses occurring within the pores of the alcogel have been identified as the most important factor affecting the integrity of the material during drying.[2] We have in previous papers[3,4] shown that controlled hyper-critical drying of silica alcogels leads to crack-free gels named aerogels. This technique allowed us to obtain aerogels having at least one dimension > 15 cm with a very high reliability. The entire operation alcogel → aerogel is < 10 hr and is the fastest procedure known to produce monolithic silica gels.[3]

Silica aerogels are characterized by an "open" texture (70–90% porosity), with a low density (0.1–0.5 g/cm^3) and are hydrophobic.

These characteristics make aerogels different in behavior from typical silica xerogels dried in air or under vacuum.

The purpose of this paper is to study the thermal densification of silica aerogels under various experimental conditions. Isothermal and constant heating rate measurements of the dimensional changes of the gel are studied from ambient to 1050°C.

MATERIALS AND EXPERIMENTAL PROCEDURES

Monolithic silica aerogels (MSA) were prepared by stoechiometric hydrolysis of tetramethoxisilane (TMS). Experimental details of the MSA synthesis and characterization have been described in previous papers.[3,4] The gels used in this study are represented by AxS where "x" denotes the volume percentage of TMS and AS means silica aerogels. Typical textural properties of the gels are $S \simeq 360 \text{ m}^2/\text{g}$ (specific surface area) and $\rho \simeq 0.25 \text{ g/cm}^3$ (apparent density).

RESULTS AND DISCUSSION

Densification of Monolithic Silica Aerogels

A60S aerogels were prepared as cylinders of 20-mm diameter and 10-mm height. They were heated from ambient to a given temperature with a constant heating rate and cooled rapidly. Porosity and apparent density were measured either by a mercury volumeter (porosity $>20\%$) or by the Archimedes method (porosity $<20\%$).

Densification in air was made using two thermal cycles (Fig. 17.1). Cycle A: linear increase in temperature. Cycle B: Samples were previously treated at various temperatures between 500 and 1000°C for 15 hr.

The last procedure was used to examine indirectly the effect of the hydroxyl content on the densification of the gel.

Figure 17.2 shows the typical evolution of apparent density and total porosity of the MSA through cycle A with a heating rate of 3.3°C/min. There is an insignificant variation of the density from ambient to 1020°C. At higher temperatures and up to 1070°C the gel starts to densify slowly and then rapidly between 1070 and 1120°C. At 1120°C the material is fully densified with a density 99.7% of the density of vitreous silica.

The total linear shrinkage is about 50% and in good agreement with that calculated (Fig. 17.3). Figure 17.4 shows the densification curve of the same aerogel using cycle B with a plateau at 650°C. If we compare Figs. 17.2 and 17.4 there is a significant displacement of the densification curve by about 50°C. A lower magnitude shift (15°C) occurs when the heating rate is varied (Fig. 17.4).

In both cases (cycles A and B) silica aerogels show a remarkable dimensional stability up to 1000°C. The temperature at which the densification starts is

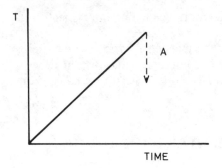

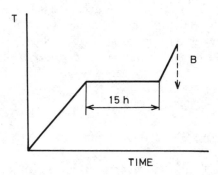

Figure 17.1 Thermal cycles used to densify MSA in air.

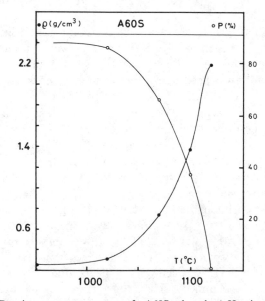

Figure 17.2 Density versus temperature of a A60S gel, cycle *A*. Heating rate 3.3. °C/min.

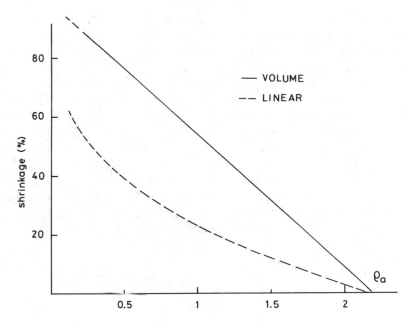

Figure 17.3 Calculated total volume and linear shrinkage versus apparent density of the original gels.

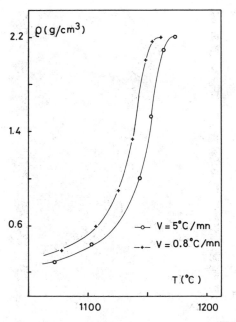

Figure 17.4 Density versus temperature of a A60S gel (cycle *B*).

strongly dependent on the previous heat treatment of the gel and to a lesser degree on the heating rate. This behavior can be explained if we consider that the chemical composition of the aerogels and especially the hydroxyl content is different depending on the thermal treatment.

In fact a long heat treatment at temperature between 500 and 1000°C before densification was effective in removing residual OH groups. Since sintering at high temperatures occurs by viscous flow, and viscosity depends on the water content it is expected that the lower this quantity the higher will be the temperature at which the gel will densify. Table 17.1 indicates the water content of the densified gel versus treatment temperature before densification.

The samples corresponding to Fig. 17.4 are shown in Fig. 17.5. The material is initially opaque and becomes progressively transparent. Their monolithicity is maintained during the heat treatment regardless of heating rate. Also, regardless of the heat treatment schedule, densification occurs very rapidly in a short range of temperature (30–40°C). Surface layers are first densified whereas the

TABLE 17.1 Hydroxyl Content[a] of Vitreous Silica Obtained by Sintering a Silica Aerogel Using a Thermal Cycle B (see Text)

$T(°C)$	—	650	920	1005	1050
OH (ppm)	3500	2870	2600	2400	1200

[a]The hydroxyl content is measured on small slide of glass using the intensity of the IR band situated at 2.73 μm.

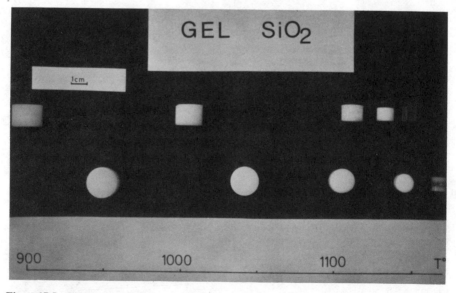

Figure 17.5 Photo of aerogels as a function of temperature. Heating rate: upper series 0.8°C/min; lower series 5°C/min.

bulk is still porous. This premature densification inhibits the escape of the hydroxyl groups and consequently provokes the well-known phenomena of bloating usually observed at higher temperatures. A more drastic dehydroxylation can be conducted by treating the gel in a chlorine atmosphere[5] or in vacuum which prevents bloating.

Densification in vacuum (10^{-1}–10^{-2} torr) was carried out in a tube furnace with graphite-heating elements. The following schedules were compared: (1) in air at 650°C for 15 hr; (2) under vacuum at 950°C for 30 min. Then they are heated up with a rate of 60°C/min to the densification temperature where they are kept for 10 min.

Figure 17.6 shows the apparent density evolution of A50S and A60S MSA versus temperature for vacuum densification. Comparison of Fig. 17.6 and Fig. 17.4 (A60S gel) shows that the densification curves are quite similar. However in vacuum the gel starts to densify at a temperature which is 40°C higher than in air.

A higher shift ($\simeq 70$°C) in the densification temperature is also observed for gel A50S in comparison with A60S. In fact A50S gel has a higher specific surface area and a lower density than A60S. Thus dehydroxylation is more effective (Table 17.2) and consequently since densification occurs in the same viscosity range[6] the densification is shifted to a higher temperature (1280°C). In this case sintering overlaps the beginning of crystallization and gel-to-glass conversion cannot be completed at this temperature.

Vitreous silica synthesized by sintering of MSA under vacuum does not show any bloating phenomena even at temperatures as high as 1800°C.

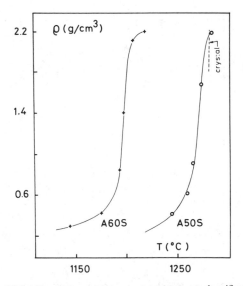

Figure 17.6 Density versus temperature (vacuum densification).

TABLE 17.2 **Average Hydroxyl Content (in ppm) of Glass-Derived Gels Versus Densification Atmosphere**

Atmosphere	Air	Vacuum (10^{-1} torr)
A60S	2500	1000
A50S	2000	600

Sintering Mechanism

We have shown in previous papers that thermal shrinkage is strongly dependent on the chemical composition, thermal history, and textural properties of the gel.[5,6,7] Figure 17.7 shows shrinkage curves of MSA having different textural properties (Table 17.3). With one exception (gel No. 2 Table 17.3) the higher the specific surface area the higher is the relative shrinkage of the gel at a given temperature.

Shrinkage curves of MSA are in a general manner characterized by two distinctive parts: (1) low shrinkage rate up to 800–900°C; (2) high shrinkage rate for temperatures higher than 1000°C. This behavior is similar to the density temperature evolution and both measurements can be correlated. However the shrinkage measurements are more accurate and they are used to identify the sintering mechanism.

Isothermal shrinkage data can usually be described by the following simple

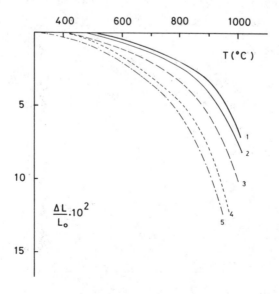

Figure 17.7 Thermal shrinkage of different silica aerogels (see Table 17.3).

TABLE 17.3   Textural Properties of Various MSA Appearing in Fig. 17.7

MSA Number	Specific Surface Area (m²/g)	Apparent Density (g/cm³)
1	320	0.23
2	250	0.26
3	369	0.25
4	374	0.19
5	448	0.34

relationship

$$\frac{\Delta L}{L_0} = Kt^n \tag{1}$$

where $\Delta L/L_0$ is the relative shrinkage, K is a constant which depends on the temperature and the nature of the material, and n is a coefficient related to the sintering mechanism.

Using this relationship and comparing experimental values of the coefficient n with the theoretical models we can identify the sintering mechanism. For a viscous flow mechanism the value of n is 1[8] whereas for a diffusional process n is between 0.3 and 0.5.[9]

Such kinetics data for a A70S gel are presented in Fig. 17.8. These data also plotted in a log–log scale allow us to calculate n from 500 to 1000°C (Fig. 17.9).

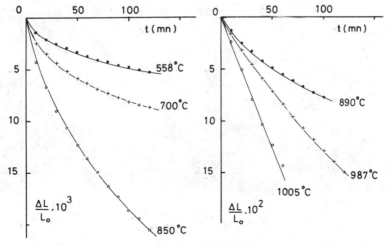

Figure 17.8   Shrinkage versus time for different temperatures of a A70S gel.

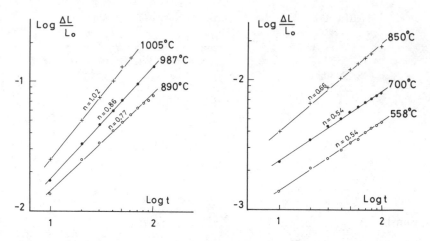

Figure 17.9   Densification curves at various temperatures.

From 500 to 850°C n is equal to 0.55. n progressively increases and reaches the value 1 at 1005°C. The gel texture, as followed by TEM, shows that up to 900°C the particle radii remain unchanged whereas at temperatures higher than 1000°C ($n \simeq 1$) a particle growth occurs and interpenetration zones appear between particles (Fig. 17.10). This agrees with both kinetics and densification data suggesting the presence of two distinctive mechanisms as is evident from calculated n values.

Apparent activation energies of the low- and high-temperature processes can be calculated if we assume that K in Eq. (1) obeys an Arrhenian relationship;

850 °C   900 °C   1050 °C

0.1μm

Figure 17.10   Transmission electron micrographs of a A70S as a function of temperature.

that is

$$K = K_0 \exp\left(\frac{-Q}{RT}\right) \qquad (2)$$

where K_0 is a constant, Q the activation energy, and T the absolute temperature.

A log K versus $1/T$ plot is a straight line with $-0.43Q/R$ slope. However, to apply Eq. (2) n must be constant, that is, one sintering mechanism should be dominant. This is the case from 500 to 850°C. The calculated activation energy is 5 to 10 kcal/mole which is of the same order of magnitude of the activation energy of the reaction:

$$-\overset{|}{\underset{|}{Si}}-OH + HO-\overset{|}{\underset{|}{Si}}- \quad \longrightarrow \quad -\overset{|}{\underset{|}{Si}}-O-\overset{|}{\underset{|}{Si}}- + H_2O \qquad (3)$$

The activation energy for the high-temperature ($>860°C$) process was estimated by using constant heating rate experiments. In fact, since n reaches the value 1, sintering occurs by a viscous flow and the Frenkel's model can be applied.

$$\frac{\Delta L}{L_0} = \frac{\sigma}{2r\eta} t \qquad (4)$$

where σ is the interfacial energy, r the particle radius, and η the viscosity.

Assuming that $\eta = A \exp(Q/RT)$ and expressing the heating rate by $v = dT/dt$, Eq. (3) becomes

$$\frac{d}{dt}\left(\frac{\Delta L}{L_0}\right) = \left(\frac{\sigma}{2rAv}\right) \exp\left(\frac{-Q}{RT}\right) \qquad (5)$$

Assuming σ and A are independent of temperature, the last equation can be integrated and becomes[10]

$$\frac{\Delta L}{L_0} = \left(\frac{\sigma RT^2}{2rAvQ}\right) \exp\left(\frac{-Q}{RT}\right) \qquad (6)$$

Thus a plot of $\log(\Delta L/L_0)(1/T^2)$ versus $1/T$ will be linear and have a slope of $-0.43Q/R$.

Such a plot was constructed for a A70S sample (Fig. 17.11). The gel was treated at 850°C until a constant weight and then heated at a rate of 3.3°C/min up to 1050°C. The calculated activation energy is 88 kcal/mole. This apparently low value is in good agreement with previously reported results.[11] In the case of vitreous silica, values ranging from 120 to 170 kcal/mole were calculated when the hydroxyl content is 1300 to 3 ppm, respectively. By extrapolating the plotted data of Ref. 11 at about 4000 ppm, corresponding to the hydroxyl content of the fully densified A70S gel, we found 84 kcal/mole.

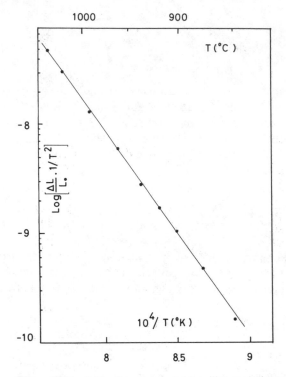

Figure 17.11   Activation energy plot for silica aerogels.

CONCLUSION

Sintering behavior of monolithic silica aerogels depends on the sintering atmosphere. The water content and the initial texture of the aerogel also play an important role in sintering.

The study of isothermal shrinkage for an A70S gel shows that densification results from several mechanisms. Between 500 and 700°C sintering is by a diffusional process. In this temperature range, material weight losses are important.[4] The observed shrinkage is due to the coalescence of gel particles as a consequence of thermal condensation of silanol groups. In fact, the low-temperature sintering is due to a chemical reaction. It is likely that particles do not interpenetrate because the decrease of the specific surface area in this range of temperature is low.

In the temperature range 750–1000°C sintering is due both to a diffusional mechanism and viscous flow. For temperatures above 1000°C sintering is by viscous flow and the activation energy of this process is related to the hydroxyl content of the material. Activation energies as low as 88 kcal/mole have been obtained.

REFERENCES

1. L. C. Klein and G. J. Garvey, in *Ultrastructure Processing of Ceramics Glasses and Composites*, L. L. Hench and D. R. Ulrich, Eds., Wiley-Interscience, New York, 1984, p. 88.

2. J. Zarzycki, M. Prassas, and J. Phalippou, *J. Mat. Science* 17, 3371 (1982).

3. M. Prassas, J. Phalippou, and J. Zarzycki, *J. Mater. Sci.* 19, 1656 (1984).

4. M. Prassas, J. Phalippou, and J. Zarzycki, *Glastechn. Ber.* 56K, 542 (1983).

5. J. Phalippou, T. Woignier, and J. Zarzycki, in Ref. 1, p. 70.

6. M. Prassas, J. Phalippou, and L. L. Hench, *J. Non-Cryst. Solids* 63, 375 (1984).

7. M. Prassas and L. L. Hench, in Ref. 1, p. 100.

8. J. Frenkel, *J. Phys.* 5 (9), 385 (1945).

9. D. L. Johnson and I. B. Cutler, *J. Am. Ceram. Soc.* 46(11), 541 (1963).

10. I. B. Cutler, *J. Am. Ceram. Soc.* 52(1), 11 (1969).

11. G. Hetherington, K. H. Jack, and J. G. Kennedy, *Phys. Chem. Glass.* 5, 130 (1964).

18

PHYSICAL PROPERTIES OF PARTIALLY DENSIFIED SiO$_2$ GELS

S. C. PARK AND LARRY L. HENCH
Ceramics Division
Department of Materials Science and Engineering
University of Florida
Gainesville, Florida

INTRODUCTION

Low-temperature sol–gel processes have been reported for several years as having enormous potential for the manufacture of unique ceramics, glasses, and composites.[1-5] Commercial applications of sol–gel derived powders and coatings have in fact been achieved.[2] However, difficulties in controlling the drying of large monolithic shapes have been the major restriction in the use of sol–gel processes to make solid objects. This is due to lack of understanding of the basic changes in ultrastructure during sol–gel–glass transformations. Therefore, complete physical property data on silica gel *monoliths* in as-dried and densified conditions are not now available in the literature. However, with the aid of drying control chemical additives (DCCA), large crack-free, transparent, high purity silica gel monoliths have been prepared by using sol–gel processing. In this paper, properties such as density, microhardness, flexural strength, and compressive strength are reported for pure silica gel samples as-dried and densified up to 950°C.

EXPERIMENTAL PROCEDURE

A flowchart of the general procedure for sol–gel processing is shown in Fig. 18.1. The steps include: (1) sol preparation, (2) casting, (3) gelation, (4) aging, (5) drying, and (6) densification.

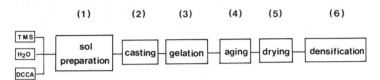

Figure 18.1   Flowchart of silica gel–glass processing steps.

The silica sol was prepared by adding tetramethylorthosilicate (TMOS) into a mixture of an organic acid drying control chemical additive (DCCA) and water using a (17:1) molar ratio of water and TMOS.

The sol was then cast into a clean styrene acrylonitrile mold and the mold was closed tightly. After 12-hr aging each at 60°C and 80°C, the gel was removed from the mold and dried in an oven at preprogrammed temperature and time schedule. The gel was dried up to 150°C, with a heating rate of 10°C/hr, after which the gel becomes a rigid monolithic body retaining the shape of the mold. The rigid fully dried silica gel (9 cm × 6 cm × 1 cm) was placed inside a high-temperature furnace and heated from 150 to 950°C with a 5-min hold at temperature followed by rapid cooling.

Density was determined using a simple mercury displacement technique. Microhardness was measured using the 136° diamond pyramid indenter on the Kentron Tester with a 500-g load. Flexural strength test specimens were pre-pared according to ASTM C158-80.[6] The polished specimens with a span: width:thickness ratio of 7:2:1, were loaded in three-point bending at a strain rate of about $3.5 \times 10^{-3} \, s^{-1}$. Rectangular specimens of 7.5 mm × 5 mm × 14 mm were loaded in order to determine compressive strength at a strain rate of $3 \times 10^{-4} \, s^{-1}$ using an Instron Testing Machine. More than five measurements of samples from each process condition were made in order to determine reproducibility. All specimens were dried at 150°C for 3 hr immediately prior to measurements.

RESULTS

Density of the silica gel was measured and plotted as a function of densification temperature in Fig. 18.2. Note that the dotted horizontal line represents the value of vitreous silica. Density increases as densification temperature increases with no significant change below 600°C, where the density remains about 52% of that of vitreous silica. However, the density of the sample heated to 950°C is about 2.04 g/cm^3, approximately 93% of that of vitreous silica.

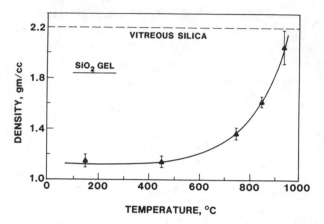

Figure 18.2   Variation of density with densification temperature.

Variations of microhardness (DPN), compressive strength, and flexural strength with densification temperature are shown in Fig. 18.3. It is also observed that there is no significant change in microhardness or flexural strength below 600°C. However, a sharp increase in these mechanical properties was observed after heating to 750°C and above. The 950°C specimens reached a DPN about 470, which is approximately 75% of that of vitreous silica, which is almost twice as high as the 850°C data. The compressive strength of the silica

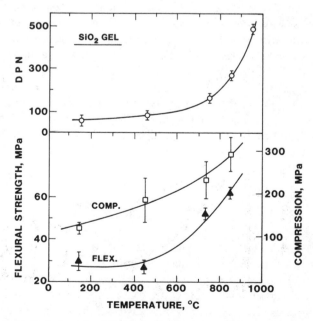

Figure 18.3   Variation of mechanical properties with densification temperature.

gel samples is also sensitive to firing temperature as low as 450°C, even though the density is unchanged (Fig. 18.2). In spite of the large variation in the 450°C data, there is a statistically significant increase in the compressive strength over the as-dried samples.

DISCUSSION

Both the density and the mechanical properties of the gel monoliths increase with densification temperature and are simply related. In Fig. 18.4, the experimental data of microhardness, compressive strength, and flexural strength are plotted as a function of density. The solid symbols at a density of 2.2 g/cm³ represent the values of vitreous silica. Within the experimental error, the mechanical properties are linearly related to the density. Dotted lines drawn in this figure were obtained from the simple relationship as follows

$$\sigma_{gel} = \sigma_s \left(\frac{\rho_{gel}}{\rho_s} \right) \tag{1}$$

where σ_{gel} and σ_s are the mechanical strength of the gel and vitreous silica, and ρ_{gel} and ρ_s are the density of the gel and vitreous silica, respectively.

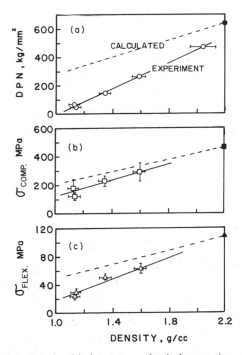

Figure 18.4 Relationship between mechanical properties and density.

The experimental data are generally lower than the calculated values. The differences become larger at lower density and smaller as the density of the gel increases. This may be due to the presence of moisture within the open pore channels in the gel, which will react with the silica matrix and weaken the strength of the gel. As the gel becomes denser, the number of pores decreases and accordingly the sites for a reaction to occur are reduced. Therefore, in the higher density region, the experimental data become close to the values the above relation predicts. In fact, the above relationship assumes that pores present in the gel are isolated and that no reactions between the matrix and moisture in the pores occur. The results shown in Fig. 18.4 indicate otherwise.

CONCLUSIONS

1. Reliable production of large transparent high purity silica gel monoliths is possible by using low-temperature sol–gel processing with a drying control chemical additive (DCCA).

2. There is no significant change in density and mechanical properties of the silica gel below 600°C, above which the properties change markedly with densification temperature.

3. There is a linear relationship between mechanical properties and density of the dried and densified gels.

4. The mechanical properties of the partially densified gels are less than predicted from a simple dependence on porosity.

ACKNOWLEDGMENT

The authors gratefully acknowledge the financial support of the AFOSR contract No. F49620-83-C-0072.

REFERENCES

1. M. Nogami and Y. Moriya, *J. Non-Cryst. Solids* **37**, 191–201 (1980).

2. J. D. Mackenzie, in *Ultrastructure Processing of Ceramics, Glasses and Composites*, L. L. Hench and D. R. Ulrich, Eds., Wiley, New York, 1984, pp. 15–26.

3. D. R. Uhlmann, B. J. J. Zelinski, and G. E. Wnek, in *Better Ceramics Through Chemistry*, C. J. Brinker, D. E. Clark, and D. R. Ulrich, Eds., North-Holland, New York, 1984, pp. 59–70.

4. S. Wallace and L. L. Hench, in *Better Ceramics Through Chemistry*, C. J. Brinker, D. E. Clark, and D. R. Ulrich, Eds., North-Holland, New York, 1984, pp. 47–52.

5. R. W. Rice, in *Better Ceramics Through Chemistry*, C. J. Brinker, D. E. Clark, and D. R. Ulrich, Eds., North-Holland, New York, 1984, pp. 337–345.

6. Annual Book of ASTM Standard, *Flexural Testing of Glass*, Section 15, American Society of Testing Materials, Philadelphia, Pennsylvania, 1983, pp. 25–34.

19

KINETIC PROCESSES IN SOL–GEL PROCESSING

D. R. UHLMANN, B. J. ZELINSKI, L. SILVERMAN,
S. B. WARNER, B. D. FABES, AND W. F. DOYLE
Department of Materials Science and Engineering
Massachusetts Institute of Technology
Cambridge, Massachusetts

INTRODUCTION

The development of improved sol–gel techniques for preparing ceramics (including glasses) is an objective of considerable interest and importance. Such processing embraces a wide range of precursors and kinetic phenomena. In cases where metal alkoxides are employed as precursors, the kinetic phenomena range from hydrolysis and condensation through drying and relaxation to densification and crystallization. In all cases, our present state of knowledge leaves much to be desired; and the lack of insight serves as an impediment to efficient development of the technology.

This paper discusses several of these phenomena; hydrolysis, crystallization, and densification by viscous sintering. The discussion of hydrolysis kinetics focuses attention on the reliability of techniques for obtaining kinetic data, presents new data on the hydrolysis of TEOS and models of kinetic behavior, and considers the difficulty of inferring reaction order from single, simple sources of data. The discussion of crystallization and densification focuses on the competition between the two processes and on defining regions of time and temperature where densification can be achieved without significant crystallization, with examples being provided for typical glass-forming systems.

173

HYDROLYSIS KINETICS

Tetraethylorthosilicate (TEOS) has been more extensively investigated than any other alkoxide used for the preparation of ceramics; and for this reason, it will be the subject of most attention here. Analysis of the hydrolysis kinetics of TEOS is complicated by the occurrence of multiple hydrolysis reactions as well as reesterification and condensation; and many studies have treated the sets of reactions as a single overall reaction (where the individual reactions are treated as having the same kinetics). Experimental investigations have employed a range of techniques, and in many cases have yielded different and even contradictory results. Comparison is difficult in other cases because of differences in the alkoxide/H_2O/solvent/acid concentration employed. Possible variations in factors such as purity of the ingredients and chemistry of the acid further complicate interpretation of results, as do changes in effective pH during the course of reaction.

Among experimental techniques, use of IR spectroscopy to monitor the free water band at 1638 cm^{-1} seems *a priori* attractive. Unfortunately, while the kinetics of disappearance of H_2O often represent empirically useful information, such data by themselves cannot provide critical insight into the phenomena. Complications specific to the IR technique are known, for example, the change in form of the H_2O absorption band with acid concentration. These do not, however, represent the principal matter of concern.

In treating kinetic data such as those in Fig. 19.1 consider the reaction: $A + B \rightarrow C + D$. The rate of disappearance of A can be expressed in terms of the concentrations [   ] as

$$-\frac{d[A]}{dt} = K_f[A]^m[B]^n - K_b[C]^k[D]^p \qquad (1)$$

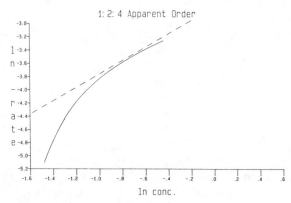

Figure 19.1 ln($-d[H_2O]/dt$) versus ln($[H_2O]$) as monitored by IR spectrometry for 1TEOS: 2H_2O:4EtOH molar initial compositions (solid line). Slope of line is apparent order of reaction with respect to water. Dashed line has a slope of one.

where m, n, k, and p are the orders of the reaction with respect to each component and K_f and K_b are the rate constants for the forward and back reactions.

Reaction rate data are usefully acquired with components (as B) in great excess and data obtained for the initial stages of reaction. Where this is not done, the analysis becomes much more complex. For example, even neglecting the back reaction, when $[B] = [A]$ initially, if the reaction is first order with respect to both components, the reaction will appear to be second order with respect to $[A]$, since $[B]$ is changing at the same rate. For the more general case where $[B] = Y + [A]$, assuming all reactions are first order,

$$YK_f t = \ln \frac{[A](Y + [A]_0)}{[A_0](Y + [A])} \qquad (2)$$

Taking $A = H_2O$ and $B = $ TEOS, $[H_2O]$ has been calculated as a function of time. The concentrations were normalized by scaling $[H_2O]_0 = 1$ and selecting K_f so that all curves pass through another common point at long times. This procedure eliminates the effects of absolute concentration and makes the $[A]/[B]$ ratio the only variable.

For such normalized $[H_2O]$ versus time curves, plots of $\ln(-d[A]/dt)$ versus $\ln[A]$ have been generated. An initial ratio $[B]/[A]$ of at least 16 is needed for the initial reaction to appear first order in $[A]$. With decreasing $[B]/[A]$, the initial apparent order of reaction increases. The apparent order at later stages of the reaction also increases with decreasing $[B]/[A]$, but not as rapidly as at the initial stages. For $[A] = [B]$, the apparent order of reaction is constant at 2. Hence the apparent initial order of reaction can have any value between 1 and 2, depending on the ratio of $[B]$ to $[A]$ in the starting mixture.

The above discussion assumed irreversibility, identical reaction kinetics for all hydrolysis steps, and all first order kinetics. It has been suggested, however, that subsequent hydrolysis steps are faster than the first in the case of TEOS. To explore the effects of this on the observed kinetics, consider the first two reactions: (1) $A + B \rightarrow C + D$; (2) $A + C \rightarrow E + D$, where A is H_2O, B is TEOS, C is triethoxysilanol, D is EtOH, and E is diethoxysilane diol. Since only A and B are present initially, the initial reaction rate would fall on the lower line in Fig. 19.2. If the rate of reaction (2) were equal to that of reaction (1), the reaction rate would continue along the line. If reaction (2) were instantaneous, the overall rate would be double that of reaction (1), as shown by the upper line in Fig. 19.2. For intermediate ratios of reaction rates, a transition is expected, characterized by orders of reaction less than unity. Computer simulations with various ratios of reaction rates show that as rate (2)/rate(1) exceeds unity, the initial apparent reaction order is less than one and increases toward unity as the reaction proceeds. When rate (2)/rate (1) is less than unity, the initial reaction order exceeds one and decreases toward unity during the reaction. Other factors that can lead to apparent orders of reaction less than unity include temperature rises during the course of reaction, increases in H^+ ion activity as the H_2O

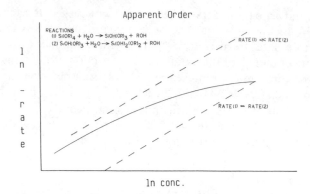

Figure 19.2 Anticipated $\ln(-d[H_2O]/dt)$ versus $\ln([H_2O])$ results for consecutive TEOS hydrolysis reactions. When rate (2) is only moderately greater than rate (1), as shown by the solid line, the apparent order changes.

concentration decreases, or autocatalytic effects of the partially hydrolyzed silane.

The effect of reesterification (the reverse of hydrolysis) is to increase the apparent order of reaction as the reaction proceeds. Condensation of two silanol groups, which also generates H_2O, should have a similar qualitative effect on the kinetic behavior.

Combining these considerations, it is not surprising that monitoring the apparent reaction order during hydrolysis of TEOS often yields complex results. The relative concentrations of the reacting species, the relative rates of the various hydrolysis steps for multifunctional alkoxides, reesterification, and condensation can all affect the apparent order of reaction. Even when the reactions are first order in all reactants, apparent reaction orders less than unity or much greater than unity can readily be observed. It is clear therefore that for the conditions employed in many investigations, it is not possible to obtain reliable insight into the order of reaction from simple measurements of $[H_2O]$ as a function of time.

In cases where the solvent is an alcohol different from the alkoxide ligands, ester exchange can further complicate the kinetic analysis. This exchange is promoted by and may require the presence of H_2O, suggesting that the exchange mechanism preferably involves a partially hydrolyzed transition state. When TEOS is solvated in MeOH or iPOH, no exchange can be detected using gas chromatography (GC). In contrast, when hydrolysis of TEOS is carried out in iPOH, the iPOH concentration initially decreases (reflecting ester exchange) and then increases toward its original value as hydrolysis proceeds (see Fig. 19.3). The overall rate of hydrolysis is appreciably slower than that of TEOS in EtOH under similar conditions, reflecting the slower rate of hydrolysis of isopropoxide ligands compared with that of ethoxide ligands. In contrast, TEOS in MeOH hydrolyzes much faster than in EtOH, reflecting ester exchange and the faster rate of hydrolysis of methoxide ligands. The rate of hydrolysis of TMOS in MeOH is in turn faster than that of TEOS in MeOH.

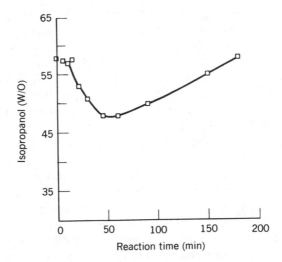

Figure 19.3   Concentration of isopropanol versus time for hydrolysis of TEOS in isopropanol.

However the [H_2O] is monitored, it does not seem possible to obtain the desired kinetic insight save under rather unconventional conditions (initial stages of reaction, greatly understoichiometric [H_2O]). For compositions and conditions of frequent concern, one must monitor not just [H_2O] versus time, but the concentrations of all important species as a function of time. Several species can be monitored using IR spectroscopy, although band overlap can present significant problems.

Gas chromatography (GC) is a highly attractive complementary technique which has been used with success in a number of investigations. In our work, use of MEK as an internal standard was found to cause problems in reactions involving TMOS (the MEK disappears), and acetone has a retention time which overlaps with EtOH. THF in concentrations of 5% was found satisfactory. Use of a hot wire (thermal conductivity) detector led to positions which depended on pH, TEOS concentration, and so on, and in general is not recommended for substances heavier than MeOH or under highly acidic conditions. A flame ionization detector, when suitably calibrated, gave satisfactory results, although nonlinearity at large concentrations is an issue.

Other techniques that appear promising for investigating hydrolysis and condensation kinetics include ^{29}Si NMR mass spectrometry, and HPLC. The first technique has been used to great advantage by Jonas and his coworkers, and is discussed in Chapters 5 and 6. As yet, the last two techniques have been relatively little utilized.

CRYSTALLIZATION VERSUS VISCOUS SINTERING

Using sol–gel techniques, it is possible to produce fine amorphous powders or dried gels having distributions of fine pores. To obtain dense, glassy bodies or coatings from such materials, it is useful to define regions of time and temperature (or heating rates) where the kinetics of viscous sintering are faster than those of crystallization—assuming that such regions exist for a given material.

In describing viscous sintering, we have used the cylindrical pore model of Scherer.[1] This model is very attractive for bodies of the type produced by sol–gel processes. It employs the basic assumptions of the Frenkel model for the viscous sintering of spheres and considering these assumptions, it is remarkable that the model predicts experimental data as well as it does.

The model predicts that the time required for appreciable sintering is proportional to the viscosity and inversely proportional to the surface energy. Among microstructural features, the most important feature is the pore radius. The initial particle size and packing density are less significant (particles of different sizes can be packed differently to give the same pore size, and exhibit similar times for sintering). In the present paper, we have taken pore sizes of 50 and 5000 Å to represent the effective limits obtained from sol–gel processing.

The process of crystallization has been described using the approach described previously.[2] Under isothermal conditions, the times at various temperatures required to reach a just-detectable degree of crystallinity are represented by time–temperature–transformation (TTT) curves; while crystallization under conditions of constant heating rate is described using the analysis of crystallization statistics. Viscous sintering and crystallization are viewed here as competitive processes and the analysis of this competition is applied to a number of representative glass-forming liquids.

Considering isothermal conditions, Fig. 19.4 indicates the calculated behavior for a representative organic glass former, o-terphenyl. The regions of time at any temperature between the appropriate sintering curve and the crystallization curve for a just-detectable volume fraction crystallized of 10^{-6} indicate the conditions where densification can be achieved without detectable crystallization. In some cases, the regions of time may be inaccessibly short; but in the

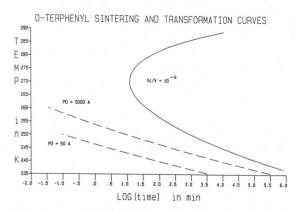

O-TERPHENYL SINTERING AND TRANSFORMATION CURVES

LOG (time)   in min

Figure 19.4 Time versus temperature relationships for the sintering (dashed lines) and crystallization (solid line) of o-terphenyl. At a given temperature in all cases, significant sintering occurs before crystallization.

case of *o*-terphenyl, there is a range of conditions under which densification can be obtained without crystallization.

The magnitude of the viscosity at any temperature affects both the sintering rate and the crystallization rate. To take account of the dependence of the viscosity, results such as those shown in Fig. 19.4 may conveniently be plotted as a function of t/η, where η is the viscosity. Such a plot is shown in Fig. 19.5 for anorthite.

For crystallization under conditions of constant heating rate, the competition between crystallization and viscous sintering is conveniently represented in terms of a heating rate versus temperature relation, such as that shown in Fig. 19.6 for anorthite. As shown there, there exist windows in temperature, over a range of heating rates, in which substantial sintering can take place without the occurrence of detectable crystallization.

All of this discussion has used bulk material properties and measured or estimated nucleation barriers, together with the assumption of steady-state nucleation rates. The effect of different nucleation barriers on the competition between sintering and crystallization is shown in Fig. 19.7 for anorthite-like materials having various nucleation barriers (represented as *BkT*). With decreasing barriers to crystal nucleation, the processing window in which sintering can be achieved without detectable crystallization is expected to decrease. For sufficiently small nucleation barriers, such as *BkT* = 40 for a material with the properties of anorthite, crystallization will take place before sintering at all heating rates evaluated. The measured nucleation barrier of anorthite is about 82 kT.

The effect of changes in the degree of crystallinity is shown in Fig. 19.8 for $Na_2O \cdot 2SiO_2$. Larger processing windows are suggested for conditions where larger volume fractions crystallized are acceptable. Also plotted in Fig. 19.8 are the experimental results of Hench et al.,[3] who observed sintering without crystallization in 1-hr treatments at 753–793°C, and crystallization at tempera-

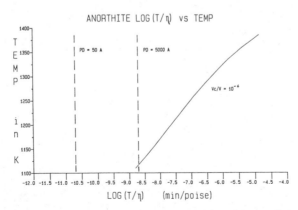

Figure 19.5 Sintering (dashed lines) and crystallization (solid lines) curves for anorthite with the effect of viscosity factored out. For compacts with large pore sizes, detectable crystallization occurs before sintering only at temperatures less than ~1115°K.

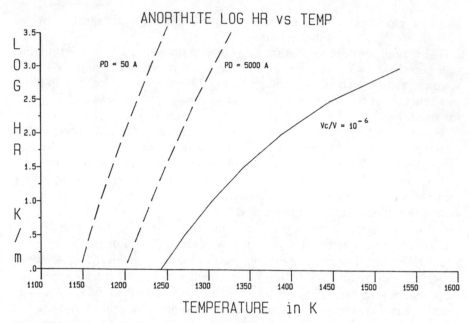

Figure 19.6 Anorthite log (heating rate) versus temperature relationships for sintering (dashed lines) and crystallization (solid lines). For all heating rates shown (>1°K/min) the temperature reached when sintering is complete is lower than the temperature where crystallization to $V_c/V = 10^{-6}$ occurs.

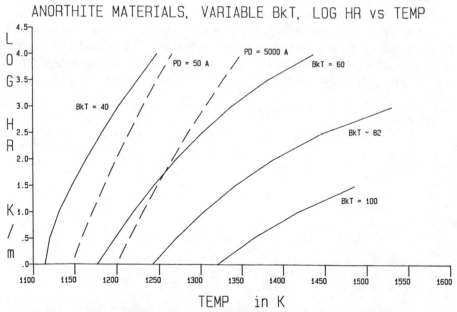

Figure 19.7 Sintering (dashed lines) and crystallization (solid lines) log (heating rate) versus temperature curves for anorthite-like materials with different nucleation barriers, *BkT*.

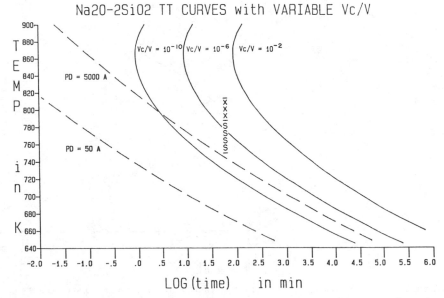

Figure 19.8 $Na_2O \cdot 2SiO_2$ sintering (dashed lines) and crystallization (solid lines) curves for various degrees of crystallinity. Experimental results of Hench et al.[3] for gel derived glasses at treatment times of 1 hr are also shown. S marks the range of temperatures where sintering alone is observed and X marks the range where both sintering and crystallization are observed.

tures of 793–813°C. Their condition for sintering corresponds to a degree of densification beyond that contemplated in the present calculations, and hence the agreement between predictions and experiment seems quite reasonable.

The assumption of steady-state nucleation rates can be relaxed by considering the incubation time during which steady-state distributions of subcritical embryos are established. The effect of such transient nucleation is to shift the crystallization curves to higher temperatures and thereby enlarge the processing window. Of the materials considered in the present study, the effect is most pronounced for $Li_2O \cdot 2SiO_2$, for which the results shown in Fig. 19.9 were obtained.

It is well recognized that sol–gel derived materials are generally characterized by properties, such as liquid viscosities, which differ from those of melted glasses. In applying the present analysis to SiO_2, use has been made of the data of Sacks et al.[4] for the viscosity of the "wet" SiO_2 employed in his studies. The viscosities reported there were appreciably smaller than those of conventional fused silica, and the lower viscosities have an appreciable effect on the expected rates of sintering and crystallization. These are shown in Fig. 19.10. Also shown in the figure are the other results of Sacks, who observed sintering to an amorphous state in 24 hr at 1274°C, but crystallization in 48 hr at 1273°C or 3 hr at 1323°C. The pore size in Sacks' experiments were a few thousand Å. From the results shown in Fig. 19.10, remarkable agreement is seen between the predicted behavior and experimental results.

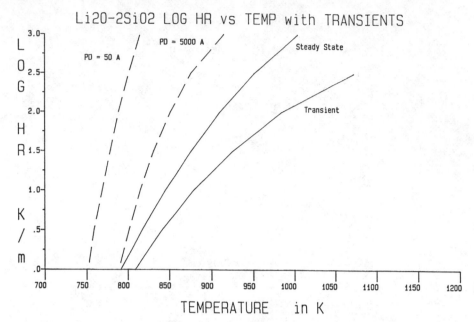

Figure 19.9 $Li_2O \cdot 2SiO_2$ log (heating rate) versus temperature sintering (dashed lines) and crystallization (solid lines) relationships showing the effect of transient nucleation.

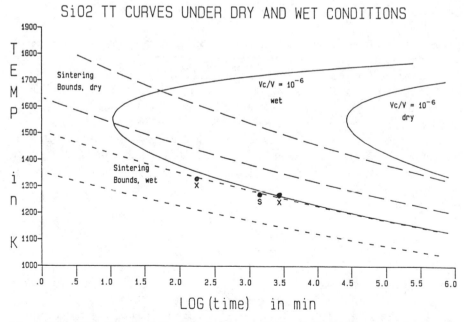

Figure 19.10 SiO_2 sintering (dashed lines) and crystallization (solid lines) curves under dry and "wet" conditions. The lower viscosity observed by Sacks et al.[4] shifts all curves to lower times at any temperature. Sacks data for sol–gel derived glasses are also included. *X* marks treatments which resulted in crystallization with little sintering and *S* marks treatments which produced sintered glasses free of crystallinity.

The effects of other factors on the competition between crystallization and viscous sintering are presently being explored in our laboratory. These include the effects of heterogeneous nucleation as well as the effects of time-dependent properties.

CONCLUSIONS

The kinetics of hydrolysis are best monitored using a combination of experimental techniques, together with thoughtful design of experimental conditions and critical appraisal of possible reaction sequences. Some studies carried out to date have pointed out effective approaches to obtaining appropriate kinetic data; but at present it must be concluded that, even for the most widely studied alkoxide, TEOS, our understanding of kinetic behavior leaves much to be desired. The situation with respect to alkoxides of other cations is even worse.

The competition between viscous sintering and crystallization can be described analytically, and processing windows can be defined for various materials where dense amorphous bodies or coatings can be obtained. Analyses carried out to date work surprisingly well considering the assumptions used; and we are optimistic that such analyses can provide useful guidance for processing. A note of caution must, however, be introduced in light of the difference in basic thermodynamic properties observed between gel-derived glass and melted glasses.[5]

ACKNOWLEDGMENTS

Financial support for the present work was provided by Emhart Corporation, Rogers Corporation, and the Air Force Office of Scientific Research. This support is gratefully acknowledged.

REFERENCES

1. G. W. Scherer, Sintering of Low-Density Glasses: I, Theory, *J. Am. Ceram. Soc.* **60** (5–6), 236–239 (1977).

2. D. R. Uhlmann and H. Yinnon, The Formation of Glasses, in *Glass: Science and Technology*, D. R. Uhlmann and N. J. Kreidl, Eds., Vol. 1, Academic Press, New York, 1983, pp. 1–47.

3. L. L. Hench, M. Prassas, and J. Phalippou, Preparation of 33 Mol % Na_2O—67 Mol % SiO_2 Glass by Gel-Glass Transformation, *J. Non-Cryst. Solids* **53**, 183–193 (1982).

4. M. D. Sacks and T.-Y. Tseng, Preparation of SiO_2 Glass from Model Powder Compacts: I, Sintering, *J. Am. Ceram. Soc.* **69** (8), 532–537 (1984).

5. M. C. Weinberg and G. F. Neilson, Elevation of Liquidus Temperature in a Gel-Derived Na_2O–SiO_2 Glass, *J. Am. Ceram. Soc.*, **60** (2), 132–134 (1983).

PART 2

Applications of Sol—Gel Processing

20

DIELECTRIC PROPERTIES OF SILICON DIOXIDE AND SILICON-OXYNITRIDE SOL–GEL THIN FILMS

LEE A. CARMAN AND CARLO G. PANTANO
Department of Materials Science and Engineering
The Pennsylvania State University
University Park, Pennsylvania

INTRODUCTION

There has been considerable interest in the development of the sol–gel process for deposition of thin films. Although the optical properties required for anti-reflection or high-reflection coatings have received the most attention,[1] there is also possible use of the sol–gel process for preparation of scratch resistant coatings, diffusion and oxidation barriers, and dielectric films. In fact, this method may be especially applicable in low-temperature processing and fabrication of microelectronic devices, for example, to deposit field and gate oxides, passivation coatings, or interlayer dielectrics. Nevertheless, there is little concerning the dielectric properties of these sol–gel films and that which is available is scattered throughout the microelectronic-processing literature. This chapter reviews the literature on dielectric properties of sol–gel films, describes potential applications of sol–gel thin films in microelectronics, and presents results on the thermal processing of silica sol–gel thin films in ammonia.

There are at least three reasons for the interest in the thermal treatment of silica sol–gel films in ammonia. First, a series of publications show [2–4] that

5–40 at% nitrogen can be chemically incorporated in silica sol–gel films during thermal densification in ammonia at 900–1300°C; the nitrogen is uniformly distributed throughout the film due to the initially microporous structure of the gel. This nitridation is expected to increase the dielectric constant, and probably the breakdown strength of the silica film, but the electrical properties of these oxynitride films have not yet been reported. The high dielectric constant and breakdown strength of the (oxy)nitride would permit the use of thinner field insulators in metal–insulator–semiconductor (MIS) devices. Second, it is well known that the thermal treatment of very-thin silica films in ammonia is an effective way to "anneal-out" the defects which limit their reliability in thin film capacitors.[5] The initial measurements on these sol–gel films revealed their susceptibility to low-voltage breakdowns, so the ammonia treatments were investigated simply to enhance the yield and reliability of the dielectric films. Third, the (oxy)nitride structure is a barrier to the transport of alkali ions, water, hydrogen, oxygen, and other impurities; it can provide a more effective dielectric film material than the pure oxide in both passive and active applications.

The behavior and properties of these sol–gel deposited silica films are contrasted here with those of the thermal-oxide of silicon because it serves as a convenient and challenging benchmark. More specifically, the sol–gel films have been used to fabricate metal–oxide–silicon (MOS) capacitors which can be used to test and evaluate the electrical characteristics of sol–gel films versus thermal oxide films in these simple diodes. Of course, one should recognize that neither these sol–gel films, nor any other deposited films, can surpass the electrical and interface properties of the thermal oxide in MOS or MIS devices. However, there are many situations where a thermal oxidation step at temperatures in the range 900–1200°C cannot be used and/or where a silica or doped silica film must be deposited (rather than grown) on a nonsilicon substrate or device. In the case of oxynitride films, the development of optimal processing methods is still being actively pursued.

BACKGROUND

In the deposition of glass silicate films for microelectronic applications, sol–gel materials are usually referred to as spin-on glasses (SOG), and virtually all of the data in the literature refers to a commercial product called Silicafilm® (manufactured by Emulsitone Co., Whippany, NJ, USA). Although the exact composition of this commercial sol(ution) has not been published, it is essentially a silicon–oxide/alcohol emulsion. Lam and Lam published two papers[6,7] which addressed the characteristics and electrical properties of this spin-on oxide. Their infrared studies showed that the spinning solution contained both silicon alkyl ($\equiv$Si—R) and silicon alkoxy ($\equiv$Si—O—R) groups, as well as ketones. The deposited films exhibited both siloxane ($\equiv$Si—O—Si$\equiv$) and silanol ($\equiv$Si—OH) groups, but after a heat treatment in N_2 at 1200°C, the

IR spectrum was characteristic of the thermal oxide of silicon. They concluded that the heat-treated film was not oxygen deficient, but other studies[8] show that although the oxide stoichiometry approaches SiO_2, it is typically in the range $SiO_{1.9}$ to $SiO_{1.98}$. Lam and Lam[6,7] concluded that the electrical characteristics of these films were clearly inferior to those of thermal SiO_2 when heat treated in N_2 at less than 900°C, but were only slightly inferior if heat treated at 1200°C. This conclusion was based primarily upon the measurement of dielectric constant and conductance, which approached that of the thermal oxide (3.9) for treatments above 900°C. In general, though, they found the dielectric breakdown strengths to be poor, even after eliminating the weak spots in each capacitor using a nonshorting technique. Thus, the films heat treated at low temperature ($<900°C$) exhibited very low breakdown fields ($\ll 0.1$ MV/cm), and the films heat treated at 900 to 1200°C showed acceptable breakdown fields (~ 2–8 MV/cm). Nevertheless, they found that even the films treated at temperatures greater than or equal to 900°C adsorbed water in their outermost surface, and this led to increased capacitance and hysterisis effects. We will show later in this chapter that a heat treatment at 900°C or higher in dry N_2 is sufficient to oxidize the underlying silicon substrate. Thus, the dielectric constant and conductance of their MOS devices were really due to an underlying thermal oxide and not to the SOG film; the capacitance behavior of the devices, however, was being influenced by the overlying spin-on glass.

More recently, Ma and Miyauchi[9] used Silicafilm® SOG to "cap" the surface of GaAs wafers during their thermal oxidation. The growth of a stoichiometric GaAs–oxide layer is needed to fabricate gate dielectrics in MIS GaAs devices, but typically this oxide exhibits high leakage currents and unacceptable capacitance characteristics. Here, Silicafilm® caps were used to prevent excessive vaporiziation of As and to limit the oxygen activity at the GaAs surface during thermal oxidation; this was expected to enhance the formation of a stoichiometric GaAs–oxide at the Silicafilm®/GaAs interface. Interestingly, these investigators found that the electrical properties of the spin-on glass film itself might be sufficient for the gate dielectric in GaAs devices (i.e., as a substitute for the GaAs–oxide). They showed that a 700°C heat treatment in dry N_2, followed by a 500°C heat treatment in forming gas, yielded SOG films with low leakage currents and the desired capacitance–voltage characteristics (surface inversion); the dielectric constant was found to be 4.4 and the breakdown field about 4.5 MV/cm. The densification heat treatment was limited to 700°C due to the generation of interfacial gas pockets and crystalline arsenic at higher treatment temperatures; these effects were believed to be nucleated by moisture in the spin-on glass film. The role of the forming gas treatment was not clearly identified, although it was suggested that the hydrogen might tie up the "trap sites" associated with dangling bonds and excess arsenic at the oxide–GaAs interface. Later, Hayashi et al.[10] tried to reproduce this fabrication of GaAs MIS diodes but found that Ga diffusion into the spin-on glass layer prevented the inversion type capacitance–voltage characteristic required in these devices. They showed that the spin-on glass (presumably, Silicafilm®, although the authors do not state clearly) could be doped with $GaCl_3$ to suppress the out-diffusion of Ga.

The spin-on glass or sol–gel process is also being evaluated for the deposition of barrier films, interlevel dielectric films, and planarizing layers which are required for bilevel or trilevel resist processing. Gupta et al.[11,12] have evaluated spin-on silica glasses for both stand-alone and planarizing interlevel dielectric layers. These layers were prepared from Accuglass ® 203 SOG (manufactured by the Electronic Chemical Products Division of Allied Chemical, Buffalo, NY, 14210, USA) which is a commercial solution of hydrolyzed alkoxysilanes in an alcoholic solution. Although these authors have not published the details of their electrical property measurements (since for this application the electrical characteristics are not especially critical), they claim that the dielectric proper-ties of the spin-on glass films become comparable with the thermal oxide after heat treatment to 800°C. They report a dielectric constant of 4.2, a breakdown field >1 MV/cm, and a resistivity $>1 \times 10^{16}$ Ω-cm; the heat-treatment atmosphere used at 800°C was not reported.

Most recently, Dosch,[16] Livage,[17] and others in this volume, have studied the dielectric properties of nonsilicate sol–gel films (e.g., barium–titanate and vanadium–oxide) on substrates other than single crystal semiconductor wafers. In these cases, the influence of the heat-treatment temperature upon crystalliza-tion and grain growth must be considered.

EXPERIMENTAL PROCEDURE

Dilute solutions of tetraethoxysilane (TEOS), ethanol, and deionized water (~ 15 MΩ/cm) were used to prepare these silica sol–gel films. The water-to-TEOS molar ratio was varied from 4 to 24. A small amount of HCl was added to the solution in all cases to promote hydrolysis. The materials were combined in a closed reaction vessel equipped with a reflux condenser, heated under constant stirring to 60°C, and held for 30 min. The solutions were then cooled to approximately room temperature and diluted with ethanol to yield an oxide content of about 5%. The solutions were subjected to a spectrochemical im-purity analysis which showed the total alkali content (Na and K) to be ~ 2–3 ppm, and the boron content to be ~ 1 ppm. The total alkali content in the TEOS was ~ 4 ppm and the boron content in the TEOS was ~ 10 ppm.

Polished n-type phosphorous-doped Si wafers having resistivities of 1–2 Ω-cm were used for the substrates. The wafers were cleaned using electronic grade reagents and the standard "RCA procedure" used in the semiconductor industry. A photoresist spinner was employed to spin-coat the wafers at 3000 rpm for 15 s. The solutions were deposited through filtered syringes in a normal laboratory atmosphere; some sets of wafers were prepared in a clean-room environment, but since their dielectric breakdown characteristics were not significantly improved, this practice was not continued.

The heat treatments were performed in a fused-silica tube furnace under flowing O_2, dry N_2, and/or anhydrous NH_3 gases. All of the samples were heated at $10°C/min$, held at temperature for 2 hr, and then cooled to room temperature in the flowing gas. A vacuum system was also available to heat treat several samples at pressures of $10^{-5}-10^{-6}$ torr. A pretreatment to $400°C$ in O_2 was performed with all samples to burn off residual organics. The thicknesses and refractive indices of the films were measured before and after each processing step using a Rudolph Automated Ellipsometer.

An array of Al electrodes was evaporated onto the treated films through a metal mask; the electrodes were approximately $2000 Å$ thick and $\sim 7 \times 10^{-3}$ cm^2. This essentially produces an array of MOS (or MIS) diode capacitors whose electrical properties can be obtained very conveniently. Usually, about 30 of the individual diodes on each wafer were tested. The electrical properties measured included current–voltage ($I–V$) or current–field ($I–E$), current–time ($I–t$), dc breakdown strength (E_{max}), and capacitance–voltage ($C–V$) characteristics. The breakdown field was measured by applying a ramp voltage (~ 0.5 V/s), and then recording the maximum voltage drop across the diode. The $I–V$ curves were obtained using step voltages and then recording the steady-state current after a few minutes stabilization. The time increment required to attain a steady-state current was determined by first monitoring the $I–t$ characteristic of each diode. Dielectric constants were determined using the maximum capacitance obtained in the accumulation regime of the high-frequency $C–V$ curve.

RESULTS AND DISCUSSION

The primary objective was to correlate the dielectric properties of the silica sol–gel films with the spin-on solution composition and heat-treatment condition. In contrast to the commercial Silicafilm® emulsion, the sol–gel solutions used here are quite simple. Only hydrolyzable silicon–alkoxides were present in solution, whereas the commercial emulsion contains silicon–alkyl groups. The effects of annealing or nitriding the film in ammonia were of particular interest.

It should be noted that most of these films were deposited in the range 500–$1000 Å$, and most importantly, that the thickness was monitored after each heat treatment. In general, the film thickness had the biggest effect upon the behavior and properties of the MOS devices; thus, it was important to account for the effect of any thermal treatment or process variation upon film thickness *per se* before its influence upon the intrinsic properties of the film could be determined. The "self-healing" (or nonshorting electrode) method was not used to eliminate weak spots in each device, and so, numerous devices on each wafer were tested until a statistically significant average was obtained.

The $I–E$ curves in Fig. 20.1 are for films prepared with a $H_2O/TEOS$ ratio of 4 and treated at $1000°C$ in vacuum or dry nitrogen. It is quite clear that the dielectric properties of the nitrogen-treated films are considerably better than

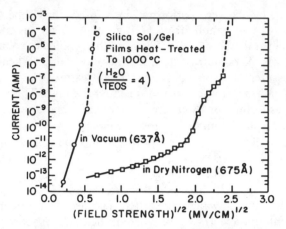

Figure 20.1 Dependence of leakage current on applied field for nitrogen and vacuum treated films. Data obtained at room temperature (Al positive) via a step voltage with a 2-min current stabilization period.

the vacuum-treated films. This dependence upon the atmosphere used in thermal treatment was also observed at other $H_2O/TEOS$ ratios. In general though, the $I-E$ curves shift to the right (higher fields) with increasing $H_2O/TEOS$ ratio for a given heat-treatment condition. The solutions whose $H_2O/TEOS$ ratio was 10 seemed to produce films whose properties and reliability were best in this thickness range.

The behavior of the films treated in dry N_2 was qualitatively similar to that reported by Lam and Lam[6,7] for the commercial Silicafilm ®. The films treated at temperatures less than 900°C were susceptible to low voltage breakdowns and high leakage currents, whereas the higher treatment temperatures produced films whose properties approximated those of the thermal oxide. Lam and Lam attributed this behavior to the more effective annealing and densification of the spin-on glass at temperatures in excess of 900°C, but the observations made in this study lead to a different interpretation. Here, it was found that the most important consequence of heat treatment in nitrogen at >900°C is an increase in the oxide film thickness.[2] This is due to oxidation of the underlying silicon substrate even in dry N_2; the dry N_2 used in this study was found to contain ~500 ppm O_2. Secondary ion mass spectroscopy (SIMS) provided in-depth concentration profiles which verified that substrate oxidation occurs at >900°C in N_2.[3] Thus, the $I-E$ characteristics of films treated in N_2 at >900°C are probably due to an underlying thermal oxide, and not to the intrinsic properties of the heat-treated sol–gel or SOG films. It is not surprising, then, that films treated in this way can exhibit some of the properties of the thermal oxide. This also explains the observations of Lam and Lam concerning the moisture sensitivity of their high-temperature films whose behavior and properties were, otherwise, comparable to the thermal oxide. Presumably, their spin-on glass film still contained microporosity which could adsorb enough water to influence the $C-V$ behavior, even though the underlying thermal oxide enhanced the resistivity and dielectric strength. All of the films described in the remainder of this chapter were heat-treated in either vacuum or NH_3 where the P_{O_2} is negligible (at least for oxidation of Si at these temperatures).

Figure 20.2 shows the effects of heat-treatment temperature and atmosphere upon the $I-E$ characteristics of films prepared with a $H_2O/TEOS$ ratio of 10. All these films have a thickness of 500 ± 20 Å after heat treatment, and therefore, the differences in $I-E$ characteristics are intrinsic to the film and are not due simply to changes in their thickness with thermal processing. One notes that the conductance of the film is not much affected by the low-temperature thermal treatment in O_2 used to eliminate organic residuals. The higher temperature treatments in vacuum and NH_3 lead to substantial decreases in the leakage current. The vacuum treatment has resulted in lower leakage currents than the NH_3 treatments, and this is consistent with the $I-E$ behavior of sputtered and pyrolytic oxide and oxynitride films;[18-20] that is, nitrides and oxynitrides usually exhibit higher conductances than the pure thermal oxide.

Figure 20.3 shows a comparison of $I-E$ curves for sol–gel oxide and oxynitride films, sputtered Si_3N_4, and pyrolytic Si_3N_4, oxynitride, and SiO_2 films. The sol–gel films were prepared with an $H_2O/TEOS$ ratio of 10 and treated to 1000°C in vacuum and 1000°C in NH_3, respectively. It can be seen that the nitrided sol–gel exhibits a higher leakage current than the vacuum-fired silica sol–gel. This is consistent with the higher leakage currents of the pyrolytic silicon–nitride and silica–oxynitride when compared to the pure thermal oxide of silicon. However, the leakage currents through the sol–gel films are higher in

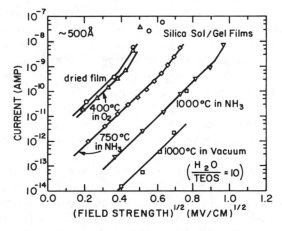

Figure 20.2 Current–field behavior of sol–gel films as a function of treatment temperature and atmosphere. All data were obtained on films with thicknesses of 500 ± 20 Å after treatment.

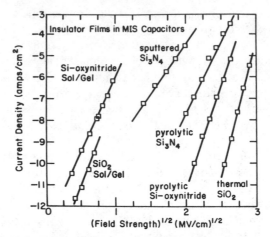

Figure 20.3 Current density versus field characteristics for insulator films with varying compositions and methods of preparation. Data are for films deposited on $1-2\,\Omega$-cm n-type silicon, (Al positive).

general, than the vacuum-deposited nitrides and oxynitrides. In fact, both the nitrided sol–gel and pyrolytic oxynitride films represented in Fig. 20.3 have refractive indices of 1.71 which corresponds to approximately 30–40 at% nitrogen. The higher conductance through the sol–gel films is probably related to the higher ionic impurity levels associated with the spin-on film deposition versus vacuum film deposition.

The sol–gel films exhibit ohmic behavior at very low fields (not shown in Fig. 20.2); this is presumably an ionic conductivity due to the presence of alkali and other mobile impurities. Perhaps of more significance, though, is the observed linear relationship between log I and $E^{1/2}$ (Fig. 20.2). This dependence is also observed for thermal silica films. It indicates that the electrons which give rise to these leakage currents are injected into the conduction band of the oxide from the silicon substrate. Considering the low applied fields at which these currents are observed, it implies that there are high local electric fields at the silicon/sol–gel oxide interface. This situation may arise due to the presence of heterogeneities and protuberances at the interface that focus the field lines applied across the film, or due to a lowering of the interface barrier by the interfacial adsorption or immobilization of positively charged ionic impurities (see below).

Figure 20.4 shows the influence of film thickness upon the IV characteristics of films treated to 1000°C in NH_3; all of these films were prepared from solutions with a H_2O/TEOS ratio of 10. The thicker films exhibit lower leakage currents and can support much higher voltages (even though they breakdown at lower fields), and not surprisingly, one can prepared more reliable dielectric insulators with the sol–gel method by depositing thicker layers. In fact, the films with thicknesses greater than 1000 Å exhibit low leakage currents and reasonable

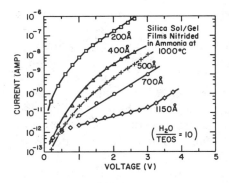

Figure 20.4 Influence of film thickness on the *I–V* characteristic with all other processing conditions remaining constant. Film thickness was determined using a Rudolph Automated Ellipsometer ($\lambda = 6328$ Å) at an incident angle of $70°$.

dielectric strengths after the 400°C treatment in O_2 only, and thus, they are reasonable dielectric insulators even without complete densification at high temperature. Unfortunately, there is a limit to the thickness that can be deposited with the sol–gel method, and where thickness uniformity is critical, the maximum film thickness is further reduced by the need to use dilute solutions. Hence, silica sol/gel films—whether fired or unfired—will be useful insulators only where leakage currents of the order nanoamps to picoamps can be tolerated. Of course, the most challenging applications for these materials in microelectronics are within the realm of very thin films, and thus, it is necessary to understand and control the properties of films <1000 Å in thickness. In these applications, the dielectric breakdown strength is, perhaps, the most limiting characteristic.

Figure 20.5 shows the dependence of the breakdown field upon film thickness and heat-treatment temperature in NH_3 and in vacuum. The oxide content of the solutions used to deposit each of these films was carefully adjusted to produce sets of films whose initial thickness varied, but whose thickness after densification were comparable. The breakdown field increases with decreasing film thickness and increasing heat treatment temperatures which is consistent with breakdown behavior for thermal silica.[21] On the other hand, defect density increases with decreasing thickness,[22] thus giving rise to higher probabilities of low field breakdown in very thin insulator films. This is consistent with

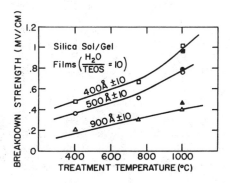

Figure 20.5 Dependence of breakdown strength on film thickness and treatment temperature. Approximately 30 MOS diodes were tested per wafer, with 95% confidence limits determined to be negligible (± 0.015). Darkened symbols represent vacuum-treated films.

the behavior of most systems in the thickness range less than 1000 Å.[23] The higher dielectric strengths in the films heat treated to higher temperatures correlate directly with the $I-E$ behavior; that is, a reduction in the conductance of the film leads to an increase in the breakdown strength. Higher treatment temperatures can also result in degradation in the breakdown strengths, due to crystallization or other localized defects. Here, the film properties increase continuously with treatment temperature, at least up to 1000°C.

It is surprising, though, that the dielectric strengths of the films treated at 1000°C in NH_3 are indistinguishable from those of the vacuum-treated films since the oxynitride is expected to exhibit better dielectric strength than the pure oxides. The refractive indices measured for the 1000°C ammonia-treated films indicated that these films should contain 30–40 at% nitrogen.[17] One interpretation of their comparable dielectric strengths is that the breakdown of these MOS devices is not intrinsic to the dense oxide or oxynitride material itself, but rather, is limited by microstructure in the film or defects in the device. This interpretation is consistent with the low values of dielectric strength for any sol–gel deposited film ($E_{max} = 0.2–1.0$ MV/cm), at least when compared with the thermal oxide ($E_{max} = 10$ MV/cm).

Although the nitridation treatment in NH_3 did not measurably enhance the dielectric strength of the film over that of the pure oxide densified in vacuum, it was found that the distribution of breakdown strengths was much narrower for the NH_3-treated films. The MOS devices were examined with optical microscopy and SEM after testing, and the mode of breakdown depended upon the heat-treatment atmosphere. The micrograph in Fig. 20.6a shows numerous small pits in the Al electrode on the vacuum treated film where breakdowns have occurred. In contrast, the micrograph in Fig. 20.6b shows a single breakdown cavity in the electrode for an NH_3-treated film. This observation was noted for films treated to 750°C and 1000°C in NH_3, and was reproducible for various treatment temperatures and H_2O/TEOS ratios. Ito et al.[5] observed a similar difference in the breakdown behavior of thermal oxides which had been annealed in NH_3 versus those which had not. They concluded that the NH_3 treatment passiviated weak spots in the oxide so that current flowed more evenly over the electrode surface until a single catastrophic breakdown occurs. Thus, it seems likely the nitridation of these sol–gel films in NH_3 has, in fact, influenced the film structure and dielectric behavior. Further improvements in the dielectric properties can be expected once the "extrinsic" defect density present in these laboratory films is more efficiently controlled. It is likely that these nitrided films would exhibit substantially better properties than the pure oxides if the "bad diodes" were eliminated via "self-healing" breakdown (a method of testing not used in this study).

Finally, the effect of heat-treatment temperature and atmosphere upon the dielectric constant is shown in Fig. 20.7. The films heated to 400°C in O_2 exhibit a high permittivity (~ 10) due to the presence of extensive microporosity, and probably, a considerable quantity of adsorbed water. After heating to 750°C in vacuum or NH_3, the permittivity drops to about 4. Since a 750°C treatment is

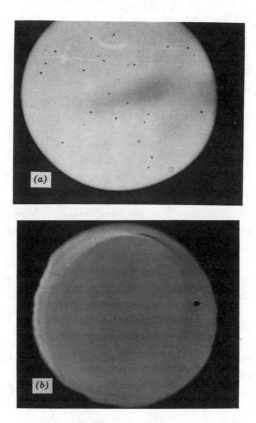

Figure 20.6 Micrographs of breakdown effects on Al electrode surfaces. (*a*) Multiple breakdown sites observed for vacuum treated films (80X). (*b*) Single breakdown pit observed for ammonia treated films (80X).

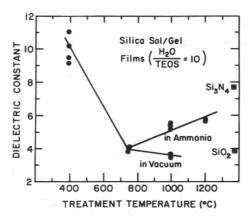

Figure 20.7 Dependence of the dielectric constant on heat treatment temperature and atmosphere. Dielectric constant determined using the maximum capacitance of the high-frequency CV curve in the accumulation regime.

insufficient to nitride the film, this average value—as well as the slightly lower value of 3.8 for the film treated to 1000°C in vacuum—are probably characteristic of the pure oxide; the dielectric constant of the thermal oxide of silicon is 3.9. In contrast, heating the film to 1000°C or 1200°C in NH_3 leads to an increase in the dielectric constant. This is due to nitridation of the film at the higher treatment temperature. The film treated at 1000°C contains ~ 15 at% nitrogen ($SiO_{1.4}N_{0.4}$) while the film treated to 1200°C contains ~ 35 at% nitrogen ($SiO_{0.7}N_{0.9}$).[4]

SUMMARY

It has been found that silica sol–gel films in the thickness range $\leqslant 1000 \text{Å}$ exhibit relatively high leakage currents and low dielectric strengths when compared with the thermal oxide of silicon. Although the distribution of dielectric strengths was considerably narrower for the NH_3-treated films—and thus the yield of MOS capacitors is improved—the average dielectric strength was not measurably improved due to the nitridation. Of course, the nitrided films exhibited a higher dielectric permittivity than the pure oxides and there were clear indications that the breakdown mechanism was being influenced by the incorporation of nitrogen. In general, the films were more reliable dielectrics when prepared using solutions of higher $H_2O/TEOS$ ratios (~ 10–16). It was also shown that a silicon (or other metal) substrate may be oxidized during a high-temperature thermal densification of sol–gel or SOG films in dry nitrogen or other atmospheres where the oxygen activity is sufficiently high. Although this substrate oxidation under a sol–gel "cap" might itself be a useful processing step in some microcircuit fabrications, it must be accounted for when MOS capacitors are used to measure or study the intrinsic dielectric properties of the sol–gel oxide film.

The high leakage currents and low dielectric strengths reported here are probably enhanced by the presence of impurities and defects in the films. The spectrochemical analysis of the sol–gel solutions, and the SIMS depth profiles of the deposited films have verified the presence of sodium, potassium, and boron impurities. The TEOS was also subjected to spectrochemical analysis (on four separate occasions), with results indicating that the TEOS itself may be acting as a source of impurities in the films, particularly the boron. The presence of these impurities in the oxide film can enhance the injection of charge carriers into the oxide film due to a lowering of the interface barrier. Using a bias-temperature stress test, the CV characteristics of these films (not reported here) verify the presence of mobile positively charged impurity species. The injection of electrons into the conduction band of the oxide would be further enhanced due to the high local fields at or near the semiconductor/oxide interface. These high local fields could be associated with microstructural features intrinsic to the film/substrate interface (e.g., micropores, impurity crystallites, pyrolysis products, stresslines, etc.), as well as with any extrinsic defects (e.g., pinholes or dust). Altogether, this suggests that these high leakage currents are due to electron injection, and thus the low dielectric strengths may be triggered by the thermal instabilities brought about by the resultant Joule heating. The other data reported in the literature for SOG films (with the exception of those SOG films treated under conditions where the underlying substrate has been oxidized) also lead one to conclude that sol/gel films are "leaky insulators," although the dielectric strengths can be much better than those obtained in this study. Thus, it is likely that the laboratory MOS devices used here did, in fact, contain many impurities and extrinsic defects.

Although an objective of this study was to define the dielectric properties of sol–gel films relative to their process chemistry, these data more closely reflect the influence of alkali impurities and extrinsic defects. That is, the effects of sol(ution) chemistry and nitridation have either been masked by, or superimposed upon, the rather low values of resistivity and dielectric strength (Fig. 20.3). Unfortunately, it will not be possible to define dielectric properties which are intrinsic to sol–gel or SOG films, and their thermal processing, until the metal–ion impurity levels and extrinsic defect densities are significantly reduced. This evaluation would require the commitment to high purity chemicals and solvents, air quality, liquid handling and thermal processing, which characterizes the technology of microelectronics materials processing.

ACKNOWLEDGMENTS

The authors gratefully acknowledge Sandia National Laboratories for financial support of this study.

REFERENCES

1. See Chapters 13, 14, and 15 in *Ultrastructure Processing of Ceramics, Glasses, and Composites*, L. L. Hench and D. R. Ulrich, Eds., Wiley New York, 1984, for reviews and complete lists of references.

2. C. G. Pantano, P. M. Glaser, and D. H. Armbrust, Nitridation of Silica Sol/Gel Thin Films, in *Ultrastructure Processing of Ceramics, Glasses, and Composites*, L. L. Hench and D. R. Ulrich, Eds., Wiley, New York, 1984, pp. 161–177.

3. P. M. Glaser and C. G. Pantano, Effect of the H_2O/TEOS Ratio Upon the Preparation and Nitridation of Silica Sol/Gel Thin Films, *J. Non-Cryst. Sol.* **63**, 209 (1984).

4. R. K. Brow and C. G. Pantano, Composition and Chemical Structure of Nitrided Silica Gel, in *Better Ceramics Through Chemistry*, Brinker et al., Eds., *Mater. Res. Soc. Symp.* **32**, Elsevier (1984), pp. 361–368.

5. T. Ito, H. Arakawa, T. Nozaki and H. Ishikawa, Retardation of Destructive Breakdown of SiO_2 Films Annealed in Ammonia Gas, *J. Electrochem. Soc.* **127**, 2248 (1980).

6. Y. W. Lam and H. C. Lam, Dielectric and Interface-State Measurements of Metal-Spin-On-Oxide-Silicon Capacitors, *J. Phys. D: Appl. Phys.* **9**, 1477 (1976).

7. H. C. Lam and Y. W. Lam, Physical and Conduction Measurements of Spin-On Oxide, *Thin Solid Films* **41**, 43 (1977).

8. J. N. Smith, S. Thomas, and K. Ritchie, Auger Electron Spectroscopy Determination of the Oxygen/Silicon Ratio in Spin-On Glass Films, *J. Electrochem. Soc.* **121**, 827 (1974).

9. T. P. Ma and K. Miyauchi, MIS Structures Based on Spin-On SiO_2 on GaAs, *Appl. Phys. Lett.* **34**, 88 (1979).

10. H. Hayashi, K. Kikuchi, Y. Yamaguchi, and T. Nakahama, Study of the Properties of Spin-On SiO_2/GaAs Interfaces, *Inst. Phys. Conf. Ser.* **56** (Chp. 5), 275 (1981).

11. S. K. Gupta, R. L. Chin, and S. A. Ferguson, Characterization of Spin-On Glass Dielectric Films for Use in IC Fabrication, presented at ACS meeting, St. Louis, Missouri (1984). To be published.

12. S. K. Gupta and C. G. Audain, Trilayer Resist Processing Using Spin-On Glass Intermediate Layers, *Proc. SPIE* **469**, 179 (1984).

13. Y. C. Lin, S. Jones, and G. Fuller, Use of Antireflective Coating in Bilayer Resist Process, *J. Vac. Sci. Technol.* **B1**(4), 1215 (1983).

14. G. W. Ray, S. Peng, D. Burriesci, M. M. O'Toole, and E. D. Liu, Spin on Glass as an Intermediate Layer in a Tri-Layer Resist Process, *J. Electrochem. Soc.* **129**, 2152 (1982).

15. B. Singh, G. C. Chern, and I. Lauks, Spin-Coated As_2S_3: A Barrier Layer for High Resolution Tri-Layer Resist System, *J. Vac. Sci. Technol.* **B3**(11), 327 (1985).

16. R. G. Dosch, Preparation of Barium Titanate Films Using Sol-Gel Techniques, in *Better Ceramics Through Chemistry*, Brinker et al., Eds., *Mater. Res. Soc. Symp. Proc.* **32**, Elsevier (1984), pp. 157–162.

17. J. Livage and J. Lemerle, Transition Metal Oxide Gels and Colloids, *Ann. Rev. Mater. Sci.* **12**, 103 (1982).

18. B. E. Deal, P. J. Fleming, and P. L. Castro, Electrical Properties of Vapor-Deposited Silicon Nitride and Silicon Oxide Films on Silica, *J. Electrochem. Soc.* **115**, 300 (1968).

19. D. M. Brown, P. V. Gray, F. K. Heumann, H. R. Philipp, and E. A. Taft, Properties of $Si_xO_yN_z$ Films on Silicon, *J. Electrochem. Soc.* **115**, 311 (1968).

20. V. A. Gritsenko, N. D. Dikavskaja, and K. P. Mogilnikov, Band Diagram and Conductivity of Silicon Oxynitride Films, *Thin Solid Films* **51**, 353 (1978).

21. S. K. Lai, Dependence of Thin-Gate Oxide Properties on Processing, *Silicon Processing*, ASTM STP 804, D. C. Gupta, Ed., American Society for Testing and Materials, 1983, pp. 260–272.

22. A. C. Adams, T. E. Smith, and C. C. Chang, *J. Electrochem. Soc.* **127**, 1787 (1980).

23. P. Solomon, Breakdown in Silicon Oxide—A Review, *J. Vac. Sci. Technol.* **14**, 1122 (1977).

21

SOL–GEL DERIVED SILICA OPTICAL FILTERS

S. H. WANG AND LARRY L. HENCH
Ceramics Division
Department of Materials Science and Engineering
University of Florida
Gainesville, Florida

INTRODUCTION

Large, monolithic pure silica gels have been made rapidly and reliably from tetramethylorthosilicate (TMOS) using drying control chemical additives (DCCAs).[1-4] Recently, attention has been shifted to the TMOS–DCCA method to make silica gels with optical filter characteristics by introducing transition or rare earth elements into the transparent and colorless silica gel matrix. Such a transparent and colorless matrix has fully filled electronic energy levels in all of its bands. When the impurity is added, incomplete d or f orbitals are introduced.[5] Ions with incomplete inner electron shells will absorb light in characteristic ranges of wavelength in the gel matrix. These electron energy levels contribute to color formation for the gels containing transition or rare earth elements.

The colors of materials with transition elements having unfilled shells are particularly subject to change in coordination numbers and the nature of adjacent ions and therefore the colors are described as resulting from specific chromophores, which are complex ions that produce a particular optical absorption effect. In contrast, the rare earth colorants, such as Nd, which depend on transitions in the inner f shell are much less subject to environmental changes.[6] These differences in d-shell and f-shell colorants are observed for the doped silica gels as shown in this work.

201

PROCEDURE

Six steps are generally used in making the monolithic silica gels and glasses containing transition or rare earth elements: (1) mixing, (2) casting, (3) gelation, (4) aging, (5) drying, and (6) densification. In the first mixing stage, use of suitable drying control chemical additives such as formamide, glycerol, or an organic acid, make it possible to make monoliths rapidly without: (1) precipitation, (2) formation of an inhomogeneous gel, or (3) crystallization. Thus, it is now possible to produce non-crystalline homogeneous optical filter silica gels and glasses with a wide variety of transition metal or rare earth elements and optical absorption bands. To our knowledge, monolithic gels containing the transition and rare earth elements mentioned below have not previously been described.

One example is the production of Co colored silica monoliths. The first step involves mixing from 0.01 mole to 0.1 mole of an organic acid DCCA with 16 moles of distilled water for 5 minutes at room temperature, followed by addition to the DCCA solution of 1 mole of TMOS and mixing for a further 5 min. A range of concentrations from 0.001 mole to 0.5 moles, of transition metal (Co) acetate (TMAc) is added to the DCCA–TMOS solution, keeping the ratio of TMAc/DCCA < 1.0 to avoid the formation of a precipitate. The molar concentration of TMAc and thereby the DCCA is selected to produce the required intensity of optical absorption in the gel-glass. The TMAc is added to the solution while the temperature is increased during 30 min mixing at 85°C to accelerate hydrolysis of the TMOS. The Co ions are incorporated homogeneously into the silica sol network as it is being formed. The thoroughly mixed homogeneous silica solution is cooled then cast into molds to form the required shape of an optical component. The molds are maintained at temperature until gelation and for an additional hour after gelation, then are heated for approximately 15 hr to age the gels and increase their strength and density. The drying procedure is as previously described[4] for pure silica gels with various DCCAs. In this case, densification can be achieved by firing in an ambient atmosphere between 600–1300°C depending on the concentration of transition or rare earth elements present and the density required. The monolithic optical filter gels described herein are heated and taken from an oven after the gel is completely dried, or at 850°C, where the gel is partially densified after heating at 10°C/hr from 160 to 850°C.

The transmission spectra are obtained by using a Perkin-Elmer UV-Visible-IR spectrophotometer model 552.

RESULTS AND DISCUSSION

The UV-Visible spectra of monolithic pure silica gels shown in Fig. 21.1 (sample 21.2E) heated to different temperatures, are compared with a pure melt silica glass (Dynasil). The percentage of transmission decreases gradually with a decrease of the wavelength from 900 to 200 nm for the four gel samples. Increas-

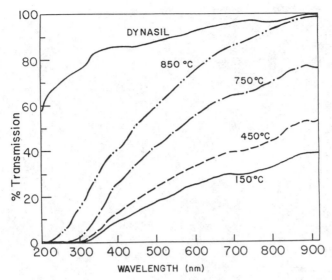

Figure 21.1   Transmission curve for pure silica gel monoliths heated to various temperatures.

ing the thermal treatment increases the optical transmission for the pure silica gels. The greyness of the silica spectra may be due to incompletely pyrolysed carbon residuals from the methanol or DCCAs left after drying the gels. Increasing the temperature in an O_2 atmosphere gradually eliminates the residual impurities.

Addition of 1% of Nd into the silica gel results in the two samples shown in Fig. 21.2D, I. The colors change from light pink at 160°C (21.2I) to light purple at 850°C (21.2D), with the optical spectra shown in Fig. 21.3. The rare earth colorant, Nd, has only a moderate sensitivity to the variation of densification temperature since the electron transitions take place in the inner f shell which are, to a first order, independent of their chemical environment. The major declination of the spectrum from 900 to 200 nm in Fig. 21.3 is similar to that of pure silica in Fig. 21.1. However, significant sharp absorptions occur in three frequencies at 562, 725, and 782 nm. Three minor absorptions at 335, 502, and 860 nm are also observed. These spectral features of Nd in a silicate melt glass is shown for comparison.[12] The melt glass absorption bands are ~25 nm higher than the bands in the silica gel. The spectrum of the Nd–silica gel heated to 850°C is also shown in Fig. 21.3. There are no significant deviations in the location of the characteristic absorption peaks of the gel with heat treatment. The only two changes with heat treatment are the transmission intensities and the cut off frequencies between 200 and 300 nm.

The silica gel samples containing 0.25% Co are shown in Fig. 21.2B, F, G. The color of the 160°C gel (F) is reddish orange. The 850°C sample's color is deep blue (G), and the 900°C sample has a greenish black color (B). The UV-Visible spectra characteristic of these three Co–silica gel samples are shown in Fig. 21.4 There is a totally different absorption curve for the 160°C orange sample than for

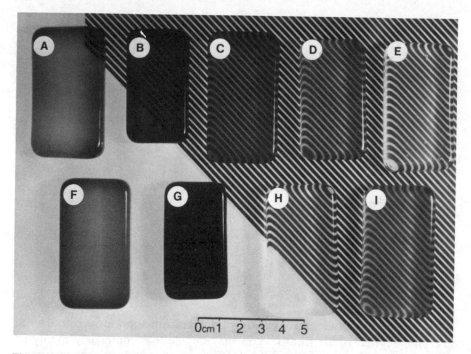

Figure 21.2 Silica gel monoliths with various additives and thermal histories: A—0.25% Co at 160°C; B—0.25% Co at 900°C; C—1.0% Cu at 160°C; D—1.0% Nd at 850°C; E—pure silica; F—0.25% Co at 160°C (DCCA II); G—0.25% Co at 850°C; H—1.0% Ag at 160°C; I—1.0% Nd at 160°C.

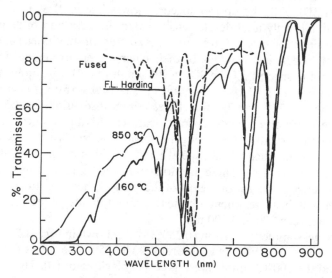

Figure 21.3 Transmission curves for silica gels containing 1.0 mol% Nd heated to 160°C and 850°C compared with melt cast Nd silicate glass.

204

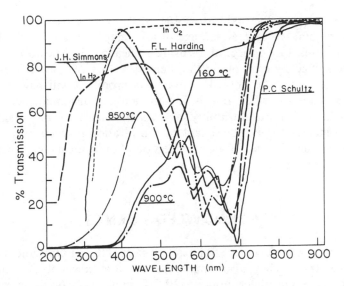

Figure 21.4 Transmission curves for silica gels containing 0.25 mol% Co heated to 160°C, 850°C, and 900°C compared with porous vitreous silicas doped with Co and heated in H_2 or O_2 (J. H. Simmons) or melt cast vitreous silica containing Co (P.C. Schultz).

the 850°C blue sample and the 900°C green sample. Since the color of transition ions such as cobalt in silicate glasses depends primarily on the outer d valence orbitals,[12] it means that the color and absorption spectra depends on the oxidation state and coordination number of the ion. The temperature sensitivity of the Co-silica gel absorption spectra indicates a shift in oxidation state and coordinate number. The low-temperature gel shows evidence of a sixfold CN similar to that reported for Co in phosphate glasses.[12] In contrast the high-temperature gel appears to have a CN of 4 more equivalent to that of a standard vitreous silicate glass,[12-14] that is,

$$Co^{II}O_6(\text{orange}) \xrightarrow{\Delta T} Co^{II}O_4(\text{blue}) \tag{1}$$

However, none of the spectra for the Co doped silica gels is equivalent to the silicate melt glass spectrum in detail. The primary absorption between 550 and 650 nm is present for the high-temperature gels but other features are also present.

CONCLUSIONS

Thermal history can alter the chemical environment or ligand-field around a transition metal ion in a silica gel and have a marked effect on its optical absorption characteristics and hence on the color produced. Rare earth colorants are much less sensitive to thermal history effects. This is the first effort to show that it may be possible to take advantage of low temperature sol–gel–glass techniques to manufacture optical filters using a silica matrix. The absorption spectra can be shifted by controlling the thermal history of silica gels containing transition elements. The optical components produced will have the unique physical properties of silica, that is, low thermal expansion coefficients, extraordinarily high chemical durability, and superb thermal shock resistance. In addition, depending upon the extent of densification reached during thermal processing the density and index of refraction of the optical component can be varied over wide ranges.[7,11]

ACKNOWLEDGMENTS

The authors are grateful for the financial support of the AFOSR under Contract #F49620-83-0072. They also appreciate the assistance of Dr. Goldberg's group for the Perkin-Elmer model 552 UV-Visible Spectrophotometer.

REFERENCES

1. S. H. Wang and L. L. Hench, 8th Annual Conference on Composites and Advanced Ceramic Materials, Cocoa Beach, Florida, Jan. 15–18, 1984.
2. S. H. Wang and L. L. Hench, Processing and Properties of Sol–Gel Derived 20 mol% Na_2O–80 mol% SiO_2 (2ON) Materials, in *Better Ceramics Through Chemistry*, C. J. Brinker, D. E. Clark, and D. R. Ulrich, Eds., Materials Research Society, Vol. 32, North Holland, New York, 1985, pp. 71–78.
3. M. Prassas and L. L. Hench, Physical Chemical Factors in Sol–Gel Processing, in *Ultrastructure Processing of Ceramics, Glasses and Composites*, L. L. Hench and D. R. Ulrich, Eds., Wiley, New York, 1984, pp. 100–125.
4. S. H. Wang and L. L. Hench, "Drying Control Chemical Additives for Rapid Production of Large Sol–Gel Monoliths Containing Transition and Rare Earth Elements," patent pending.
5. Kingery, *Introduction to Ceramics*, 2nd ed., Wiley New York, 1976.
6. L. H. Van Vlack, *Physical Ceramics for Engineers*, Addison Wesley, Reading, MA, 1964.
7. L. L. Hench, S. H. Wang, and S. C. Park, "SiO_2 Gel Glasses," SPIE's 28th Annual International Technical Symposium on Optics and Electro-Optics Aug. 19–24, 1984, San Diego, California.
8. Jenkins and White, *Fundamentals of Optics*, 3rd ed., McGraw-Hill, New York, 1957.
9. J. Zarzycki, Monolithic Xero- and Aerogels for Gel-Glass Processes, in *Ultrastructure Processing of Ceramics, Glasses and Composites*, L. L. Hench and D. R. Ulrich, Eds., Wiley, New York, 1984, pp. 27–42.
10. J. Phalippou, T. Woignier, and J. Zarzycki, Behavior of Monolithic Silica Aerogels at Temperatures Above 1000°C, in *Ultrastructure Processing of Ceramics, Glasses and Composites*, L. L. Hench and D. R. Ulrich, Eds., Wiley, New York, 1984, pp. 70–87.
11. S. Park and L. L. Hench, "Physical Properties of Partially Densified SiO_2 Gels," to be published, 1985.

12. Foster L. Harding, "The Development of Colors in Glass," Brockway Glass Co., Inc., Brockway, PA, 15824.

13. J. H. Simmons, Private discussion, Department of Materials Science and Engineering, Universite of Florida, Gainesville, FL 32611.

14. P. C. Schultz, Optical Absorption of the Transition Elements in Vitreous Silica, *J. Am. Ceram. Soc.* **57**, 7 (July 1974).

22

GEL-DERIVED GLASSES FOR OPTICAL FIBERS PREPARED FROM ALKOXIDES AND FUMED SILICA

E. M. RABINOVICH, J. B. MacCHESNEY,
AND D. W. JOHNSON, Jr.
AT&T Bell Laboratories
Murray Hill, New Jersey

INTRODUCTION

There are two basic methods of silica gel glass preparation: (1) hydrolysis and polymerization of alkoxides,[1-3] and (2) gelation of sols formed from colloidal oxide dispersions.[4-9] Each of these methods has advantages and disadvantages. Alkoxides provide an easy way to make uniform multicomponent gels and, due to very small pores (2–7 nm), these gels can be sintered at temperatures lower than typically necessary to form glasses. Thus the problem of crystallization can be avoided because sintering takes place much below the temperature range where nucleation and growth of crystals occur. However drying of alkoxide monolithic gels is a difficult problem; it has been solved by way of hypercritical evacuation in an autoclave,[10,11] but this method is cumbersome, requires heavy equipment, and would be expensive if implemented commercially.

On the other hand, the colloidal methods do not have such a severe drying problem. The gels are readily dehydrated but preparation of multicomponent glasses in this way is more difficult. These methods consist of dispersion of colloidal particles in liquids, and these suspensions readily form gels. Shoup[7] prepared gels from dispersed colloidal silica Ludox ®* and sodium or potassium silicates. To reduce the alkali content to <0.02% it was necessary to wash these gels in weak acid or NH_4NO_3 solutions. These gels, when dried, could be sintered to a monolithic transparent glass.

We previously reported[4 – 6,12] a method of preparation of rods and tubes up to 40 cm long and up to 3.8 cm in diameter from colloidal "fumed" silica (Cab-o-Sil®†) with BET surface area near 200 m^2/g. This silica, when dispersed in water (60 g solids per 100 g water), forms a sol capable of gelation in several hours. However this gel also cracked during drying. To prepare large dried and sintered glass articles, the "double processing" method was developed. In this method, Cab-o-Sil® first gelled and then dried forming pieces having practically the same surface area. These pieces were redispersed in water, cast in molds, released upon gelation, and dried on a laboratory table without breaking. The porous bodies (about 0.6 g/cm^3 in bulk density) could be sintered to a transparent glass in He atmosphere at 1260–1500°C. This glass (95–100 wt% SiO_2, 0–5% B_2O_3)) had properties similar to that of commercial fused silica or Vycor®‡ glass. Addition of Cl_2 to the sintering atmosphere allowed removal of OH^- in the glass to <1 ppm by weight.

If fumed silica with a lower surface area is used, larger amounts of solids should be added to water to ensure gelation; still this sol is slow to gel, although the "double processing" can be eliminated in this case. To facilitate gelation of a powder with lower surface area (27–45 m^2/g) and to eliminate the "double processing", Scherer and Loung[8] used organic liquids such as chloroform or n-decanol instead of deionized water.[4 – 6,12] This is reported to result in somewhat simpler processing, but these liquids are much more expensive than deionized water, and necessitate safety precautions (chloroform is a suspected carcinogen[13]) which may significantly increase the cost.

GELS IN OPTICAL FIBER FABRICATION

There is as yet no commercial production of optical fibers by sol–gel methods, but Table 22.1 shows a summary of the intensive research in this field. Data on seven experimental fibers have been published during 1977–1985. Fleming et al.[14] prepared GeO_2—B_2O_3—SiO_2 glass particles (100 to 800 μm in size) by hydrolysis of silicon, boron, and germanium alkoxides and sintering at temperatures between 700° and 1000°C in an oxygen atmosphere. This technique provided an intimate mix of the oxides and prevented loss of volatile constituents (Ge and B) during subsequent melting. A colloidal sol of Cab-o-Sil® doped with $(NH_4)_2B_{10}O_{16} \cdot 8H_2O$ was also employed for the fabrication of B_2O_3—SiO_2 glass powders. Melting used a dry oxygen plasma torch operating at

*Ludox ® is a trademark of E. I. DuPont de Nemours, Inc. Wilmington, Delaware.
†Cab-o-Sil ® is a trademark of Cabot Corp.
‡Vycor ® is a trademark of Corning Glass Works.

TABLE 22.1 Optical Fibers Made of High-Silica Gels

No.	Brief Description of Method	Dopants in		Optical Loss		Authors
		Cladding	Core	Value (dB/km)	Wavelength (µm)	
1	Glass particles prepared from alkoxides and colloidal silica sols. These particles were plasma fused.	B_2O_3	$GeO_2—B_2O_3$	—	—	Fleming et al.[14]
2	Monolithic gel rods were prepared by hydrolysis of $Si(OCH_3)_4$ and used as cores; claddings were deposited by the MCVD process.	B_2O_3	None	6	0.85	Susa, Matsuyama et al.[15]
3				4	0.80	Matsuyama, Susa et al.[16]
4	A whole gel fiber prepared from alkoxides: $Si(OC_2H_5)_4$, $Ge(OC_2H_5)_4$.	None	8% GeO_2	22	0.85	Puyane et al.,[17] Harmer et al.[18]
5	Glass particles were prepared from alkoxides: $Si(OC_2H_5)_4$, $Ge(OC_4H_9)_4$. These particles were melted at ~1800°C.	None	GeO_2	7	0.80	Sudo et al.[19]
6	A substrate tube for ordinary MCVD process was made of colloidal gel.	P_2O_5, F	GeO_2	0.7	1.15	Rabinovich et al.[12]
7	Fluorinated colloidal gel substrate tubes were used to match the cladding index. This allowed to reduce the cladding thickness in four times.	P_2O_5, F	GeO_2	0.28	1.55	MacChesney et al.[20]

atmospheric pressure. Transparent glass boules were formed at 4–8 g/min. The germanium-containing glass was used for the core and B_2O_3—SiO_2 for the cladding in a step index lightguide made by rod-in-tube method. No loss results were reported on fibers made by this technique.

Susa, Matsuyama, et al.[15,16] were the first to report loss results. They prepared alkoxide-derived rods of pure silica and used them to form a core by placing the rod in a tube which had been deposited with a B_2O_3—SiO_2 glass layer by MCVD to provide cladding. Original dry gels had bulk density from 0.55 to 1.2 g/cm³, but gels with higher density, although they formed clear pore-free glass on sintering at 1100°C, tended to foam at higher temperatures. The gels with lower density could be dehydrated in chlorine and sintered in helium without foaming. The authors[16] attributed the foaming to hydroxyl groups and to chlorine; the latter was responsible for foaming at 1800° to 2000°C, the temperature range needed to draw fibers. To reduce the amount of residual chlorine, the gel was treated in oxygen at 1000–1150°C. The process of drying was rather slow, up to 1 week. Two fibers prepared by this method have been reported: the minimal optical loss in one was 6 dB/km and 4 dB/km in the other.

Payane, Harmer, and Gonzalez-Oliver[17,18] reported a method of preparing a gel fiber derived from alkoxides. In this method, a solution is introduced into a vertical silica tube and drained to form a thin film on the inside wall. During a 30-min thermal treatment this layer is gelled, and the next layer, containing a dopant (8 wt% GeO_2), is deposited. The authors[18] compared this process to deposition of successive layers in the MCVD process, although the gel process is essentially cold. No details or conditions of drying were given. Apparently only a small preform could be fabricated by this method because only 500 m of fiber were drawn. The loss of 22 dB/km was attributed to transition metal impurities in $Ge(OC_2H_5)_4$.

Sudo et al.,[19] using a method similar to Fleming et al.,[14] prepared glass particles (100–300 μm in size) by hydrolysis of silicon and germanium ethoxides and subsequent sintering at 1300–1400°C in presence of an inert gas. Glass particles were melted in an oxy-hydrogen flame at $\sim$1800°C forming a transparent glass preform at a rate higher than 5 g/min. This preform was apparently used as a core, because the authors[19] write that a fiber of 125/50 μm in diameter was prepared by the rod-in-tube method. The lowest loss of this fiber was 7 dB/km.

We have reported two fibers prepared by MCVD in substrate tubes derived from colloidal gels.[12,20] In both cases the tubes were prepared from fumed silica Cab-o-Sil ® as described by Rabinovich et al.[12] In the first case reported in Wurzburg in 1983[2,12] the full cladding formed from a mixture of $SiCl_4$, $POCl_3$, CF_2Cl_2, and O_2 was deposited; the thickness of the deposit was $\sim$0.5 mm. A core consisting of SiO_2 and GeO_2 was then deposited and the preform was collapsed as usual. The drawn fiber was 150/8.6 μm in outside/core diameter, had $\Delta n \sim 0.0045$, and showed a minimal loss of 0.7 dB/km. However the gel part was not optically active.

Our second fiber was reported by MacChesney et al.[20] at a 1985 meeting on

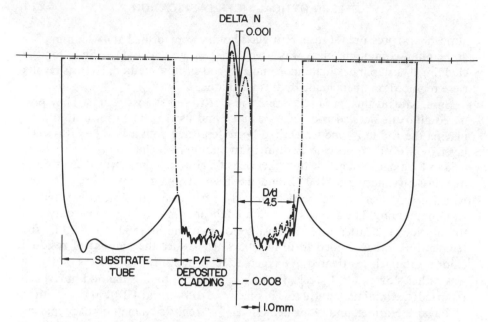

Figure 22.1 Refractive index profile of an MCVD preform for single mode fibers prepared simultaneously: in the fluorinated gel tube (solid curve) and in a commercial silica tube (dash–dot curve).

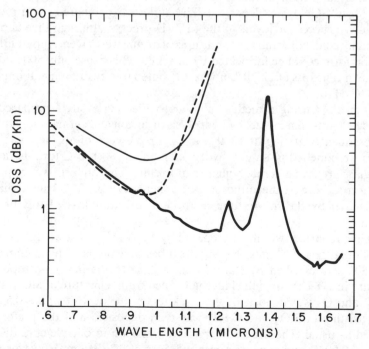

Figure 22.2 The spectral loss curve for the fibers prepared from the preforms described in Fig. 22.1.: fluorinate tube fiber—bold curve, commercial tube fiber, experiment—light curve, theory for fiber wound on 7.5-cm diameter drum—dashed curve.

212

Optical Fiber Communications in San Diego, Ca. The tubes were prepared as described,[12] except that the tube was down-doped by fluorine. Usually fibers based on a decreased cladding index require thick deposits to overcome leaky-mode losses resulting from the higher refractive index of the substrate tube. However such losses have been eliminated by using a fluorinated colloidal gel tube with the index matched to that of the fluorine-doped deposited cladding. In this case the cladding thickness could be reduced to one quarter of that normally necessary.

One of these tubes was joined with a commercial fused silica tube, so the MCVD could be done simultaneously on both. After collapse, profiles of the preforms were determined and are presented in Fig. 22.1, while loss data are shown in Fig. 22.2. As seen, a deposited cladding/core diameter ratio 4.5 gives a loss of 0.28 dB/km at 1.55 μm for the gel tube substrate, but the loss is significantly higher for the same ratio in case of the commercial tube with a nondepressed index. The mechanism of this phenomenon is discussed in detail in our paper.[20] The experimental results are in good agreement with theoretical predictions.[20,21] The same low loss with the commercial tube would have been achieved if a cladding four times thicker had been deposited.

COMBINED ALKOXIDE–COLLOIDAL METHOD

Commercial Cab-o-Sil® is a relatively pure material and, as such, is resistant to crystallization; this explains the success at sintering it to a transparent glass at 1300–1500°C, that is, at temperatures of possible crystallization. However, it is not pure enough for low loss optical fibers. It contains several ppm Fe and Cr, and fiber technology requires these metals to be on the ppb level. Another problem with fumed silica is the difficulty of doping because any salt introduced into the water used for dispersion will form a layer on the silica particle. If this layer does not form an amorphous oxide layer during calcination (as B_2O_3, P_2O_5, GeO_2) it will cause unavoidable crystallization during sintering. Not many elements can be incorporated in fumed silica's structure during processing, fluorine is a fortunate exception.

Powders from other sources are widely used in glass and ceramic processing.[22,23] Johnson[22] reviewed different techniques of powder preparation including gel methods. Scherer[9] prepared flame-generated particles of oxide powders. Fleming et al.[14] and Sudo et al.,[19] as described above, used similar methods for preparation of powders for melting. We also employed hydrolysis of alkoxides for the preparation of silica particles which could be substituted for Cab-o-Sil® in our process. A brief chart of the process is given in Fig. 22.3. Glass specimens prepared by this method from $Si(OC_2H_5)_4$ are shown in Fig. 22.4. They are of much better quality than the fumed silica samples with respect to bubbles and other defects. They could be easily dried to form large bodies and fluorinated. As in case of Cab-o-Sil®, the fluorination results in reduction of the refractive index by as much as 0.005 compared with pure silica. Sintering

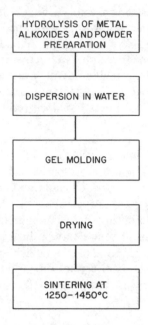

Figure 22.3   Schematic for the combined alkoxide–colloidal method of glass preparation.

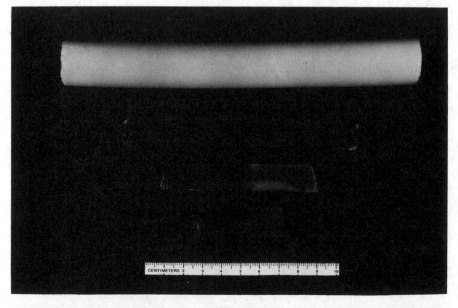

Figure 22.4   Specimens of silica dry gel and glass prepared by combined alkoxide–colloidal method.

at 1300–1450°C in He with dehydration in Cl_2 results in glass with undetectable amounts of OH^-. The first 150 m of fiber have been drawn from one of the rods (15 mm in diameter) prepared by this method. We expect that this glass, if prepared in clean conditions, can serve as an optically active part in the fibers.

SUMMARY

Gel methods of glass preparation have started to make their way into fiber optics technology. Two possible applications can be considered: gel glass as optically inactive material (substrate tubes) and as optically active core and cladding. The tubes can be also used for any other applications of fused silica and Vycor® glasses. We have shown that fluoridation of gel substrate tubes for MCVD permits a fourfold reduction of the deposited cladding thickness.

ACKNOWLEDGEMENTS

We are grateful to K. A. Jackson and S. R. Nagel for reading the manuscript and their valuable comments.

REFERENCES

1. S. Sakka, Gel Method of Making Glass, in *Treatise on Materials Science and Technology*, Vol. 22, *Glass III*, M. Tomozawa and R. H. Doremus, Eds., Academic Press, New York, 1982, pp. 129–167.

2. Glasses and Glass Ceramics for Gels, Proc. 2nd Internat. Workshop, H. Scholze, Ed., *J. Non-Cryst.* **48**(1–2) (1984).

3. Materials Research Society Symposia Proc., Vol. 32, *Better Ceramics Through Chemistry*, C. J. Brinker, D. E. Clark, and D. R. Ulrich, Eds., North-Holland, New York, (1984).

4. E. M. Rabinovich, D. W. Johnson, Jr., J. B. MacChesney, and E. M. Vogel, Preparation of Transparent High-Silica Glass Articles from Colloidal Gels, *J. Non-Cryst. Solids* **47**(3), 435–439 (1982).

5. E. M. Rabinovich, D. W. Johnson, Jr., J. B. MacChesney, and E. M. Vogel, Preparation of High-Silica Glasses from Colloidal Gels: I, Preparation for Sintering and Properties of Sintered Glasses, *J. Am. Ceram. Soc.* **66**(10), 683–688 (1983).

6. D. W. Johnson, Jr., E. M. Rabinovich, J. B. MacChesney, and E. M. Vogel, Preparation of High-Silica Glasses from Colloidal Gels: II, Sintering, Ref. 5, 688–693.

7. R. D. Shoup, Controlled Pore Silica Bodies Gelled from Silica Sol-Alkali Silicate Mixtures, in *Colloid and Interface Science*, Vol. III, M. Kerker, Ed., Academic Press, New York, 1976, pp. 63–69.

8. G. W. Scherer and J. C. Luong, Glasses from Colloids, see Ref. 2, pp. 163–172.

9. G. W. Scherer, Glasses and Ceramics from Colloids, see Ref. 3, pp. 205–211.

10. S. Henning and L. Svensson, Production of Silica Aerogel, *Phys. Scripta* **23**, 697–702 (1981).

11. M. Prassas, J. Phalippou, and J. Zarzycki, *J. Mater. Sci.* **19**, 1656–1665) (1984).

12. E. M. Rabinovich, J. B. MacChesney, D. W. Johnson, Jr., J. R. Simpson, B. W. Meagher, F. V. DiMarcello, D. L. Wood, and E. A. Sigety, Sol-Gel Preparation of Transparent Silica Glass, see Ref. 2, pp. 155–161.

13. *Casarett and Doull's Toxicology*, 2nd ed., J. Doull, C. D. Klaassen and M. O. Amdur, Eds. Macmillan, New York, 1980, p. 471.

14. J. W. Fleming, R. E. Jaeger, and T. J. Miller, GeO_2–B_2O_3–SiO_2 Optical Glass and Light-guides, U.S. Pat. 4,011,006 (1977).

15. K. Susa, I. Matsuyama, S. Satoh, and T. Suganuma, New Optical Fiber Fabrication Method, *Electron. Lett.* **18**(12), 499 (1982).

16. I. Matsuyama, K. Susa, S. Satoh, and T. Suganuma, Syntheses of High-Purity Silica Glass by the Sol-Gel Method, *Ceram. Bull.* **63**(11), 1408 (1984).

17. R. Puyane, A. L. Harmer, and C. J. R. Gonzalez-Oliver, Optical Fibre Fabrication by the Sol-Gel Method, European Conf. on Optic. Communication, Cannes, 21–24 Sept. 1982, p. 623, Sitecmo Dieppe Paris.

18. A. L. Harmer, R. Puyane, and C. Gonzalez-Oliver, The Sol-Gel Method for Optical Fiber Fabrication, *IFOC*, p. 40 (Nov./Dec. 1982).

19. S. Sudo, M. Nakahara and N. Inagaki, A Novel High-Rate Fabrication Process for Optical Fiber Preforms, 4th Intern. Conf. on Integrated Optics and Optical Fiber Communication, Main Conf. Techn. Digest, paper 27A3-4, Tokyo, Japan (June 27–30, 1983).

20. J. B. MacChesney, D. W. Johnson, Jr., P. J. Lemaire, L. G. Cohen, and E. M. Rabinovich, Fluorosilicate Tubes to Eliminate Leaky-Mode Losses in MCVD Single Mode Fibers with Depressed Index Cladding, Conference on Optical Fiber Communication, OFC/OFS 185 (February 11–13, 1985), Techn. Digest, San Diego, California, paper WH2, pp. 98–100.

21. L. G. Cohen, D. Marcus, and W. L. Mammel, *IEEE J. Quantum Electron.*, **QE-18**, 1467–1472 (1982).

22. D. W. Johnson, Nonconventional Powder Preparation Techniques, *Ceram. Bull.* **60**(2), 221 (1981).

23. K. S. Mazdiyasni, Chemical Synthesis of Single and Mixed Phase Oxide Ceramics, see Ref. 3, p. 175–186.

23

STRONGER GLASS VIA SOL–GEL COATINGS

B. D. FABES, W. F. DOYLE, L. S. SILVERMAN,
B. J. J. ZELINSKI, AND D. R. UHLMANN
Department of Materials Science and Engineering
Massachusetts Institute of Technology
Cambridge, Massachusetts

INTRODUCTION

It is widely accepted that the strengths of glass bodies are limited by the presence of microflaws in their surfaces. These flaws may be introduced during processing, handling, or even simple exposure to atmospheric contaminants. For an applied stress σ_a, the stress σ_m at the tip of a flaw of radius ρ and length c is

$$\sigma_m \approx 2 \left(\frac{c}{\rho} \right)^{1/2} \sigma_a \qquad (1)$$

The present approach to increasing the strength of glass bodies is based on the use of sol–gel coatings which can be applied in the form of low-viscosity solutions to glass surfaces and thereby fill in the microflaws, and which on firing will form primary chemical bonds to the glass. The coatings should further have a higher strength than the glass, and also a higher hardness (so that the effects of subsequent abrasion can be reduced).

The high-temperature ammonolysis of porous gels has been shown to produce oxynitride glasses with attractive properties.[1-4] While the nitridation of silica gels does not take place until about 1000°C,[1] Brinker et al.[2-4] showed

217

that the temperature of nitrogen incorporation could be lowered to 600°C with the addition of boron and other 3^+ cations, which act as preferential sites for the nucleophilic attack of the ammonia. The preparation of high boron content gels offers the possibility of producing oxynitride glass coatings at low temperatures. This chapter reports an initial investigation of such coatings applied to fused silica substrates.

EXPERIMENTAL PROCEDURE

Borosilicate gels were made by partially hydrolyzing tetraethyl orthosilicate (TEOS) under acidic conditions (pH = 2.5), allowing the TEOS to react for a predetermined amount of time, adding trimethyl borate (TMB), mixing the solution for 1 hr, and then adding the rest of the water to complete the stoichiometric hydrolysis of the TEOS. To maximize the amount of boron remaining in the final coatings, the effects of solvent and the time before borate addition were investigated by coating 0.5 mm n-type silicon wafers with various solutions. These coatings were then densified at 800°C for 2 hr, and their compositions examined by comparing the 1340 cm^{-1} B—O vibration with the 1070 cm^{-1} Si—O vibration.

For mechanical testing, polished silica slides (25 mm × 75 mm × 1 mm) were preflawed with a 2.5 cm long scratch, made with a 100-gr Vicker's diamond. The slides were dip coated in the gel solutions and fired at 650°C for 2 hr in either ammonia or air, followed by densification under N_2 at 1000°C. Coating thickness was measured with a double-beam interferometer before firing; strength was determined with the flaw side in tension under four-point bending in an Instron machine.

RESULTS

Figure 23.1 shows the ratio of the boron to silicon peaks as a function of the time of boron addition and the solvent composition. The maximum boron retention was obtained when the boron was added 10 min after the TEOS was partially hydrolyzed and when the coating solution contained 25 mL of THF. The incorporation of nitrogen in the final ammonia-fired coatings is confirmed by difference spectra, where the B—N vibration at[3] 1510 cm^{-1} has appeared. Coating thicknesses were found to average 1200 Å. SEM micrographs (Fig. 23.2) show that these coatings do indeed help fill in the flaws, and survive uncracked over the surface of the glass. Modulus of rupture (MOR) data are plotted in Fig. 23.3, where the controlled flaw is observed to decrease the strength of the silica slides by just under a factor of 2. All subsequent treatments increased the strength to very near the initial (unflawed) value. Fracture occurred without exception along the scratch (flaw) for untreated samples, and rarely for treated specimens.

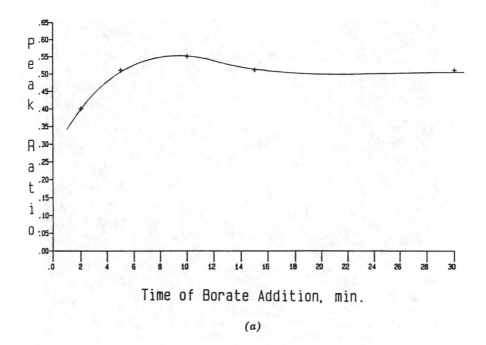

Time of Borate Addition, min.

(a)

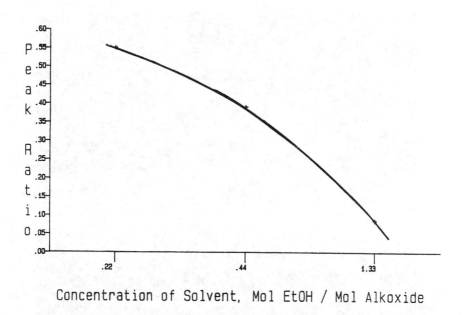

Concentration of Solvent, Mol EtOH / Mol Alkoxide

(b)

Figure 23.1 B–O to SiO IR peak ratios versus (a) time of borate addition and (b) concentration of solvent including ethanol evolved during hydrolysis. ● indicates the effect of adding 0.25 mole THF to aid in mixing.

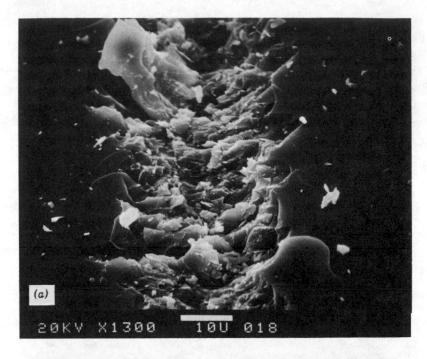

20KV X1300   10U 018

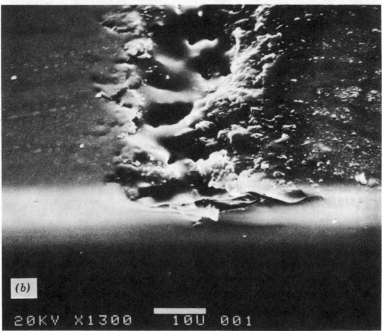

20KV X1300   10U 001

Figure 23.2 Scanning electron micrograph of (*a*) freshly scratched SiO$_2$ slide and (*b*) scratched, coated, and fired slide.

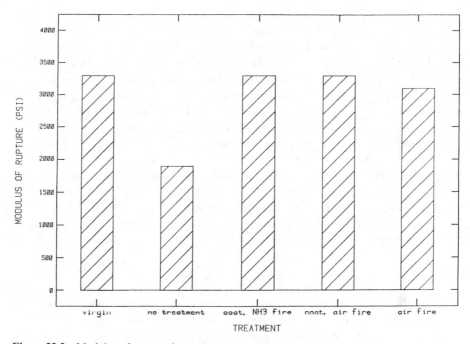

Figure 23.3 Modulus of rupture for various treatments, where "virgin" refers to the as-received quartz slides, "no treatment" to slides scratched with the Vicker's Diamond, "coat, NH₃ fire" to slides scratched, coated with borosilicate gel and fired under ammonia, "coat, air fire" to slides scratched, coated with borosilicate gel and fired in air, and "air fire" to slides scratched and fired in air.

DISCUSSION

The boron content in borosilicate gels has been largely limited by two problems, the slow hydrolysis kinetics of TEOS and the volatility of TMB. The slow hydrolysis of TEOS has traditionally been compensated by partially hydrolyzing the TEOS before the other components are added to the solution. Attention must be paid, however, to the point at which this second species is added. Addition too soon may result in the inhomogeneous formation of boric acid, $B(OH)_3$, and addition too late may result in fewer OH groups available for boron–silicon condensations. In this work, the retained boron was maximal when added 10 min after hydrolysis commenced, with the most dramatic effect seen at short times Fig. 23.1a.

The problem of borate volatility is exacerbated by the reversibility of the hydrolysis reaction in the acidic conditions necessary for the production of high surface area gels. This reesterification

$$(OR)_2B(OH) + ROH \rightarrow B(OR)_3 + H_2O \qquad (2)$$

results in the reformation of the volatile borate. It is evident, however, that minimizing the use of alcoholic solvents should retard this effect (Fig. 23.1b).

Comparison of the MOR data on thermally annealed and oxynitride coated samples with data on the unscratched specimens suggests that in all cases failure occurred in the substrate, as in all cases the minimum critical flaw size is larger than the coating thickness. The similarity of these data with those for the un-nitrided coatings supports this suggestion, as a change in the coating properties showed no effect on the MOR. The SEM micrographs, showing no cracks in the coating before or after fracture, combined with our inability to scratch the coating off the slide after firing, indicate that primary chemical bonds have formed between the coating and the substrate. The lack of cracking also implies that the thermal stress imposed by the differences in expansion coefficients $(\alpha_{coating} - \alpha_{SiO_2} \sim 4 \times 10^{-6\circ}C^{-1})$ does not result in failure of the coating. If the elastic properties of the coating and substrate are assumed to be similar, the thermal stress can be approximated by

$$\sigma = E(T - T_0)(\alpha_{coating} - \alpha_{SiO_2})(1 - 3_j + 6_j^2) \qquad (3)$$

where E is Young's modulus, $T - T_0$ is the difference between ambient temperature and the temperature at which the coating can no longer relax, and j is the ratio of coating to substrate thickness. Assuming that relaxation effectively ceases about 100°C below T_g, and taking E as 10×10^6 psi Eq. (3) indicates a thermal stress of about 22 ksi in the coating. When the additional 10-ksi bending stress is added, we conclude that, without any special handling techniques, coatings are produced with strengths exceeding 30 ksi.

CONCLUSIONS

Simple kinetic and thermodynamic considerations, such as the rate of hydrolysis and the equilibrium conditions of reesterification, can be used to optimize the composition and properties of borosilicate gel coatings. The full potential of these coatings has yet to be determined; but strengths above those commonly developed after processing have been demonstrated. Exploration of greater coating thicknesses and a broader range of firing conditions is indicated, as is determination of the effectiveness of oxynitride coatings in resisting abrasion and fatigue.

ACKNOWLEDGMENTS

Financial support for the present work was provided by the Emhart Corporation, Rogers Corporation, and the Air Force Office of Scientific Research. This support is gratefully acknowledged.

REFERENCES

1. D. N. Coon, J. G. Rapp, R. C. Bradt, and C. G. Pantano, Mechanical Properties of Silicon-Oxynitride Glasses, *J. Non-Cryst. Solids* **56** (1–3), (1983).

2. D. M. Haaland and C. J. Brinker, In Situ FT-IR Studies of Oxide and Oxynitride Sol-Gel Derived Thin Films, *Mater. Res. Soc. Symp.* **32** (1984).

3. C. J. Brinker, Formation of Oxynitride Glasses by Ammonolysis of Gels, *J. Am. Ceram. Soc.* **65** (1), (1982).

4. C. J. Brinker, and D. M. Haaland, Oxynitride Glass Formation from Gels, *J. Am. Ceram. Soc.* 66 (11), (1983).

24

PROCESSING AND STRUCTURAL PROPERTIES OF Li$_2$O—Al$_2$O$_3$—TiO$_2$—SiO$_2$ GELS

GERARD ORCEL AND LARRY L. HENCH
Ceramics Division
Department of Materials Science and Engineering
University of Florida
Gainesville, Florida

INTRODUCTION

Over the past several years, there has been an increasing interest in the gel route for the preparation of glasses and ceramics. Even though different methods can be used for the manufacture of gels,[1] the difficulties of preparing samples on a commercial basis are not yet overcome. Monolithic gels larger than 10 cm can be produced, but with considerable precautions, which are generally costly. The price of the raw materials limits the possible commercial exploitations to very special devices or applications, such as coatings, optical fibers, and electronic materials.[2] The chief advantages of the sol–gel route for these applications are very high purity and homogeneity on a molecular scale. Knowing this, it is surprising that very little attention had been paid to glass–ceramics. The main efforts in this domain have been to achieve the preparation of zero thermal expansion coefficient materials.[3-5]

The purposes of this preliminary study were: (1) to demonstrate the possibility of producing monolithic gels in the system SiO$_2$—Al$_2$O$_3$—TiO$_2$—Li$_2$O using drying control chemical additives (DCCA)s,[6-8] and (2) to study the evolution of the structure of the gels with temperature.

PROCEDURE

The composition studied was: 70 mole% SiO_2–19 mole% Al_2O_3–6 mole% TiO_2–5 mole% Li_2O. This composition potentially can yield a zero thermal expansion coefficient glass–ceramic by crystallization of phases with negative expansion coefficients. The procedure followed for the manufacture of the gels is shown schematically in Fig. 24.1. The sols were cast in Teflon® containers and aged in sealed vials at room temperature. A photo of a monolithic gel produced by this method is given in Fig. 24.2.

The surface area, pore volume, and average pore radius were determined by the nitrogen adsorption–desorption method, using an automated Quanta-chrome Autosorb-6 sorption system. The infrared spectra were recorded on a Nicolet MX-1 FTIR Spectrometer. A hot-stage diffusion reflection spectroscopy system was used for in-situ analysis of the crystallization of the gels. X-ray diffraction (XRD) analysis was obtained using Philips spectrometers, one with hot stage capabilities.

$$CH_3OH + H_2NCHO$$

$$\downarrow \quad 5'$$

$$Si(OCH_3)_4 + H_2O$$

$$\downarrow \quad 30'$$

$$Ti(OC_3H_7)_4 + H_2O$$

$$\downarrow \quad 30'$$

$$Al(OC_4H_9)_3$$

$$\downarrow \quad 30'$$

$$Li(OCH_3)$$

$$\downarrow \quad \text{heat to } 60°C \text{ for } 30'$$

$$H_2O$$

Figure 24.1 Schematic diagram of the sol–gel process for preparing 70S-19A-6T-5L monoliths.

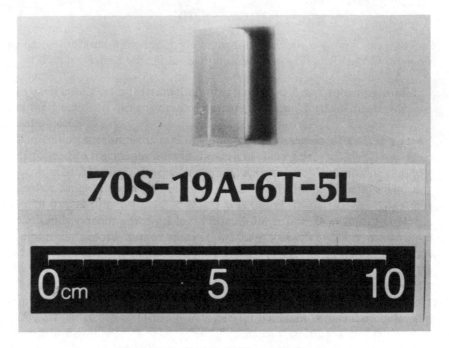

Figure 24.2   70S-19A-6T-5L monolithic gel.

RESULTS

Structure

It is well known that a large number of variables affect the pore characteristics of gels.[9] This work was limited to studying the influence of gelling temperature, aging time and temperature. Three sols, GT 0, GT 25, and GT 80 were prepared according to the procedure outlined in Fig. 24.1, and were allowed to gel at 0°C, 25°C, and 80°C, respectively. After 8 hr at the gelling temperature (which is slightly more than the time required for GT 0 to gelate) the samples were analyzed. The results are given in Table 24.1. The differences between the pore characteristics are rather small. Thus, when the gelation time is short, the gelling temperature is not a determinant factor.

Three aging temperatures (0°C, 25°C, 90°C) and three aging times (1 day, 1 week, 1 month) were also investigated. The sample code is 'A' followed by the aging temperature and then by a letter which indicates the aging time (A 25 W was aged at 25°C for one week). The results are reported in Table 24.2. At 0°C aging, no time dependence is observed while a slight change (about 30%) in the pore volume and the surface area can be seen at room temperature after one month, the pore diameter remaining constant. For the aging at 90°C, the changes in properties with time are more drastic. The pore volume and the pore size increased by 3.5 times, after one month.

TABLE 24.1 Structural Characteristics (Surface Area, SA; Pore Volume, PV; and Pore Radius, PR) as a Function of Gelation Temperature

	GT 0	GT 25	GT 80
SA (m^2/g)	406	374	414
PV (cm^3/g)	0.37	0.29	0.37
PR (Å)	18	15	18

TABLE 24.2 Influence of the Aging Time and Temperature on the Surface Area (SA), Pore Volume (PV), Pore Radius (PR)

	A 0 D	A 0 W	A 0 M	A 25 D	A 25 W
SA (m^2/g)	330	334	308	314	313
PV (cm^3/g)	0.25	0.25	0.22	0.23	0.23
PR (Å)	15	15	14	15	14

	A 25 M	A 80 D	A 80 W	A 80 M
SA	414	416	499	439
PV	0.30	0.47	0.97	1.74
PR	14	39	39	80

The mechanism involved during aging is a dissolution of the small particles which redeposit on the larger ones. This process is thermally activated, evidently is negligible at 0°C, and takes several weeks to produce a change in pore volume at 25°C. Since the average pore size does not vary with time at 25°C, the particles which dissolve at this temperature should be in the 14-Å range. At 90°C this process is more active and larger particles can dissolve. This may explain the increase in the pore diameter after short aging times. If the gel is allowed to remain in its liquor for longer periods of time, then the consolidation process of the silicate network becomes more effective and less shrinkage takes place. Much higher pore volumes and larger pore radii result.

FTIR SPECTROSCOPY

The possibility of using *in situ* FTIR analysis to follow the crystallization of gels is very attractive. The short times required for the collection of data (1 min), compared with the usual XRD methods, makes it possible the study of the kinetics of crystallization. Figure 24.3 shows the FTIR spectra recorded for the heat treatments given in the insert. The asterisks indicate when the spectra were taken. The usual silicate network vibrational modes at about $1000 \, cm^{-1}$ decrease during crystallization. Meanwhile, the peaks at 840, 770, 580, and 464 cm^{-1} grow in intensity. The band at $840 \, cm^{-1}$ can be assigned to the stretching Si—O—Al mode, where Al is in tetrahedral sites. This mode is accompanied by bands at 770 and $580 \, cm^{-1}$. The band at $464 \, cm^{-1}$ is the usual silica rocking vibration.[10] From Fig. 24.3 we deduce that the onset of crystallization for the gel is between 200 and 400°C. Isothermal runs showed that the gel crystallizes at a noticeable rate between 375 and 400°C.

Information on the volume of the crystalline phase can be obtained by plotting the intensity of a vibration mode characteristic of the crystalline phase as a

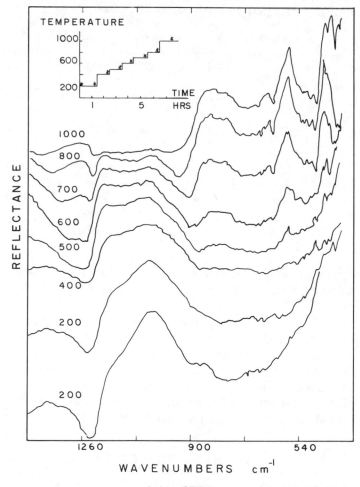

Figure 24.3 Temperature variation of FTIR spectra of the 70S-19A-6T-5L gel.

TABLE 24.3   Variation with Time of the Intensity of
the 580 cm^{-1} FTIR Band at 400°C

Time (hr)	0	0.5	1	2	3
Intensity	0.38	0.40	0.45	0.53	0.53

function of time. Table 24.3 reports such data for the gels heated at 400°C. It is not possible to compare intensities obtained at different temperatures due to the change in reflectivity of the surface of the gels with temperature.

X-RAY DIFFRACTION

We studied the evolution of the structure between 400 and 1000°C. As shown in Fig. 24.4, there is no change of the crystalline phase, synthetic β-spodumene ($Li_{0.6}Al_{0.6}Si_{2.4}O_6$), as a function of crystallization temperature. The changes are only a decrease in the amorphous scattering and an increase of the crystalline peaks as the temperature increases. The β-spodumene phase, which is the equilibrium phase for this gel composition,[11] is very desirable since it has a small thermal expansion coefficient.[12]

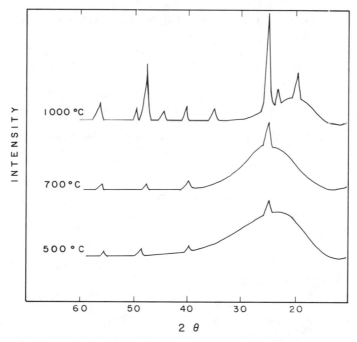

Figure 24.4   Variation of X-ray diffraction spectra with temperature.

CONCLUSION

Monolithic gels of the composition 70 mole% SiO$_2$–19 mole% Al$_2$O$_3$–6 mole% TiO$_2$–5 mole% Li$_2$O can be prepared using formamide as a DCCA. The pore size, pore volume, and surface area can be controlled through the aging process. It is possible to follow the crystallization of the gels using hot-stage FTIR and X-ray diffraction. The stable crystalline phase is β-spodumene which appears between 375 and 400°C.

ACKNOWLEDGMENT

The authors are grateful for the financial support of AFOSR contract # F49620-83-C-0072 and the assistance of Guy LaTorre (FTIR) and Dr. Stanley Bates and Wayne Acree (XRD).

REFERENCES

1. L. L. Hench and D. R. Ulrich, Eds., *Ultrastructure Processing of Ceramics, Glasses and Composites*, Wiley, New York, 1984.

2. C. J. Brinker, D. E. Clark, and D. R. Ulrich, Eds., *Better Ceramics Through Chemistry*, Materials Research Society, Vol. 32, North Holland, New York, 1985.

3. H. Dislich, *Angew. Chem. Internat. Edition* **10**(6), 363 (1971).

4. R. Roy, D. K. Agrawal, J. Alamo, and R. A. Roy, *Mater. Res. Bull.* **19**, 471 (1984).

5. B. E. Yoldas, *J. Non-Cryst. Solids* **38** and **39**, 81 (1980).

6. G. Orcel and L. L. Hench, in *Better Ceramics Through Chemistry*, C. J. Brinker, D. E. Clark, and D. R. Ulrich, Eds., Materials Research Society, Vol. 32, North Holland, New York, 1985, p. 79.

7. S. Wallace and L. L. Hench, in *Better Ceramics Through Chemistry*, C. J. Brinker, D. E. Clark, and D. R. Ulrich, Eds., Materials Research Society, Vol. 32, North Holland, New York, 1985, p. 47.

8. S. H. Wang and L. L. Hench, in *Better Ceramics Through Chemistry*, C. J. Brinker, D. E. Clark, and D. R. Ulrich, Eds., Materials Research Society, Vol. 32, North Holland, New York, 1985, p. 71.

9. M. Prassas, Ph.D. Thesis (1981), Montpellier France.

10. V. Stubican and R. Roy, *Am. Mineral.* **46**, 32 (1961).

11. R. A. Eppler, *J. Am. Ceram. Soc.* **46**, (1963).

12. P. W. McMillan, *Glass Ceramics*, 2nd ed., Academic Press, London, 1979.

25

SILICON CARBIDE/SILICA GEL MATRIX COMPOSITES

BURT I. LEE AND LARRY L. HENCH
Ceramics Division
Department of Materials Science and Engineering
University of Florida
Gainesville, Florida

INTRODUCTION

The need for tough and stable composites containing silicon carbide (SiC) for high-temperature structural applications is increasing. In the past, SiO_2/SiC composites were prepared by hot pressing SiO_2 powder with SiC fibers, particulates, whiskers, and so on. Composite materials of a Vycor®* (96% SiO_2) glass matrix reinforced with SiC fiber, prepared by Prewo et al.[1,2] using hot pressing, exhibit high toughness. However, hot pressing has limits on size, shape, and complexity of parts.

Encouraged by the progress of using sol–gel techniques to produce large monolithic silica parts,[3,4] it was decided to apply the same methods for producing composites to attempt to overcome the processing limitations of hot pressing. This chapter describes: (1) the processing of SiC/SiO_2 composites via sol–gel techniques, and (2) characteristics of composite products in their initial developmental stages. Drying control chemical additives (DCCAs)[3,4] were used to avoid cracking during processing.

---

*Vycor® is the registered trademark of Corning Glass Works.

EXPERIMENTAL

Composites of β-SiC and α-SiC in a pure silica gel matrix were prepared using Nicalon®* and Silar™† as the reinforcing filler phases.

Chopped fibers of Nicalon were pretreated to remove a polyvinyl acetate coating by successive washing in ethyl acetate, benzene, and acetone followed by firing at 500°C in air for 2 hr. The cleaned Nicalon® was crushed in a polypropylene container containing alumina balls, on a vibratory mill for 1 hr. In addition, continuous fibers and weaves of Nicalon® were cut into the sizes of molds and the polymer coating was removed by burning it on a propane burner. Silar™ SiC was heated in air at 900°C for 1 hr to obtain monolayer oxidation.

Silica sol was prepared by hydrolyzing tetraethyl orthosilicate (TEOS) with HCl in ethanol as the solvent in the mole ratio of 1:4:0.5:0.06, TEOS:water: alcohol:HCl. Ten grams of SiC was mixed with 75 ml of the silica sol containing 1 to 10 mL of glycerol and/or 1–10 mL of formamide as drying control chemical additives (DCCAs) for 1–30 min followed by ultrasonication before casting in polystyrene molds of various shapes and sizes. An organic surfactant was added to improve wettability of the SiO_2 sol on the SiC. The various configurations of the composites made in this manner are shown in Fig. 25.1.

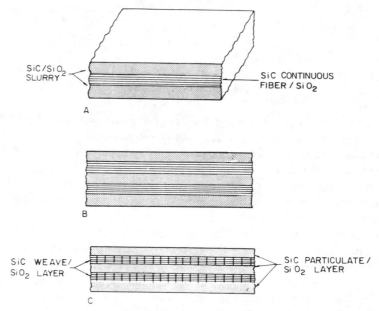

Figure 25.1 Cross sections of some Nicalon® SiC/SiO$_2$ composites prepared by sol–sel processing. (a) Continuous Nicalon fiber sandwiched unidirectionally. (b) Multilayer continuous fiber sandwich. (c) Multilayer of bidirectional weaves.

*β-SiC by Nippon Carbon Co., distributed through Dow Corning, Midland, MI.
†α-SiC by ARCO, Greer, SC.

The SiC/SiO$_2$ sol mixture was aged in an oven at 40–80°C for ~10 hr before drying in air for 5 hr at 90°C. The dried green composite bodies were impregnated with the SiO$_2$ sol for four impregnation-drying cycles. Densities of the composite bodies were measured by mercury volume displacement after each heat treatment.

Three point flexural strengths were determined by using an Instron Testing machine and the compressive modulus was measured with an MTS machine. Microhardness was measured with a Kentron Tester using a 100-g load. Porosity was measured using mercury porosimetry.

RESULTS

Porosity analyses gave ~10% open porosity for a composite sample fired at 1200°C. The densities and flexural strengths for Nicalon® composites after each treatment are given in Table 25.1. Silar™ composites under similar conditions showed lower densities and strengths.

The volume percent SiC in the resulting composite bodies was estimated from SEM micrographs to be ~60%, which is close to the intended proportion.

Heating of the composites up to 1500°C in air showed no visible change in shape or physical state. However heating composite bodies in vacuum ($\sim 10^{-2}$ torr) at 1400°C resulted in a disintegration of the composite body and an evolved gas which may have been H$_2$S.

Hardness and modulus values are given in Table 25.2.

TABLE 25.1 Density and Flexural Strength of SiC/SiO$_2$ Composites of Nicalon® After Heat Treatments in N$_2$

Temperature, °C	ρ (g/cm^3)	σ_f (psi)
80	1.6	3500
500	1.8	4000
900	2.0	4500
1100	2.1	5000

TABLE 25.2 Other Physical Properties of Nicalon®/SiO$_2$ Composites

Highest Density Obtained	Mean Green Density	Vicker Hardness	Compressive Modulus
2.2 g/mL	1.6 g/mL	360 DPN	4×10^6 psi

DISCUSSION

The high microporosity is due to the large inherent shrinkage in the silica gel matrix. The flexural strength was related to the density of the composite body. The lower density and lower strengths of the Silar™ system indicate that the SiO_2 sol is less compatible with α-SiC Silar™ than β-SiC Nicalon® (Fig. 25.2). This may be because of the amorphous nature and large amount of SiO_2 and graphite present in Nicalon®.[6]

Although density and strength can be improved by multiple impregnation of SiO_2 sol into the micropores of the composite bodies, there is a limit to the improvement since the impregnation is confined to the surface region after ~4 cycles. This problem may be rectified by vacuum infiltration but has not been successful thus far. Impregnation by silane SiC precursors is expected to improve density, strength, and toughness.[4]

Densification of the SiO_2 sol/SiC composites by heat treatment is another major problem. The higher the green density, the easier the densification by heat treatments. SEM micrographs of a polished surface of a Nicalon/SiO_2 sample (Fig. 25.3a) shows that 1200°C may be sufficient to achieve sintering.

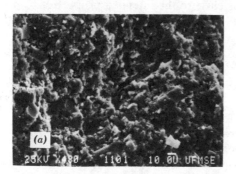 

Figure 25.2 SEM micrographs of the fractured composite surfaces. (a) Nicalon/SiO_2 after heat treatment at 1000°C in air for 2 hr. (b) Silar/SiO_2 after same heat treatment as (a).

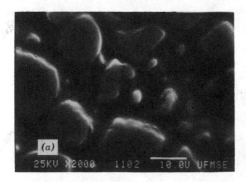 

Figure 25.3 SEM micrographs of polished surfaces. (a) Nicalon/SiO_2 after heat treatment at 1200°C. (b) Silar/SiO_2 after heat treatment at 600°C.

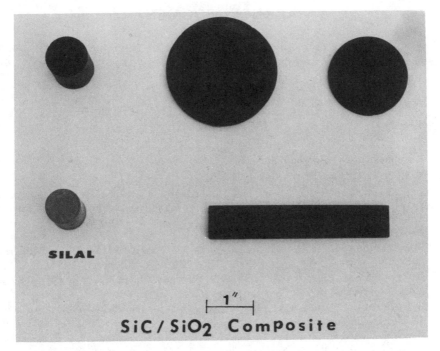

Figure 25.4 Gel-derived SiO_2 matrix SiC composites of various shapes after exposure to air at 1000°C for 2 hr.

A positive result is the oxidation resistance of the composites, which change little upon exposure to air, (Fig. 25.4). The oxidation kinetics of SiC in the SiO_2 matrix ($\rho = 1.9$ g/mL) after heating in air at 1100°C is three to five times slower than the pure SiC under the same conditions.[3] However due to the residual porosity, complete protection of SiC from oxidation in the matrix was not seen.

The disintegration of SiC/SiO_2 composite bodies observed in vacuum ($\sim 10^{-2}$ torr) at 1400°C may be due to the reduction reaction of SiO_2 as shown below.[7] $2SiO_2(s) + SiC(s) \rightleftarrows 3SiO(g) + CO(g)$

This will certainly limit the use of SiC/SiO_2 composites in vacuum at high temperatures.

CONCLUSION

Low-density SiC/SiO_2 composites can be prepared easily by a sol–gel method. Complex shapes can be easily cast using the SiC dispersed in the SiO_2 sol.

Stability at high temperature in oxygen, and moderate mechanical properties have been demonstrated. For the composites, stability in vacuum at high temperatures is a problem as is achieving high densities by this method.

ACKNOWLEDGMENT

The authors are grateful to the financial support from the U.S. Air Force Office of Scientific Research under contract AFOSR #F49620-83-C-0072 for this work.

REFERENCES

1. K. M. Prewo and J. Brennan, *J. Mater. Sci.* **17**, 1201 (1982).

2. K. M. Prewo, *J. Mater. Sci.* **17**, 3549 (1982).

3. S. H. Wang and L. L. Hench, Processing and Properties of Sol-Gel 20 Mol % Na_2O–80 Mol % SiO_2 (20N) Materials, in *Better Ceramics Through Chemistry*, Materials Research Society Symposium Proceedings, Vol. 32, J. Brinker, D. E. Clark, and D. Ulrich, Eds., North-Holland, New York, 1984, p. 71.

4. S. Wallace and L. L. Hench, The Processing and Characterization of DCCA Modified Gel-Derived Silica, Ref. 3, p. 47.

5. B. I. Lee and L. L. Hench, Oxidation Kinetics of Silicon Carbides in Sol-Gel Derived Silica Matrix, to be published.

6. B. I. Lee and L. L. Hench, Mechanical Properties of Silane Impregnated Gel Matrix SiO_2/SiC Composites, work in progress.

7. N. A. Toropov and V. P. Barzakovskii, *High-Temperature Chemistry of Silicates and Other Oxide Systems*, Consultants Bureau, New York, 1966.

26

SOL–GEL DERIVED CERAMIC MATRIX COMPOSITES

D. E. CLARK
Department of Materials Science and Engineering
University of Florida
Gainesville, Florida

INTRODUCTION

Ceramics are generally resistant to high-temperature degradation and hence provide a unique advantage in many applications over other materials. However, a major disadvantage of ceramics, and a primary reason for not using them as structural components, is their tendency to fracture in a brittle fashion. Moreover, failure often occurs without prior warning (i.e., lack of toughness). What is needed is a material that is both tough and resistant to high temperatures. Ceramic composites potentially can meet these requirements.

The mechanisms of toughening for a variety of composites are discussed in detail by Rice et al.[1] Additional studies and discussion on the reinforcement mechanism of whiskers, importance of aspect ratios and whisker packing can be found in Refs. 2 and 3. It is generally accepted that the interfacial bond strength between the matrix and reinforcement can influence the mechanical properties of the composite. Thus, there are significant efforts being directed towards controlling the interfacial strength with coatings.[4,5] Rice et al.[4] report increases in SiO_2/SiC composite strength and fracture toughness values of 4 and 100, respectively, when the SiC fibers are coated to limit interfacial bonding.

Their low densities and good chemical, thermal and mechanical properties make these materials potential alternatives to the nickel-based super alloys currently used in turbine engines.

237

COMPOSITE FABRICATION

One process for ceramic composite fabrication involves the mixing of ceramic oxides, either as dry powders or slurries, with whiskers. Upon densification these oxides form the matrices which hold together the reinforcement, protect it, and permit it to effectively carry the load. Whiskers are particularly attractive because their strengths approach theoretical values. Whisker strengths over 1 million psi have been reported by Gac et al.[6]

Recent investigations by Wei and Becker[7,8] on ceramics reinforced with silicon carbide whiskers provide evidence for the superior properties of ceramic composites. Both the fracture toughness and strength of the composites are about a factor of 2 greater than those for the ceramics (alumina and mullite) without silicon carbide reinforcement. An additional advantage of the alumina/silicon carbide composites is their better resistance to slow crack growth compared to alumina alone.[7]

Alternatively, long fibers may be coated with a powder slurry and wound to produce a tape which can then be laminated to form a 3-D structure. Prewo and Brennan[9,10] have produced glass/SiC fiber composites with fracture toughness (K_{IC}) and strength values approaching 11–18 MPa·m$^{1/2}$ and 700 MPa, respectively.

Currently, one of the major disadvantages of ceramic composite fabrication is the requirement for hot pressing in order to achieve densification. In addition to fiber damage resulting from hot pressing, complex shapes are difficult to fabricate. An objective of our program is to evaluate the feasibility of fabricating near-net shape ceramic composites without hot pressing. The approach that we have selected is based on sol–gel processing.

Sol–gel processing is a relatively new method for preparing ceramic composites. The reinforcement (in the form of whiskers, fibers, weaves and particulates) can be infiltrated with a low-viscosity sol, with or without the application of pressure, resulting in the formation of an intimate interface between the two phases after gelation. Additionally, incorporation of whiskers into the sol prior to gelation permits casting of near-net shape components with complex geometries. Although the technique appears promising, a number of processing variables must be understood, controlled, and evaluated before the advantages of sol–gel derived composites can be realized. The thrust of this chapter is to describe the sol–gel fabrication techniques that we have explored and some of the associated processing advantages and problems that have been encountered.

We have fabricated two types of ceramic composites using alumina gels: (1) preform fiber composites, and (2) whisker composites. Major problems that have been encountered are:

1. Sticking of the composite to the mold during drying.
2. Cracking of the composite during drying.
3. Segregation of the reinforcement (particularly whiskers) prior to gelation.

Solutions have been found for each of these problems. The addition of a mold release agent (Union Carbide R-272) has eliminated the sticking problem. Cracking can be controlled with glycerol which also provides flexibility and toughness to the dried composite. $Al(NO_3)_3$ can be used to accelerate gelation after the whiskers are added to prevent segregation.

SOL–GEL/PRE-FORM FIBER REINFORCED COMPOSITES

The matrix for the composites discussed in this section was produced according to the methods described by Clark and Lannutti.[11–14] The sol was boiled to reduce its volume and promote gelation.

Two techniques were used for preparing these composites: (1) casting, and (2) multiple dipping. Casting involved placing fiber pre-forms (random mats or woven cloth) into a tray and then pouring sol (near gelation) into the container. Care was taken to allow air to escape from the pre-form during casting. Multiple dipping was used for producing coatings on woven cloth. Matrix continuity on the cloths was expected to be poor with the casting technique, due to the large volume shrinkages of the gel during drying. The woven pre-forms were immersed in the sol for about 2 min, removed, and heated to 300°C. This procedure resulted in a stable gel on the pre-forms that could be dipped into the sol again without dissolving. The process was repeated 10 times.

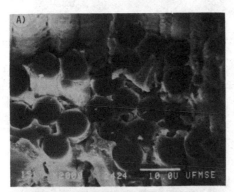

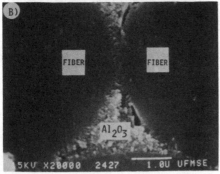

Figure 26.1 (A) Cross section of dipped Nextel composite fired at 1200°C for 1 hr in air. 2000X. (B) Dipped Nextel composite fired at 1200°C for 1 hr in air.20,000X.

Fiber coatings are fairly easy to achieve using sol–gel processing. Figure 26.1 shows a Nextel fiber bundle (tow) that has been dipped in an alumina sol and subsequently fired at 1200°C. Although the coating appears to be adequate, it is anticipated that better coatings can be achieved on single fibers.

SEMs of typical cast and dipped fracture surfaces are shown in Fig. 26.2. In general, matrix continuity was better on the dipped samples. Nicalon fibers retained their integrity after firing to 1200°C although there appears to be some interaction between the fibers and the matrix. This interaction is attributed to a thin silica layer that formed on the Nicalon fibers when they were preheated to 700°C prior to using them in composite fabrication.[14]

Two other features are worth noting on these micrographs. Fiber pullout does not seem to accompany fracture, suggesting that the fracture is brittle in nature and that the fibers do not provide toughening. Additionally, there is substantial porosity in all of the samples after firing to 1200°C. Porosity potentially can be reduced by hot pressing at higher temperatures, but as mentioned earlier, an objective of our program is to eliminate the need for hot pressing.

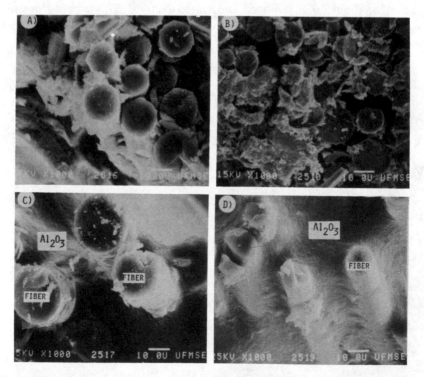

Figure 26.2   (A), (C) Cross section of Nicalon cloth composite (dipped) fired at 1200°C for 1 hr in argon. 1000X. (B), (D). Cross section of Nicalon cloth composite (cast) fired at 1200°C for 1 hr in argon. 1000X.

SOL–GEL/WHISKER REINFORCED COMPOSITES

Whiskers appear to be ideally suited for the fabrication of ceramic composites using the sol–gel process. The whiskers can be dispersed in a sol, cast into nearly any desired shape, and then fired or hot pressed to produce a ceramic with isotropic properties. Thus, the amount of machining required on the final part can be minimized. It is also thought that whisker-reinforced composites might offer better environmental stability than long fiber or pre-form fiber-reinforced composites. The reason for this is that the latter provides direct pathways from the interior of the composite to its surface, thus increasing the risk of degradation.

The addition of whiskers to the sol can significantly reduce the extent of drying and firing shrinkages.[12] For example, a pure alumina gel will experience about 25% linear shrinkage during drying and another 40% after firing to 1200°C. Total shrinkage is reduced to less than a few percent when 50 wt% of SiC whiskers are added to the sol. Unfortunately, there is considerable porosity in these composites which prevents realization of their potential strengths. However, in regions of the samples where high density is achieved, the fracture surface exhibited fiber pullout which suggests toughening.

A flow diagram of the sol–gel process for fabricating ceramic composites is shown in Fig. 26.3. One of the techniques that we have investigated for decreasing the porosity is multiple infiltrations. The sample is cast, gelled, and then fired to 600°C for stabilization. Figure 26.4 shows that a composite consisting of 8 wt% Al_2O_3–92 wt% SiC contains about 80% porosity after casting and firing to 600°C. The porosity can be reduced by vacuum infiltration with the sol and

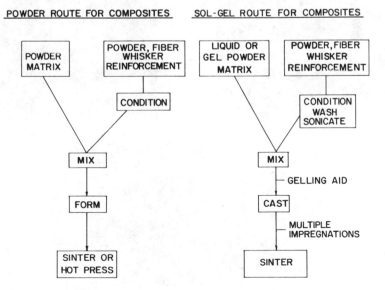

Figure 26.3   Flow diagram for the traditional and sol–gel processes used in composite fabrication.

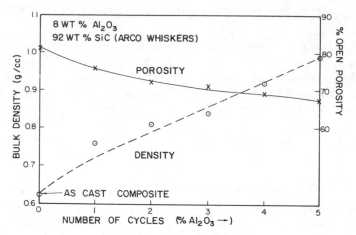

Figure 26.4   Bulk density and porosity versus the number of cycles for a cast composite containing 8 wt% Al_2O_3 + 92 wt% SiC whiskers. A cycle involves vacuum infiltration of the composite with an alumina sol and firing to 600°C—2 hr.

then firing again to 600°C. This process may be repeated many times prior to final densification at a higher temperature. Based on the trends in Fig. 26.4, about 25–30 cycles would be required to approach full densification, at which point the composite would consist of about 80% Al_2O_3–20% SiC. This calculation assumes that all of the original 80% porosity would be filled with Al_2O_3. Microstructural alterations accompanying infiltration can be seen in Fig. 26.5.

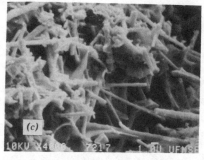

Figure 26.5   SEMs of a composite originally consisting of 8 wt% Al_2O_3–92 wt% SiC. (a) As-received Arco whiskers. (b) After casting in Al_2O_3 gel, 600°C. (c) After second infiltration, 600°C.

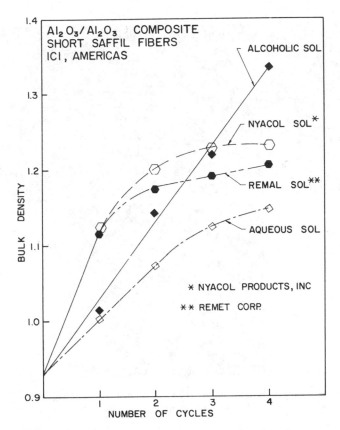

Figure 26.6 Bulk density versus cycles. Each cycle involves infiltrating a porous composite orig-inally consisting of short Saffil fibers and Nyacol sol, then firing to 500°C. Several types of sol were used as indicated in the graph. Densities of 2.6–3.3 g/cm³ were obtained by hot pressing to 1250–1450°C.

The multiple infiltration process also has been evaluated for the alumina/alumina composite. Efficiency of infiltration depends on the sol. Alcohol-based sols provided the best infiltrations while water based sol resulted in a self-limiting process (Fig. 26.6).

Two serious problems have been encountered in the processing of whisker-reinforced composites from sols: (1) whisker agglomeration, and (2) whisker segregation. Whisker agglomeration can be seen in Fig. 26.7. Some of the agglomerates are 500 μm in diameter, which can lead to their settling to the bottom of the casting prior to gelation. Ultrasonication is beneficial, but does not result in elimination of all of the agglomerates. The effects of these agglome-rates on mechanical properties of the composites are not understood. It is possible that, if controlled, such agglomerates could act as crack quenchers. Studies are currently underway to evaluate this theory. The major problem is rapid segregation (due to gravity) of these agglomerates prior to gelation.

A number of inorganic gelling aids have been evaluated in our laboratory.

Figure 26.7   (A) Silar SC-9 single crystal whiskers (as received). 10,000X. (B) Silar SC-9 single crystal whiskers (as received). 24X. (C) Composite formed with as-received whiskers (prewetted). 24X.

One gelling aid that works well is $Al(NO_3)_3$. When added during casting, it results in accelerated gelation and consequently the prevention of segregation. FTIR analysis shown in Fig. 26.8 illustrates the effectiveness of $Al(NO_3)_3$ in providing homogeneity to Al_2O_3–SiC composites.

SUMMARY

Ceramic matrix composites, although a relatively unexplored area of science, can provide a new generation of high-performance materials. Sol–gel appears to be a viable method for producing ceramic composites. The advantages of fabricating gel-derived matrices containing high-strength SiC whiskers (and other whiskers) certainly deserves further attention. In order to adequately assess the processing/property/performance relationships in these materials, much additional testing, particularly high-temperature strength and toughness, is required.

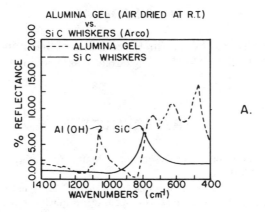

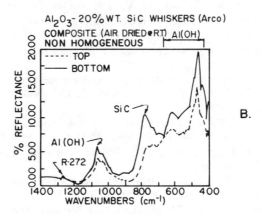

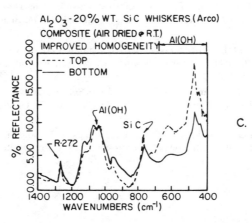

Figure 26.8 FTIR analysis of Al_2O_3—SiC composites. (A) Individual spectra of pure Al_2O_3 gel and pure SiC whiskers. (B) Spectra of top and bottom of Al_2O_3—SiC composite without gelling agent. (C) Same as (B), but with $Al(NO_3)_3$ added.

245

ACKNOWLEDGMENT

This work was supported by the Air Force Office of Scientific Research under Contract No. F49620-83-C-0072. Fig. 26.6 was provided by Jean-Claude Cavalier.

REFERENCES

1. R. W. Rice, Mechanisms of Toughening in Ceramic Matrix Composites, *Ceram. Eng. Sci. Proc.* **2**(7–8), 661–701 (1981).

2. J. V. Milewski, How to Use Short Fiber Reinforcements Efficiently, in Proceedings of the 37th Annual Conference, Reinforced Plastics/Composite Institute, The Society of the Plastics Industry, Inc., Jan. 11–15, 1982.

3. J. V. Milewski, Problems and Solutions in Using Short Fiber Reinforcements, see Ref. 2.

4. R. W. Rice, Processing of Advanced Ceramic Composites, in *Better Ceramics Through Chemistry*, C. Brinker, D. Clark, and D. Ulrich, Eds., Elsevier, 1984, pp. 337–345.

5. R. W. Rice, Ceramic Composites-Processing Challenges, *Ceram. Eng. Sci. Proc.* **2**(7–8), 493–508 (1981).

6. F. D. Gac, J. J. Petrovic, and J. V. Milewski, Short Fiber Reinforced Structural Ceramics, *Quarterly Progress Report of Advanced Research & Technology Development*, Los Alamos National Laboratory, April 1983.

7. P. R. Becker and G. C. Wei, *Fracture Toughness and Crack Growth in SiC Whisker Reinforced Oxide Ceramics*, Oak Ridge National Laboratory, ORNL Report No. WS-32253, 1984.

8. G. C. Wei and P. F. Becker, Development of SiC-Whisker-Reinforced Ceramics, *Am. Ceram. Soc. Bull.* **64**(2), 298–304 (1985).

9. K. M. Prewo and J. J. Brennan, High-Strength Silicon Carbide Fibre-Reinforced Glass-Matrix Composites, *J. Mater. Sci.* **15**, 463–468 (1980).

10. K. M. Prewo and J. J. Brennan, Silicon Carbide Yarn Reinforced Glass Matrix Composites, *J. Mater. Sci.* **17**, 1201–1206 (1982).

11. D. E. Clark and J. J. Lannutti, Phase Transformations in Sol-Gel Derived Aluminas, *Ultrastructure Processing of Ceramics, Glasses, and Composites*, L. L. Hench and D. R. Ulrich, Eds., 1984, pp. 126–141.

12. J. J. Lannutti and D. E. Clark, Sol-Gel Derived Ceramic-Ceramic Composites Using Short Fibers, in *Better Ceramics Through Chemistry*, C. Brinker, D. Clark, and D. Ulrich, Eds., Elsevier, New York, 1984, pp. 369–374.

13. J. J. Lannutti and D. E. Clark, Long Fiber Reinforces Sol-Gel Derived Al_2O_3 Composites, in *Better Ceramics Through Chemistry*, C. Brinker, D. Clark, and D. Ulrich, Eds., Elsevier, 1984, pp. 375–382.

14. J. J. Lannutti and D. E. Clark, Sol-Gel Derived Coatings on SiC and Silicate Fibers, Proceedings of the 8th Annual Meeting on Advanced Ceramics and Composites, Cocoa Beach, Fla., Jan. 1984.

27

NUCLEATION AND EPITAXIAL GROWTH IN DIPHASIC (CRYSTALLINE + AMORPHOUS) GELS

RUSTUM ROY, YOSHIKO SUWA,
AND SRIDHAR KOMARNENI

Materials Research Laboratory,
The Pennsylvania State University,
University Park, Pennsylvania

INTRODUCTION

SG, the Route to Homogeneous Ceramics

The sol–gel (SG) process as a route to homogeneous ceramics and glasses has been investigated by the senior author and his colleagues and students since July 1948. In the decade following we reduced to routine practice the mixing in solution (starting with both inorganic and organic precursors many of which were rare chemicals in those days) and thence the formation of sols, and then gels, and xerogels. We covered the most common ceramic oxide compositions (involving Al, Si, Ti, Zr, etc.)[1-5] in both simple (one-component) and complex (up to five- and six-component) systems. This discovery of making ultra-*homogeneous* oxide solids via the SG route instead of the oxide powder mixing route became widespread in the community of experimental geochemists and petrologists. Literally hundreds of papers were published employing the SG method to make homogeneous multicomponent oxide compositions. Nearly two decades elapsed before industrial research laboratories and the ceramics community became aware of its potential.

247

Technologically, till the midsixties these papers stimulated very little interest. The catalyst industry (e.g., the Filtrol Corporation, which had supported some of our work) where aluminosilicate gels were used extensively, and DuPont where the pioneering work of R. K. Iler led to new silica sol products under the trade name Ludox ®*, were exceptions.

Between 1965 and 1980 the pace of research in industry on *technological applications* of the SG process was intensified in a few corporations and new products were successfully introduced or partially developed and abandoned:

1. Nuclear fuel pellets at Oak Ridge.[6]

2. Ceramic fibers by 3M and Carborundum,[7] including nonoxide fibers.[8]

3. Abrasive grain by 3M.[9]

4. Glass-melting research done on a small-pot scale by Bausch and Lomb was extended to full-tank scale in a very substantial development effort by Owens-Illinois.[10] This was terminated just before the oil crisis, just when its energy-conserving advantage might have helped.

5. Two obvious areas of applications: large-area thick (+1 mil) coatings and bulk ceramics received considerable industrial attention, but they had not been successfully translated into major products. The more recent successful work of Yoldas[11] and Dislich[11] in the coatings area showed not only that one could make excellent coatings from organic precursors, but also that these could not be made very thick ($\gg 1$ μm).

The products of the science, and of all these technologies, were *maximally homogeneous* ceramics and glasses. The recent voluminous literature on SG structures and processes which has accumulated in the last four or five years is likewise exclusively devoted to the same goal—*homogeneous materials*.

The New Direction: Maximally Heterogeneous Nanocomposites

In 1982 the senior author reported[12] that the SG route which was used to make such *maximally homogeneous* ceramic materials, could be turned around to make *maximally* (i.e., in the degree of interpenetration of phases) *heterogeneous* materials. We have now demonstrated a wide range of uses of the SG route for making this new class of materials which we call *nanoscale composites* derived from *di* (or multi) *phasic xerogels*.

A nanoscale composite (including the di-phasic xerogel) is a material which has two (or more) phases with the physical dimensions of the phases lying in the range 1–10 nm. The two phases may differ in either composition or structure or both. Thus there can be a nanocomposite of 10-nm SiO_2 and 10-nm Al_2O_3 particles, or 10-nm $AlOOH$ and 2.5-nm $AgCl$, or 10-nm rutile crystals + 100-nm noncrystalline TiO_2, or 10-nm $AlOOH$ + 20-nm α-Al_2O_3. The sig-

---

*Registered trademark E. I. DuPont de Nemours, Inc., Wilmington, Delaware.

nificance of some members of this class of nanocomposite materials is just beginning to be appreciated. In the metals area Gleiter and Marquardt,[13] for instance, title their 1984 paper "Nanokristalline Strukturen ein Weg zu neuen Materiales". Schechtman, Blech, Gratias, and Cahn;[14] and Levine and Steinhardt[15] in *Phys. Rev. Letters* show, for example, that in certain rapidly quenched alloys (A186:Mn14) extraordinary crystallographic structures are seen or quasicrystals with the fivefold icosahedral symmetry disallowed to crystals with large sizes and translational symmetry. We had already approached the making of such very finely heterogeneous material from *the vapor phase* by cosputtering of ceramics + ceramics, ceramics + metals, and ceramics + polymers some years ago in this laboratory. The papers by Messier, Roy, and Cowley, and Roy and Cowley[16,17] described, for example, the extraordinary structure of the $\approx$2-nm gold crystals dispersed in a Al_2O_3 matrix. However, both the methods referred to in the referenced papers—the very recent rapid solidification of a melt or earlier cosputtering—have severe limitations on compositional flexibility, volume of samples, and so on. The success of the experiments summarized herein is in the synthesis of nanoscale composites via the sol–gel route. This method provides enormous flexibility with respect to variation of composition and structure and metastable energy storage combined with relative ease of preparation.

Epitaxial Growth in S–S Systems

There has been little work in the case of pure S $\rightarrow$ S epitaxy, that is, overgrowth of a crystalline phase by solid-phase reaction in the presence of an epitaxial substrate.[18] In a manner of speaking, various topotactic reactions[19] in complex silicates may be thought of as a self-seeded epitaxy, where one part of the structure provides a persisting structural element which controls the formation of the final product. Martensitic phase transitions may also be treated as examples of solid-phase epitaxy[20] in this limited sense.

As we embarked on our new emphasis of studying the science of heterogeneous or diphasic gels the role of seeding became important. Two radically different mechanisms of seeding of the process can exist. The first is the mutual interaction of charged colloidal particles and the possible resulting mutual orientation of the two particles. From the classical work of Biltz summarized by Zsigmondy in 1905 and 1925[21,22] one notes that "protective colloids" have a particular relation to each other. Figure 27.1 illustrates Zsigmondy's picture of the specific "protection" of a Au sol particle by gelatine particles, with a reversal of phases occurring when the ratio of sizes is reversed. Such mutual interactions which are at least (in part) controlled by the charge on the particles can only occur in mixed sols of relatively low viscosity. While the role of the charge on the particle is clear, what is less clear is whether the structure of the solid phase can cause a specific mutual orientation between particles.

That crystal "structure" plays some role is evident from the well-known technique of "decorating" clay particles with colloidal gold. Here the diphasic sol

Figure 27.1   Reversal of interrelationships of phases in protective colloids as a size function.[21,22]

on dehydration results in a solid with gold particles being adsorbed on the *edges* of the dominantly hexagonal clay platelets.[23]

Our purpose in the work on diphasic xerogels was very different. It was to catalyze reactions to attain equilibrium within the final solid assemblage. In our early studies with gels to get reaction at the lowest temperatures we introduced the practice of seeding of gels with the same structure as we expected in the final product. It was found that the phase diaspore (AlOOH) could not even be formed without diaspore seeds in an alumina gel.[24] We were later able to override this requirement of seeds by using higher pressure and no seeds.[25] Similar results were obtained with α-spodumene and other systems.[26] The mechanism of such seeding was assumed to be transport of Al^{3+} and O^{2-} via the solution phase. The final element relevant to development of nanocomposites was our research on crystal growth in gels. For instance, in McCauley and Roy's work[27] on the formation of $CaCO_3$ crystals in SiO_2 gel it was possible to control the *structure* of the crystalline phase (whether calcite, aragonite or vaterite) by adjustment of ion activities. We demonstrated the epitaxial control of the final phase (i.e., aragonite growing on Sr or Mg-rich seeds) by finding the original

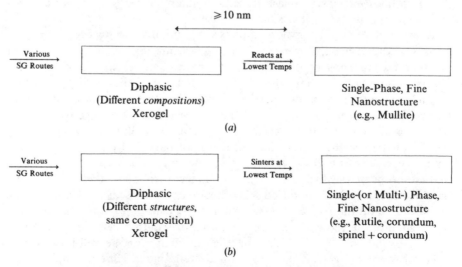

Figure 27.2   Alternative routes for making nanocomposites.

nucleus inside the final crystal and analyzing it in the electron microprobe. In this case, although the gel network is critically controlling the diffusion, we are dealing with a L → S crystallization reaction.

Our present experiments use the SG route to make two classes of diphasic xerogel ceramic materials which are nanocomposites and which can be precursors for making at lower temperatures, nanocomposite ceramics (or cermets) or single-phase ceramics with a fine nanostructure, as shown in Fig. 27.2.

Most of the results deal with the route (b); and the novelty is the positive effects of the second (seed) phase on the reaction temperatures which prove the importance of epitaxial effects in purely solid-state reactions.

EXPERIMENTAL

As described earlier,[28,29] we have developed two different methods for making such materials (Fig. 27.3). For most experiments we have used method II. We

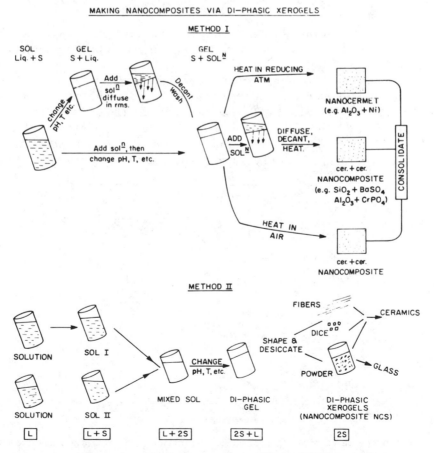

Figure 27.3 Methods of making nanocomposites via diphasic xerogels.

thoroughly mix two preexisting sols at low viscosities and assume "perfect mixing." A gel is then formed from the mixed sol, dried, and in part heated to an anhydrous (xerogel) condition, and used for further characterization and other reaction studies.

Data on the thermodynamics of the xerogel → ceramics transformation was obtained by DTA, Harrop (Model TA700), and Perkin-Elmer (Model DTA1700). X-ray and SEM characterization has been done on most samples at various stages of the heating cycle. Optical microscopy proved to be significant in describing the extraordinarily large corundum crystals obtained under certain conditions.

RESULTS

Solid-State Epitaxial Effects in One-Component Diphasic Systems

The System TiO_2

A TiO_2 gel was made by the method described elsewhere.[30] XRD and TEM show that it is an anhydrous noncrystalline oxide, with no crystallinity at <200°C.

By adding a rutile-sol and mixing thoroughly prior to gelation of titania a series of diphasic gels are formed so that the total solid phases contained 0.2, 0.5, . . . , 5% by weight of 0.1-μm rutile crystals. The unseeded and seeded gels were dried at 400°C to remove most of the organics prior to the DTA runs (Fig. 27.4). The DTA pattern shown at the top in Fig. 27.4 shows that the titania gel crystallizes to rutile (via an anatase stage) at approximately 900°C

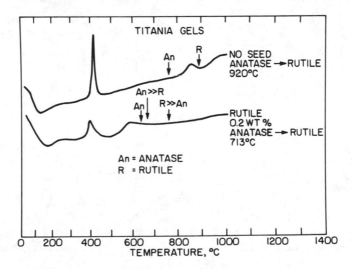

Figure 27.4 Effect of rutile seeding on the transformation temperatures in titania xerogels.

under the dynamic conditions of the DTA experiment. With 0.2% of rutile seeds, the diphasic xerogel transforms to rutile nearly 250°C lower than the unseeded TiO_2 xerogel, as is shown in the bottom curve of Fig. 27.4.

The DTA curves for higher concentrations of seeds show no further effect beyond 0.2%.

The System Al_2O_3

In this case a boehmite (AlO·OH) sol (~20-nm particle), not precisely isoplethal with the sol of α-Al_2O_3 seeds (~100 nm) was used. However, the diphasic xerogel obtained at approximately 500°C when the boehmite has dehydrated to a "γ-Al_2O_3" phase is truly isoplethal. The results comparing the unseeded and seeded gels with increasing weight percents of corundum seeds are shown in Fig. 27.5. These data are parallel to the TiO_2 data with a marked (~150°C) lowering of the θ–α transformation exotherm from 1280° to 1150°C with the addition of only 0.1% seeds. Higher concentrations do not significantly lower the θ–α monotropic transformation further.

In a separate series of experiments a similar boehmite sol was quickly gelled by evaporation and thin layers of the xerogel heated to 1150°C for 30 min. The α-Al_2O_3 crystals which formed from what were originally 20-nm grains were 10–15 μm in diameter. Figure 27.6a shows them in polarized light. While detailed examination showed that there are low-angle grain boundaries within the 10-μm grains, this is remarkable grain growth. When the sol was seeded with

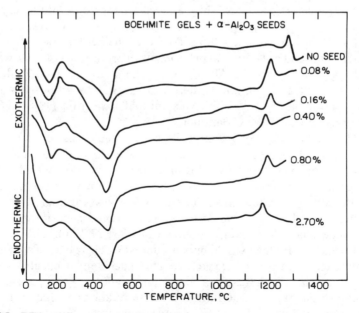

Figure 27.5 DTA results comparing the transformation temperatures of unseeded and seeded boehmite gels with increasing wt%'s of α-Al_2O_3 seeds.

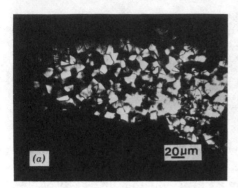
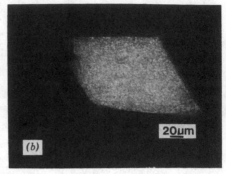

Figure 27.6   Photomicrographs of boehmite xerogels heated to 1150°C. (A) Unseeded gel showing 10-μm grains of α-Al$_2$O$_3$. (B) α-Al$_2$O$_3$ seeded gel showing submicrometer grains.

the 0.2% α-Al$_2$O$_3$ seeds and the experiment repeated, no large crystals could be found at all (Fig. 27.6b), the grain size was submicrometer, and could not be discerned optically. Details are reported elsewhere.[31]

Epitaxial Effects in Two-Component Diphasic Systems

Addition to Alumina and 93% Al₂O₃–7% MgO Gels

We have also explored hetero–homo and hetero–hetero epitaxy effects in diphasic xerogel-derived ceramics. This has been attempted by adding crystalline seeds isostructural with α-Al$_2$O$_3$ (i.e., Fe$_2$O$_{3[\text{corundum}]}$ and Cr$_2$O$_{3[\text{corundum}]}$) to boehmite sols, and subsequent gelling. To check whether the effects found were structural epitaxy and not colloidal phenomena of protection colloids, we added also SiO$_{2[\text{quartz}]}$ seeds, and MgAl$_2$O$_4$ seeds. The quartz (Fig. 27.7) shows no effect on the θ–α transition, and while Fe$_2$O$_3$ does show a lowering of 30°C, the Cr$_2$O$_3$ shows hardly any. The lattice mismatch with Fe$_2$O$_3$ is 5.86% and with Cr$_2$O$_3$ is 4.56% for the C parameter.

The 93% Al$_2$O$_3$–7% MgO composition was studied as an example of a binary system in which both α-Al$_2$O$_3$ and spinel would crystallize at equilibrium at 1150°C. It is of technological significance since the composition of 3M Regal abrasive grains is near this composition. Due to the known topotactic relations between the corundum and spinel structures, it was also considered worthwhile to investigate whether the spinel structure seeds would nucleate the spinel or the corundum phase. The data are shown in Figs. 27.7 and 27.8. The effect appears to be slight. Epitaxy would only occur on the (111) plane of spinel, and in general the cubic close-packed oxygen array of spinel is not an effective substrate for the hcp face-shared octahedra of corundum. Figure 27.8 compares the effect of spinel and corundum seeds on the θ–α transformation as a function of composition. About 0.2 wt% of seeds is a saturation level. This would appear to correspond to the growth of grains of approximately 2 μm in diameter on seeds 0.2 μm in diameter (assuming spherical morphologies).

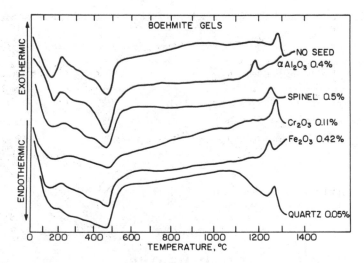

Figure 27.7 Effect of structurally and compositionally different seeds on the θ- to α-Al$_2$O$_3$ transition in boehmite gels.

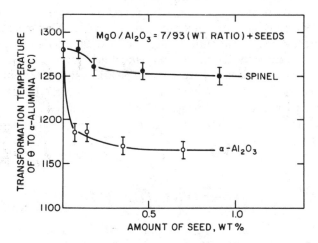

Figure 27.8 Transformation temperature of θ- to α-Al$_2$O$_3$ as affected by the concentration of spinel and α-Al$_2$O$_3$ seeds.

Compositionally Diphasic Nanocomposites Derived from Diphasic Xerogels

This paper has focused on precursor ceramic powders where the two phases are structurally different. However, earlier work[29] shows that compositionally diphasic xerogels differ even more radically in reaction rates and patterns, for example, the mullite formation exotherm at 980°C is completely eliminated. These diphasic xerogels have been shown to sinter much better than the single-phase xerogels as measured by the densities (Table 27.1) attained under identical conditions. We believe that the exceptional results obtained by Prochaszka[32]

TABLE 27.1   Sintering Behavior of Single and Diphasic Mullite Xerogels as Measured by Densities

Starting Materials	Density at 1200°C	Density Relative to Mullite (%)
Single-Phase—$Al(NO_3)_3 \cdot 9H_2O$ + $Si(OC_2H_5)_4$	2.71	85.4
Diphasic—$AlOOH$ + $Si(OC_2H_5)_4$	2.92	92.0
Diphasic—$AlOOH$ + Ludox (SiO_2)	3.05	96.2

on transparent mullite ceramics may be attributed to this probable use of diphasic gels.

CONCLUSIONS

We have presented data that show a generalized process to make a new class of materials, nanocomposites, using structural epitaxy where even 0.1% of such seeds can affect reaction rates and sintering behavior. The temperature and microstructure differences in sintering of the α-Al_2O_3 seeded boehmite have been reported recently by Kumagai and Messing.[33]

At least of equal significance are the compositionally heterogeneous nanocomposite precursor mixtures made by the diphasic xerogel route. By designing binary ceramic powders as heterogeneous nanocomposites it should be possible to achieve very substantial (200–300°C) lowering of reaction and sintering temperatures.

ACKNOWLEDGMENT

This work draws on the research supported by two grants: AFOSR Grant 83-0212 and NSF, DMR-8119476.

REFERENCES

1. R. Roy, Aids in Hydrothermal Experimentation II, *J. Am. Ceram. Soc.* **39**, 145–146 (1956).
2. D. M. Roy and R. Roy, An Experimental Study of the Formation and Properties of Synthetic Serpentines and Related Layer Silicate Minerals, *Am. Mineralogist* **39**, 957–975 (1954).
3. R. C. DeVries, R. Roy, and E. F. Osborn, Phase Equilibria in the System CaO-TiO_2-SiO_2, *J. Am. Ceram. Soc.* **38**, 158–171 (1955).
4. R. Roy, Gel Route to Homogeneous Glass Preparation, *J. Am. Ceram. Soc.* **52**, 344 (1969).
5. G. J. McCarthy, R. Roy, and J. M. McKay, Preliminary Study of Low-Temperature Glass Fabrication from Noncrystalline Silicas, *J. Am. Ceram. Soc.* **54**, 637 (1971).

6. M. E. A. Hermans, Importance of Sol-Gel Processes for Ceramic Nuclear Fuels, *Sci. Ceram.* **5**, 523–538 (1970).

7. K. Miyahara and N. Nakayama, Process for Producing Polycrystalline Oxide Fibers, U.S. Patent 4,159,205 (1979).

8. M. A. Leitheiser and H. G. Sowman, Non-Fused Alumina Based Abrasive Material, U.S. Patent Appl. 145,383 (1980).

9. S. Yajima et al., Synthesis of Continuous Silicon Carbide Fibre with High Tensile Strength and High Young's Modulus, *J. Mater. Sci.* **13**, 2569–2576 (1978).

10. I. M. Thomas, Metal-Organic-Derived (MOD) Glass Compositions. Preparation, Properties, and Some Applications, *Abstracts, Annual Meeting of the Materials Research Society*, Boston, MA, p. 370 (Nov. 1982).

11. B. E. Yoldas, Deposition and Properties of Optical Oxide Coatings from Polymerized Solutions, *Appl. Opt.* **21**, 2960–2964 (1982); H. Dislich and E. Hussmann, *Thin Solid Films*, **77**, 129 (1981).

12. R. A. Roy and R. Roy, New Metal-Ceramic Hybrid Xerogels, *Abstracts, Annual Meeting of the Materials Research Society*, Boston, MA, p. 377 (Nov. 1982).

13. H. Gleiter and P. Marquardt, Nanokristalline strukturen-einweg zu neuen materialien?, *Z. Metallkunde* **75**, 263 (1984).

14. D. Schechtman, I. Blech, D. Gratias, and J. W. Cahn, *Phys. Rev. Lett.* **53**, 1951 (1984).

15. D. Levine and P. J. Steinhardt, *Phys. Rev. Letters* (submitted, Oct. 31, 1984).

16. R. A. Roy, R. F. Messier, and J. M. Cowley, Fine Structure of Gold Particles in Thin Films Prepared by Metal-Insulator Co-Sputtering, *Thin Solid Films* **79**, 207–215 (1981).

17. J. M. Cowley and R. A. Roy, Microdiffraction from Small Au Particles, pp. 143–152, SEM (1981).

18. C. S. Fang and R. Roy, Heteroepitaxial Growth of KDP and TGS Crystals on Muscovite, *J. Cryst. Growth* **60**, 182–184 (1982).

19. G. W. Brindley, Role of Crystal Structure in Solid-State Reactions of Clays and Related Minerals, *Intl. Clay Conf., Stockholm, Sweden* **1**, 37–44 (1963).

20. R. Roy, A Syncretist Classification of Phase Transitions, in *Proc. Conf. on Phase Transitions and Their Applications in Materials Science*, pp. 13–27, L. E. Cross, Ed., Pergamon Press, New York, 1973.

21. R. Zsigmondy, *Zur Erkenntnis der Kolloide*, Fischer Verlag, Jena, New York, 1905.

22. R. Zsigmondy and P. A. Thiessen, *Das Kolloide Gold*, Akad. Verlag, Leipzig, 1925, p. 229.

23. H. Van Olphen, *An Introduction to Clay Colloid Chemistry*, Wiley, New York, (1978), p. 318.

24. G. Ervin, The System Al_2O_3-H_2O, Ph.D. Thesis in Ceramic Technology, The Pennsylvania State University, 1949.

25. D. M. Roy, R. Roy, and E. F. Osborn, The System MgO-Al_2O_3-H_2O and Influence of Carbonate and Nitrate Ions on the Phase Equilibria, *Am. J. Sci.* **251**, 337–361 (1953).

26. R. Roy, D. M. Roy, and E. F. Osborn, Compositional and Stability Relationships Among the Lithium Aluminosilicates: Eucryptite, Spodumene and Petalite, *J. Am. Ceram. Soc.* **33**, 152–159 (1950).

27. J. W. McCauley and R. Roy, Controlled Nucleation and Crystal Growth of Various $CaCO_3$ Phases by the Silica Gel Technique, *Am. Mineralogist* **59**, 947–963 (1974).

28. D. Hoffman, R. Roy, and S. Komarneni, Diphasic Ceramic Composites Via a Sol-Gel Method, *Mat. Lett.* **2**, 245–247 (1984).

29. D. Hoffman, R. Roy, and S. Komarneni, Diphasic Xerogels, A New Class of Materials: Phases in the Al_2O_3-SiO_2 System, *J. Am. Ceram. Soc.* **67**, 468–471 (1984).

30. S. Komarneni, E. Breval, and R. Roy, Structure of Solid Phases in Titania and Zirconia Gels, *J. Non-Cryst. Solids* (submitted).

31. W. A. Yarbrough and R. Roy, Microstructural Evolution in Sintering of AlOOH Gels, Abstracts of the American Ceramic Society meeting, 1985.

32. S. Prochaszka, personal communication, 1984.

33. M. Kumagai and G. L. Messing, Enhanced Densification of Boehmite Sol-Gels by α-Al_2O_3 Seeding, *Comm. Am. Ceram. Soc.* **67**, C230–C231 (1984).

28

SEEDED TRANSFORMATIONS FOR MICROSTRUCTURAL CONTROL IN CERAMICS

GARY L. MESSING, MASATO KUMAGAI,
RICHARD A. SHELLEMAN, AND JAMES L. McARDLE
Department of Materials Science and Engineering
The Pennsylvania State University
University Park, Pennsylvania

INTRODUCTION

It is well known that many transformations in ceramics proceed by a nucleation and growth process.[1] To initiate the transformation sufficient energy must be supplied to the system to exceed the nucleation barrier. After nucleation the transformation occurs rapidly, by growth. Usually the high surface area product of transformation sinters to form an aggregated mass because of the temperature requirements for nucleation. To diminish powder aggregation the transformation can be controlled by using lower temperatures. However, at the lower temperatures excessively long times (i.e., days) are required for nucleation and growth.

Recent research in this laboratory[2] indicates that the nucleation step may effectively be eliminated by supplying nuclei to the system. This process, known as seeding, involves adding ceramic particles of the high-temperature phase to the ceramic matrix to be transformed. By eliminating the nucleation step less energy is required for the transformation and it can occur at a lower temperature. By increasing the number of nucleation sites in the system the rate is increased. There are certain physical requirements of the seed particles, relative to the matrix phase to be transformed, before seeding is effective. Nevertheless, if properly practiced, seeding offers unique control over the transformation process and thus microstructural control and control over densification. In this chapter recent results on the controlled transformation of a boehmite (γ-AlOOH) gel to α-Al$_2$O$_3$ by either α- or γ-Al$_2$O$_3$ seeding will be presented to illustrate the potential that seeding offers for controlled transformation in ceramics.

EXPERIMENTAL

To ensure the homogeneous distribution of the seed particles, a colloidal boehmite* with an ultimate crystallite size of 10 nm was used. When heated boehmite transforms to α-Al$_2$O$_3$ by the following sequence

$$\text{boehmite} \rightarrow \gamma\text{-} \rightarrow \delta\text{-} \rightarrow \theta\text{-} \rightarrow \alpha\text{-Al}_2\text{O}_3 \tag{1}$$

with the final transformation from θ occurring at $\sim 1200°$C. As nucleation of α-Al$_2$O$_3$ is commonly reported[3] to be difficult, it was decided to add α-Al$_2$O$_3$ particles to the boehmite gel to seed the θ- to α-Al$_2$O$_3$ transformation. Prior to seeding, agglomerates were removed from the boehmite hydrosol by first dispersing it with nitric acid at pH = 2.5 and then settling for extended times. The seed particles were also dispersed at pH = 2.5 and settled to obtain an agglomerate-free α-Al$_2$O$_3$ slurry having a median particle size of 0.1 μm. After adding seed particles to a 20 wt% boehmite sol, it was gelled by dehydration and air dried for 10 days. Since the gels cracked during drying, only fragments of ~ 1.0 cm were used in the sintering studies. For transformation studies the dried gel was hand-ground to pass a 325 mesh screen (-44 μm). Isothermal transformation kinetics were determined by quantitative X-ray analysis. All densities were measured by Archimedes' technique.

ALPHA ALUMINA SEEDING

The nominal 0.1-μm α-Al$_2$O$_3$ seeds were added to the hydrosol at concentrations of 0.05 to 10 wt% on a dry weight basis. As seen from the DTA data in Fig. 28.1, only the θ to α-Al$_2$O$_3$ transformation temperature was affected by the α-Al$_2$O$_3$ seeding. The transformation peak temperature was reduced by $\sim 170°$C, supporting the claim that seeding reduces the transformation temperature. The plot of the transformation peak temperature as a function of seeding concentration in Fig. 28.2 shows that 1.5 wt% α-Al$_2$O$_3$ seeding reduced the transformation by 150°C. This clearly demonstrates that excessive quantities

*Catapal SB, Vista Chemical, Ponca City, OK.

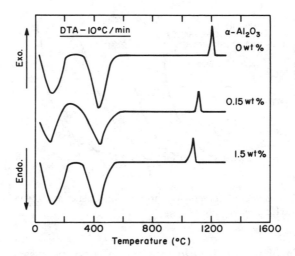

Figure 28.1   Differential thermal analysis of the α-Al$_2$O$_3$ seeded and unseeded boehmite gels.

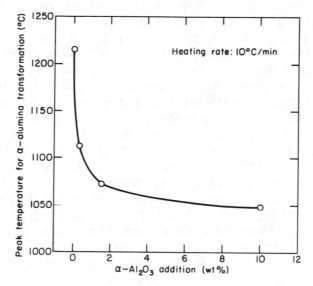

Figure 28.2   Change in the theta to α-Al$_2$O$_3$ peak transformation temperature as a function of α-Al$_2$O$_3$ seeding.

of seed particles are not necessary to affect significantly the boehmite to α-Al$_2$O$_3$ phase transformation.

In Fig. 28.3 the isothermal transformation kinetics between seeded and unseeded boehmite gels at 1050 and 1150°C are compared. The incubation period at 1150 is 37 s for the unseeded gel, versus 9 s for the seeded gel. At 1050°C the incubation period increases to 270 s whereas for the seeded gel it is 40 s. The transformation kinetics of the seeded systems are enhanced because

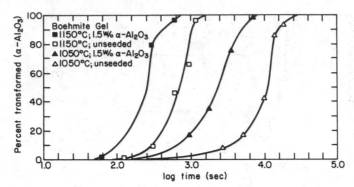

Figure 28.3   Comparison of isothermal transformation kinetics at 1050 and 1150°C for α-Al$_2$O$_3$ seeded and unseeded gels.

the longer incubation period for nucleation is not necessary. From the DTA temperature shift and the increased kinetics, it is clear that the α-Al$_2$O$_3$ particles do indeed act as nuclei for the θ- to α-Al$_2$O$_3$ transformation.

The series of micrographs in Fig. 28.4 follows the microstructural evolution of the seeded and unseeded boehmite gels at 1025°C. The unseeded sample was heated for 2 hr before significant transformation occurred. The area of larger grain size in Fig. 28.4a is an α-Al$_2$O$_3$ colony that is surrounded by the θ-Al$_2$O$_3$ matrix. Similar structures having a contiguous grain structure and no grain boundaries between what appear to be individual α-Al$_2$O$_3$ grains are commonly reported[4-6] for other precursors (e.g., gibbsite, alum, aluminum nitrate) that undergo the α-Al$_2$O$_3$ transformation. It is suggested that this colony has grown from a single nucleus. In Fig. 28.4b a 0.15 wt% seeded sample is shown after heating for 15 min at 1025°C. This sample is $\sim 70\%$ transformed and is characterized by multiple α-Al$_2$O$_3$ colonies that are smaller than those observed in the unseeded sample. The number of colonies is a function of the seeding concentration, whereas the smaller colony size is a result of the reduced time and temperature conditions for transformation. That is, the colony size would increase until impingement with adjacent colonies if heated for longer times at this temperature, assuming that there is no additional nucleation in the sample. The 1.5 wt% seeded sample shown in Fig. 28.4 is $\sim 95\%$ transformed after 10 min at 1025°C. The α-Al$_2$O$_3$ grains are 0.1 μm and do not show the contiguous pore and α-Al$_2$O$_3$ structures seen in Figs. 28.4(a) and 28.4(b). Comparison of this microstructure to the θ-Al$_2$O$_3$ matrix in Fig. 28.4(a) shows that the α-Al$_2$O$_3$ grains are significantly larger than the matrix material. This suggests that the α-Al$_2$O$_3$ grains have grown from a single nucleus. Indeed, the observed grain size is in agreement with the calculated α-Al$_2$O$_3$ grain size if spherical geometry and uniform distribution of boehmite per nucleus is assumed. From these micrographs the number of nucleation sites is seen to effect whether the deleterious α-Al$_2$O$_3$ aggregates or individual α-Al$_2$O$_3$ grains are formed. By increasing the seeding concentration the total volume transformed per nucleus is

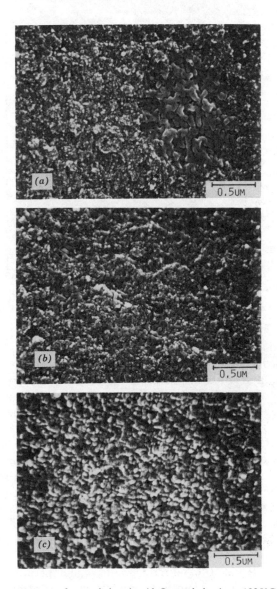

Figure 28.4 Microstructures of unseeded and α-Al_2O_3 seeded gels at 1025°C. (a) 2 hr, unseeded, (b) 15 min. 0.15 wt%, (c) 10 min, 1.5 wt%.

decreased such that the α-Al_2O_3 transformed per nucleus is much less than observed in the unseeded boehmite gel. Consequently, the large vermicular α-Al_2O_3 colonies that characterize unseeded Al_2O_3-transformations cannot develop.

The surface area changes during transformation of an unseeded and a 1.5 wt% seeded sample are compared in Fig. 28.5 as a function of the degree of transformation. As expected, the surface area decreases continuously with

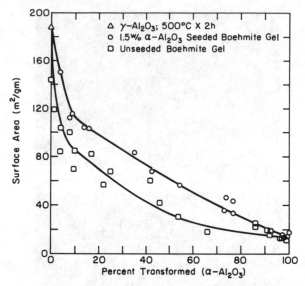

Figure 28.5 Surface area change as a function of percent transformation to α-Al$_2$O$_3$ for unseeded and α-Al$_2$O$_3$ seeded gels.

increasing transformation. At the same degree of transformation, the surface area of the seeded sample is greater than the unseeded sample, which is due to the shorter time and lower temperature required for the seeded samples to achieve the same degree of transformation as the unseeded samples. The relative distribution of porosity after transformation may influence these surface areas but has not yet been investigated. Interestingly, the surface areas after 100% transformation are essentially the same despite the fact that the microstructures are considerably different. The major microstructural differences appear to be the relative distribution of the porosity and solid phase. That is, the unseeded microstructure has a continuous pore phase that is intertwined with the interconnected α-Al$_2$O$_3$ grain colonies, whereas the pore phase is evenly distributed between individual, solid α-Al$_2$O$_3$ grains in the seeded sample.

The sintering kinetics of the unseeded and 1.5 wt% seeded fragments is given in Fig. 28.6 at sintering temperatures of 1150 to 1260°C. In these experiments the sample was heated at 50°C/min to the sintering temperature. The unseeded samples do not significantly densify during this experiment and reach a maximum relative density of 75%. In contrast the seeded sample is fully dense after 40 min at 1220°C. At 1185°C 150 min is needed to achieve full density and 6 hr is required to reach 97% of theoretical density at 1150°C. It should be noted that no MgO was added to these samples to inhibit exaggerated grain growth. However, they do contain ~0.2% TiO$_2$ as the commercial boehmite xerogel is produced by the Zeigler process, which uses titanium as a catalyst. While it is known[7] that TiO$_2$ increases diffusion in Al$_2$O$_3$ it is not known how important this is at the temperatures used in these studies. Experiments with high-purity boehmite will examine this effect. The excellent densification behavior of the

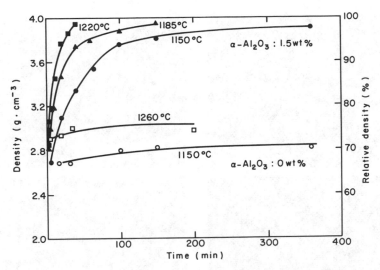

Figure 28.6 Sintering kinetics for the unseeded and α-Al₂O₃ seeded gels.

seeded samples is clearly a result of the uniform, fine-grained microstructure that is developed during transformation. It is important to note that the major difference between the seeded and unseeded samples before sintering is the connectivity and distribution of the pore and grain phases. That is, an aggregated structure forms during transformation in the unseeded samples whereas an aggregate-free microstructure is developed in the seeded sample. A similarly dramatic decrease in sintering temperature has been reported by Rhodes[8] when he sintered a centrifugally cast submicrometer, aggregate-free zirconia.

Microstructural evolution of the 1.5 wt% α-Al$_2$O$_3$ seeded sample during sintering at 1185°C is shown in Fig. 28.7. The microstructure in Fig. 28.7a shows the $\sim$0.1–0.2 μm α-Al$_2$O$_3$ grains developed upon transformation after 2 min. This sample is 71% dense, indicating that there has already been some densification. After 20 min at 1185°C (Fig. 28.7(b)) the microstructure is characterized by clusters that are composed of 0.05–0.1 μm particles of α-Al$_2$O$_3$. The "cracks" are probably related to the shrinkage within the clusters as a result of the volume change during the θ- to α-Al$_2$O$_3$ transformation and/or as a result of differential shrinkage during sintering of the θ-Al$_2$O$_3$ prior to its transformation. Interestingly, densification of the sample involves both intercluster and intracluster sintering. The sample is 99% dense after 100 min at 1185°C and shows in Fig. 28.7c that the grain size ranges from 0.1 to 0.5 μm with an average of 0.45 μm. Some "cracks" are still present in this microstructure but after 150 min at 1185°C the sample is 99% dense and there is no evidence of the crack structure.

The low-temperature sintering of these samples is a result of the exceptionally fine grain size of the α-Al$_2$O$_3$ that is formed as a result of the sol–gel process and the controlled transformation to α-Al$_2$O$_3$. Actually the active grain size for sintering is <0.1 μm, as this is the α-Al$_2$O$_3$ grain size within the clusters after

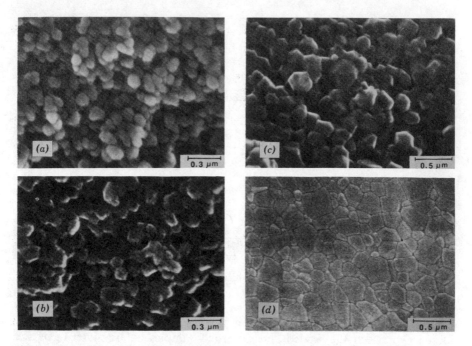

Figure 28.7   Microstructural development at various times at 1185°C for the α-Al$_2$O$_3$ seeded gels. (a) 2 min, (b) 20 min, (c) 100 min, (d) 150 min.

transformation. Therefore, initial stage densification is controlled by local densification at unit contacts and within the α-Al$_2$O$_3$ clusters, whereas final stage densification is controlled by the sintering of the almost fully dense α-Al$_2$O$_3$ clusters or grains.

These results demonstrate that the transformation to α-Al$_2$O$_3$ can be accelerated by nucleating it with α-Al$_2$O$_3$ seed particles. However, they also indicate that part of the transformation control depends on having numerous nucleation sites available for transformation. Other methods to increase the number of active nucleation sites should have a similar effect to that of α-Al$_2$O$_3$ seeding.

It was known from our transformation studies with a particular γ-Al$_2$O$_3$ powder that the θ to α transformation temperature was lower than observed for the boehmite used in these studies. Therefore, a series of experiments was designed to determine whether seeding the boehmite hydrosol with γ-Al$_2$O$_3$ powder would also enhance the transformation process and in turn yield microstructure control and enhanced sintering.

γ ALUMINA SEEDING

The γ-Al$_2$O$_3$ (CR-125, Baikowski International) was first dispersed at pH = 3.0 and settled to yield a 0.1-μm median grain size powder. This powder was mixed with the boehmite hydrosol up to 8 wt%. Thermal analysis by DTA revealed that the transformation peak temperature was reduced from 1220 to 1130°C at 6.5 wt% γ-Al$_2$O$_3$ (Fig. 28.8). The magnitude of the temperature shift with seeding concentration is considerably less than observed with the α-Al$_2$O$_3$ seeding. It is possible to explain the difference by noting that the γ-Al$_2$O$_3$ powder has a transformation peak temperature of 1150°C and therefore is probably not fully transformed to α-Al$_2$O$_3$ until the sample reaches this temperature. Consequently, the α-Al$_2$O$_3$ seed for nucleating the gel transformation does not develop in the sample until it is sufficiently heated. Nevertheless, the transformation of the gel is affected because of the large number of nuclei present in the sample when it does transform. Note that there is a shift of the boehmite to γ-Al$_2$O$_3$ transformation peak temperature as a result of seeding. However, we have not yet examined this part of the transformation process in any detail.

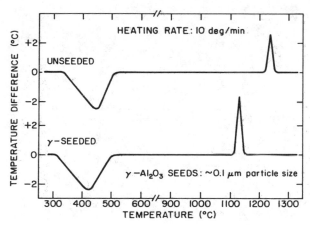

Figure 28.8 Change in the peak transformation temperature as a function of γ-Al$_2$O$_3$ seeding.

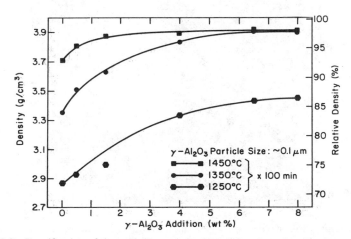

Figure 28.9 Densification of the γ-Al$_2$O$_3$ seeded gels as a function of seeding concentration.

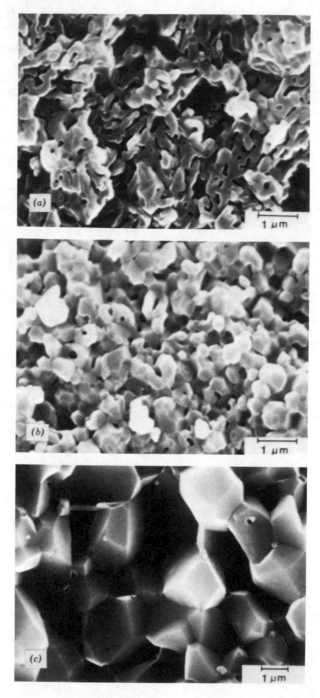

Figure 28.10 Microstructural development for the unseeded and γ-Al$_2$O$_3$ seeded gels after heating for 100 min. (a) 1250°C, unseeded, (b) 1250°C, 6.5 wt%, (c) 1350°C, 6.5 wt%.

Figure 28.9 shows the densification of the γ-Al$_2$O$_3$ seeded samples as a function of seeding concentration. At 1250°C a maximum density of 87% is obtained with the 8 wt% sample. This suggests that γ-Al$_2$O$_3$ is not as effective as α-Al$_2$O$_3$ for seeding. However, the samples with >1.5 wt% γ-Al$_2$O$_3$ sinter better than the unseeded samples. Furthermore, when these samples are heated to 1350 and 1450°C there is a significant increase in densification, relative to the unseeded samples. In general the γ-Al$_2$O$_3$ seeding yields better densification than the unseeded samples but it is not as effective as α-Al$_2$O$_3$.

Microstructures of the γ-Al$_2$O$_3$ seeded samples are compared with the unseeded samples in Fig. 28.10. The unseeded sample, heated at 1250°C for 100 min is seen in Fig. 28.10a to consist of the characteristic vermicular micro-structure commonly reported for precursors of α-Al$_2$O$_3$ after calcination. The 6.5 wt% γ-Al$_2$O$_3$ seeded sample (Fig. 28.10b) fired at 1250°C for 100 min consists of individual, submicrometer grains of α-Al$_2$O$_3$ with $\sim$15% pore phase. When heated to 1350°C for 100 min the unseeded sample develops irregular α-Al$_2$O$_3$ grains with a large pore phase within and between the grains. In contrast, the γ-Al$_2$O$_3$ seeded sample (Fig. 28.10c) is 97% dense and consists of 2–5 μm grains that have sintered to form a uniform dense structure.

The micrographs in Fig. 28.10 show that γ-Al$_2$O$_3$ seeding may also be used to control microstructural development during transformation of a boehmite gel to α-Al$_2$O$_3$. Although these results are preliminary and do not represent any attempt at optimization, it appears that γ-Al$_2$O$_3$ particles are not as effective as α-Al$_2$O$_3$ particles in promoting transformation, controlling microstructural development, or enhancing densification. This may be a result of morphological or crystallographic differences in the γ- and α-Al$_2$O$_3$ seed particles or the greater suppression of the θ- to α-Al$_2$O$_3$ transformation temperature by α-Al$_2$O$_3$. Additionally, the volume contraction associated with the γ- to α-Al$_2$O$_3$ transformation may disrupt the local surroundings of each transformed γ-Al$_2$O$_3$ particle, resulting in the loss of continuity with the surrounding metastable matrix.

SEEDED TRANSFORMATIONS

A boehmite sol–gel system was used in these studies because a gel offers the advantages of uniform dispersion of individual seed particles in the matrix as well as uniform packing of the surrounding adjacent gel matrix. Evidently, it is important that the seed particles are in intimate contact with numerous other particles to maintain continuity in the transforming structure during growth. Powder systems in which the seed and matrix particles differed little in size would not be strongly influenced by seeding. Also, it is known[9] that the number of seeds is critical to the successful control of the transformation. It has been shown that 10^{12} seeds/cm^3 has little influence on transformation of boehmite to α-Al$_2$O$_3$ whereas 5×10^{13} seeds/cm^3 (or 1.5 wt% 0.1 μm particles) represents an optimum concentration for the θ to α transformation, based on sintering results. In contrast, Dynys and Halloran have reported[5] a nucleation density of $10^8/cm^3$ for an alum-derived γ alumina. The reason for the strong influence of seed concentration is the degree of refinement (volume per seed) that is attained upon transformation.

The seeding of the α-Al_2O_3 transformation has a dramatic effect on the transformation temperature, kinetics, microstructure after transformation, and the sintering process. That seeding can have such a large effect on these processes suggests that a variety of other ceramics, which transform by a nucleation and growth process, may equally benefit by seeding their transformations. Lange[10] showed that by adjusting the relative proportion of α to β silicon nitride powder in a MgO-doped sample, the microstructure and fracture toughness could be controlled. In a later publication he[11] attributed this to control of the α to β silicon nitride transformation during liquid-phase densification. It is evident that by seeding the preferred phase of silicon nitride upon densification he was able to control microstructural development. Inoue et al.[12] showed that they could control the phase, powder size, and shape of silicon nitride by seeding its gas-phase synthesis in mixtures of SiO_2, carbon, and silicon nitride.

Seeding offers a unique and nonintrusive means of microstructural control in the above silicon nitride examples and in Al_2O_3. Additional examples where seeding has been applied or may be useful, include powder synthesis by decomposition, solid-state, vapor-phase, or liquid-phase reactions, since these are often nucleation and growth processes. Furthermore, seeding may play an important role in the adaptation of these synthetic processes to *in situ* ceramic fabrication such as has been shown here with boehmite.

REFERENCES

1. C. N. Rao and K. J. Rao, Phase Transformations in Solids, in *Progress in Solid State Chemistry*, 131–185 (1967).

2. M. Kumagai and G. L. Messing, Enhanced Densification of Boehmite Sol-Gels by Alpha Alumina Seeding, *J. Am. Ceram. Soc.* 67(11), C230–231 (1984).

3. F. W. Dynys and J. W. Halloran, Alpha Alumina Formation in Al_2O_3 Gels, in *Ultrastructure Processing of Ceramics, Glasses, and Composites*, L. L. Hench and D. R. Ulrich, Eds., 1984, pp. 142–151.

4. P. A. Badkar, J. E. Baily, and H. A. Barker, Sintering Behavior of Boehmite Gels, in *Sintering and Related Phenomena*, G. C. Kuczynski, Ed., Materials Science Research, Vol. 6, Plenum, New York, 1973, pp. 311–322.

5. F. W. Dynys and J. W. Halloran, Alpha Alumina Formation in Alum-Derived Gamma Alumina, *J. Am. Ceram. Soc.* 65(9) 442–448 (1982).

6. K. A. Morrissey, K. K. Czanderna, C. B. Carter, and R. P. Merrill, Growth of Alpha Alumina Within a Transition Alumina Matrix, *J. Am. Ceram. Soc.* 67(5), C88–90 (1984).

7. R. J. Brook, Effect of TiO_2 on the Initial Sintering of Al_2O_3, *J. Am. Ceram. Soc.* 55(2), 114–115 (1972).

8. W. H. Rhodes, Agglomerate and Particle Size Effects on Sintering Yttria-Stabilized Zirconia, *J. Am. Ceram. Soc.* 64(1), 19–22 (1981).

9. M. Kumagai and G. L. Messing, Controlled Transformation and Sintering of a Boehmite Sol Gel by Alpha Alumina Seeding, *J. Am. Ceram. Soc.* **68**(9), 500–505 (1987).

10. F. F. Lange, Fracture Toughness of Si_3N_4 as a Function of the Initial α-Phase Content, *J. Am. Ceram. Soc.* **62**(7–8), 428–430 (1979).

11. F. F. Lange, Fabrication and Properties of Dense Polyphase Silicon Nitride, *Bull. Am. Ceram. Soc.* **62**(12), 1369–1374 (1983).

12. H. Inoue, K. Komeya, and A. Tsuge, Synthesis of Silicon Nitride Powder from Silica Reduction, *J. Am. Ceram. Soc.* **65**(12), C205 (1984).

29

PROCESSING OF SOL–GEL DERIVED DOPED POWDERS AND SUBSTRATES

K. W. WISTROM AND D. E. CLARK
Department of Materials Science and Engineering
University of Florida
Gainesville, Florida

INTRODUCTION

One of the advantages of sol–gel processing is that it permits intimate mixing of multiple phases. This can be accomplished by any of several methods. For example, transition metal elements can be dissolved in a solution and then added to an alumina sol. Vigorous stirring of the sol prior to gelation will result in molecular level mixing of the two materials. After gelation and pyrolysis, the transition element will be homogeneously dispersed throughout the alumina matrix on a scale difficult to achieve with standard doping procedures, we have shown for neodymium-doped alumina.

Homogeneous mixing of a reinforcing solid phase in a ceramic matrix can also be done using sol–gel processing. The reinforcement phase is usually more dense than the liquid sol in which it is dispersed. Consequently, segregation due to settling can occur unless careful control of gelation is exercised. One way of ensuring uniform dispersion of the reinforcement is to use a gelling agent.[1] Upon gelation and drying, the material is pulverized to form a fine powder consisting of both gel and reinforcement phase. The powders can then be pressed into substrates and fired to form dense ceramic composites with well-dispersed reinforcement. This process has been used to fabricate an alumina/ yttria-stabilized zirconia composite. Such composites appear promising for applications requiring high-strength, high-fracture toughness and resistance to high-temperature degradation.[2]

In this chapter we compare the properties of two composites, identical in composition, but prepared by different methods. In the first, an 80 wt% alumina–20 wt% zirconia composite was made using commercially available alumina and yttria-stabilized zirconia. In the second case, the same composite was made by substituting a gel-derived alumina for commercially available alumina.

PROCEDURE

A flow diagram of the steps involved in preparing the gel-derived alumina zirconia composites is shown in Fig. 29.1. Alumina sol was made by reacting aluminum sec-butoxide (1 mole) with excess hot (85°C) distilled water (100 moles) and nitric (0.0233 mole) and hydrochloric (0.070 mole) acids. The mixture was gently boiled in a round-bottom flask under a condenser column for several days until the solution cleared.

Thirty-six hundred milliliters of alumina sol was reduced to 1200 ml by evaporation and allowed to cool under slow stirring. Yttria-stabilized zirconia (21.95 g) was added to the concentrated alumina sol. The sol and zirconia were

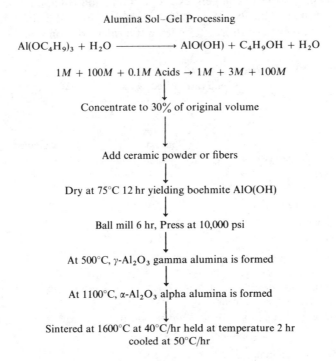

Alumina Sol–Gel Processing

$$Al(OC_4H_9)_3 + H_2O \longrightarrow AlO(OH) + C_4H_9OH + H_2O$$

$$1M + 100M + 0.1M \text{ Acids} \rightarrow 1M + 3M + 100M$$

Concentrate to 30% of original volume

Add ceramic powder or fibers

Dry at 75°C 12 hr yielding boehmite AlO(OH)

Ball mill 6 hr, Press at 10,000 psi

At 500°C, γ-Al_2O_3 gamma alumina is formed

At 1100°C, α-Al_2O_3 alpha alumina is formed

Sintered at 1600°C at 40°C/hr held at temperature 2 hr
cooled at 50°C/hr

Figure 29.1 Flow diagram of process used for preparing gel-derived composites.

mixed well, by hand, then 19.4 g of $Al(NO_3)_3 \cdot 9H_2O$, dissolved in 40–50 ml of hot distilled water, was added to gel the sol. A small amount of $Mg(NO_3)_2$ dissolved in water (0.22 g) was added as a sintering aid. The mixture was stirred for 5 min, gelled and dried at 75°C for 12 hr. The dried gel (about 120 g) was ball milled for 6 hr and the resulting powder was sieved (−48 mesh), and dried at 60°C. This procedure produced a gel powder consisting of 80 wt% alumina–20 wt% zirconia.* Rectangular bars weighing approximately 7 g each and with dimensions of 7 × 1.5 × 0.5 cm were uniaxially pressed at 70 MPa without a binder. Samples of 80% calcined alumina–20% zirconia were also pressed at 70 MPa. Calcined alumina powders† are much more dense than gel powders. Therefore, a larger amount of calcined alumina was required (10 g) to yield equivalent sample sizes to those pressed from gel powders.

Samples were fired to 1600°C at 40°C/hr, held at 1600°C for 2 hr, and cooled at 50°C/hr. Flexural strength was measured using four-point loading according to ASTM C 651-70. Span length was set at 3.81 cm and cross-head speed at 0.0254 cm/min. Apparent porosity, specific gravity, and bulk density of the samples were determined using boiling water according to ASTM C 20-80a.

RESULTS

The physical properties of the gel-derived alumina and calcined alumina composites are given in Table 29.1. Composites prepared from the alumina gel were considerably weaker than those made from the calcined alumina. This is prob-

TABLE 29.1 Physical Properties of 80 wt% Al_2O_3–20 wt% ZrO_2 Composites Fired to 1600°C

	Material	
Property	Gel-Based Al_2O_3 Composite[a]	Calcined Al_2O_3 Composite[b]
Strength (MPa)	52.1	138.2
Bulk density (g/cm³)	2.91	3.95
Apparent density (g/cm³)	4.21	4.11
Porosity (%)	31.1	3.8
Linear shrinkage (%)	48.0	29.7

[a]Yttria-stabilized zirconia from Cerac. This material contains 9.2% and had an average particle size of 2.76 μm.
[b]Calcined alumina from Alcoa. Superground with average particle size of 0.6 μm.

*Yttria-stabilized zirconia from Cerac.
†Alumina powders from Alcoa.

ably due to the higher porosity of the former. It is worth noting that the variation in strengths were much less for the gel-derived composites.

Although the gel-derived composite experienced more shrinkage during firing, its prefired density was considerably less (about 30% less) than that of the calcined alumina composite. The large shrinkages observed in these samples after firing are characteristic of gel-derived ceramics. These shrinkages usually result in cracking, making it difficult to cast monolithic bodies. This problem has been eliminated by pulverizing the cast gels, pressing, and then firing. Although large shrinkages did occur for the gel-derived composites, cracking did not occur on any of the fabricated samples.

Scanning electron micrographs of fracture surfaces are shown in Fig. 29.2 for each of the two composites. Differences can be seen in the microstructures

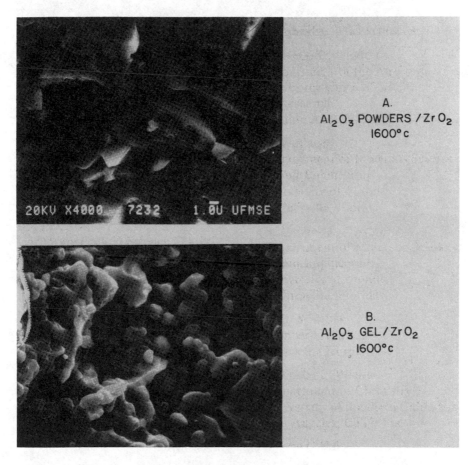

Figure 29.2 SEMs of $Al_2O_3/20$ wt% ZrO_2 bodies fired at 1600°C. (a) Prepared from calcined Al_2O_3 powders from Alcoa and ZrO_2 powders from Cerac. (b) Prepared from alumina gel and ZrO_2 powders from Cerac.

of the composites. Although the gel-derived composite is more porous, its average grain size is significantly smaller and less variable than the calcined alumina composites.

CONCLUSION

1. Al_2O_3–ZrO_2 composites made from alumina gel powders and sintered at 1600°C exhibited a large shrinkage (48.0%), high porosity (31.1%), fine grain structure, and relatively low strength. The low density of alumina gel powder (0.67 g/cm³) makes it difficult to sinter to full density.

2. Al_2O_3–ZrO_2 composites made from calcined alumina exhibited low porosity (3.8%), high bulk density (3.95 g/cm³), high strength (138.2 MPa), and

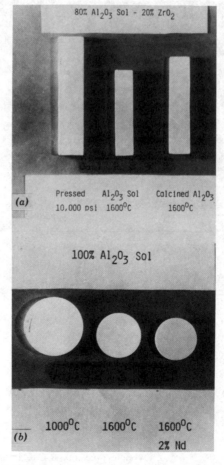

Figure 29.3   Compacts of (A) Sol-gel Al_2O_3-ZrO_2 composite and (B) Al_2O_3; as a function of processing history.

large grain size. The high reactivity (e.g., particle size of 0.6 μm) and relatively high density of the alumina powder enable it to sinter to a dense composite at 1600°C (Fig. 29.3). Although shrinkage was high, it was less than for the gel-derived composite.

ACKNOWLEDGMENT

This work was supported by the Air Force Office of Scientific Research under Contract No. F49620-83-C-0072

REFERENCES

1. R. Krabill, K. W. Wistrom, and D. E. Clark, Chemically Derived Ceramic Composites, (to be published.

2. R. W. Rice, Processing of Advanced Ceramic Composites, in *Better Ceramics Through Chemistry*, C. J. Brinker, D. E. Clark, and D. Ulrich, (Eds.), Elsevier, New York, 1984, pp. 337–346.

30

INTERFACE PROPERTIES OF TRANSITION METAL OXIDE GELS

J. LIVAGE, Ph. BARBOUX, AND E. TRONC
Spectrochimie due Solide
Université Pierre et Marie Curie
Paris, France

J. P. JOLIVET
Chimie des Polymères Inorganiques
Université Pierre et Marie Curie
Paris, France

INTRODUCTION

The "sol–gel process" has been widely studied during the last few years offering new approaches to the preparation of glasses and ceramics. Some of the main advantages of this process come from the great flexibility of the chemical reactions and the possibility of processing materials at lower temperatures.[1] In all cases, gels result from a polycondensation reaction and are only intermediates in the process. A thermal treatment always follows to remove the solvent and to densify the material. It has been shown that gels could be considered as materials and lead to new applications: antistatic coatings,[2] switching devices,[3] or electrochromic displays.[4] All these devices are based on coatings deposited from colloidal solutions and dried at low temperatures (below 80°C) so that a xerogel is obtained.

Such layers are actually biphasic materials. They contain a liquid phase (water) trapped into a solid network (transition metal oxide). They then exhibit electronic properties arising from the mixed valence oxide together with ionic properties arising from the ionization of water molecules.

In this chapter, we discuss the fundamental role of interfacial phenomena at the oxide–water interface. We will show that the electronic properties of the oxide depend on the solvent composition while the ionic properties of the liquid depend on redox reactions inside the transition metal oxide.

REDOX REACTIONS AND HYDRATION OF VANADIUM PENTOXIDE GELS

Vanadium pentoxide gels $V_2O_5 \cdot nH_2O$ can be obtained through the poly-condensation of vanadic acids. They are made of entangled fibres that look like flat ribbons about 10^3 Å long and 10^2 Å wide.[5] When deposited onto a substrate, these ribbons are stacked one upon the other giving rise to a turbo-stratic one-dimensional order in a direction perpendicular to the plane of the ribbons and the surface of the substrate. This can be very easily shown by X-ray diffraction experiments where a series of 001 peaks is observed.[6] The basal spacing, d, between the ribbons, depends on the amount of water in the $V_2O_5 \cdot nH_2O$ gel: $d = 11.5$ Å for a xerogel dried under ambient conditions and containing about $1.8H_2O$ per V_2O_5. Under vacuum another phase, corresponding to $V_2O_5 \cdot 0.5H_2O$, is obtained, in which the basal spacing $d = 8.7$ Å. By comparison with sheet silicates, the 2.8 Å shortening of the d-spacing can be attributed to the loss of one intercalated water layer.[7] Hydration and dehydration of $V_2O_5 \cdot nH_2O$ gels remain reversible down to $n = 0.5$, the n value depending mainly on the partial water pressure above the sample.

Further dehydration can only be obtained by heating. This process is no longer reversible and leads to the crystallization of orthorhombic V_2O_5.[8]

Vanadium pentoxide gels always contain some vanadium ions in a reduced state, namely V(IV). They are therefore mixed-valence compounds and exhibit semiconducting properties arising from an electron hopping between V(IV) and V(V) ions.[2] Such electronic conductivity can be only observed on V_2O_5–$0.5H_2O$ xerogels, when most of the water has been removed. It depends on the V(IV) amount, $\sigma = 4.10^{-5} \, \Omega^{-5} \, cm^{-1}$ at room temperature for a xerogel containing 1% of V(IV) ions.[9] Electronic properties of V_2O_5 xerogels can be described following the small polaron model.[2]

Ionic conductivity prevails as soon as some water is intercalated between the V_2O_5 ribbons. ac conductivity then depends on the amount of water in the xerogel and therefore on the water pressure above the sample, as shown in Fig. 30.1. Note that $V_2O_5 \cdot 1.8H_2O$ xerogels appear to be very good proton conductors. Their room temperature conductivity is quite high, $\sigma = 1.5$ $10^{-2} \, \Omega^{-1} \, cm^{-1}$.[9] Such xerogels can be described as particle hydrates as defined by Goodenough.[10] They are negatively charged particles held together by water and hydronium ions and exhibit all the properties typical of such hydrous oxides, that is, high proton conductivity[9] ion-exchange behavior[11] and easy

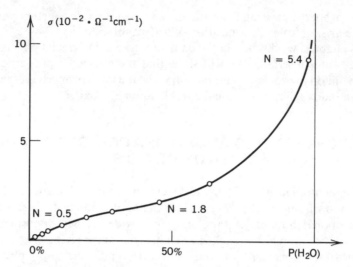

Figure 30.1   Dependence of ionic conductivity in V_2O_5 on water vapor pressure.

densification by cold pressing. The high proton conductivity presumably arises from the ionization of water molecules at the oxide–water interface.

ESR and ENDOR experiments performed on $V_2O_5 \cdot nH_2O$ xerogels suggest some couplings between the electronic charge carriers, V(IV) ions, and the water molecules or protons at the interface of the colloidal particles.[12] We may then expect that increasing the amount of reduced vanadium ions should lead to some modification of the water layers between the V_2O_5 ribbons. There-fore, partial reduction, using either organic or inorganic reducing agents, has been done. The red color of the gel gradually turns to green. This is due to the development of an optical intervalence band around 1430 nm, in the near infra-red part of the spectrum.[13]

Thermal analysis of the reduced $V_2O_5 \cdot nH_2O$ xerogels shows that the amount of water trapped in the oxide network has increased. Under ambient conditions ($pH_2O = 13$ torr), $n = 1.8H_2O$ for a ratio V(IV)/V $= 0.01$. It increases up to $n = 2.4H_2O$ when V(IV)/V $= 0.16$.[13]

X-ray diffraction experiments on a xerogel deposited on to a substrate, show that the layered structure is conserved as long as the amount of V(IV) ions is less than 20%. Beyond this value, flocculation occurs. Moreover, the basal spacing arising from the stacking of the colloidal particles increases to $d = 14.3$ Å, under ambient conditions ($pH_2O = 13$ torr), for a xerogel containing more than 10% of reduced V(IV) ions. Such a structure is quite stable and can be kept for months without much variation of the water amount and the d spacing.

The increase of the basal spacing, $\Delta d = 2.8$ Å, together with the increase of the amount of water, $\Delta n = 0.6$, suggest that another water layer has been intercalated between the ribbons.[7] A two-phase X-ray diagram is observed when the amount of V(IV) ions lies between 5 and 10%, corresponding to the two basal spacings $d = 11.5$ Å and $d = 14.3$ Å.

EXCHANGE AND REDOX REACTIONS AT THE INTERFACE OF IRON OXIDE COLLOIDS

Colloidal solutions of magnetic iron oxide can be obtained through the precipitation of an aqueous mixture of $FeCl_2$ and $FeCl_3$. The precipitate is then peptized either by $N(CH_3)_4OH$ or $HClO_4$ leading to anionic or cationic sols. The stability of such aqueous colloidal solutions is due to surface charges on the oxide that can be either negative or positive depending on the pH. Flocculation occurs at the point of zero charge, around pH 8–9.[14]

Electron microscopy and X-ray diffraction studies show that cationic sols are made of roughly spherical particles with a mean diameter of about 100 Å.[15] They exhibit a spinel structure that stands between Fe_2O_3 and Fe_3O_4, depending on their Fe(II) content.[16] Unlike anionic sols which are stable under anaerobic conditions, cationic sols release Fe(II) ions spontaneously, so that they are close to γ-Fe_2O_3. However, their composition can be modified by adding Fe(II) to the solution.[17]

Figure 30.2 shows the titration curves of the sol, for various Fe(II) amounts added to the solution prior to titration. Two types of curve are observed, with a single stage around pH 6–7 (a-type) or with another one around pH 8.5–9 (b-type). This latter stage corresponds to the $Fe(OH)_2$ precipitation from free Fe(II) ions no longer interacting with the colloid. The main result comes from the fact that free Fe(II) cannot be seen as long as the amount of added Fe(II) remains smaller than a limited value. This suggests that Fe(II) ions should be sorbed by the colloid. Taking into account the colloid's own Fe(II) content, we deduce that cationic colloids can adsorb Fe(II) species up to an overall Fe(II)/Fe(III) ratio of about 0.5, as in Fe_3O_4. Figure 30.2 also shows that whatever the amount of Fe(II), twice that amount of OH^- is needed to get complete neutralization, suggesting a net exchange reaction between one Fe(II) in the solution and two protons at the surface of the colloid.

Structural analysis, by powder diffraction, performed on protonated (H) and exchanged (E) colloids shows that only a spinel-like phase could be detected, and any variation in the unit–cell parameter (0.835 nm) could not be seen. Iron distribution among the spinel lattice, determined by measurement of integrated intensities, is reported in Table 30.1. It leads to an overall unit cell content of $Fe_{21.2}O_{32}$(H) and $Fe_{21.8}O_{32}$(E), respectively. Such a small variation

TABLE 30.1   Iron Site Population based on 32 Oxygens Per Average Unit Cell

	Colloid H	Colloid E
O_h sites	12.9	13.7
T_d sites	8.3	8.1
Total	$Fe_{21.2}O_{32}$	$Fe_{21.8}O_{32}$

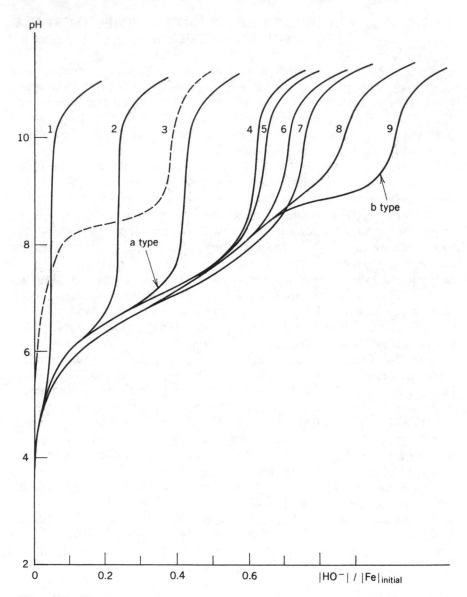

Figure 30.2   Titration curves of iron oxide sol as a function of Fe(II) added prior to titration.

cannot be interpreted in terms of iron diffusion toward the vacancies inside the core of the colloid. It suggests an extension of the oxygen-ordered array. Therefore, colloid E should be described as a coherently scattering two-phase system, made of a core H, surrounded by an ordered outer shell. Mass balance, based on X-ray and chemical data allows determination of its composition. The epitaxial shell is shown to have the same composition and structure as Fe_3O_4, its thickness being about 0.5 nm.[17] In the average cell E, the Fe(II) amount is just half the overall octahedral site population, suggesting an inverse spinel-type distribution of the cations in the outer shell and in the core, despite the existence of iron vacancies.

CONCLUSION

Transition metal oxide gels exhibit both electronic and ionic properties arising either from the mixed valence colloidal particles or from the aqueous solvent. Experiments described in this chapter suggest that electronic and ionic behaviors are related and can only be fully understood by taking into account interfacial phenomena occurring at the oxide–water interface.

Vanadium pentoxide gels, $V_2O_5 \cdot nH_2O$, exhibit a layered structure in which water molecules can be intercalated. The amount of water, n, and therefore the number of intercalated water layers depends on the partial water pressure above the sample.[7] It also depends on the ratio of reduced V(IV) ions inside the colloidal particles. It has been shown that n increases upon reduction of V_2O_5. A stable phase containing two intercalated water layers is then obtained. $V_2O_5 \cdot nH_2O$ gels belong to the so-called particle hydrate group[10] in which charged particles are held together by water and either H_3O^+ or OH^- ions. We may then assume that redox reactions could modify the charge equilibrium and therefore the liquid-phase composition trapped into the solid network.

Cationic iron oxide colloids whose structure and composition are close to γ-Fe_2O_3 are able to adsorb a large amount of Fe(II) ions coming from the aqueous solution. This uptake leads to an extension of the spinel framework and the formation of a Fe_3O_4 outer shell. Moreover, Fe(II) adsorption can only proceed via an electron transfer from Fe(II) at the surface of the colloid toward Fe(III) in the core. Adsorption stops as soon as the octahedral sites, inside the core, become equally populated by Fe(II) and Fe(III) ions as in the inverse spinel structure. Fe(II)–Fe(III) electron transfer thus appears to be the driving force governing Fe(II) adsorption.

REFERENCES

1. H. Dislich, Glassy and Crystalline Systems from Gels, *J. Non-Cryst. Solids* **57**, 371–388 (1983).
2. C. Sanchez, F. Babonneau, R. Morineau, J. Livage, and J. Bullot, Semi-Conducting Properties of V_2O_5 Gels, *Philos. Mag. B* **47**, 279–290 (1983).
3. J. Bullot, O. Gallais, M. Gauthier, and J. Livage, Threshold Switching in V_2O_5 Layers Deposited from Gels, *Phys. Status Solidi A* **71**, K1-4 (1982).
4. A. Chemseddine, R. Morineau and J. Livage, Electrochromism of Colloidal Tungsten Oxide, *Solid State Ionics* **9–10**, 357–162 (1983).
5. J. J. Legendre and J. Livage, V_2O_5 Gels: 1. Structural Study by Electron Diffraction, *J. Colloid Interface Sci.* **94**, 75–83 (1983).
6. J. J. Legendre, P. Aldebert, N. Baffier, and J. Livage, V_2O_5 Gels: II. Structural Study by X-ray Diffraction, *J. Colloid Interface Sci.* **94**, 84–89 (1983).

7. P. Aldebert, H. W. Haesslin, N. Baffier, and J. Livage, V_2O_5 Gels: III. X-ray and Neutron Diffraction Study of Highly Concentrated Systems: One-Dimensional Swelling, *J. Colloids Interface Sci.* **98**, 478–483 (1984).

8. P. Aldebert, N. Baffier, N. Gharbi, and J. Livage, Layered Structure of Vanadium Pentoxide Gel, *Mater. Res. Bull.* **16**, 669–676 (1981).

9. P. Barboux, N. Baffier, R. Morineau, and J. Livage, Diffusion protonique dans les xerogels de V_2O_5, *Solid State Ionics* **9–10**, 1073–1080 (1983).

10. W. A. England, M. G. Cross, A. Hamnet, P. J. Wiseman, and J. B. Goodenough, Fast Proton Conduction in Inorganic Ion-Exchange Compounds, *Solid State Ionics* **1**, 231–249 (1980).

11. P. Aldebert, N. Baffier, J. J. Legendre, and J. Livage, V_2O_5 Gels: A Versatile Host Structure for Intercalation, *Rev. Chim. Minérale* **19**, 485–495 (1982).

12. P. Barboux, D. Gourier, and J. Livage, ESR and ENDOR Study of V_2O_5 Gels, *Colloids Surf.* **11**, 119–128 (1984).

13. F. Babonneau, P. Barboux, F. A. Josien, and J. Livage, Reduction of V_2O_5 Gels, *J. Chim. Phys.* (in press).

14. R. Massart, Preparation of Aqueous Magnetic Liquids in Alkaline and Acidic Media, *IEEE Trans. Magn.* **17**, 1247–1248 (1981).

15. J. P. Jolivet, R. Massart, and J. M. Fruchart, Synthèse et étude physicochimique de colloides magnétiques non surfactés en milieu aqueux, *Nouv. J. Chim.* **7**, 325–331 (1983).

16. E. Tronc, J. P. Jolivet, and R. Massart, Defect Spinel Structure in Iron Oxide Colloids, *Mater. Res. Bull.* **17**, 1365–1369 (1982).

17. E. Tronc, J. P. Jolivet, J. Lefebvre, and R. Massart, Ion Adsorption and Electron Transfer in Spinel-Like Iron Oxide Colloids, *J. Chem. Soc. Faraday Trans. I* **80**, 2619–2629 (1984)..

31

PREPARATION OF BULK AND THIN FILM Ta$_2$O$_5$ BY THE SOL–GEL PROCESS

H. C. LING
AT&T Technologies, Inc.
Princeton, New Jersey

M. F. YAN AND W. W. RHODES
AT&T Bell Laboratories
Murray Hill, New Jersey

INTRODUCTION

The use of tantalum pentoxide, Ta$_2$O$_5$, for the storage capacitor dielectric in future high-density dynamic RAM devices has been proposed.[1] With a reported dielectric constant of about 25,[2] Ta$_2$O$_5$ provides an approximately sixfold increase in capacitance density as compared to SiO$_2$. Thus a sufficiently large stored charge per memory cell can be maintained without reducing the dielectric thickness to impractical values, as required if SiO$_2$ continues to be used as the dielectric.

Various methods of forming tantalum pentoxide thin films for use as gate dielectrics have been investigated. Generally tantalum films were deposited by electron-beam evaporation or by rf sputtering. The Ta film was anodized[1] or thermally oxidized[3] below 600°C to form amorphous Ta$_2$O$_5$. Low-pressure chemical vapor deposition (LPCVD) of Ta$_2$O$_5$ has also been attempted.[4] The common problem encountered is the relatively low crystallization temperature, between 600 and 700°C, of thin film Ta$_2$O$_5$. Unless the crystallization temperature is raised to $\geqslant 900°C$, polycrystalline Ta$_2$O$_5$ will form as a result of current memory cell design and processing temperature specification for IC fabrication.

As an alternative to the deposition techniques described above, we have investigated the formation of thin film metal oxides using the sol–gel process. Starting with the use of metal–organic compounds, particularly metal alkoxides, an oxide network can be formed by chemical polymerization at low temperatures by controlled hydrolysis. Metal alkoxide is first mixed with a solvent such as the corresponding alcohol, and water is added. As the polymerization progresses and as the solvent gradually evaporates, the solution becomes more viscous and eventually forms a gel. This process has been studied extensively[5–8] as a method of forming monolithic oxide glass instead of the usual solidification of oxide melts from elevated temperatures. An application relevant to the telecommunication industry may be the fabrication of glass preforms for optical fibers.[8] Yoldas[9] had also studied the use of solution-prepared TiO$_2$ film as optical coatings. Our interest in the sol–gel process stems from the possibility of formulating multicomponent metal oxides such as BaTiO$_3$ as sols and the subsequent use of spin-on technique to form thin film oxides. It would be much more difficult to prepare thin films of these complex oxides using vapor deposition or sputtering techniques. In this chapter we discuss the preparation and characterization of Ta$_2$O$_5$. Some of the spin-on characteristics, such as thickness uniformity on silicon wafers, should be applicable to other oxides prepared with the sol–gel technique.

EXPERIMENTAL PROCEDURES

We used ethoxides of the respective metals as the source of alkoxides, as they are the most readily available. In the concentrated form, the ethoxides are easily hydrolyzed when exposed to atmospheric air. Thus we first prepared a large batch of the diluted solution of the ethoxide in alcohol in a dry box. This diluted ethoxide solution became the source of subsequent chemical manipulation.

Tantalum ethoxide,* Ta(C$_2$H$_5$O)$_5$, was mixed with 100% pure ethyl alcohol, C$_2$H$_5$OH, and stirred thoroughly to form a 0.5M solution. Similarly, triple-distilled water was mixed with ethyl alcohol to form a 5M water solution. These solutions were stored in polyethylene bottles to eliminate possible contamination by Na. When equal volume of the two solutions are mixed together, the water to metal molar ratio equals 10. In general, an acid is also added to prevent rapid hydrolysis and precipitation of the metal oxide.[11] We had investigated the effect of two acids: HCl, and CH$_3$COOH. The hydrochloric acid was first prepared as 0.1M HCl solution in pure ethyl alcohol while acetic acid was used in its pure form. Selected amounts of the acid or acid solution was first mixed thoroughly with 30 g of 5M water solution, which was then added slowly to an

*Alfa Products, Danvers, Ma.

equal amount of the 0.5M Ta ethoxide solution under constant stirring. Tables 31.1 and 31.2 list the acid concentrations investigated for HCl and acetic acid, respectively. The acid concentrations which resulted in a clear sol and the longest gelation time were selected for subsequent experiments. The optimum acid to water molar ratio is 0.005 for HCl and 1.33 for acetic acid.

Whenever thin-film samples were to be prepared, new Ta_2O_5 sol was made by mixing equal amounts of the 0.5M ethoxide solution with the 5M water solution. Within 1 hr of preparation, we used a pipette to deposit 1 or 2 mL of sol to cover the entire surface of 51 mm (2 in.) or 76.2 mm (3 in.) Si wafers, which had already been cleaned using a HCl—HF solution and dried. The wafers were then spun at different speeds on a photoresist spinner. The remaining sol was poured into petri dishes and allowed to gel and dry at room temperature. We used Si wafers with either (100) or (111) orientations. They are p-doped with a specific resistivity of 0.001 $\Omega \cdot cm$.

Thermogravimetric analysis (TGA)* was used to measure the weight loss of bulk pieces of gel as a function of temperature. Continuous weight measurement was recorded between 25 and 1000°C at a heating rate of 20°C/min. However, this method does not have sufficient sensitivity to measure the weight loss of Ta_2O_5 gel spun on the Si substrate.

TABLE 31.1   Gelation Time Versus HCl Concentration (H_2O/Ta = 10)

Mixture	HCl/Alkoxide Molar Ratio	At Mixing	Gelation Start (hr)	Gelation Complete (hr)
1	0	White precipitate	—	—
2	0.025	Clear liquid	2.0	4.25–5
3	0.0375	Clear liquid	2.5	4.25–5
4	0.050	Clear liquid	3.0	5+
5	0.117	Clear liquid	0.8	1.00–1.25

TABLE 31.2   Effect of Acetic Acid (H_2O/Ta = 10) on Mixture

Mixture	$\dfrac{\text{Acetic Acid}}{\text{Alkoxide}}$ (Molar Ratio)	After Mixing
1	2.66	Cloudy after 3 min
2	5.33	Cloudy after 15 min
3	13.32	Clear liquid after 30 min gelation complete after 4 hr

*DuPont Model 1090 thermal analysis system.

Bulk pieces of the dried gel were fired at different temperature between 100 and 1200°C in flowing O$_2$. Then they were crushed into fine powders for powder X-ray diffraction (XRD) analysis. Similarly, cut pieces of Ta$_2$O$_5$ gel-coated Si wafers were fired in O$_2$ and in vacuum (<1 torr). The thin-film thickness and refractive index were measured with an ellipsometer* before and after firings. For XRD analysis, we used an automated Philips XRD system with Cu radiation filtered through Ni to study the crystallization process on powder and thin film samples.

RESULTS AND DISCUSSION

General Consideration of the Sol–Gel Process

The formation of oxides from metal alkoxides is believed to proceed through two simultaneous reactions: hydrolysis and polymerization. This is shown schematically in Fig. 31.1 for tantalum pentoxide. The polymerization reaction forms bridging oxygens and leads to localized oxide networks. The extent of this localized polymerization depends on the rates of reaction, which in turn are determined by the type of alkoxide, dilution of the system by the addition of a solvent, availability of water relative to the metal alkoxide content, the reaction temperature, and so on. It was found[11] that in SiO$_2$, Al$_2$O$_3$, and TiO$_2$, the oxide content of the hydrolysis product reaches a saturation value when the water to metal alkoxide mole ratio is between 10 and 20. Therefore in this study, we prepared Ta$_2$O$_5$ gel at room temperature with the water to alkoxide ratio fixed at 10.

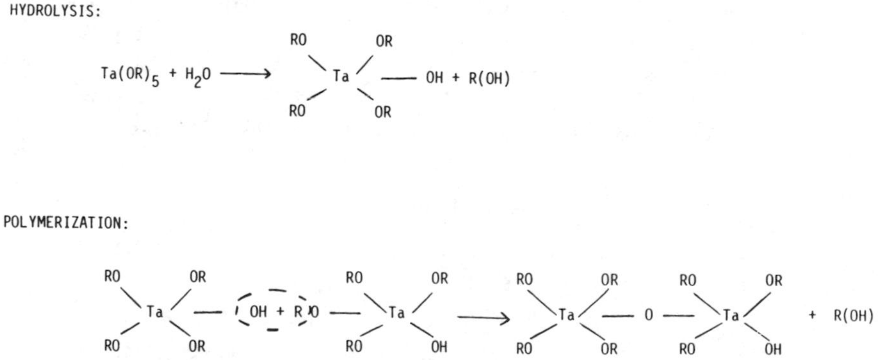

Figure 31.1   Hydrolysis and polymerization reactions in the sol–gel process.

*AutoEL-II automatic ellipsometer, Rudolph Research, Flanders, NJ.

Effect of Acid Addition

When equal amounts of the tantulum ethoxide solution and water solution were mixed together, a very fine, white precipitate of Ta_2O_5 formed immediately. This is an irreversible step; neither heat nor more alcohol solvent will redissolve the precipitates.

In general, a critical amount of acid is introduced in the sol–gel process to prevent the precipitation of metal oxide powders. While acid additions are generally specified in terms of pH, it was observed that in the preparation of alumina sol,[11] the type of acid played a much more important role than the pH of the system. The specification of the acid to alkoxide ratio was a more sensitive designation to ensure reproducibility of the resultant sols. We shall use the same parameter to describe our results.

We investigated the effect of HCl and CH_3COOH on the gelation process. It was found that when the acid concentration exceeded a certain minimum value, precipitation would not occur and the mixture of alkoxide and water solution remained as a clear liquid at the time of mixing. After an incubation time, which varied with the acid concentration, the liquid gradually became more viscous and sometimes a little hazy in appearance. This time was noted as the start of gelation. When the gelation was completed, the mixture no longer flowed like a liquid; it moved as a whole piece. Because the gelation times were determined visually, they can only be treated as rough estimates. The total time between the start and finish of the gelation process was also dependent on the acid concentration. In Tables 31.1 and 31.2, we list the approximate time spans estimated from visual observation of the gelation process. With HCl addition, we observed that the alkoxide and water solution remained as clear liquid when the HCl/alkoxide molar ratio was greater than 0.025. The incubation time to the start of gelation ranges between 2 and 3 hr when the HCl/alkoxide ratio was between 0.025 and 0.05. This offers a reasonable period of time to perform the spin-coating operation. Then it took approximately another 2 hr for the gelation to complete. With higher HCl to alkoxide ratio, both the incubation time and the time to complete gelation decreased. Thus the longest incubation and gelation times occur at an intermediate HCl/alkoxide molar ratio of 0.05. In the case of acetic acid, a much higher concentration was needed to prevent precipitation throughout the gelling process.

Based on these results, we used solutions prepared with a HCl to alkoxide molar ratio of 0.05 or an acetic acid to alkoxide ratio of 13.32 for subsequent studies.

The extent of localized polymerization during formation of oxide network can be considered as the size of the "particle," which depends on the rate of reaction. The role of the acids is to provide a charged electrolyte surrounding the "particles," thus controlling the coalesence of these "particles" into forming visible precipitates while the polymerization process is progressing to form larger and larger oxide networks. Even though acetic acid has a high $[H^+]$ concentration and a low pH value, it is a weak electrolyte. Therefore a much

CONTROLLED HYDROLYSIS

1. DILUTE BOTH THE METAL PRECURSORS AND WATER WITH A MUTUAL SOLVENT

2. CONTROL THE ELECTROLYTE CONCENTRATION TO INCREASE SURFACE CHARGE ON "PARTICLES" TO AVOID PRECIPITATION AND ENHANCE GELATION

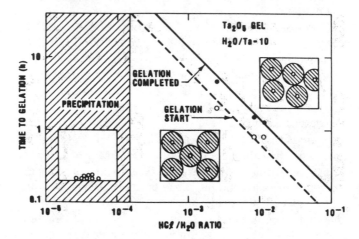

Figure 31.2   Effect of HCl addition on the gelation process.

higher concentration of acetic acid is needed to control the hydrolysis step. Figure 31.2 shows schematically the concentrations where HCl would be effective in preventing precipitation in the solution.

Studies on Bulk Ta$_2$O$_5$ Gel

Thermogravimetric analysis in flowing O$_2$ of dried gels containing HCl is shown in Fig. 31.3. There are three distinct regions of weight loss. The first region occurs below 100°C and is associated with the evaporation of ethyl alcohol, which has a boiling point of 78.5°C. The weight loss ranges between 9 and 11% for different samples. The major weight loss occurs in region II, between 400 and 500°C, with a loss equivalent to 33% of the total sample weight. A third, minor weight loss of less than 1% occurs between 700 and 750°C. By 800°C only 55% of the total weight is retained. The weight loss in regions II and III is probably related to the loss of C and H from the gel complex.

A TGA curve for gels containing acetic acid in flowing O$_2$ is shown in Fig. 31.4a. This run was conducted at a heating rate of 10°C/min. There are also three regions of weight loss in approximately the same temperature ranges as the gel containing HCl. The slightly lower transition temperatures are probably a result of the slower heating rate. While regions I and III exhibit similar weight loss percentages as in the case of HCl, region II shows a much reduced value of 5.68%. After firing to 800°C, approximately 84% of the total weight is retained. Changing the atmosphere to flowing N$_2$ in the TGA run does not change the

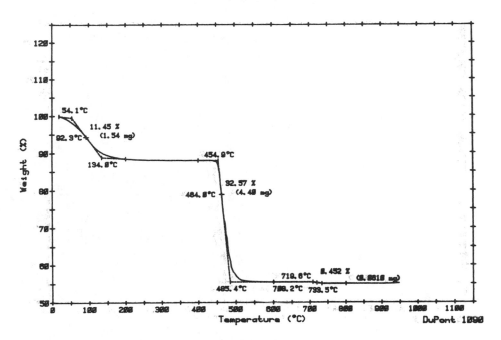

Figure 31.3 Thermogravimetric analysis of dried Ta$_2$O$_5$ gel with HCl addition (HCl/Alkoxide = 0.05), O$_2$ atmosphere. Heating rate = 20°C/min.

temperature ranges or percentages of weight loss. However, continuous change in weight is observed above 750°C, indicating the reduction of Ta$_2$O$_5$ (Fig. 31.4b).

To study the phase transitions with temperature, small pieces of the dried gel, typically less than 2mm in any dimensions, were heated in Pt crucibles to temperatures between 100 and 1200°C for 2 hr in flowing O$_2$. Then the pieces were crushed into very fine powders for XRD analysis. The results are from gels containing HCl; we did not observe significant differences in gels containing acetic acid. The color of the gel changed from clear and whitish to slightly brownish after the 300°C heat treatment. Between 350 and 500°C the gel turned black, indicating that some carbonization of the gel may have occurred. Above 500°C, the gel was again whitish in color.

The powder XRD patterns of the Ta$_2$O$_5$ gel heat treated at various temperatures are shown in Fig. 31.5. The intensity has been normalized to the same scale of 100 counts/s, thus comparing the peak intensities at different temperatures gives a realistic indication of the progressing crystallization with increasing temperature. We also include the standard pattern* of the orthorhombic Ta$_2$O$_5$ phase for comparison.

*JCPDS file # 25-922.

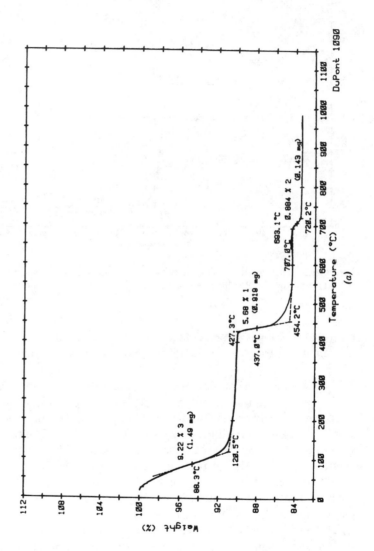

(a)

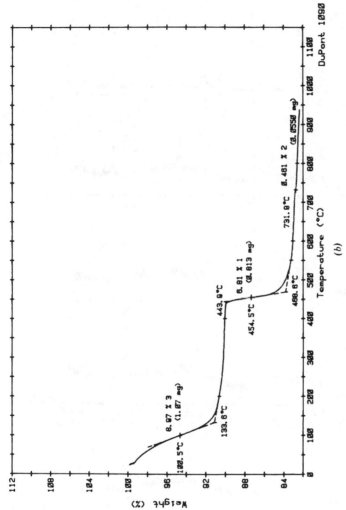

Figure 31.4 (*a*) Thermogravimetric analysis of dried Ta_2O_5 gel with acetic acid addition (acetic acid/alkoxide = 13.32), O_2 atmosphere. Heating rate = 10°C/min. (*b*) Thermogravimetric analysis of dried Ta_2O_5 gel with acetic acid addition (acetic acid/alkoxide = 13.32), N_2 atmosphere. Heating rate = 20°C/min.

293

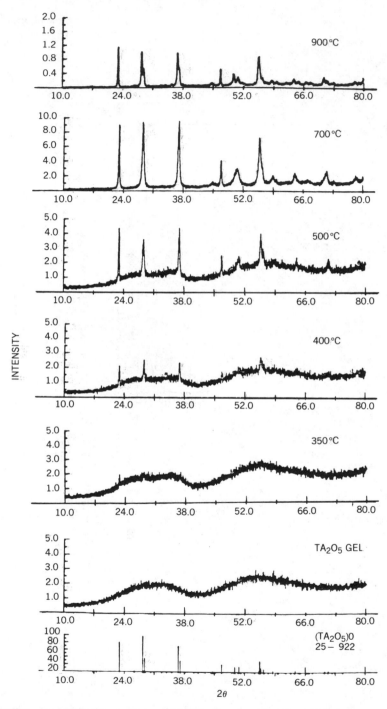

Figure 31.5 X-ray diffraction patterns of dried Ta_2O_5 gel as a function of firing temperature (HCl/alkoxide = 0.05).

294

Before heat treatment, Ta_2O_5 gel shows two very broad X-ray diffraction peaks superimposed on a background intensity which increases with the 2θ angles. Both broad peaks encompass 2θ ranges where the most intense peaks are found in crystalline Ta_2O_5. Thus, while the as-formed gel is amorphous, short-range order of the Ta_2O_5 crystalline structure is already present at room temperature. At 350°C, very faint peaks can be seen at the 2θ positions where the three most intense crystalline peaks are expected. Their intensities increase with firing temperature while the background intensity decreases. At 700°C, the background intensity has reached the normal level while the crystalline peaks continue to evolve towards the orthorhombic form. At 900°C both the $(11\overline{1}0)$ peak $(2\theta = 28.3°)$ and the (200) peak $(2\theta = 28.9°)$ are revealed, with $(11\overline{1}0)$ having the higher intensity. Similarly, the $(11\overline{1}1)$ peak at $2\theta = 36.7°$ has a higher intensity than the (201) peak at $2\theta = 37.2°$. The diffraction pattern at 900°C is identical to the standard JCPDS pattern #25-922.

The crystallization of Ta_2O_5 is a nucleation and growth phenomenon as demonstrated in Fig. 31.6 where the XRD patterns of samples heated at 400°C for 2, 8, and 16 hr are compared. The intensities of the peaks increase with annealing time, indicating a growing volume fraction of crystalline Ta_2O_5.

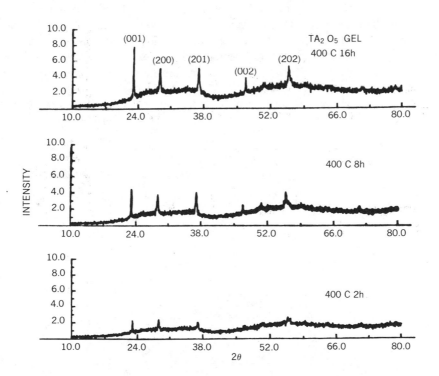

Figure 31.6 X-ray diffraction pattern of dried Ta_2O_5 gel as a function of time at 400°C (HCl/alkoxide = 0.05).

Thin-Film Ta$_2$O$_5$

Physical Characteristics of Spin-On Gel

Since the spinning process depends on the viscosity of the sol, there might be differences in using sols prepared with the addition of HCl or acetic acid. Visual inspection cannot distinguish Si wafers coated with the two different kinds of sols. In general, the color was quite uniform across the entire wafer surface, however, there were often radial streaks of colors different from the uniform background color. Under optical microscope, we did not find small particles which might give rise to streaking. It is more likely that these streaks arise from uneven evaporation of the solvent. A preliminary experiment using butyl alcohol instead of ethyl alcohol as the solvent seemed to eliminate the streaks. In any case, nearly all results are with ethyl alcohol as solvent. Furthermore, we report spin-on data on Ta$_2$O$_5$ sol prepared with addition of acetic acid, since TGA results on bulk gel indicated that nearly 84% of the total weight was retained up to 800°C.

Figure 31.7 shows the average as-spun thickness as a function of spinning speed. Thickness measurements were made after the sol had been deposited on the Si wafer for at least one day, so that gelation was complete. Further evaporation from the gel at room temperature was assumed to be negligible. The spinning time was fixed at 20 s since we observed that longer spinning time did not change the thickness. For three different lots of identically prepared sols, we found that at the same spinning speed, the maximum variation in thickness was less than 10% whether 51-mm (2-in.) or 76-mm (3-in.) Si wafers were used or whether 1 or 2 ml of sol was initially deposited to cover the wafer surface.

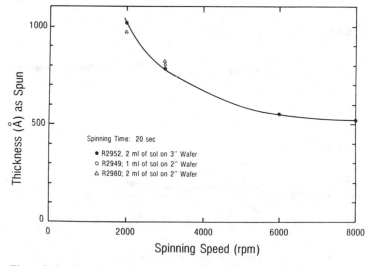

Figure 31.7   Variation of thickness as-spun on Si wafer with the spinning speed.

Furthermore, the thickness depends only on the spinning speed. This result is well known in photoresist deposition, where an empirical relationship has been established[12]:

$$t = \frac{k}{\sqrt{w}} \tag{1}$$

where t is the thickness in (Å), k is a constant combining the characteristics of a particular spinner and the viscosity of the photoresist, and w is the spinning speed in revolutions per minute divided by 1000. Using the data from Fig. 31.7, we replotted the thickness in Å versus the square root of w. This is shown in Fig. 31.8. Similar to the case of photoresist, a linear relationship is obtained, with $k = 1400$.

Typical thickness variation across a single wafer is less than $\pm 2.5\%$. Figure 31.9 shows the variation cross a 76-mm (3-in.) wafer spun for 20 s at a speed of 3000 rpm. On this wafer the deposition process was repeated twice, so that an average thickness of 1750 Å is nearly twice the thickness shown in Fig. 31.6 for a single deposition step. The refractive index is nearly constant at 1.75 across the entire wafer. This value is the same as the thin-film gel after one deposition. Thus thick layers of Ta_2O_5 gel can be built up with repeated deposition steps. Alternatively, one can envision a series of operations involving deposition, firing to form a dense Ta_2O_5 layer, followed by a second deposition and firing. This will also reduce the likelihood of pinholes, as the second deposition will fill in the holes while building up a thicker layer at the same time.

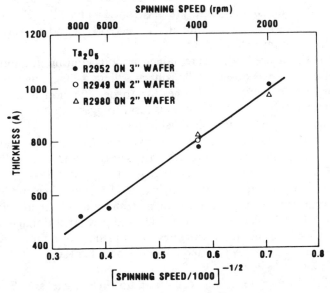

Figure 31.8 Variation of spin-on thickness with the square root of (spinning speed/1000).

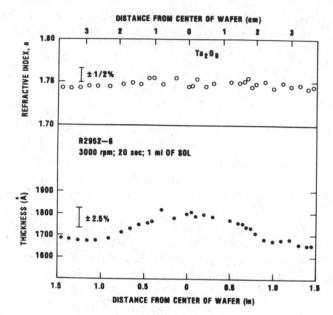

Figure 31.9   Variation of thickness and refractive index across a 76-mm (3-in.) Si wafer.

Heat Treatment

The spin-coated Si wafers were cleaved into pieces approximately 10 mm × 5 mm in size. Individual pieces were heated in flowing O_2 or vacuum (<1 torr) at a rate of 100°C/hr to a temperature between 100 and 1200°C. At the higher temperatures, this would have eliminated excess solvent and hydrocarbons and densified the resultant Ta_2O_5 film.

Up to 800°C, the fired surfaces remained smooth, continuous, and shiny. No breakup of the film was observed. This means there is no change in the lateral dimensions. Any dimensional change due to densification is limited to the direction normal to the Si substrate. This result suggests that bonding between the Si substrate and the as-deposited gel is quite strong.

Figure 31.10 shows the thickness shrinkage and changes in the refractive index with the firing temperature. The thickness decreases rapidly between 100 and 500°C, reaching a total shrinkage of 50% at 500°C. Higher firing temperatures do not cause further shrinkage. The refractive index increases with the firing temperature, attaining a plateau value of 2.1 at temperatures higher than 500°C. Both the refractive index and percent shrinkage were found to be independent of the initial thickness, t_0. Furthermore, they did not differ in either O_2 or vacuum environment. The use of either (100) or (111) Si wafers also did not affect the results.

X-ray diffraction patterns between 10 and 55° 2θ angles are shown in Fig. 31.11 for thin film samples on (100) Si substrates heated to a temperature range

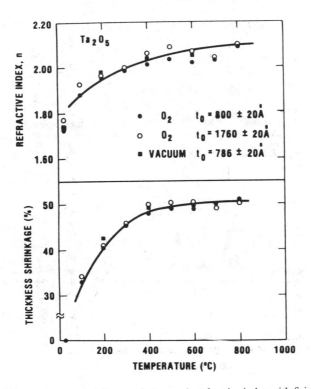

Figure 31.10 Thickness shrinkage and changes in refractive index with firing temperature.

between 500 and 700°C for 2 hr in flowing O_2. The 500°C XRD pattern is representative of samples heat treated below 500°C. The single peak at $2\theta = 33°$ is the (200) Si reflection. Below 635°C, there is no crystalline peak in the XRD pattern. Even annealing for a longer time, such as 12 hr at 600°C does not result in any crystalline phase, hence the film is amorphous. Crystallization of Ta_2O_5 begins between 635 and 650°C. Furthermore, there are only four discrete diffraction peaks in the 2θ range between 10 and 100°. Even at temperatures as high as 1200°C, the peaks at $2\theta = 28.9°$ and 37.2° did not split up into two reflections, as seen in the 900°C diffraction pattern of the bulk gel (Fig. 31.5). The four discrete peaks seen in the thin film are identified as the (001), (200), (201), and (002) reflections. Missing were the peaks of $(1\overline{1}10)$ and $(1\overline{1}11)$, which normally have the higher diffraction intensities. Thus thin film Ta_2O_5 crystallizes on (100) Si substrates with preferred orientations. Because the (001) to (200) intensity ratio observed in Fig. 31.11 is approximately the same as in the standard XRD pattern, both types of grains must be present in approximately equal volume fractions. Consequently, it is unlikely that an epitaxial layer of Ta_2O_5 can be formed on a Si single-crystal substrate. This is also obvious from the large lattice mismatch between Si ($a_0 = 5.431$ Å) and orthorhombic Ta_2O_5 ($a_0 = 6.198$ Å, $b_0 = 40.29$ Å, $c_1 = 3.888$ Å).

The crystallization temperature of thin film Ta_2O_5 is reported to be between

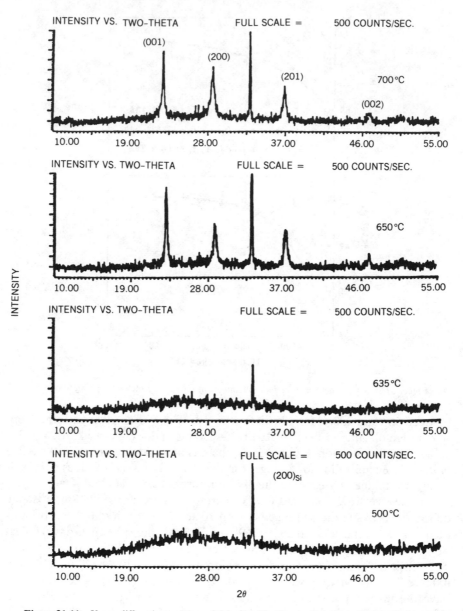

Figure 31.11   X-ray diffraction pattern of thin film Ta$_2$O$_5$ as a function of firing temperature.

600 and 700°C independent of deposition technique. We identify the onset of crystallization in thin film Ta$_2$O$_5$ to a temperature between 635 and 650°C. Compared with a value of 350°C in bulk gel, the crystallization temperature of thin film Ta$_2$O$_5$ is almost 300°C higher. This increase may be due to a reduced number of nucleation sites for crystallization.

Thin Film Ta₂O₅ Doped with Al

It is believed that polycrystalline Ta_2O_5 has a high leakage current due to grain boundary diffusion. Thus in applications where low leakage current is required, amorphous Ta_2O_5 with a high crystallization temperature is desired. Since thin film Ta_2O_5 has a crystallization temperature $\sim 650°C$ it is too low for applications related to IC fabrication. To address this problem, we doped the tantalum ethoxide solution with aluminum butoxide so that Al constituted 0.1

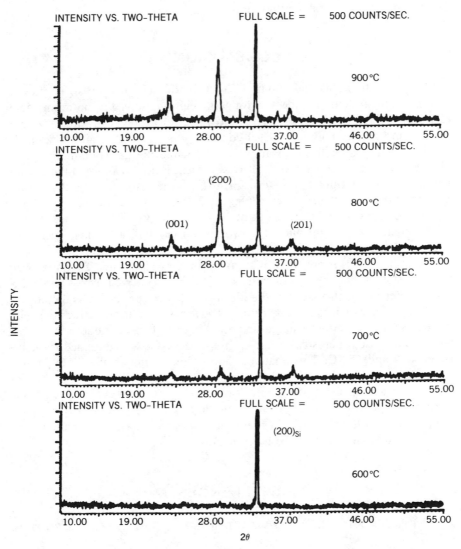

Figure 31.12 X-ray diffraction pattern of thin film Al-doped Ta_2O_5 as a function of firing temperature.

mole of the nominal metal content. The mixture was heated at 85°C for 3 hr before hydrolysis took place with the addition of acetic acid. The sol was spun onto (100) Si wafers and heat treated between 300 and 1000°C. Figure 31.12 shows the XRD patterns at different firing temperatures. The striking feature is the reduced crystalline peak intensities at 700°C compared to the undoped Ta$_2$O$_5$ film (Fig. 31.11). There is also a change in the relative intensities; that is, (200) reflection is now dominant while the (001) and (002) peak intensities are greatly diminished. From these results it appears that Al is effective to some extent in raising the crystallization temperature and in changing the preferred orientation of the crystalline thin film.

CONCLUSION

We have prepared bulk and thin film Ta$_2$O$_5$ by the sol–gel process. To prevent precipitation and to provide the longest gelation time, the optimum acid/alkoxide molar ratio is 0.05 for HCl and 13.32 for acetic acid. After drying at room temperature, bulk Ta$_2$O$_5$ gel containing acetic acid retains 84% of the original weight after firing to above 800°C while the gel containing hydrochloric acid retains only 55 wt%. However, there is no difference in the crystallization behavior. Bulk Ta$_2$O$_5$ transforms from the amorphous to the crystalline orthorhombic structure beginning at 350°C, with an increasing volume fraction of the crystalline phase at higher temperatures. The volume fraction also increases with firing time, indicating that crystallization is a nucleation and growth phenomenon.

Thin film Ta$_2$O$_5$ is prepared by spin-coating Si wafers when Ta$_2$O$_5$ is in the form of sols. A thickness uniformity of $\pm 2.5\%$ is achieved across the entire wafer. After firing, shrinkage of the film is limited to the thickness dimension, with a shrinkage of 50% at temperatures above 500°C. The refractive index also increases with the firing temperature, reaching a saturated value of 2.1 above 500°C. The onset of crystallization in thin film Ta$_2$O$_5$ is determined to be between 635 and 650°C. This is almost 300°C higher than bulk Ta$_2$O$_5$ and is attributed to the reduced number of nucleation sites. Thin film Ta$_2$O$_5$ also crystallizes with preferred orientations relative to the Si substrate. The crystallization temperature can be increased by doping the Ta$_2$O$_5$ with Al. Further improvement will be essential in order for amorphous Ta$_2$O$_5$ to be used as an insulator material in current IC fabrication.

REFERENCES

1. K. Ohta, K. Yamoda, K. Shimiru, and Y. Tarui, *IEEE Trans. Electron Devices* **ED-29**, 368 (1962).

2. L. I. Maissel and R. Glang, Eds., *Handbook of Thin Films*, McGraw-Hill, New York, 1970, Chapter 16, p. 17.

3. G. S. Oehrlein and A. Reisman, *J. Appl. Phys.* **54**, 6502 (1993).

4. S. B. Desu, D. W. Johnson, Jr., and E. M. Vogel, private communication.

5. B. E. Yoldas, *J. Mater. Sci.* **14**, 1843 (1979).

6. B. E. Yoldas, *J. Non-Crystal. Solids* **38**, 81 (1980).

7. C. J. R. Gonzalez-Oliver, P. F. James, and H. Rawson, *J. Non-Crystal. Solids* **48**, 129 (1982).

8. E. M. Rabinovich, D. W. Johnson, Jr., J. B. MacChesney, and E. M. Vogel, *J. Ceram. Soc.* **66**, 683 (1983).

9. B. E. Yoldas, *Appl. Opt.* **21**, 2960 (1982).

10. H. M. Quek and M. F. Yan, private communication.

11. B. E. Yoldas, *Ceram. Bull.* **54**, 289 (1975).

12. G. F. Damon, Collected papers from Kodak Seminars on Microminiaturization 1965–66, Kodak Publications, p. 195, 1965.

32

USE OF MIXED TITANIUM ALKOXIDES FOR SOL–GEL PROCESSING

WILLIAM C. LaCOURSE AND SUNUK KIM
Institute of Glass Science and Engineering
New York State College of Ceramics
Alfred University
Alfred New York

INTRODUCTION

A major problem in forming homogeneous multicomponent gels is the unequal hydrolysis and condensation rates of the metal alkoxides. Tetraethoxysilane (TEOS), for instance, can require up to several days for gelation while titanium and zirconium alkoxides tend to precipitate almost immediately under similar conditions.[1]

Under conditions of unequal reaction rates there will be a strong tendency for the condensation reaction to occur between molecules of the faster reacting species. In the case of TiO_2—SiO_2 sols this means that a greater than random number of Ti—O—Ti bonds will form at the expense of Ti—O—Si bonds and a microheterogeneous sol results.

Several methods for increasing the homogeneity have been attempted.[2–4] For binary TiO_2—SiO_2 sols some workers prehydrolyze the slower reacting silicon ethoxides under acid conditions prior to adding the monomer the faster titanium alkoxides species. However, in acid-catalyzed TEOS one tends to form Si—O—Si linkages and sol particles of pure SiO_2 during hydrolysis. When the titanium alkoxide is added a preferential reaction still occurs between titania but some reaction with the surfaces of preformed silicate sol particles also occur. Thus some improvement is expected but nonrandom structures still result.

Yamane[4] attempted to form homogeneous gels by increasing reaction rates of silicate to match more closely those of the titanate. Here silicon methoxide was employed and gelation was accomplished under base conditions. The homogeneity of glasses prepared was found to be superior to those formed using prehydrolysis, but neither exhibited the homogeneity of melt-formed glasses.

Sakka and Kamiya[3] have prepared acid-catalyzed binary silicate sols of titanium or zirconium alkoxides by premixing the alkoxides in the absence of water then allowing an extremely slow hydrolysis by absorption of moisture from the air. The technique appears to avoid problems of precipitation and fibers could be drawn from the sols. However, the preparation time was quite long.

In the present paper we present an approach in which the reaction between hydrolyzed titanium alkoxides is slowed considerably, while reactions between silicon and titanium monomers is enhanced.

EXPERIMENTAL PROCEDURE

The material chosen as a source for titanium is a complex alkoxide, "titanium acetylacetonate" ($Ti(C_5H_7O_2)_2(OC_3H_7)_2$, hereafter referred to as TIACA. This material is known to be stable in alcohol–water solutions with some compositions stable for days. Several preparation procedures and gel compositions were investigated. In most cases parallel reactions, using conventional titanium alkoxides, were also carried out. Specific preparation procedures are listed below.

Pure Titania Sols and Gels

Monoliths

A TIACA–alcohol solution (18% TIACA in propanol) was mixed with an equal amount of ethyl alcohol and allowed to mix for several minutes. The desired amount of water was then added dropwise. Depending on the amount of water added the solution became cloudy but would clear when the pH was decreased to below about 6.0 or above 8.0, or when the alcohol to TIACA ratio was increased to about 2 to 1 by volume. It is assumed that this effect is due to flocculation (immiscibility) of the sol. The zero point of charge on titania is near pH 6. Ionization of hydrolyzed bonds on either side of the ZPC would cause deflocculation and clearing of the sol.

Both acid and base catalysis was employed and gels could be formed using either condition. However, acid-catalyzed sols required several weeks for gelation and the gels tended to be weak. Base-catalyzed sols were prepared by addition of 2 mL of $15N$ NH_4OH to the solution. Mixing was continued for 2 hr. The sol was then stored in a closed container for final gelation.

TiO₂ Fibers

Fibers could be drawn from TIACA–alcohol–water sols using the following preparation procedures: 5 mL of TIACA, 10 mL alcohol, and 0.5 mL (2.67 mole ratio of water to TIACA) water were comixed for 5 min at room temperature. One mL of $15N$ NH_4OH was then added slowly and the sol was stirred for 30 min in air. The sol was then covered and allowed to age at room temperature. Three small pinholes were placed in the cover to allow slow evaporation of alcohol and excess water. The sol viscosity increased slowly and the system became fiberizable after 45–55 hr.

Binary TiO₂ —SiO₂ Gels

Titanium silicate gels were produced with up to 90% TiO_2. Here approximately 5 mL of the TIAC solution was mixed with the appropriate amount of TEOS and enough alcohol to provide a clear sol. Water was then added to provide the desired water contents (generally more than 5 moles of water per mole of alkoxide for clear gels). After 10 min mixing 0.6 mL of $0.5N$ HCl was added. Mixing was continued for 30 min to permit initial hydrolysis and condensation. For sols containing 10 or 20% TiO_2 1.5 mL of $0.5N$ NH_4OH was added to raise the pH and accelerate the final gelation. $5N$ NH_4OH was was used for 40% TiO_2 sols and $15N$ was used for the 60, 80, and 90% TiO_2 sols. Mixing was continued for an additional hour, after which the sols were covered and stored for gelation.

RESULTS AND DISCUSSION

The gelation behavior of TIACA-based sols was found to be quite different from those of the simpler alkoxides. TIACA sols were stable in alcohol solutions for several hours with water/TIACA mole ratios of up to 40, while those produced with conventional titanium alkoxides precipitated immediately under similar conditions. Prassas and Hench[1] have also found that titanium isopropoxide sols tend to be inhomogeneous and to crystallize partially to anatase in the gel state, if the water content is above a 3/1 mole ratio. Gel times were decreased to about 1 hr for a 2.7 ratio.

In the present study clear, amorphous gel monoliths of pure titania have been produced under a variety of conditions. Gelling times for representative compositions are provided in Table 32.1. The gels are amorphous and remain so to temperatures between 400 and 500°C. Figure 32.1 shows the X-ray diffraction spectra obtained on a sample that was heat treated for 1 hr at 300, 400, 500, and 700°C. A very broad anatase peak appears at 500°C, indicating extremely small grain size. Further heating to 700°C caused a transformation

**TABLE 32.1 Gelation Times for
TIAA-Derived TiO$_2$ Sols**

R^a	Gel Time (hr)	Appearence
4	11	Clear
10	6.33	Clear
20	5	Clear
40	3	Cloudy

^aR = moles H$_2$O/moles TIAA.

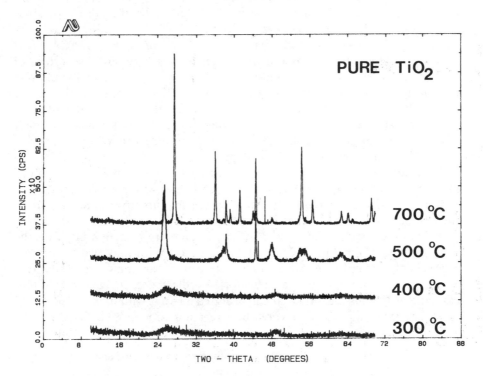

Figure 32.1 X-ray diffraction patterns for TiO$_2$ gels after various heat treatments.

to the rutile structure and a sharpening of the peaks. The sample remained transparent throughout.

It might be noted that the amorphous structure appears to be somewhat more stable in TIACA-derived sols than in similar material produced from titanium isopropoxide. Mukherjee[5] found that crystallization of small anatase particles occurs at temperatures as low as 350°C after slow heating but with no soak at this temperature. Present samples were soaked for 1 hr at 100°C intervals and yet showed no crystallinity below 400°C.

Binary TiO_2–Silicates

Monolithic gels of binary TiO_2–SiO_2 were formed with up to 90% TiO_2. Table 32.2 provides gel and fired glass characteristics for several preparation conditions for a 40% TiO_2–60% SiO_2 composition. Gel times ranged from 5 hr with base-catalyzed samples and a water to alkoxide ratio of 5.5 to 31 hr when samples were first hydrolyzed in acid, with a base-catalyzed gelation step. Again, these times are considerable and exceed those for pure TEOS sols prepared under similar conditions.[6]

Samples of the 40% TiO_2 composition were fired to temperatures between 200 and 700°C. All compositions remained clear and transparent to 700°C. X-ray analysis of the 40% gel prepared using 0.5N HCl and final gelation with 0.6 mL of 5N NH_4OH indicated no crystallization over the full temperature range studied (Fig. 32.2).

Perhaps the most interesting result is the ability of fiberize TIACA-based sols when prepared using base catalysis. Diameters between several tens of micrometers and 1 mm were handdrawn. Fiberizing from base-catalyzed sols is unusual. Silicate sols cannot be fiberized under these conditions due to the growth of noninteracting spherical particles. The ability to form molecular structures suitable for fiberizing in base-catalyzed TIACA is a strong indication of a linear polymerization mechanism. Indeed, the structure of TIACA would indicate that linear chain growth is expected.

Infrared spectra obtained on the sol[7] indicated almost immediate hydrolysis of the propoxide groups on the TIACA which suggests that the slower reaction is due to a reduced condensation rate. Reasons for this can be seen by noting the monomer structure of TIACA (Fig. 32.3).

Two of the organic groups on TIACA (C_3H_7O) are hydrolyzable while the chelate groups ($C_5H_7O_2$) are quite stable in water. With only two hydrolyzed species polymerization is forced to occur in a linear chain structure. This explains the ability to fiberize these sols. Furthermore, in binary or multicomponent systems reaction of the hydrolyzed TIACA with hydrolyzed TEOS or other species is probably favored over reaction between two TIACA mole-

TABLE 32.2 Preparation Conditions and Gelation Times
for 40 TiO_2–60 SiO_2 Sols

Sample	R	HCl (mL)	NH_4OH (mL)	Gel Time (hr)
AB40-1	2.2	0.60	0.60	31
AB40-2	4.4	0.60	0.60	11
AB40-3	11.0	0.60	0.60	6.5
B40-1	3.3		0.60	7
B40-2	5.5		0.60	5

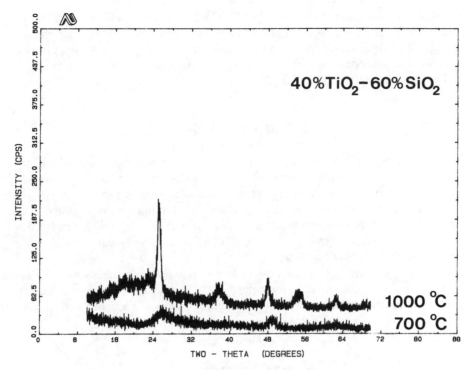

Figure 32.2 X-ray diffraction patterns for 40 TiO₂–60 SiO₂ gel after heat treatment.

```
          H   H   H
        HCH  C  HCH
          \ //\  \ /
           C      C
           |      ||
           O      O
            \    /
             \  /
H  H  H       \/            H  H  H
|  |  |       Ti            |  |  |
H-C-C-C-O -Ti- O-C-C-C-H
|  |  |      /\             |  |  |
H  H  H     /  \            H  H  H
           O    O
           ||   |
           C    C
          / \\// \
        HCH  C  HCH
         H   H   H
```

Figure 32.3 Structure of the TIAA monomer.

309

cules. Approach and reaction of small TEOS molecules would be much easier than two TIACA due to the steric henderance provided by the large chelate groups. This suggests that binary silicate gels formed using TIACA might have a *greater* than random number of Ti—O—Si linkages, particularly in the high TiO_2 content material.

CONCLUSION

Use of complex alkoxides such as TIACA, and similar materials which are available for zirconia, allows much greater control over the resulting sol structures. One can generate linear chain molecules by use of totally unreactive side groups, or randomly crosslinked chains by having organic groups with alkoxides which react at different rates. Similar crosslinking might result from use of two different alkoxides, one of which contains only two hydrolyzable species while the other has four.

In binary or complex systems use of large unhydrolyzable species can cause preferential reaction between molecular units which might otherwise not react. Our work indicates that extension to multicomponent systems such as titanium-nucleated lithium aluminosilicate glass–ceramics is possible.[7]

REFERENCES

1. B. E. Yoldas, Formation of Titania-Silica Glasses by Low Temperature Polymerization, *J. Non-Cryst. Solids* **38/39**, 81–86 (1980).

2. K. Kamiya, S. Sakka, and I. Yamanaka, Non-crystalline Solids of the TiO_2 and Al_2O_3 Systems Formed from Alkoxides, *10th International Glass Congress Proceedings*, Kyoto, Japan, 1974, pp. 13-44–13-48.

3. M. Yamane, S. Aso, and T. Sakaino, Preparation of a Gel from Metal Alkoxide and its Properties as a Precursor of Oxide Glass, *J. Mater. Sci.* **13**, 865–870 (1978).

4. M. Prassas and L. L. Hench, Physical Chemical Factors in Sol-Gel Processing, in *Ultrastructure Processing of Ceramics, Glasses and Composites*, L. L. Hench and D. R. Ulrich, Eds., Wiley-Interscience, New York, 1984, pp. 100–125.

5. S. Mukherjee, Inorganic Oxide Gels and Gel Monoliths: Their Crystallization Behavior, in *Emergent Process Methods For High Technology Ceramics*, R. F. Davis, H. Palmour III, and R. L. Porter, Eds., Plenum Press New York, 1984, pp. 95–110.

6. W. C. LaCourse, M. Md. Akhtar, S. Dahar, R. Sands, and J. Steinmetz, Factors Controlling the Sol-Gel Conversion in TEOS, *J. Can. Ceram. Soc.* **52**, 18–23 (1983).

7. T. Mattson, Glass-Ceramics from Sol-Gel Processing, M.S. Thesis, N.Y.S. College of Ceramics, Alfred University, Jan. 1985.

8. S. Sakka and K. Kamiya, The Sol-Gel Transition in the Hydrolysis of Metal Alkoxides in Relation to the Formation of Fibers and Films, *J. Non-Cryst. Solids* **48**, 31–46 (1982).

9. W. C. LaCourse, S. Kim, T. Mattson, and B. Kilinski, submitted to *J. Am. Ceram. Soc.*

33

THE EFFECTS OF PROCESSING CHEMISTRY ON ELECTRICAL PROPERTIES OF HIGH FIELD ZnO VARISTORS

R. G. DOSCH
Sandia National Laboratories
Albuquerque, New Mexico

The nonohmic properties of doped, polycrystalline ZnO, first reported in the early 1970s,[1] have led to its increasing use in electronic devices known as varistors for applications such as transient voltage suppressors and voltage regulators. At low voltages, varistors behave like high-value ohmic resistors. As applied electric fields are increased, varistors switch to a nonlinear conducting condition in which $J = KE^\alpha$ where J and E are current density and electric field, respectively, and K is a constant. The exponent α, defined as $d(\log J)/d(\log E)$, is called the nonlinearity coefficient.

From the work of numerous investigators, it is known that additives or dopants are necessary for producing ZnO varistors with the desirable properties of high insulating resistance at low voltages and large α's at high voltage. The addition of Bi_2O_3 to polycrystalline ZnO increases the low-voltage resistivity and also acts as a sintering aid, presumably due to a $ZnO–Bi_2O_3$ eutectic which forms at $\sim 750°C$. Since Bi_2O_3 is essentially insoluble in ZnO at ambient temperature, it segregates either at grain boundaries or in second phases at triple points.[2–4] Because of the former location, it is thought to contribute to the large electrostatic barriers which form at the grain boundaries.[5] Average voltage drops of 2–4 V per grain boundary are generally reported for ZnO

311

varistors operating in the nonlinear regime. Other dopants, such as Co and Mn oxides, are used to enhance specific electrical properties such as increasing α. These dopants are generally reported to be homogeneously distributed in ZnO grains.[3,4]

Commercial varistors are generally prepared for low-voltage applications ($E < 5$ kV/cm) by conventional mixed oxides techniques requiring sintering temperatures of 1000–1300°C for densification. The application resulting in this work required varistors with E in the 40–60 kV/cm range at $J = 5$ A/cm^2, $\alpha > 30$ at $2.5 \leqslant J \leqslant 5$ A/cm^2, and densities greater than 90% of theoretical. These properties could not be achieved by mixed oxide techniques due to large grain sizes (> 3 μm) and microstructural heterogeneity. The addition of inter-granular phases[6] and sol–gel methods[7] have been used to increase E; however, the electrical properties reported did not meet the above criteria.

In this work, a chemical preparation method based on the above model was used for Bi, Co, and/or Mn doped varistors containing $\geqslant 98.4$ mole% ZnO. In the baseline method, Zn, Co, and/or Mn were coprecipitated from a solution of their chloride salts as hydrous oxides which were then converted to oxalates. The oxalates were calcined and Bi was precipitated on the surface of the resulting oxides by a localized hydrolysis reaction. After drying, the precursor powder was uniaxially pressed and sintered in air in the range of 675–740°C. This method produced varistors with submicrometer average grain sizes, E in the 20–60 kV/cm range at $J = 5$ A/cm^2, $\alpha > 30$ at $2.5 \leqslant J \leqslant 5.0$ A/cm^2, and greater than 90% densification. By reducing density requirements to 70%, the E range was extended to approximately 100 kV/cm. Details concerning the chemistry, effects of sintering conditions, dopant combinations, and dopant levels on varistors properties are published elsewhere.[8]

In this chapter, two processing variables related to precursor powder preparation, which in turn affected varistor electrical properties, are discussed. These include the zinc salt used as the starting material and the aging time of the hydroxide coprecipitate prior to conversion to the oxalate form.

EXPERIMENTAL PROCEDURE

Precursor and Varistor Preparation

A baseline process was developed from parametric studies using the physical and electrical properties of varistors as guidelines to produce 200-g batches of material. Hydrous oxides of Zn, Co, and/or Mn were precipitated from a 1.2M ZnCl$_2$ solution containing desired dopant levels using a 1% excess of NaOH. After stirring for 2 min, a solution containing an equimolar amount of H$_2$C$_2$O$_4 \cdot 2$H$_2$O based on the Zn, Co, and Mn content was added. The oxalates were separated by filtration and calcined to oxides at 600°C. Bismuth addition was done by contacting the oxide mixture with a HNO$_3$ solution of Bi. After filtration, the powder was dried at 400°C, pressed at 68.9 MPa into 1.25 cm diameter, 6-g pellets, and sintered in air at 675–740°C. No mixing/milling steps or binder materials were used.

The deviations from the baseline process to be discussed include the use of the sulfate, nitrate, or acetate salts rather than chlorides and different aging times for hydrous oxide precipitates where periods of 1 min to 24 hr between the addition of NaOH and $H_2C_2O_4 \cdot 2H_2O$ to the chloride salt solutions were used. The effects of these parameters were evaluated with respect to the morphology of the hydrous oxide precipitates and to the physical/electrical and microstructural properties of varistors.

Electrical Measurements

Three circular electrodes of 0.2–0.32 cm^2 were applied to each varistor sample using Ag paint (DuPont conductor composition 4817). The varistors referred to in this paper were tested in concordance with the previously outlined goals. Thus, values of E reported were measured at 5 A/cm^2 and α values were determined for the 2.5–5.0 A/cm^2 range. Current pulses approximately 12 μs in duration with a rise time of 1 μs between 10 and 90% amplitude levels were used. Voltages corresponding to the 2.5 and 5.0 A/cm^2 current pulses were read by a 2-channel, 9-bit digitizer. Voltage readings were taken at about 8 μs where voltage change with time was very small. Each electrode on a given varistor was tested three or more times and the E and α values represent the mean of nine or more measurements unless otherwise stated. Standard deviations were typically 0.5–2% and 5–25% of the reported values of E and α, respectively.

RESULTS AND DISCUSSION

The electrical properties, E and α, of varistors doped with 0.56 mole% Bi_2O_3 and different levels of Co and Mn, prepared by the baseline method are shown in Figs 33.1 and 33.2, respectively. Ninety percent or greater densification was obtained for all samples in the 710–725°C range. The general trends of decreasing E with increased temperature and doping level and the increase in α with increasing doping levels were also typical for varistors doped with Co and/or Mn alone. The average grain sizes of the varistors referred to in Figs 33.1 and 33.2 were all submicrometer.

Microstructural characterization was done by transmission electron microscopy (TEM) and X-ray diffraction (XRD). In addition to zincite (ZnO), all varistors examined contained γ-Bi_2O_3. An additional phase, $ZnMnO_3$, was identified in Mn-doped materials, however, the presence of Co compounds could not be confirmed in Co-doped varistors. A second Bi-rich phase which could not be identified was also observed. Both Bi-rich phases rarely occurred at triple points and were primarily present as randomly dispersed agglomerates containing approximately 50 vol% of ZnO grains. There was no evidence of a second phase(s) associated with grain boundaries although a slight Bi enrich-

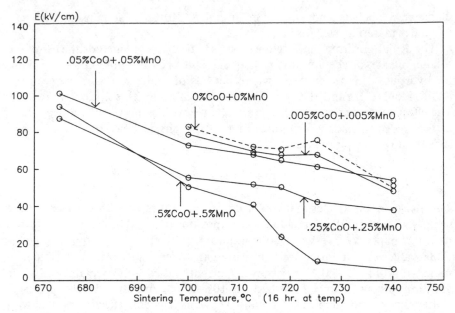

Figure 33.1 Electric field, E, as a function of sintering temperatures for ZnO varistors doped with 0–1 mole% CoO + MnO and 0.56 mole% Bi_2O_3. E, kV/cm, was measured at $J = 5$ A/cm^2.

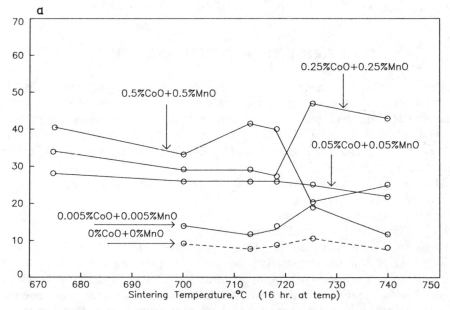

Figure 33.2 Nonlinearity coefficients, α, as a function of sintering temperature for ZnO varistors doped with 0–1 mole% CoO + MnO and 0.56 mole% Bi_2O_3. Determined for $2.5 \leqslant J \leqslant 5.0$ A/cm^2.

ment was detected at grain boundaries. No evidence of residual Cl was detected.

The electrical properties of varistors doped with 0.56 mole% Bi_2O_3 and 0.5 mole% MnO prepared using the baseline method and the sulfate, acetate, nitrate, and chloride salts of zinc are given in Table 33.1. With the exception of the varistor prepared from $Zn(Ac)_2$ at 725°C, the varistors prepared from non-chloride salts and sintered in the 700–725°C range evidenced complete or partial dielectric breakdown under standard test conditions. Varistors prepared from nonchloride salts and sintered at 725–740°C which did not show dielectric breakdown had unacceptably low α values. As a group, varistors prepared from nonchloride salts had higher densities and average grain sizes compared to those prepared from $ZnCl_2$ under comparable conditions.

Anion analyses, Cl^-, NO_3^-, and SO_4^{2-}, were done by ion chromatography on precursor powders prepared from the different zinc salts. These anions were detected in all precursors, however, Cl^- and SO_4^{2-} were slightly higher in precursors prepared from their Zn salts. Considering all samples, it was not possible to correlate residual anion concentrations with varistor properties. No differences in morphology could be seen in scanning electron microscope (SEM) micrographs of the precursor powders. The BET surface areas of the precursors were also similar where Cl^-, SO_4^{2-}, NO_3^-, and Ac^- salts produced powders with surface areas of 13.3, 12.6, 12.8, and 13.6 m^2/g, respectively, after the Bi loading and final 400°C calcine. Fracture surfaces of varistors prepared from Cl^- and SO_4^{2-} salts were examined by low-energy ion scattering spectroscopy (ISS) and no significant differences could be detected.

Differences in microstructures of varistors prepared from the Cl^- and SO_4^{2-} salts were detected by TEM and were related to Bi distribution. The microstructure of the SO_4^{2-}-prepared varistor differed from that of Cl^--prepared varistor described previously in that only one Bi-rich phase, γ-Bi_2O_3, was

TABLE 33.1 Electrical Properties of Varistors Prepared with Different Zinc Salts

Zinc Salt[b]	700/16[a]		718/16		725/16		740/16	
	E	α	E	α	E	α	E	α
Chloride	69.5	40	42.7	33	18.9	22	8.4	21
Nitrate	BD[c]		BD		49.7[d]	6	22	6
Sulfate	BD		BD		BD		45.8	8
Acetate	BD		BD		36.8	8	15.8	6

[a]Sintering temperature, °C/time, hr.

[b]Zinc salt used with baseline method. All varistors doped with 0.5 mole% MnO and 0.56 mole% Bi_2O_3.

[c]Dielectric breakdown.

[d]Tests obtained on single electrode. Two other electrodes showed dielectric breakdown.

observed and it was randomly dispersed at triple points rather than as agglomerates.

Early work of Feitknecht[9] showed that precipitation of hydrous Zn oxides from dilute solution ($\sim 10^{-4}M$) could produce up to five crystalline forms in addition to an amorphous form(s) depending on reaction conditions including aging times. The more concentrated solutions used in the baseline method were dictated by a need for relatively large quantities of material. However, the hydroxide aging time used in the baseline method could be easily varied with no significant impact on the overall process and the effect of this parameter on varistor properties was investigated.

Varistors doped with 0.5 mole% MnO and 0.56 mole% Bi_2O_3 were prepared using hydroxide aging times in the range of 2–240 min. The surface areas of the final precursor powders were independent of aging times. However, differences were observed in electrical properties of varistors prepared using different aging times, Table 33.2. At lower sintering temperatures (675–700°C), increases in aging times resulted in a general decrease in E. Increased aging times also resulted in a general decrease in open porosity and greater densification under comparable sintering conditions. At the higher sintering temperatures, no definitive trends were observed. The comparatively sharp decline in E at 718 and 740°C in varistors prepared using 2-min aging times is an anomaly as the physical properties of these samples were comparable to those of the varistors prepared using the longer aging times.

SEM photomicrographs of hydroxide precipitates obtained from a 1.2M $ZnCl_2$ solution and a 1.2M $ZnCl_2$–0.006M $MnCl_2$–0.006M $CoCl_2$ solution after 2- and 10- min aging times are shown in Figs. 33.3 and 33.4, respectively. Differences in morphology due both to aging time and Co–Mn presence are

TABLE 33.2   Effect of Hydroxide Precipitate Aging Time on Varistor Electrical Properties

Aging Time[a] (min)	675/16[b]		700/16		718/16		740/16	
	E	α	E	α	E	α	E	α
2	105	48	70	44	27	17	11	17
5	BD[c]	—	72	41	58	38	46	45
10	BD	—	72	39	55	37	45	38
20	BD	—	67	36	55	32	44	37
60	102	42	66	39	53	29	No data	
240	82	51	63	41	51	27	44	38

[a]Time elapsed between NaOH and $H_2C_2O_4$ addition in the baseline method. All varistors were doped with 0.5 mole% MnO and 0.56 mole% Bi_2O_3.
[b]Sintering temperature, °C/time, hr.
[c]Dielectric breakdown under normal test conditions.

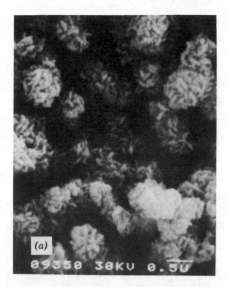

Figure 33.3 Hydrous oxide precipitates separated from 1.2M ZnCl$_2$ solution (a) 2 min and (b) 10 min, after addition of NaOH.

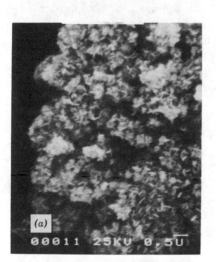
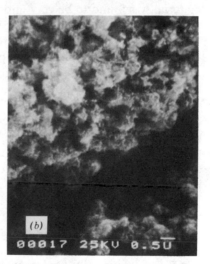

Figure 33.4 Hydrous oxide precipitate separated from 1.2M ZnCl$_2$–0.006M CoCl$_2$–0.006M MnCl$_2$ solution (a) 2 min and (b) 10 min, after addition of NaOH.

readily apparent. Intermediate and longer aging times produced further changes in morphology. However, XRD done on nine samples of the coprecipitate of Zn, Co, and Mn with aging times ranging from 1 min to 1 day identified all samples as zincite (ZnO). Extra lines or evidence of amorphous material were not observed.

The results of these experiments suggest that hydroxide precipitate aging times should be considered as a variable in the baseline preparation. The differences in precipitate morphology resulting from Co and Mn additions suggest that different dopants or dopant levels may cause further, unpredictable changes. An aging time of 2 min was used in the baseline method as it produced the most consistent results with a wide range of dopant concentrations. However, anomalies such as those described above probably exist for other compositions/doping levels and were not identified in this work.

CONCLUSION

A chemical preparation method was developed which produced doped ZnO varistors with E in the 40–60 kV/cm range, $\alpha > 30$, and greater than 90% densification. The primary contributions of the chem-prep techniques were in the area of microstructural control where homogeneity and small grain size were desirable. The latter was achieved due to relatively low sintering temperatures, 675–740°C, needed for densification.

Two parameters common to nearly all chem-prep techniques used for ceramic precursors were identified as being important with respect to both electrical and physical properties of varistors. The anion of the zinc salt used as a starting material was found to be very important in this application. Acceptable varistors were prepared from Cl^- salts; however, those prepared under comparable conditions using SO_4^{2-}, NO_3^-, or Ac^- salts had unacceptable electrical properties even though they showed comparable or superior sintering characteristics. Although the differences in varistor properties could not be correlated with the presence of anion impurities, anions apparently have some influence on the varistor microstructure relative to the Bi distribution. Precursor morphology developed during precipitation reactions was also found to influence both electrical and physical properties of the varistors. Although the effects were not as pronounced as those produced by use of different zinc salts, the results indicated that this property should be considered, and preferably quantified, for each composition or set of conditions for this application.

ACKNOWLEDGMENTS

Significant and appreciated contributions to this work were made by Sandia coworkers K. Kimball, B. Tuttle, R. Brooks, and G. Pike in the areas of precursor preparation and varistor testing, electrical properties and testing, thermal processing, and theoretical aspects of varistor behavior, respectively. Supported by U.S. Dept. of Energy Contract #DE-AC-04-76DP00789.

REFERENCES

1. M. Matsuoka, Nonohmic Properties of Zinc Oxide Ceramics, *Jpn. J. Appl. Phys.* **10**(6), 736 (1971).

2. W. D. Kingery, J. B. VanderSande, and T. Mitamura, A Scanning Transmission Electron Microscopy Investigation of Grain Boundary in a $ZnO-Bi_2O_3$ Varistor, *J. Am. Ceram. Soc.* **62**(3–4), 221–222 (1979).

3. D. R. Clarke, Grain Boundary Segregation in a Commercial ZnO-based Varistor, *J. Appl. Phys.* **50**(11), 6829–6832 (1979).

4. J. P. Gambino, Ph.D. Thesis, MIT (1984), unpublished.

5. G. E. Pike, Electronic Properties of ZnO Varistors: A New Model, in *Grain Boundaries in Semiconductors*, H. J. Leamy, G. E. Pike, and C. H. Seager, Eds., Vol. 5, Elsevier, New York, 1982, pp. 369–379.

6. G. S. Snow, S. S. White, R. A. Cooper, and J. R. Armijo, Characterization of High Field Varistors in the System ZnO-CoO-PbO-Bi_2O_3, *Am. Ceram. Soc. Bull.* **59**(6), 617–622 (1980).

7. R. J. Lauf and W. D. Bond, Fabrication of High-Field Zinc Oxide Varistors by Sol-Gel Processing, *Am. Ceram. Soc. Bull.* **63**(2), 278–281 (1984).

8. R. G. Dosch, and K. M. Kimball, SAND85-0195 (Sept.) 1985.

9. W. Feitknecht, *Helv. Chim. Acta* **13**, 314–345 (1930).

34

SOL–GEL PROCESSING OF LEAD ZIRCONATE TITANATE FILMS

R. A. LIPELES, N. A. IVES, AND M. S. LEUNG
Chemistry and Physics Laboratory
The Aerospace Corporation
El Segundo, California

INTRODUCTION

High dielectric constant lead zirconate titanate (PZT) films are needed for capacitors, optical modulators, and optical storage. Transparent PZT films have been reported recently with bulklike electrical properties. These PZT films have been produced by rf sputtering,[1,2] sol–gel processing[3] and ion sputtering.[4] In these studies platinum is the preferred substrate[1–3] although indium tin oxide (ITO) coated fused quartz,[2] fused silica,[3] Ni–Cr–Au on Pyrex*, and Invar[4] have been used. In an effort to find a substrate that is inexpensive, easy to prepare, and can be used for large area devices, polished stainless steel substrates are investigated in this study.

Homogeneous PZT films are produced from organometallic presursors by first dissolving the starting materials in a common solvent. This solution is prehydrolyzed to form an intimately mixed metal–oxygen–metal polymer. This solution can be sprayed, spun, or dip-coated onto substrates. A crystalline oxide is formed by annealing this film to remove organics and water, and to grow the desired grain structure. In this chapter, we first examine the sol–gel processing of PZT in the absence of any substrate. Next, we examine the metal–

---

*Registered trademark of Corning Glass Works.

organic solution deposition (MOSD) of the organometallic precursors on stainless steel substrates.

EXPERIMENTAL

The starting solution consisted of lead acetate [$Pb(C_2H_3O_2)_2 \cdot 3H_2O$], zirconium n-tetrapropoxide [$Zr(OC_3H_7)_4$], and titanium n-tetrabutoxide [$Ti(OC_4H_9)_4$] in methanol with 0 to 10% excess lead by weight to form $Pb_1Zr_{0.58}Ti_{0.42}O_3$ after hydrolysis. The solution contained about 0.3 mole/L of PZT. This solution was prehydrolyzed by refluxing for 2 hr using a 2.5 mole ratio of water to total moles of metal. Films were formed by applying the solution to polished 304 stainless steel substrates and spinning off the excess. The solvent was driven off at about 80°C in air and the substrates were recoated resulting in a 0.5 μm thick film. The films were typically heated to 275°C to remove organics and water and then annealed for $\frac{1}{2}$ hr to 4 hr from 450 to 600°C in air.

RESULTS AND DISCUSSION

The reaction rate of PZT with stainless steel substrates is a strong function of the processing temperature. Processing temperatures were determined for the PZT gel using thermogravimetric analysis (TGA) in the absence of a substrate in order to select processing temperatures for films. The initial weight loss due to the removal of the methanol and water occurs between 25 and 240°C as shown in Fig. 34.1. Organics are removed between 240 and 325°C. This is similar to the TGA curves of Ibraimov et al using carbonate and hydroxide starting materials.[5] Oxidation plays a role in pyrolysis as shown by the difference in the rate of weight loss in air and argon atmospheres.

The loss of the alkoxide and formation of metal–oxygen bonds were followed using FTIR spectroscopy. PZT gels were dried in air to remove methanol. The alkoxide was pyrolyzed in air at 450°C and then the resulting powder was annealed at 600°C. Some evidence of carbon was found in both the 450 and the 600°C samples. The FTIR spectrum of air-dried PZT gel in Fig. 34.2a, shows evidence of alkoxides similar to that found for $SiTiO_3$ precursors.[6] As shown in Fig. 34.2b, pyrolysis at 450°C resulted in a metal–oxygen shoulder at about 600 cm^{-1} and a metal–oxygen band at about 450 cm^{-1}. After annealing at 600°C, the spectrum in Fig. 34.2c shows a very strong metal–oxygen band at 600 cm^{-1} and a spectrum very similar to lead titanate and lead zirconate. Crystalline compounds were identified using X-ray powder diffraction. The air-dried sample was amorphous. The sample annealed at 450°C contained lead oxide[7] and a very broad peak at $d = 2.90$ Å due to PZT. The sample annealed at 600°C consists of some lead oxide and major peaks at $d = 2.897$, 1.673, 2.047 Å indicating perovskite PZT.[2] Thus, temperatures greater than

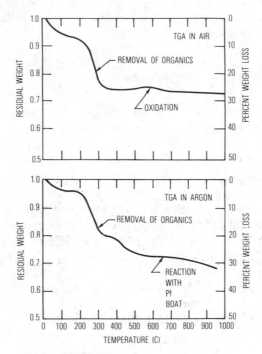

Figure 34.1 Thermogravimetric analysis of a PZT gel heated at 4°C/min in (*a*) air and (*b*) argon flowing at 100 cm^3/min.

450°C are needed to form perovskite PZT. Using these data as a guide, PZT films on stainless steel substrates were examined.

The thickness and grain size of the films were measured. The films on stainless steel were about 0.5 μm thick as determined by a Dektak profilometer. The electron micrograph in Fig. 34.3 confirms the thickness and shows a lack of grain structure in the PZT film. This result is expected because chromium is known to suppress the formation of grain boundaries in PZT.[8]

The effect of annealing temperature on the structure of the film was studied by annealing 0.5 μm thick films at 450, 500, 550, and 600°C for $\frac{1}{2}$ hr. X-ray diffraction measurements showed three lines due to stainless steel and a broad line at about $d = 3.00$ Å that narrowed with increasing temperature. The grain size estimated from the width of this line ranged from 130 Å at 450°C to 450 Å at 550°C. X-ray diffraction of PZT films annealed at 600°C showed the formation of a lead chromate phase.[9] A trace of lead chromate was also observed in the 550°C film.

The composition of the film on a stainless steel substrate annealed at 600°C was measured using Auger electron spectroscopy (AES). The sample was coated with about 600 Å of carbon in order to avoid charging. The carbon was sputtered off prior to making composition measurements. The sensitivity coefficients of Okada for lead, titanium, and zirconium were used to convert the Auger peak-to-peak heights to atomic concentration.[2] The peak heights and relative atomic

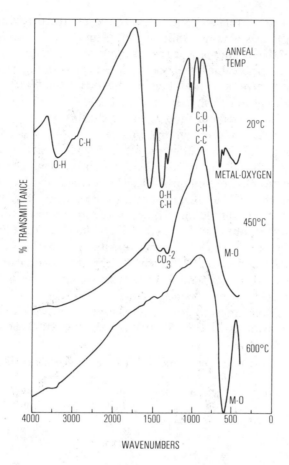

Figure 34.2 The FTIR spectrum of (a) PZT gel dried in air; (b) annealed at 450°C; and (c) annealed at 600°C.

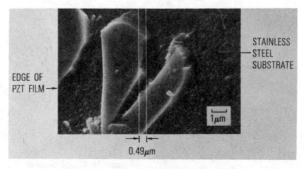

Figure 34.3 Electron micrograph of PZT annealed at 600°C at the edge of a polished stainless-steel substrate. Grains are not visible in the PZT. Uniform crack-free material was obtained over most of the sample.

concentrations are given in Table 34.1. Based on these results the formula $Pb_1Zr_{0.5}Ti_{0.5}O_3$ was obtained which agrees with the expected formula within the accuracy of the technique. The results indicate very little loss of lead during annealing. Furthermore, the line shape of the lead peak in Fig. 34.4 and the ratio of lead to titanium peak-to-peak heights show that the film has a perovskite structure according to line shapes given by Okada.[2] The spectrum also shows chromium and iron contamination from the substrate.

The distribution of chromium and iron in the film annealed at 600°C was examined using AES. The sputter profile in Fig. 34.5 shows a diffusion-like profile of chromium and iron in the PZT. (Some chromium and iron background is probably due to pin holes in the PZT.) The correlation of the chromate concentration with that of lead is consistent with the a stable lead chromate phase observed using X-ray diffraction. The ratio of zirconium to titanium remains constant throughout the film. Due to the diffuse interface, the PZT adheres very well to the substrate. Based on these results, stainless steel cannot be used as a substrate for PZT annealed at 600°C due to chromium and iron diffusion.

The electrical properties of the films were measured for the films deposited on stainless steel using 1-mm diameter conductive silver paint electrodes. The capacitance and loss tangent were measured using an impedance bridge at 1 kHz. The dielectric constants of these 0.5-μm films were about 400 and the loss tangent

TABLE 34.1   Composition of a PZT Film on Stainless Steel

Peak Energy (eV)	Peak-to-Peak Height	Correction Factor[2]	Calculated Atomic Concentration
Pb, 94	410	1	0.2
Zr, 116	80	2.44	0.095
Ti, 387	184	1.06	0.095

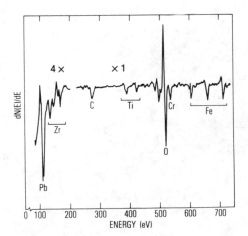

Figure 34.4   AES sputter profile of PZT on stainless steel annealed at 600°C. The sputtering rate measured on tantalum oxide was about 100 Å/min.

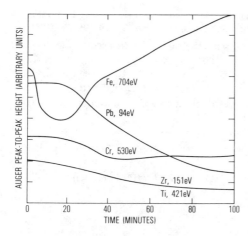

Figure 34.5 Auger spectrum of PZT on stainless steel annealed at 600°C. The spectrum was obtained using a 3 keV, 3 μA electron beam, 3 eV modulation, and 1 eV/s scan rate.

was 0.4 for films annealed over 450°C. The dc resistivity was greater than 6×10^8 Ω-cm. A low value of the dielectric constant of 30 was found in PZT annealed at over 550°C and is attributed to the presence of lead chromate.

CONCLUSION

PZT films can be made from lead acetate, zirconium *n*-propoxide, and titanium *n*-butoxide starting materials. The hydrolysis and pyrolysis of PZT powders and films occur between 240 and 350°C and annealing at higher temperatures is required to form the perovskite form of PZT. Annealing temperatures below 550°C must be used in order to avoid the formation of lead chromate on stainless steel substrates.

ACKNOWLEDGMENTS

The authors thank P. A. Adams for the X-ray work, G. A. To for the FTIR spectra, and S. E. Harris and D. J. Coleman for preparing the films. Work performed, in part under Aerospace Sponsored Research.

REFERENCES

1. S. B. Krupanidhi, N. Maffei, M. Sayer, and K. El-Assai, rf Planar Magnetron Sputtering and Characterization of Ferroelectric Pb(ZrTi)O₃ Films, *J. Appl. Phys.* **54**, 6601–6609 (1983).

2. A. Okada, *J. Appl. Phys.* **48**, 2905, (1978); A. Okada, Electrical Properties of Lead-Zirconate-Lead-Titanate Ferroelectric Thin Films and Their Composition Analysis by Auger Electron Spectroscopy, *J. Appl. Phys.* **49**, 4495–4499 (1978).

3. J. Fukushima, K. Kodaira, and T. Matsushita, Preparation of Ferroelectric PZT Films by Thermal Decomposition of Organometallic Compounds, *J. Mater. Sci.* **19**, 595–598 (1984).

4. R. N. Castellano and L. G. Feinstein, "Ion-Beam Deposition of Ferroelectric Lead Zirconate Titanate (PZT), *J. Appl. Phys.* **50**, 4406–4411 (1979).

5. N. S. Iraimov, N. E. Troshin, I. A. Silvant'eva, Zh. A. Dolgay, and V. A. Golovnin, Formation of Lead Titanate-Zirconate, *Izv. Akad. Nauk SSSR Neorg. Mater.* **14**, 276–279 (1978).

6. S. Smith II, R. T. Dolloff, and K. S. Mazdiyani, Preparation and Characterization of Alkoxy-Derived $SrZrO_3$, *J. Am. Ceram. Soc.* **53**, 91–95 (1970).

7. ASTM Powder Diffraction File 5-561 (PbO).

8. B. Jaffe, W. R. Cook Jr., and H. Jaffe, *Piezoelectric Ceramics*, Academic Press, New York, 1971, pp. 160–162.

9. ASTM Powder Diffraction File 29-767A (Pb_2CrO_5) and 29-769 ($Pb_2[CrO_4]O$).

35

OXY-ALKOXY POLYMER NETWORKS

P. E. D. MORGAN, H. A. BUMP, E. A. PUGAR,
AND J. J. RATTO
Rockwell International Science Center
Thousand Oaks, California

INTRODUCTION

We report in this chapter the synthesis of oxidic polymers with compositions intermediate between oxides and alkoxides (viz: oxy-alkoxides). Polymers with silicon–oxygen backbones, the silicones, are well known but little effort has been reported of extending the chemistry to other metal–oxygen linkages. These latter types would have potential for further processing to oxides, as fibers, films, or bulk solids, with some of the advantages of alkoxides, that is, the control of viscosity, decomposition temperature, oxidation behavior, and so on, via the built-in variability of OR groups with more or less bulky, cross-linked, or polyfunctionality of the R groups. Additionally, mixed polymers with both two or more different metals and differing R groups seem to be feasible, leading to a new, large chemistry.

RESULTS AND DISCUSSION

The simplest way to polymerize an alkoxide is to react an alkoxide monomer or oligomer, such as $Al(O-sC_4H_9)_3$, with water just sufficient to hydrolyse all the —OR groups and allow alcohol produced to be given off.

A simple way to establish a dense polymer network by this method is to spread single or mixed, undiluted, liquid alkoxides out on a thin sheet, for example,

327

on aluminum foil, so that it can slowly react with air/moisture. After several days, no smell of alcohol is detectable in the resulting powder, which is X-ray amorphous, and has, for example, a formula in the case of $Al(O—sC_4H_9)_3$, of close to $AlO(OC_4H_9)$. This method, reported in more detail elsewhere,[1] leads to an oxide network which crystallizes with difficulty to, for example, a γ-Al_2O_3 like product, with a Scherrer X-ray line broadening grain size of $<100 \text{Å}$ at 600°C, Fig. 35.1. When the network also contains, for example, zirconium as from $Zr (O-nC_3H_7)_4$, to produce ultimately $Al_2O_3/10$ vol% ZrO_2, crystallization is inhibited in a pronounced way so that the "γ" alumina does not appear until $\sim 900°C$, with a grain size of $<250 \text{Å}$, Fig. 35.2. The lattice parameter of this γ alumina is the same as for the pure γ case. All firings are overnight. No XRD evidence exists for by-product cubic or tetragonal zirconia being thrown out of solution until between 947 and 1000°C. This result parallels what had earlier been seen with the use of γ precursors for β aluminas or for radwaste hosts.[2,3]

Conversion of the γ aluminas to α-Al_2O_3 is also somewhat unusual. For the pure Al_2O_3 case, Fig. 35.1, α-Al_2O_3 appears as low as 850°C. Seeding with α does not seem to have any enhancing effect. By 950°C, substantial conversion to $\alpha + \theta$ occurs and progressive conversion to α is complete by $\sim 1100°C$. For the $Al_2O_3/10$ vol% ZrO_2 case, Fig. 35.2, the γ form produces some θ with difficulty by 974°C, much more and better crystalline at 1054°C, where α-Al_2O_3 also appears (Fig. 35.3). The crystallization of zirconia, barely visible at 974°C, rapidly accelerates in a coupled manner with the $\gamma + \theta$ and $\gamma \rightarrow \alpha$-$Al_2O_3$ conversions.

Treatment of the polymer network powders with excess water at 60°C for 3 days causes pseudoboehmite to be produced. This, for the $Al_2O_3/10$ vol% ZrO_2 case, leads to results of Fig. 35.4, where the course of crystallization is very similar to that seen if starting hydroxides are used. The fact that this result (Fig. 35.4) is quite different from that of Fig. 35.1 leads to the corollary that the network established by the thin film air/moisture method must be quite different from that of the hydroxide cases, including also gels where water is used in excess. A summary of grain sizes for the $Al_2O_3/10$ vol% ZrO_2 case is contained in Table 35.1.

The possibility of making new types of ultramicrostructures with good alkoxide mixing led to testing a method of polymerization related to one of earlier workers.[4] If single or mixed alkoxides are refluxed in dry inert atmosphere (N_2 or Ar) at up to $\sim 300°C$, a mixture of alkenes, alcohols, and traces of aldehydes, corresponding to the original R group, are given off, for example, as follows:

$$Ti(O—nC_4H_9)_4 \xrightarrow{\Delta} TiO_{1.35}(O—nC_4H_9)_{1.3} + \text{distillate at } \sim 100°C \quad (1)$$

$$Al(O—sC_4H_9)_3 \xrightarrow{290°C} AlO_{0.93}(O—sC_4H_9)_{1.15} + \text{distillate at } \sim 90°C \quad (2)$$

$$Zr(O—nC_3H_7)_4 \xrightarrow{300°C} ZrO_{1.35}(O—nC_3H_7)_{1.3} + \text{distillate at } \sim 90°C \quad (3)$$

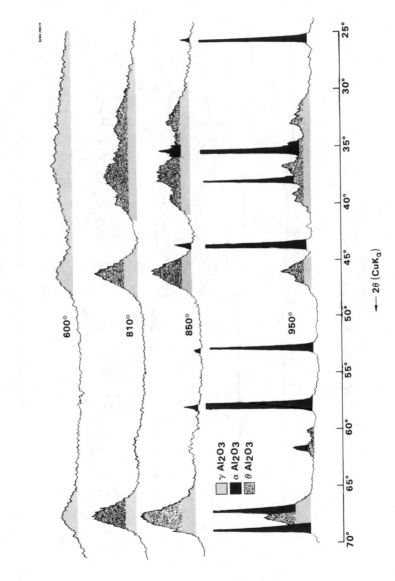

Figure 35.1 Al₂O₃ low-temperature phase behavior.

329

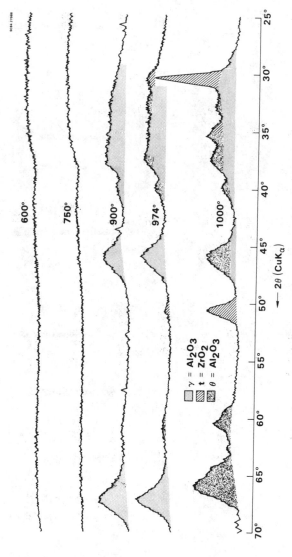

Figure 35.2 Al$_2$O$_3$/10 vol% ZrO$_2$ low-temperature phase behavior.

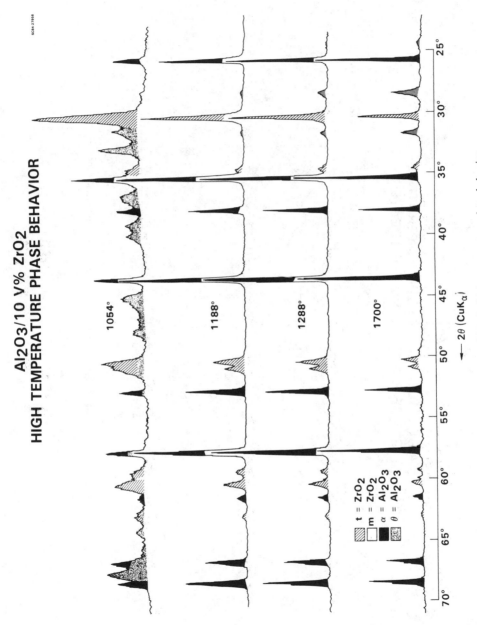

Figure 35.3 Al$_2$O$_3$/10 vol% ZrO$_2$ high-temperature phase behavior.

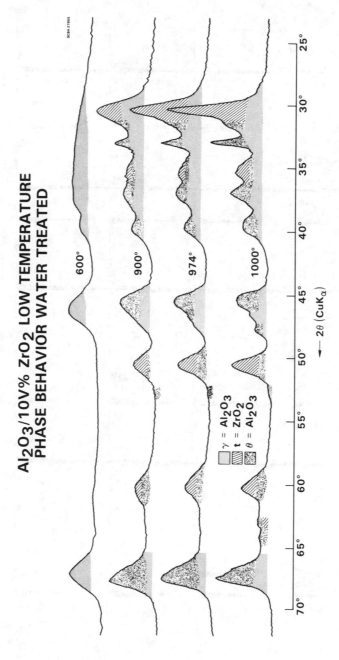

Figure 35.4 Al$_2$O$_3$/10 vol% ZrO$_2$, water-treated, low-temperature phase behavior.

332

TABLE 35.1   Scherrer Line Broadening Particle Size,[a]
$$t = 0.9\lambda/(B \cos \theta_B)$$

| Temperature (°C) | ZrO$_2$ | | Al$_2$O$_3$ | Size |
	Form	Size[b] (Å)	Form	(hkl)	(Å)
600	—	—	γ	(400)	82
900	tet	96	θ	(200)	236
974	tet	121	θ	(200)	236
1000	tet	164	θ	(200)	331
1054	tet	411	θ	(200)	552
			α	(113)	>1000

[a] $\lambda = 1.54178$ Å, $B =$ width at half-height in radians, $\theta =$ Bragg angle.
[b] Measured from tetragonal (101) line.

The liquid in the flask slowly becomes more viscous and with sufficient loss of the distillate, solidifies to a glass or a powder. The latter, in the Ti case, has an XRD pattern of poorly crystalline anatase, others are amorphous. If the reaction is stopped just before solidification and cooled, a glassy or resinous solid results. NMR and IR analysis show that the R groups left on the polymeric material have not undergone any isomerization or other change. The empirical formulae of these is established by weighing, hydrolyzing with water, drying slowly, and firing to 1000°C in air.

$$\frac{\text{wt after treatment}}{\text{wt before treatment}} \equiv \frac{Al_2O_3}{Al_2(OR)_{6-2x}O_x} \tag{4}$$

Solving for X, with the appropriate atomic weights, leads to the empirical formulae shown.

It is observed that the viscous liquids (at $\sim 300°C$) persist up to about the halfway stage of transformation of the alkoxide to the oxide. These represent a new group of polymeric oxy-alkoxides.

The glassy or resinous solids are surprisingly soluble in organic solvents; alcohols, ketones, and even chloroform. This may come about as a result of the metal unit having coordination of 4 or 6 for Al, 6 usually for Ti, and >6 for Zr, as is known in simple alkoxide compounds. The inward facing oxygen atoms then leave the organic groups on the outside of the polymer molecule, leading to the ready solubility. Fibers and films can be made, both from the molten oxy-alkoxides, and from organic solutions. The use of these matrices for ceramic composites or extrusion aids seems possible.

The sequence of crystallization reactions of these new polymer networks, as they are oxidized on firing appears to be similar to that of the air/moisture hydrolyzed networks. However, there are important differences in ceramic formability and unprecedented new microstructures.

CONCLUSION

A new class of tractable oxy-alkoxy polymers has been established which promise new ultramicrostructural types and may have a role in a variety of ceramic forming routes.

ACKNOWLEDGMENT

Work supported by Rockwell International Internal Research and Development Fund and the Rockwell International Youth Motivation Fund.

REFERENCES

1. E. A. Pugar and P. E. D. Morgan, Coupled (Inhibited) Grain Growth Effects in $Al_2O_3/10v/o$ ZrO_2, in press, *J. Am. Ceram. Soc.*

2. P. E. D. Morgan, Preparation of Powders Suitable for Conversion to Useful β-Aluminas, U.S. Patent No. 4,339,511, July 1982.

3. P. E. D. Morgan, D. R. Clarke, C. M. Jantzen, and A. B. Harker, High Alumina Tailored Nuclear Waste Ceramics, *J. Am. Ceram. Soc.* **64**, 249–258 (1981).

4. G. Carturan, V. Gottardi, and M. Graziani, Physical and Chemical Evolutions Occurring in Glass Formation from Alkoxides of Silicon, Aluminum and Sodium, *J. Non-Cryst. Solids* **29**, 41–48 (1978).

Note added in proof. A most important reference to earlier work is D. C. Bradley and M. M. Faktor, "The Pyrolysis of Metal Alkoxides," *Trans. Faraday Soc.* **55**, 2117 (1959). This we were (perhaps fortunately) unaware of; important new findings emerge in that the normal and secondary alkoxides used by us appear to allow the isolation of the intermediate oxy-alkoxides, whereas the tertiary cases treated earlier apparently did not.

PART 3

Materials from Organometallic Precursors

36

CROSSLINKING OF POLYSILANES AS SILICON CARBIDE PRECURSORS

ROBERT WEST, XING-HUA ZHANG,
PETER I. DJUROVICH, AND HARALD STÜGER
Department of Chemistry
University of Wisconsin
Madison, Wisconsin 53706

INTRODUCTION

Polysilanes are macromolecules in which the backbone of the polymer consists entirely of silicon atoms. Synthesis of polysilanes is accomplished by dechlorination of diorganodichlorosilanes with sodium metal; the reaction is conveniently carried out in toluene at 110°C, but other inert solvents can also be used. Both homopolymers (1) and copolymers (2) can be made:

$$R^1R^2SiCl_2 \xrightarrow[110°]{Na} \begin{matrix} R^1 \\ | \\ -(Si-)_n \\ | \\ R^2 \end{matrix} \qquad \begin{matrix} R^1R^2SiCl_2 \\ R^3R^4SiCl_2 \end{matrix} \xrightarrow[110°]{Na} \begin{matrix} R^1 & R^3 \\ | & | \\ -(Si)_n(Si)_m- \\ | & | \\ R^2 & R^4 \end{matrix}$$

$$(1) \qquad\qquad\qquad\qquad (2)$$

The substituent groups can be widely varied, as can the ratio of different kinds of silicon groups, so that the possible range of polymer compositions is very large. Several of these polymers have been described in the chapter, "Polysilane

337

Precursors to Silicon Carbide," in the volume preceding this one,[2] as well as in other publications.[3-5]

Depending on the nature of the side groups, the polysilanes can be crystalline polymers, glassy solids, waxes, or elastomers. These polymers have potential industrial utility, not only as precursors to silicon carbide as described below, but as resist materials for photolithography in the manufacture of microelectronic components,[6] and as photoinitiators for the radical polymerization of acrylates and other vinyl monomers.[7]

Polysilane polymers are by now well known as precursors for silicon carbide ceramic. In the original process of Yajima and Hayashi,[8] permethylpolysilane (3) is first thermolyzed at moderate temperature to form a polycarbosilane, (4) in which the backbone consists mainly of alternating silicon and carbon atoms. The carbosilane is heated *in vacuo* to remove volatile constituents, fractionated, drawn into fibers which are air-oxidized on the surface to provide rigidity, and finally pyrolyzed to silicon carbide.

$$
\underset{\textbf{(3)}}{\underset{\overset{|}{CH_3}}{-(\underset{\overset{|}{CH^3}}{Si})_n-}} \xrightarrow{>400°} \underset{\textbf{(4)}}{\underset{\overset{|}{CH_3}}{-(\underset{\overset{|}{H}}{Si}-CH_2)_n-}} \xrightarrow{>800°} SiC + CH_4 + H_2
$$

Other polymer precursors to silicon carbide and silicon carbide–nitride are known. Workers at Dow-Corning Corporation,[9] Union Carbide Corporation,[10] and MIT[11] have described several more complex polymer systems useful in such ceramic conversions, and we have shown that certain polysilane polymers can be pyrolyzed directly to silicon carbide, without prior conversion to polycarbosilanes.[12]

Linear polysilane polymers are most desirable for purposes of characterization and fabrication. However, it is important that the polysilane be crosslinked before pyrolysis, as otherwise the ceramic yield is very small. Initially crosslinking was brought about by photolysis of phenyl-substituted polysilanes such as poly(dimethylsilylene-co-phenylmethylsilylene), "polysilastyrene," (2 where $R_1 = R^2 = R^3$ = methyl, R^4 = phenyl, and $n/m \sim 0.7$ to 2.0). However this photocrosslinking is always accompanied by undesirable chain scission.[13]

Recently we have developed several new crosslinking systems for polysilane polymers, which are described below. Experiments to test these as silicon carbide ceramic precursors are being carried out in other laboratories.

AIR-CURING POLYSILANES: OXYGEN CROSSLINKING OF CYCLOTETRAMETHYLENESILANE POLYMERS

Using cyclotetramethylenedichlorosilane as one of the starting monomers, we have synthesized polymers (5a–5c) containing silicon in a five-membered heterocyclic ring:

$$\text{(ring)}SiCl_2 \xrightarrow[110°]{Na, toluene} -(Si)_n- \qquad (5a)$$

$$\text{(ring)}SiCl_2 + R^1R^2SiCl_2 \xrightarrow[110°]{Na, toluene} -(Si)_n-(Si)_m- \quad \overset{R}{\underset{CH^3}{|}}$$

(5b), R = methyl
(5c), R = phenyl

These polymers resemble other linear polysilanes, except that their ultraviolet spectrum is abnormal. They absorb at unusually short wavelengths, 280–290 nm for (5a) and (5b), whereas most polysilanes have absorption maxima from 300 to 350 nm. The abnormal λ_{max} for (5a–b) probably results from conformational effects due to the strained rings in these polymers.

Ring strain in (5a–c) is also manifested in their abnormally high reactivity toward oxygen. Simply heating in air to a moderate temperature ($\sim 80°C$) leads to rapid oxygenation. The products are crosslinked, insoluble polymers which appear from their infrared spectra to contain both Si—O—C and Si—O—Si bonds. We envision the crosslinking to take place as shown; the dotted lines indicate the crosslinks to other polysilane chains.

$$-(Si)_n-(Si)_m- \quad \overset{R^1}{\underset{R^2}{|}} \xrightarrow[80°C]{O_2} -(Si)_n-(Si)_m- \quad \overset{O}{\underset{O}{\overset{R^1}{\underset{R^2}{|}}}}$$

Air-curing of cyclotetramethylene polysilanes is very easy to carry out, but the crosslinked polymer contains some oxygen, which may be undesirable for some purposes. There is another minor practical disadvantage: air oxidation and consequent crosslinking takes place slowly even at room temperature, so the linear polymers must be stored out of contact with the atmosphere.

ROOM TEMPERATURE VULCANIZING POLYSILANES

Because hydrogen bonded to silicon is quite reactive, Si—H bonds would provide another possibility for crosslinking reactions of polysilanes. However, solvolysis of the Si—H links may take place during workup of the polymer, which must therefore be done with special care.

We have synthesized polymers containing various amounts of phenylsilylene units, PhSiH. The results are summarized in Table 36.1. The polymers were

TABLE 36.1 Polysilane Polymers Containing Si—H Groups, Ph_2MeSi—$[PhSiH$—$(R^1R^2Si)_{\overline{m}}]_nSiMePh_2$

	R^1	R^2	m	$\bar{M}_w{}^a$	Appearance
(6a)	—	—	0	1000	White powder
(6b)	Me	Me	0.63	1200	White powder
(6c)	Me	Ph	0.69	1000	White powder
(6d)	Me	Ph	4.5	1100	White powder
(6e)	Me	PhC_2H_4	1.15	1000	Resinous solid
(6f)	Me	PhC_2H_4	5.4	1200	Resinous solid
(6g)	Me	Cy-Hex[b]	0.7	1100	White powder
(6h)	Me	Cy-Hex[b]	5.3	1300	White powder
(6i)	Me	n-Hex	0.91	1000	Grease
(6j)	Me	n-Hex	4.8	1500	Fluid

[a] From GPC elution volume, relative to polystyrene.
[b] Cyclohexyl.

prepared in the usual way by sodium coupling of dichlorosilanes, but the reaction was carefully monitored to ensure that the chlorosilanes were completely consumed. At this point the silyl anion-terminated, "living" polymer was capped with a monochlorosilane, $Ph_2MeSiCl$:

$$2Ph_2MeSiCl + {}^-Si\!\sim\!\sim\!Si^- \rightarrow Ph_2MeSi—Si\!\sim\!\sim\!Si—SiMePh_2$$

The end-blocked polymers (**6a–j**) were then isolated under strictly neutral conditions to prevent solvolysis of Si—H bonds. The isolated polymers had relatively low molecular weights, and were meltable solids, or in some cases flowable liquids (Table 36.1).

Crosslinking of Si—H containing polysilanes might be brought about by a variety of reactions. In practice we used trivinylphenylsilane as the crosslinking agent, with a trace of chloroplatinic acid as catalyst. The mixture of components is, in some cases, a viscous liquid which gradually sets to a firm crosslinked polymer. This is the first example of a room temperature vulcanizing polysilane, analogous to the well-known RTV silicone elastomers. The crosslinking reaction can be outlined as:

THERMAL OR PHOTOCROSSLINKING OF ALKENYLPOLYSILANES

Carbon–carbon double bonds are reactive toward free radicals, which are produced in the photolysis of polysilanes. Therefore polysilanes with C=C unsaturation in the side chains are promising candidates for photocrosslinking. However, the double bond must be at a sufficient distance from the silicon atoms in the dichlorosilane monomer. If the double bond is immediately adjacent to the polysilane chain (vinylsilanes), crosslinking takes place during the synthesis. Likewise when the double bond is removed by one carbon atom (allylsilanes), the yield of linear polymer containing allyl groups is very low, because of side reactions leading to crosslinking.

Good results are however obtained using 3-cyclohexenylethyl substituents. Three polymers, (7a–c), containing this group were synthesized in the usual way in satisfactory yield:

(7a) homopolymer

(7b) R = n-propyl

(7c) R = phenyl

These functionalized polymers may be converted to other derivatives through reaction at the side chain. For example, addition of HBr or HCl takes place to convert (7a–c) into polymers containing halogen atoms bound to the cyclohexane ring. To our knowledge, this is the first example of side-chain derivatization of a polysilane polymer.

Crosslinking of (7a–c) takes place upon exposure of thin films to ultraviolet radiation of $\lambda > 300$ nm. The crosslinked polymers are insoluble in solvents which dissolve (7a–c) such as toluene, THF, hexane, and chloroform, and show greater rigidity. Films of these polymers also undergo crosslinking when heated to 200°C for 4 hr in vacuum. The crosslinking evidently involves reactions of the C=C double bond, but the structures of the crosslinked polymers have not yet been studied in detail.

THERMAL OR PHOTOCHEMICAL CROSSLINKING OF POLYSILANES WITH POLYVINYL COMPOUNDS

The final crosslinking system developed in our studies is a decidedly simple one. A polysilane polymer is mixed with a polyunsaturated compound containing several carbon–carbon double bonds; either photolysis, or thermal generation of radicals from an added initiator, leads to crosslinking of the polysilane. Polyunsaturated compounds used in this process included tetravinylsilane, 1,2-bis(trivinylsilyl)ethane, 1,6-bis(trivinylsilyl)hexane, methylvinylcyclotetrasiloxane, 1,4-cyclooctadiene, 1,9-decadiene, and triallyl benzene-1,3,5-tricar-boxylate. Both alkyl and arylsilane polymers are crosslinked under these conditions; typical polymers studied were poly(phenylmethylsilylene) and poly(cyclohexylmethylsilylene-co-*n*-hexylmethylsilylene).

The photochemical crosslinking probably takes place through initial cleavage of the polysilane chain to form radicals. The photolysis of polysilanes has been investigated in our laboratories and shown to take place by a combination of some or all of the reactions shown here:[14]

$$\sim\!\!\overset{|}{\underset{|}{Si}}\!\!-\!\!\overset{|}{\underset{|}{Si}}\!\!\sim\;\xrightarrow{h\nu}\;\sim\!\!\overset{|}{\underset{|}{Si}}\cdot\;+\;\cdot\overset{|}{\underset{|}{Si}}\!\!\sim$$

$$\sim\!\!\overset{|}{\underset{|}{Si}}\!\!-\!\!\overset{|}{\underset{|}{Si}}R_2\!\!-\!\!\overset{|}{\underset{|}{Si}}\!\!\sim\;\xrightarrow{h\nu}\;\sim\!\!\overset{|}{\underset{|}{Si}}\cdot\;+\;R_2Si\;+\;\cdot\overset{|}{\underset{|}{Si}}\!\!\sim$$

$$\sim\!\!\overset{|}{\underset{|}{Si}}\!\!-\!\!\overset{|}{\underset{|}{Si}}R_2\!\!-\!\!\overset{|}{\underset{|}{Si}}\!\!\sim\;\xrightarrow{h\nu}\;R_2Si\;+\;\sim\!\!\overset{|}{\underset{|}{Si}}\!\!-\!\!\overset{|}{\underset{|}{Si}}\!\!\sim$$

The silyl radicals probably add to the C=C double bonds of the polyunsaturated additives, causing formation of crosslinks and generating new carbon radicals which may lead to further crosslinking reactions:

$$\sim\!\!\overset{|}{\underset{|}{Si}}\cdot\;+\;\diagup\!\!\diagdown R\diagup\!\!\diagdown\;\longrightarrow\;\sim\!\!\overset{|}{\underset{|}{Si}}\diagdown\!\!\diagup\diagdown R\diagup$$

Using thermal radical initiators, the first steps may be either addition of initiator radicals to the double bonds of the polyunsaturated compound, or abstraction of hydrogen from organic groups on the polysilane. For instance, loss of hydrogens from methyl groups on the polysilane would lead to formation of crosslinks:

$$R\cdot\;+\;\sim\!\!\overset{|}{\underset{\underset{CH_3}{|}}{Si}}\!\!\sim\;\longrightarrow\;RH\;+\;\sim\!\!\overset{|}{\underset{\underset{\cdot CH_2}{|}}{Si}}\!\!\sim$$

$$\downarrow\diagup\!\!\diagdown R\diagup$$

$$-\overset{}{\underset{}{Si}}\!\!-\!\!CH_2\diagdown\!\!\diagup\diagdown R\diagup\diagdown\;,\;etc.$$

ACKNOWLEDGMENT

This work was supported by the Air Force Office of Scientific Research, Contract No. F49620-83-C-0044 and the 3M Company.

REFERENCES

1. Visiting Scholar from the Institute of Chemistry, Academic Sinica, Beijing, People's Republic of China.

2. R. West, Polysilane Precursors to Silicon Carbide, in *Ultrastructure Processing of Ceramics, Glass and Composites*, L. Hench and D. R. Ulrich, Eds., Wiley, New York, New York, 1984, Chapter 19.

3. (a) P. Trefonas, P. I. Djurovich, X.-H. Zhang, R. West, R. D. Miller, and D. Hofer, Organosilane High Polymers: Synthesis of Formable Homopolymers, *J. Polym. Sci., Polym. Lett. Ed.* **21**, 819–822 (1983).
   (b) X.-H. Zhang and R. West, Organosilane Polymers: Formable Copolymers Containing Methyl(β-Phenethyl)-Silylene or Cyclohexyl(methyl)silylene Units, *J. Polym. Sci., Polym. Chem. Ed.* **22**, 159–170 (1984).
   (c) X.-H. Zhang and R. West, Organosilane Polymers: Formable Copolymers Containing Diphenylsilylene Units, *J. Polym. Sci., Polym. Chem. Lett.*, **23**, 479–485 (1985).
   (d) X.-H. Zhang and R. West, Organosilane Polymers: Formable Copolymers Containing Diphenylsilylene Units, *J. Polym. Sci., Polym. Chem. Lett.*, **23**, 479–485 (1985).
   (e) X.-H. Zhang and R. West, Syntheses and Properties of Some Organosilane Polymers: Polysilane and Poly(dimethylsilylene), *Polym. Commun.* (Beijing, China), in press.
   (f) P. Trefonas and R. West, Organosilane High Polymers: Poly(phenylmethylsilylene), in *Inorganic Syntheses*, H. Allcock, Ed., in press.

4. J. P. Wesson and T. C. Williams, Organosilane Polymers. II. Poly(ethyl-co-dimethylsilylene) and Poly(methylpropyl-co-dimethylsilylene, *J. Polym. Sci., Polym. Chem. Ed.* **18**, 959 (1980). J. P. Wesson and T. C. Williams, Organosilane Polymers. III. Block Copolymers, *J. Polym. Sci., Polym. Chem. Ed.* **19**, 65 (1981).

5. R. E. Trujillo, Preparation of Long-Chain Polymethylphenylsilane, *J. Organomet. Chem.* **198**, C27 (1980).

6. R. D. Miller, D. Hofer, G. C. Willson, P. Trefonas, and R. West, Soluble Polysilane Derivatives: Interesting New Radiation Sensitive Polymers, in *Materials for Microlithography: Radiation Sensitive Polymers*, L. Thompson, G. C. Willson and J. M. J. Préchet, Eds., Series 266, American Chemical Society, Washington, D.C., 1984.

7. A. R. Wolff, R. West, and D. Peterson, Polysilanes: A New Class of Vinyl Photoinitiators, presented to 18th Organosilicon Symposium, Schenectady, NY, April 4, 1984.

8. S. Yajima, K. Okamura, J. Hayashi, and M. Omori, Synthesis of Continuous SiC Fibers with High Tensile Strength, *J. Am. Ceram. Soc.* **59**, 324 (1976). S. Yajima, Special Heat-Resisting Materials from Organometallic Polymers, *Am. Ceram. Soc. Bull.* **62**, 893 (1983).

9. R. H. Baney, in *Ultrastructure Processing of Ceramics, Glass and Composites*, L. Hench and D. R. Ulrich, Eds, Wiley, New York, N.Y. 1984, Chapter 19.

10. C. L. Schilling, J. P. Wesson, and T. C. Williams, Polycarbosilane Precursors for Silicon Carbide, *Am. Ceram. Soc. Bull.* **62**, 912 (1983).

11. D. Seyferth and G. H. Wiseman, High-Yield Synthesis of Si_3N_4/SiC Ceramic Materials by Pyrolysis of a Novel Polyorganosilazane, *J. Am. Ceram. Soc.* **67**, C132 (1984).

12. R. West, L. D. David, P. I. Djurovich, H. Yu, and R. Sinclair, Polysilastyrene: Phenylmethyl-silane-di-methylsilane Copolymers as Precursors to Silicon Carbide, *Am. Cer. Soc. Bull.* **62**, 825–934 (1983).

13. P. Trefonas III, R. West, R. D. Miller, and D. Hofer, Organosilane High Polymers: Electronic Spectra and Photodegradation, *J. Polym. Sci., Polym. Lett. Ed.* **21**, 823–829 (1983).

14. P. Trefonas III, R. West, and R. D. Miller, Polyosilane High Polymers: Mechanism of Photo-degradation, *J. Am. Chem. Soc.*, **107**, 2737–2742 (1985).

37

CROSSLINKING AND PYROLYSIS OF SILANE PRECURSORS FOR SILICON CARBIDE

BURT I. LEE AND LARRY L. HENCH

Ceramics Division
Department of Materials Science and Engineering
University of Florida
Gainesville, Florida

INTRODUCTION

Several organosilicon polymers designed to yield silicon carbide (SiC)[1-5] ceramic compositions are known. The primary successful example is Nicalon® made by the Yajima process. In the Yajima[5] process, β-SiC fibers are produced from polydimethyl silane through polycarbosilane and oxidative crosslinking of polycarbosilane in air. Polysilastyrene (PSS)[1,4,6] on the other hand yields SiC without the intermediate step required in the Yajima process. Crosslinking of PSS by UV light has been studied by Sinclair et al.[1,7] However, it is found that the crosslinking of PSS is limited to the surface region and is unsuitable for bulk crosslinking.

High-energy irradiation and chemical-free radical initiators (CFRI) may be viable methods for bulk crosslinking of silane precursors for SiC and are described in this chapter.

TABLE 37.1 Characteristics of Silanes and Summary of Experimental Conditions that Induce Silane Crosslinking

Silane Type	Structure	Softening or mp (°C)	Solvent or Atmosphere	Additive	Means	Crosslinking[a] Temperature (°C)	Time (min)
Vinyl silane oligomer	$Me_3-Si-(Si-Si)_n-SiMe_3$ with $Me\ Me$ / $H\ HC\!=\!CH_2$	Liquid at RT	Vacuum	None	Thermal	~200	10
Vinyl silane oligomer	$Me_3-Si-(Si-Si)_n-SiMe$ with $Me\ Me$ / $H\ HC\!=\!CH_2$	Liquid at RT	Ar	None	Thermal	~200	20
Vinyl silane oligomer	$Me_3-Si-(Si-Si)_n-SiMe_3$ with $Me\ Me$ / $H\ HC\!=\!CH_2$	Liquid at RT	Ar	DCP	Thermal	~200	20
Vinyl silane oligomer	$Me_3-Si-(Si-Si)_n-SiMe_3$ with $Me\ Me$ / $H\ HC\!=\!CH_2$	Liquid at RT	Vacuum	None	γ-ray	RT	~10 days

346

Sample	Structure		Atmosphere	Additive	Irradiation		
SS oligomer	Me Me —(Si—Si)$_m$— Ph Me		Viscous liquid at RT	Vacuum	None	γ-ray	RT	~20 days
SS oligomer	Me Me —(Si—Si)$_m$— Ph Me	$m < n$	Viscous liquid at RT	Vacuum	DCP	γ-ray	~200	~20
PSS	Me Me —(Si—Si)$_m$— Ph Me		~200	Vacuum		γ-ray	RT	~29 days
PSS	Me Me —(Si—Si)$_n$— Ph Me		~200	Benzene/N$_2$	DCP		~200	30
PSS	Me Me —(Si—Si)$_n$— Ph Me		~200	None	DCP		~200	20

[a]Oligomers showed viscosity increases under γ irradiation: an indication of polymerization rather than crosslinking.

EXPERIMENTAL

Silane polymers and oligmers of silastyrene (SS) and a vinyl silane were obtained from 3M Company and Union Carbide, respectively. The structural formulas are given in Table 37.1 along with the experimental conditions.

For crosslinking via γ-ray irradiation, PSS was melt-coated on thin stainless steel plates in glass tubes in vacuum, Ar, air, and N_2O atmosphere. The vinyl silane was placed in evacuated borosilicate test tubes. The glass tubes containing silane samples were irradiated with γ-rays from a ^{60}Co source at 1-in. distance for various lengths of time up to 29 days.

CFRI's benzoyl peroxide (BPO), aszobisisobutyronitrile (AIBN), and dicumyl peroxide (DCP) were recrystallized from methanol before use. Few grams of silanes were dissolved in 5–10 mL of benzene in a test tube or in a 3-neck round bottom flask and the silane solution was degassed with an inert gas. After 30–60 min, a CFRI in the range of 2–15 wt% was added under an inert atmosphere and the crosslinking reaction was carried out with heating on a hot plate or by a heating mantle. The crosslinking reaction in the 3-neck flask was allowed to reflux for 12 hr before cooling to room temperature. The crosslinked product was extracted and washed with methanol. Viscosity measurements were obtained by using Haake Rotovisco RV 1000 at room temperature.

Detection and confirmation of crosslinking was via FTIR, solubility in a solvent (benzene or THF), and fusion at $\sim 200°C$. Pyrolysis of the polymers was carried out in a high-temperature furnace with a flowing inert gas and also in a thermogravimetric analyzer* with continuous N_2 flushing and $10°C/min$ heating rate.

RESULTS

Crosslinking reaction conditions are summarized in Table 37.1 for those silanes that showed a positive crosslinking reaction. Oligmers did not solidify completely under γ-ray irradiation but showed an increase in viscosity.

Silanes with a γ-ray dose greater than 200 Mrad (~ 30 days at 1-m. distance from the source at 0.3 Mrad/hr) were infusible at temperatures above $200°C$ and were insoluble in THF or in benzene indicative of crosslinking. Additions of CFRI prior to γ-ray irradiation gave no difference in the crosslinking reaction rate.

FTIR spectra of PSS before and after 29 days of γ-ray irradiation are shown in Fig. 37.1. The polymer film on a stainless steel plate after 29 days of irradiation showed insolubility in THF and infusibility upon heating up to $250°C$. γ-ray irradiation of the vinyl silane oligomer in vacuum showed an increase in viscosity within 12 days from 9 cP to $>100,000$ cP. The vinyl silane upon heating at $\sim 250°C$ for 10 min in N_2 was transformed into a light yellow translucent

---

*DuPont TGA 951.

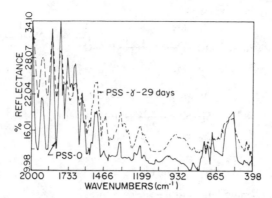

Figure 37.1   IR spectra of PSS before and after 29 days of γ-ray exposure.

solid which was insoluble in toluene. γ-ray irradiation of PSS in a N_2O atmosphere for >11 days changed the color of PSS from translucent yellow-green to bright red-brown. The viscosity of PSS decreased sharply indicating the degradation of the polymer.

Among the CFRIs (BPO, AIBN, and DCP), only DCP in the range of 2–15 wt% in PSS/benzene solution showed a positive crosslinking reaction. The FTIR spectrum of DCP crosslinked PSS compared with that of as-synthesized PSS is shown in Fig. 37.2. Figure 37.3 shows the probable mechanism of crosslinking by DCP. The DCP reacted polymers were insoluble in benzene and infusible at temperatures above 200°C.

The TGA data given in Fig. 37.4 show the char yield of SiC for different silane systems. The FTIR spectrum of DCP-reacted PSS pyrolyzed in N_2 at 1040°C is shown in Fig. 37.5 and compared with that of Nicalon®. The spectra show the characteristic absorption band of Si—C stretching at 793 cm^{-1} along with a small SiO_2 band at ~ 1040 cm^{-1}. An XRD powder pattern showed that the pyrolyzed product is amorphous which is identical with Nicalon.

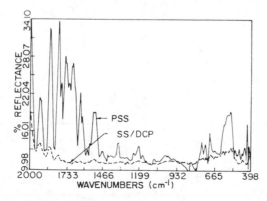

Figure 37.2   IR spectra of PSS and SS/DCP crosslinked.

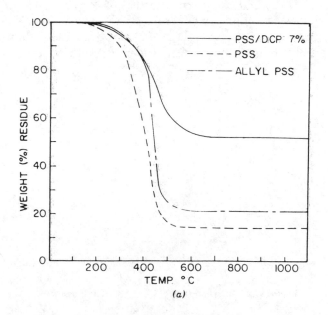

Figure 37.3 Probable crosslinking mechanism of PSS by DCP.

The char yield of PSS without crosslinking was less than 20 wt% which is close to the char yield of vinyl silane while the char yield of the PSS/DCP system shows 52 wt% SiC. The SS oligomer crosslinked with DCP shows 67% char yield.

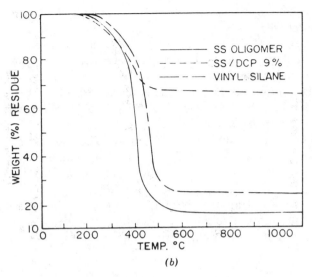

Figure 37.4 (a) TGA of Polysilanes showing char yield after pyrolyzing in N_2 at 10°C/min heating rate. (b) TGA of silane olgimers as in (a).

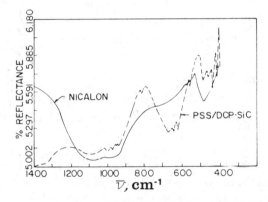

Figure 37.5 IR spectra of PSS/DCP-SiC compared with Nicalon.

DISCUSSION

An energy equivalent to 200 Mrad or greater γ-ray irradiation required to cross-link PSS is not unusual because of the phenyl group[8] on the chain and the absence of α-hydrogen.

In Figs. 37.1 and 37.2, the sharp bands at $\sim$700 cm^{-1} are completely lost and the DCP reacted SS oligomer lost most of its IR bands for Si—CH$_3$, and Si—H. This indicates that the crosslinked SS oligomer with DCP is an entirely different polymer from PSS. Crosslinking of carbon polymers in N_2O under

γ-irradiation as shown by Okada[9] did not occur with silanes but rather the opposite was observed.

Among the several CFRI studied, only DCP yielded an insoluble and infusible solid of polysilanes. This is probably due to the active methyl radical which is not present in any other CFRI used.

The TGA data show that the SS oligomer reacted with DCP yielded a higher char yield (67%) than the PSS reacted with DCP (52%). This may be due to opening up the six-membered oligomer ring which reacts with DCP in the repolymerization and crosslinking of the silane. The theoretical weight loss for PSS transformed to SiC is 55%.[1] The greater char yield of SS oligomer/DCP than the theoretical value 45% is an indication of residue of DCP in the pyrolysis products.

These results demonstrate the potential usefulness of SS oligomer as a filler for infiltrating microporous ceramic bodies with the liquid oligomer and DCP and then pyrolyzing the impregnated composite.

CONCLUSIONS

PSS and SS oligomer can be crosslinked to yield SiC by γ-rays and/or by DCP. The char yield of the crosslinked SS system is 52–67% of SiC. Two advantages of the PSS system in this paper have been demonstrated over the Yajima process for producing SiC:

1. Direct processing of the polymer to SiC is possible without the intermediate step of polydimethylsilane to polycarbosilane.
2. No oxidative crosslinking in air is necessary which evidently introduces oxygen in SiC.

Repolymerization and crosslinking of the oligomer of SS is significant since the proportion of the oligomer in PSS synthesis is larger and the oligomer may be used as an impregnant for microporous ceramic bodies.

γ-ray radiation is possible to crosslink or repolymerize oligomers but it is not practical for a polymer without more reactive functional groups on the chain. Although the vinyl silane is potentially attractive for higher char yields through the vinyl group and the α-hydrogen, it has not shown to be the case in this study.

ACKNOWLEDGMENT

The authors are grateful to the financial support from the U.S. Air Force Office of Scientific Research under contract AFOSR #F49620-83-C-0072 and the silane precursors provided by 3M Company and Union Carbide.

REFERENCES

1. R. West, L. David, P. Djurovich, H. Yu, and R. Sinclair, *Bull. Am. Ceram. Soc.* **899**, 62 (1983).

2. C. L. Schilling, Jr. and J. C. Williams, Polymer Preprints, *Am. Chem. Soc.* **1**, 25 (1984).

3. R. H. Baney, see Ref. 2, p. 2.

4. R. West, I. Nozue, X-H. Zhang, and P. Trefonas, see Ref. 2, p. 4.

5. (a) S. Yamjima, J. Hayashi, M. Omori, and K. Okamura, *Nature* **261**, 683 (1976); **273**, 525 (1978). (b) S. Yajima, *Bull. Am. Ceram. Soc.* **893**, 62 (1983).

6. R. West, in *Ultrastructure Processing of Ceramics, Glasses and Composites*, L. L. Hench and D. R. Ulrich, Eds., Wiley, New York, 1984, p. 235.

7. R. Sinclair, see Ref. 6, p. 256.

8. A. Miller, Polysiloxanes, in *Radiation Chemistry of Polymers*, Academic Press, New York, 1973, p. 179.

9. Y. Okada, Irradiation of Polymers, in *Advances in Chemistry Series*, R. F. Gould, Ed., American Chemical Society, 1969, p. 44.

38

A NOVEL POLYMERIC ORGANOSILAZANE PRECURSOR TO Si$_3$N$_4$/SiC CERAMICS

DIETMAR SEYFERTH AND GARY H. WISEMAN
Department of Chemistry
Massachusetts Institute of Technology
Cambridge, Massachusetts

INTRODUCTION

Silicon nitride has become an important ceramic.[1] It has high thermal stability (to 1800°C), high oxidative stability (to 1500°C), and, if maximum theoretical density can be achieved, it is the hardest material which can be made at atmospheric pressure. Other advantageous properties of silicon nitride are its low electrical conductivity, low coefficient of thermal expansion, excellent thermal shock and creep resistance, high strength at elevated temperatures, and corrosion resistance. As a consequence, commercial interest in silicon nitride is high.

There are three major routes for the preparation of silicon nitride:

1. The high temperature reaction of gaseous nitrogen with elemental silicon ("nitridation").
2. The gas-phase reaction of ammonia with a chlorosilane (SiCl$_4$, HSiCl$_3$, H$_2$SiCl$_2$) at higher temperatures.
3. The reaction of ammonia with SiCl$_4$ or HSiCl$_3$ in solution, followed by pyrolysis of the insoluble ammonolysis product after removal of ammonium chloride.

A newer method which we have investigated involves the reaction of ammonia with H_2SiCl_2 in an appropriate solvent to give a soluble silazane polymer which is then pyrolyzed in a nitrogen atmosphere to produce Si_3N_4.[2] The initially formed polymer, however, is not stable at room temperature (under nitrogen). It undergoes a crosslinking reaction and within a few days sets to a clear, hard glass. Added to this drawback is the fact that the starting material, H_2SiCl_2, is a rather hazardous chemical.[3]

The preceramic polymer route to silicon nitride nevertheless continued to be of interest and we have continued our research efforts in this area.

The room temperature crosslinking process which the initially formed H_2SiCl_2 ammonolysis product, $[H_2SiNH]_x$ cyclic and possibly linear oligomers, underwent may have involved chemistry of Si—H *and* N—H linkages, since cyclopolysilazanes of type $[R_2SiNH]_x$ (R = an organic group) are stable with respect to such crosslinking condensation reactions. The ideal preceramic polymer is one which will undergo a "self-curing" reaction (i.e., a crosslinking process) during the early or middle stages of its pyrolysis, so that a counterproductive decomposition to small (volatile) molecules is avoided and the yield of ceramic residue is maximized. This requires the presence of potentially reactive functionality in the polymer. In the case under discussion, the Si—H and N—H bonds are the reactive functionality. Unfortunately, this crosslinking process obviously was too facile, since it occurred even at room temperature. We reasoned that it was necessary to retain at least some Si—H bonds in the polymer for crosslinking purposes, but that some of the Si—H groups in $[H_2SiNH]_x$ could be replaced by an organic group. This is, of course, a compromise. In order to obtain a more tractable preceramic system, a carbon-containing group is added and it is probable that at least part of the carbon will end up in the ceramic product, as SiC or as free carbon. This is an acceptable compromise since silicon carbide also has very attractive properties. In order to minimize such introduction of carbon into the ceramic product, we used methyldichlorosilane, CH_3SiHCl_2, in place of H_2SiCl_2 in our subsequent work.

Methyldichlorosilane is commercially available; it is a by-product of the "Direct Process," the high-temperature, copper-catalyzed reaction of methyl chloride with elemental silicon and it is potentially inexpensive. In contrast to H_2SiCl_2, it is a liquid under ambient conditions and it appears to be "safe." The ammonolysis of CH_3SiHCl_2 has been reported to give a mixture of cyclic and (possibly) linear oligomers, $[CH_3SiHNH]_x$.[4] We have examined this reaction in some detail. It is very easily effected by introducing gaseous ammonia into a stirred solution of CH_3SiHCl_2 in an appropriate anhydrous organic solvent (diethyl ether, tetrahydrofuran, benzene, etc.). The ammonolysis product, after removal of the precipitated NH_4Cl which also is produced, can be isolated as a clear, mobile liquid in high yield. Its C, H, and N analysis and its spectroscopic (^1H NMR, IR) data are in agreement with the $[CH_3SiHNH]_x$ formulation. Molecular weight determinations (cryoscopy in benzene) of several preparations ranged from 280–320 g/mole ($\bar{x} = 4.7$–5.4). The product is quite stable at room temperature, but it is sensitive to moisture and must be protected from

the atmosphere. It is not stable to the conditions required for gas chromato-
graphic analysis, but approximately one-half of it is volatile when heated to
100°C in high vacuum. This mixture of $[CH_3SiHNH]_x$ oligomers is not suitable
for ceramics preparation without further processing. On pyrolysis to 1000°C
in a stream of nitrogen the ceramic yield is only 20%. Clearly, it is necessary to
convert these cyclic $[CH_3SiHNH]_x$ oligomers to material of higher molecular
weight.

The conversion of the cyclopolysilazanes obtained by ammonolysis of
diorganodichlorosilanes was investigated by Rochow and his coworkers[5]
some years ago when there was interest in polysilazanes as polymers in their
own right. This procedure, the ammonium salt-induced polymerization, which
in the case of hexamethylcyclotrisilazane appears to give polymers with units (**A**)
and/or (**B**), was applied in our investigations to the CH_3SiHCl_2 ammonolysis
product. It produced a very viscous oil of higher molecular weight, but the

$$
\left[
\begin{array}{c}
SiR_2 \quad\quad SiR_2 \\
\diagdown \diagup \diagdown \diagup \diagdown \diagup \\
N \quad\quad N \\
| \quad\quad | \\
SiR_2 \quad SiR_2 \\
\diagdown \quad \diagup \\
NH
\end{array}
\right]_x
\quad
\left[
\begin{array}{c}
-N-SiR_2-NH-SiR_2-N-SiR_2- \\
| \quad\quad\quad\quad\quad\quad | \\
SiR_2 \quad\quad\quad\quad SiR_2 \\
| \quad\quad\quad\quad\quad\quad | \\
-N-SiR_2-NH-SiR_2-N-SiR_2-
\end{array}
\right]_x
$$

$$
\textbf{(A)} \quad\quad\quad R = CH_3 \quad\quad\quad \textbf{(B)}
$$

ceramic yield obtained on pyrolysis was a disappointing 36%. The $Ru_3(CO)_{12}$
catalyzed ring-opening polymerization of cyclo-$[(CH_3)_2SiNH]_4$, reported
recently by Zoeckler and Laine,[6] could not be adapted to the conversion of the
$[CH_3SiHNH]_x$ cyclics to a soluble polymer. An insoluble, rubbery solid was
formed, which suggests that Si—H bonds as well as the Si—N bonds were activated by the transition metal catalyst.

The solution to our problem of converting the $[CH_3SiHNH]_x$ cyclics to a
useful preceramic polymer was provided by earlier work of the Monsanto
Company,[7] which described the conversion of silylamines of type (**1**) to cyclo-
disilazanes, (**2**), in high yield by the action of potassium in di-n-butyl ether
[Eq. (1)]. In this

$$
2R_2SiNR' \xrightarrow[Bu_2O]{K} 2H_2 + R_2Si
\begin{array}{c}
H \\
| \\
N \\
\diagup \diagdown \\
\quad\quad SiR_2 \\
\diagdown \diagup \\
N \\
| \\
H
\end{array}
\tag{1}
$$

$$
\underset{HH}{\overset{||}{}}
$$

$$
\textbf{(1)} \quad\quad\quad\quad\quad\quad \textbf{(2)}
$$

reaction, the potassium serves to metalate the NH functions to give (**3**). This then either eliminates H^- from silicon to give an intermediate with a silicon–nitrogen double bond, $R_2Si = NR'$, which then undergoes head-to-tail dimerization to form (**2**) or alternatively, reacts with a molecule of $R_2Si(H)NHR'$

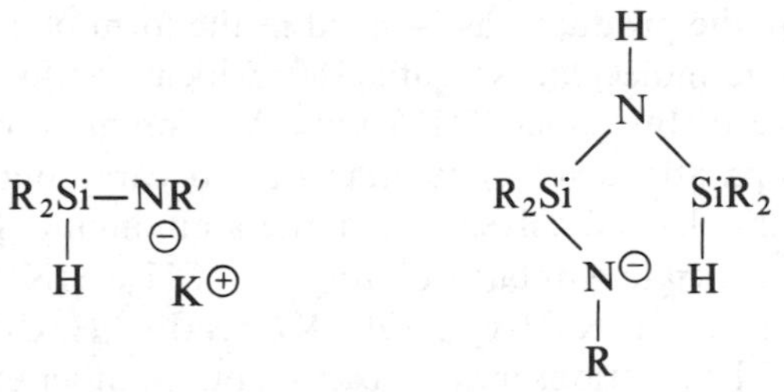

R₂Si—NR'
| ⊖
H   K⊕

(**3**)

H
|
N
/   \
R₂Si   SiR₂
\ |
N⊖   H
|
R

(**4**)

to give intermediate (**4**) which undergoes cyclization to (**2**) with displacement of H^-. This interesting mechanistic question still needs to be resolved.

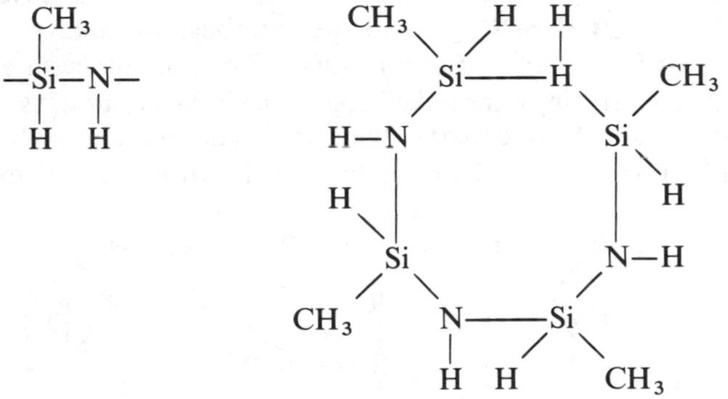

(**5**)

(**6**)

The repeating unit in the $[CH_3SiHNH]_x$ cyclics is (**5**). Thus the cyclic tetramer is the eight-membered ring compound (**6**). On the basis of Eq. (1), the adjacent NH and SiH groups provide the functionality which permits the molecular weight of the $[CH_3SiHNH]_x$ cyclics to be increased.

Treatment of the CH_3SiHCl_2 ammonolysis product, cyclo-$[CH_3SiHNH]_x$, with a catalytic amount of a base (generally an alkali metal base) strong enough to deprotonate the N—H function in a suitable solvent results in evolution of hydrogen. The resulting solution contains polymeric basic species and these are quenched by addition of methyl iodide or a monochlorosilane. After filtration to remove alkali metal halide, evaporation of the filtrate gives the product in essentially quantitative yield.

In our experiments we generally have used potassium hydride as the base, and the following is a typical experiment. The cyclosilazane, $[CH_3SiHNH]_x$

($\bar{x}$ = 4.9), 15.29 g (0.258 mole of CH_3SiHNH unit) was added under nitrogen to a slurry of 0.40 g (10 mmoles, 3.9 mole% based on the CH_3SiHNH unit) of KH in 300 mL of tetrahydrofuran (THF). After gas evolution had ceased, 2.3 g of methyl iodide was added to quench all basic species, including basic sites on the resulting polymer. After the KI and solvent were removed by the appropriate workup, the product was isolated in the form of a white powder (in 99% yield) (average molecular weight, 1180) which was found to be soluble in hexane, benzene, diethyl ether, THF, and other common organic solvents.

For larger preparations it is advantageous to carry out the CH_3SiHCl_2 ammonolysis in the desired solvent, filter the ammonium chloride, and then add the KH to the filtrate without isolating the $[CH_3SiHNH]_x$.

Other bases (e.g., NaH, $NaNH_2$, KOH, $KB(sec$-$Bu_3)H$, CH_3Li, etc.) may be used as catalysts. The reactions may be carried out in other ethers or in hydrocarbon solvents, and we have used reaction temperatures between 0 and 65°C, but usually worked at room temperature. It is imperative to exclude atmospheric moisture since the $[CH_3SiHNH]_x$ cyclics are readily hydrolyzed. The product polymer on the other hand, is of greatly diminished sensitivity to hydrolysis.

The composition of the polysilazane product of the experiment detailed above, ascertained by proton NMR spectroscopy, was $(CH_3SiHNH)_{0.39}$ $(CH_3SiHNCH_3)_{0.04}(CH_3SiN)_{0.57}$ and its combustion analysis (C,H,N,Si) agreed with this formulation. These results are compatible with a process in which $(CH_3SiHNH)_n$ rings are linked together via Si_2N_2 bridges. Thus, if for example, the silazane (6) were to be polymerized in this way, the eight-membered rings could be linked in a ladder polymer as shown in (7). The experimental

R = CH_3   (7)

$(CH_3SiN)/(CH_3SiHNH)$ ratio of ~1.3, however, indicates that further linking together of ladders via Si_2N_2 rings must have taken place. High polymers obviously are not formed, but the molecular weight is increased sufficiently so that pyrolysis proceeds satisfactorily.

A better understanding of the chemistry leading to these polysilazanes and of their structure is needed. For instance, in our various preparations the average molecular weights of the products obtained after the methyl iodide quench varied between 800 and 2000. Before the methyl iodide quench, silyl-amide functions (i.e., catalytically active functions) still were present, yet growth to higher molecular weights did not occur. Why is this so? The answer to this question must be connected with the solution structure and conformation of the polymers. If the polymerization involves linkage of the $[CH_3SiHNH]_x$ rings via Si_2N_2 rings in more than one direction, that is, if it involves formation of linked ladders as postulated above, then a sheet structure will result. Sheet structures are common in silicon chemistry, for instance, in silicates, and in this connection some work reported by Kenney and his coworkers[8] is of interest and possibly pertinent. In these studies, reaction of the silicate mineral chryso-tile with hydrochloric acid and Me_3SiCl gave a stable polymer with a silicon–oxygen sheet framework. These polymer sheets curled up into scrolls when dry. It could be that if we have sheet polymers in our case, that in solution the repulsion of the negative charges of the amide functions will result in partial "curling up" of the sheets, thus introducing steric factors which would inhibit further growth.

Whatever the structure of the silazane polymers obtained by KH treatment of the $[CH_3SiHNH]_x$ cyclics, these polymers are excellent ceramic precursors. Examination of the polymers from various preparations by TGA showed the weight loss on pyrolysis to be only between 15 and 20%. The pyrolysis appears to take place in three steps: a 5% weight loss (involving evolution of H_2) from 100–350°C; a 2% weight loss from 350–550°C; and a 9% weight loss from 550–900°C. During the 550–900°C stage a mixture of H_2 and methane was evolved. A trace of ammonia, in addition to H_2, was lost between 350 and 550°C.

In a typical bulk pyrolysis experiment, a carbon sample boat containing the polysilazane from an experiment such as that described previously was placed (nitrogen atmosphere) in a mullite tube in a tube furnace. Pyrolysis was conducted under a slow stream of nitrogen. The sample was heated quickly to 500°C and then slowly (over 8 hr) to 1420°C and was held at 1420°C for 2 hr. The ceramic powder was a single body and black. Powder X-ray diffraction (CuK$_\alpha$ with Ni filter) showed only very small, broad peaks for α-Si_3N_4. Scanning electron microscopy analysis showed little discernible microstructure with only a few very fine grains appearing at high magnification. The bulk appearance of the ceramic suggested that pyrolysis took place after the polymer had melted. There were many large holes and craters where the liquid bubbles apparently had burst. In such experiments ceramic yields usually were between 80 and 85%. The polymer used in the experiment above had been prepared in diethyl ether. It had a molecular weight of around 900 and went through a melt phase when it was heated. This could be shown when it was heated in a sealed capillary: it began to soften around 65°C, becoming more fluid with increasing temperature. The polymer prepared in THF is of higher molecular weight (MW = 1700–2000) and does not soften when heated to 350°C. It gave an 83% yield of a black ceramic material on pyrolysis under nitrogen to 1000°C.

The pyrolysis of the silazane polymer may be represented by Eq. (2). Here the ceramic yield (Si_3N_4 + SiC + C) would be 83 wt%.

$$2(CH_3SiHNH)(CH_3SiN) \rightarrow Si_3N_4 + SiC + C + 2CH_4 + 4H_2 \qquad (2)$$

An analysis of such a ceramic product gave 12.87% C, 26.07% N, and 59.52% Si. By difference, % O would be 1.54%, but it is not necessarily present. This analysis is compatible with Eq. (2) and leads to a ceramic constitution, based on the 4Si of Eq. (2), of 0.88 Si_3N_4, 1.27 SiC, 0.75 C, and (possibly) 0.09 SiO_2.

Thus the *chemistry* leading to the desired ceramic product is quite satisfactory: the starting material is readily available, not expensive, and safe to handle. Its reaction with ammonia is easily carried out and gives the cyclopolysilazane product in high yield. The conversion of the latter to the preceramic polymer is easily effected. It requires only a catalytic quantity of base and the polymer is easily isolated in essentially quantitative yield. The simple workup (filtration, evaporation of solvents) allows the recovery and recycling of the anhydrous solvent. The polysilazane product is stable on storage at room temperature and is not especially sensitive to hydrolysis. Its pyrolysis gives a high yield of ceramic product and the volatiles evolved are not toxic or corrosive. It remains to be seen whether the ceramics applications are equally advantageous, but preliminary indications are promising.

Among potential applications of preceramic polymers which are of current interest are the following:

1. The forming of the polymer into complex shapes and its subsequent pyrolysis to give a ceramic material of the same shape (but not necessarily the same size).

2. The spinning of such polymers into continuous fibers whose subsequent pyrolysis would give ceramic fibers. (Note the current importance of graphite fibers and the great interest in the Nicalon ® SiC fibers which were developed in Japan.)

3. The use of such polymers as matrix material for carbon or ceramic fibers, or as a binder for ceramic powders (with subsequent pyrolysis to form the ceramic).

4. The production of oxidation-resistant coatings on otherwise oxidizable materials (such as carbon–carbon composites). After the polymer coatings has been made, it would then be pyrolyzed to give the ceramic coating.

5. The infiltration of porous ceramic bodies such as ones obtained from reaction-sintered silicon nitride by the polymer itself (if liquid) or by a solution of the polymer, with subsequent pyrolysis to form the ceramic, resulting in better strength, oxidation resistance, and so on, of the body.

6. The formation of thin films of the ceramic material for electronics applications.

Our initial research has dealt with the first three of these potential applications. In an investigation of the conversion of a shaped body of the polymer to a ceramic body we studied the pyrolysis of polysilazane bars. The problems associated with such a conversion have been pointed out by Rice.[9] Bar-shaped bodies of the polysilazane prepared in THF solution were prepared by first uniaxially pressing a sample in a rectangular steel die and then isostatically pressing the resulting bar at 40,000 psi. Pyrolysis in a tube furnace under nitrogen (10°C/min to 1100°C) gave a coherent, rectangular ceramic bar which had not cracked or bloated and could not be broken by hand. Even more promising were experiments in which blends of the nonmelting and the meltable polysilazanes were used.

Fiber-reinforced composites as engineered structural materials are a very important area of research and development in current materials engineering.[11] The Japanese Nicalon® silicon carbide-containing fiber[12] has aroused much interest and also has prompted research to find other ceramic fibers, hopefully with better high-temperature properties. In collaboration with ceramists at the Celanese Research Company it was found that our meltable polysilazane cannot be melt-spun. Apparently, the thermal cross-linking process which is so effective in giving a high ceramic yield on pyrolysis takes place in the heated nozzle of the spinning machine and quickly gives infusible polymer. However, the infusible polysilazane which is obtained when the preparation is carried out in THF solution can be dry-spun. In this process the solid polysilazane is dissolved in an appropriate solvent and then is extruded through a spinneret into a heated drying chamber in which the solvent is volatilized, leaving the solid polymer. These polymer fibers could be pyrolyzed to give ceramic fibers. It was shown that fibers 0.3–0.6 m in length could be drawn from the sticky, waxy solid which remained when a toluene solution of the polysilazane was evaporated. Pyrolysis of these fibers under nitrogen produced long, flexible black fibers.

We also have studied ceramic powder composites using our polysilazane as a binder. The preparation of such composites (using commercial samples of fine α-SiC, β-SiC, and α-Si$_3$N$_4$ of 0.36–0.4 μm mean particle size) required use of dispersion methods. Best results were obtained using a blend of the meltable and the infusible powder. The ceramic powder was dispersed in a solution of toluene containing the appropriate weight of polysilazane and then the toluene was evaporated using a rotary evaporator, leaving a waxy residue. Vacuum distillation removed the remaining solvent to leave chunks of solid material. These were finely ground and pressed into a bar at 5000 psi. Isostatic pressing to 40,000 psi followed and then the bars were pyrolyzed in a tube furnace under nitrogen (10°C/min, to 1100°C). The maximum density ($\sim$2.4 g/cm^3) was achieved in these experiments with a polymer loading of 30%. However, all bars were relatively weak. SEM micrographs of the fracture surfaces of the bars showed that these composites had many large cracks throughout the bodies. These cracks probably resulted from nonuniform shrinkage.

CONCLUSION

Polysilazanes prepared from the base-catalyzed polymerization of the cyclo-polysilazanes obtained in the ammonolysis of methyldichlorosilane are potentially useful for the fabrication of ceramics in the form of fibers or monolithic ceramic bodies prepared by pyrolysis of the appropriately shaped polymeric precursor.[13,14] These polysilazanes also are useful as dispersants for SiC and Si_3N_4 powders. Composites can be prepared from these powders by using the polysilazanes as binders which themselves are converted to ceramic materials on pyrolysis.

ACKNOWLEDGMENT

This work was supported in part by the Office of Naval Research.

REFERENCES

1. D. R. Messier and W. J. Croft, in *Preparation and Properties of Solid State Materials*, Vol. 7, W. R. Wilcox, Ed., Dekker, New York, 1982, pp. 131–212.
2. (a) D. Seyferth, G. H. Wiseman, and C. C. Prud'homme, *J. Am. Ceram. Soc.* **66**, C-13 (1983). (b) U.S. Patent 4,397,828 (August 9, 1983). (c) D. Seyferth and G. H. Wiseman, in *Ultrastructure Processing of Ceramics, Glasses and Composites*, L. L. Hench and D. R. Ulrich, Eds., Wiley, New York, 1984, pp. 265–271.
3. K. G. Sharp, A. Arvidson, and T. C. Elvey, *J. Electrochem. Soc.* **129**, 2346 (1982).
4. S. D. Brewer and C. P. Haber, *J. Am. Chem. Soc.* **70**, 3888 (1948).
5. (a) C. R. Krüger and E. G. Rochow, *J. Polym. Sci., Part A* **2**, 3179 (1964). (b) E. G. Rochow, *Monatsh. Chem.* **95**, 750 (1964).
6. M. T. Zoeckler and R. M. Laine, *J. Org. Chem.* **48**, 2539 (1983).
7. Monsanto Company, Neth. Appl. 6,507,996 (December 23, 1965); *Chem. Abstr.* **64**, 19677d (1966).
8. (a) S. E. Frazier, J. A. Bedford, J. Hower, and M. E. Kenney, *Inorg. Chem.* **6**, 1693 (1967). (b) J. P. Linsky, T. R. Paul, and M. E. Kenney, *J. Polym. Sci., Part A-2* **9**, 143 (1971).
9. R. W. Rice, *Am. Ceram. Soc. Bull.* **62**, 889 (1983).
10. P. Beardmore, J. J. Hardwood, K. R. Kinsman, and R. E. Robertson, *Science* **208**, 833 (1980).
11. R. W. Rice, *Ceram. Eng. Sci. Proc.* **2**(7–8), 493, 661 (1981); **4**(7–8), 485 (1983).
12. S. Yajima, *Am. Ceram. Soc. Bull.* **62**, 893 (1983).
13. D. Seyferth and G. H. Wiseman, *J. Am. Ceram. Soc.* **67**, C-132 (1984).
14. D. Seyferth and G. H. Wiseman, U.S. Patent 4,482,669 (November 13, 1984).

39

ELECTROCHEMICAL AND SONOCHEMICAL ROUTES TO ORGANOSILANE PRECURSORS

PHILIP BOUDJOUK
Department of Chemistry
North Dakota State University
Fargo, North Dakota

ELECTROORGANOSILICON CHEMISTRY

Electrochemical methods of synthesis such as controlled current and controlled potential syntheses have been useful tools in the hands of inorganic chemists for many years dating from the successes of Faraday. Electrosynthesis in aqueous systems is particularly common because of the high dielectric constant of water. Thus many of the early developments in electrochemistry dealt with inorganic, water soluble systems.

In this chapter the results of attempts to bring this technique to bear on a problem in organosilicon chemistry, that is, the challenge of making silicon-to-silicon bonds to form disilanes and ultimately, polysilanes are discussed. We could not simply transpose the operations of inorganic chemistry to organosilanes because nearly all organosilicon compounds are insoluble in water, and, of greater consequence, the organosilanes of most importance react with water and with most water substitutes such as alcohols.

The solubility problem is easily overcome. Organic chemists have had excellent success with highly polar solvents such as acetonitrile, dimethylforamide, tetrahydrofuran, diglyme, and sulfolane in combination with electrolytes such

363

as tetraalkylammonium halides which have the unique property of dissolving in organic solvents yet ionizing sufficiently to permit electrical conduction.

The notable successes of organic electrochemists are shown in Table 39.1[1]:

Having prepared highly strained all-carbon structures electrochemically the same technique was applied to the synthesis of novel polysilanes. Some of those are listed in Table 39.2. At the time the study began only one class of compounds was populated: the silylenes, which are represented by the general formula, R_2Si. Recently, the chemistry of the other systems has also been developed.

While organic electrochemists circumvented the problem of solubility by using highly polar solvents in combination with a suitable electrolyte, in organosilicon chemistry hydroscopicity is a major problem especially in the presence of electrolytes which are believed to catalyze the hydrolysis of functionalized silanes.

The water problem is serious in this work and it is overcome in less than half of our electrolyses. We were not the first to electrolyze functionalized silicon

TABLE 39.1   Some Novel Organic Compounds Prepared Electrochemically

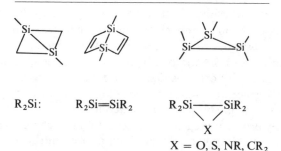

TABLE   39.2   Novel   Target   Polysilanes   via Electroreduction of Halosilanes

R_2Si: $R_2Si{=}SiR_2$ $R_2Si{-}\!\!-\!\!-SiR_2$

                                    X

                    X = O, S, NR, CR_2

compounds. As early as 1965, Dessy had reported the half-wave potential of triphenylchlorosilane.[2] Other work followed sometimes victimized by the water problem, that is, data were gathered on the hydrolysis products and not the substrate. Two important developments were reported by Hengge[3] and Corriu[4]: the former showed that a chlorosilane could be coupled to form a disilane under controlled current conditions while the latter demonstrated that a chlorosilane could be electrochemically coupled at controlled potential, even if the potential was rather high. Our approach was to determine the half-wave potentials of a broad variety of functionalized silanes as a foundation for a synthetic effort. Our plan was to investigate the obvious candidates for electrocoupling, that is, organosilanes with one chlorine atom attached to the silicon and also silanes with two or more halogen atoms on the silicon. We wished to determine if the organic part of the functionalized silane had an effect on the ease of electrochemical reduction.

Another very important variable was the functional group itself. We surveyed not only the halogens but also other groups reactive on silicon toward reducing agents in chemical systems. In Tables 39.3 and 39.4 are listed the reduction potentials for a series of halosilanes. Included are not only the monosubstituted derivatives but the more challenging dihalosilanes. The latter are more important as feedstock materials for silicones; producing those polymers upon

TABLE 39.3   Reduction
Potentials of Some
Monochlorosilanes

Me_2HSiCl	$-3.2v$
Me_3SiCl	-3.2
Et_3SiCl	-3.2
$t\text{-}BuMe_2SiCl$	-3.2
Ph_2HSiCl	-3.2
Ph_3SiCl	-2.85

TABLE 39.4   Reduction
Potentials for Some Difunctional
Silanes

Me_2SiCl_2	$-3.2v$
$t\text{-}Bu_2SiCl_2$	-3.2
Ph_2SiCl_2	-2.70
Mes_2SiCl_2	$-2.80, -3.0$
$(t\text{-}Bu_2Si)_2$	$-2.95, -3.2$
$\quad\ \ |$	
$\quad\ \ Br$	

hydrolysis. They are also likely to be intermediates in formation of polymers that do not contain oxygen.

In Table 39.5, the reduction potentials of a series of organosilanes that contain functional groups other than silicon are listed. These compounds have the advantage over chlorosilanes in that they are not as water sensitive and are therefore easier to handle.

In summary:

1. There are no distinct reduction waves for the alkyl-substituted chloro or bromosilanes.
2. When the halogens are replaced by other groups such as acetate, phenoxide, or silanolate, there is no significant effect on the reduction potial.
3. Only phenyl, hydro, and vinyl groups affected the reduction potential to a measurable extent.
4. The only leaving group that showed a significant drop in the reduction potential was the tosylate group (OTs).

In spite of the difficulties with the system we continued to develop electrochemistry as a synthetic technique for organosilanes. In Table 39.6 are listed the successful electroreduction of a number of chlorosilanes, yields of the disilanes, and the current efficiencies of the process.

These are difficult experiments because of the effect of even small amounts of water. There are two reasons why water interferes: (1) it reacts with the silane and destroys the starting material producing siloxane, a by-product that is often difficult to remove from the target compound; and (2) water and halosilanes produce acids such as hydrogen chloride and hydrogen bromide which will reduce far more easily than the silicon compound, effectively stopping the chosen reaction. Experimental runs are successful less than half the time. The apparatus must be "aged" by several runs before we can produce consistently high yields of disilane. Our best runs have about 5% of siloxane impurity in the product.

To summarize, electrosynthesis of organosilanes has promise because the reaction conditions are mild and there is the possibility not only of batch synthesis but also of flow electrolysis, an attractive feature for large-scale work.

TABLE 39.5 Reduction Potentials of Functionalized Silanes

$Me_3Si—OTs$	$-2.9v$
$Me_3Si—OCPh_3$	-3.2
$Me_3Si—OPh$	-3.2
$Et_3Si—OAc$	-3.2
$Ph_2HSi—OPh$	-3.2

TABLE 39.6   Electrosynthesis of Some Disilanes

Organosilane	Product	% Yield[a]	% Current Efficiency[b]
Me_2HSiCl	$Me_2HSiSiHMe_2$	88	82
Me_3SiCl	$Me_3SiSiMe_3$	76	60
Et_3SiCl	$Et_3SiSiEt_3$	77	67
$CH_2{=}CHMe_2SiCl$	$(Ch_2{=}CHMe_2Si)_2$	83	73
$PhMe_2SiCl$	$PhMe_2SiSiMe_2Ph$	90	79
Ph_2HSiCl	$Ph_2HSiSiHPh_2$	90	70

[a]Yields based on GLC peak integration of all products formed and unreacted starting material.

[b]Current efficiencies calculated as: $\left(\dfrac{\text{no. coulombs calculated}}{\text{no. coulombs actual}}\right) \times \%$ yield of dimer.

The reactions can also be run at high current efficiency.

The disadvantages are: sensitivity toward protic reagents and reduction potentials that are often so high that the solvent is also reduced during the reaction unless extreme care is taken. These problems are much more difficult on the small scale than on large scale; for example the silicon industry has been handling chlorosilanes in bulk for decades. More challenging is the high reduction potential problem. We believe that the answer to this problem lies in electro-catalysis.

SYNTHESIS WITH ULTRASONIC WAVES

As early as 1927, Richards and Loomis observed that the hydrolysis of methyl sulfate and the iodine clock reaction were accelerated by ultrasonic waves.[5] In the years that followed other workers discovered that ultrasound would help a variety of chemical processes resulting in a number of patents based on ultrasonic acceleration. The use of ultrasonic waves did not become widespread however because: (1) the rate accelerations published were often not very significant (5–15%), (2) the systems were not of general interest, or (3) ultrasonic generators. In spite of the limitation by the absence of cheap, simple ultrasonic units, important progress was made in this new field of "sonochemistry." As early as 1936, Freundlich reported that ultrasound promoted the liquefaction of gels. He followed this with a series of papers through the thirties, investigating iron and aluminum oxides in alkaline media.[6] At about the same time, several workers observed that the reactions of metals and some of their salts were quicker with acids in the presence of ultrasonic waves than in their absence. Zinc and calcium carbonate are two examples which react at an accelerated rate with hydrochloric and sulfuric acids in an ultrasonic field.[7]

Studies of the effect of ultrasonic waves on polymers showed that macro-molecules like polystyrene could be disintegrated readily, from 850M to 30M as measured by viscosity.[8] Generally, it was observed that larger molecules were easily degraded and the mechanism clearly involved the breaking of the carbon–carbon bond.

There is considerable energy available from ultrasonic waves. As early as 1940, ultrasound was used to degas molten aluminum[9] and was found useful in the tin plating of aluminum sheets.[10] The aluminum sheet was immersed in molten tin and ultrasonic waves were transmitted through the sheet. The rapid vibrations of the metal permitted penetration of the oxide layer, a barrier to efficient tinning, by the tin droplets.

For chemists interested in synthesis, a major step forward was made by Alfred Weissler in the late 1940s when he began a study of the effects of ultra-sound on organic compounds, demonstrating that there was sufficient energy available to break carbon–carbon, carbon–hydrogen, and carbon–chlorine bonds.[11]

In Fig. 39.1, a simplified picture of the effect of ultrasonic waves on a liquid medium is given. The chemical effects of ultrasonic waves are ascribed to cavitation, the process that releases energy. As the sound wave passes through the liquid a low-pressure region is developed of sufficient energy to cause a dis-ruption of the structure of the liquid. The result is the formation of a bubble or

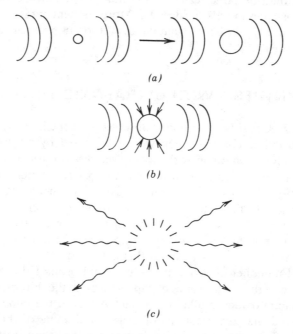

(a)

(b)

(c)

Figure 39.1 Schematic of cavitation in a liquid. (a) Expansion of bubble in low-pressure region of wave; (b) implosion by high-pressure region; (c) shock wave released following implosion giving rise to high temperatures and pressures sufficient to clean or erode surface of metals.

"cavity" that is nearly empty, containing some vapor solvent. The bubble continues to grow in the low-pressure region of the wave, and with it, the potential energy increases. The cavity is collapsed by the high pressure, or compression part of the wave. This implosion releases a large amount of energy—enough to raise temperatures to 2000°C and higher as well as develop pressures of 1 to 2 kbar. The time frame is in nanoseconds. On a macroscopic level temperatures are not that high. For the reactions discussed below, the instrument used is a $200 ultrasonic cleaner in which a reaction flask containing reagents is immersed. The temperature of the bath never exceeds 50°C.

We have focused our efforts on heterogeneous reactions involving metals because we have the most success with those systems. The major effect of ultrasound, that is, cavitation, is to clean the surface of the metal of impurities that may be present initially or formed during the course of the reaction and thus provide a fresh, reactive site for molecules to attack and react.

We have successfully produced metal powders by reacting metal halides using pure lithium as the reducing agent,[12] see eq. (1). A wide variety of powders have been generated in 5–30 min at room temperature. Typically, reaction times are much longer without ultrasound, on the order of several hours with

$$n\text{Li} + \text{MX}_n \xrightarrow{\text{)))}} \text{M(powder)} + n\text{LiX} \tag{1}$$

$$\text{M} = \text{Mg, Ca, Al, Si, C, Ge, Sn, Pb, Fe, Cr, Ni, Co, Pd, Pt, Ti,}$$

heating. Without heat or ultrasound, these reductions can require a day or longer. The problem arises, during the course of the reaction, of by-products forming on the lithium metal. Ultrasound cleans the surface, permitting the reaction to proceed. Such powders are far more reactive than commercial powders and are useful in synthesis.

Our main interest is silicon. Equation (2) is a simple reaction used to make disilanes which is much like the electroreduction reaction discussed earlier.

$$2\text{R}_3\text{SiCl} + \text{Li} \xrightarrow[\text{THF}]{\text{)))}} \text{R}_3\text{Si}-\text{SiR}_3 + \text{LiCl} \tag{2}$$

$$\text{R}_3 = \text{Me}_3, \text{Ph}_3, \text{Ph}_2\text{H}, t-\text{Bu}_2\text{H}$$

Li wire: 10 hr, 42–95%

Li dispersion: <1 hr, 85–95%

This is a simple and efficient way to make Si—Si bonds at room temperature in a very short period of time. The reaction will go well without ultrasound if the Li is very pure and very finely divided but it proceeds for only a little while before it slows down. If we intercede with ultrasonic waves the reaction increases in rate.[13]

Difunctional silanes are important precursors to silicon-containing polymers. When we irradiate a mixture of lithium and the dichlorosilanes [Eq. (3)] very satisfactory yields of cyclic polysilanes are obtained in a relatively short period of time[13]:

$$R_2SiCl_2 + Li \xrightarrow{\;)))\;} (R_2Si)_n + (R_2Si)_m \qquad (3)$$

$$R = Me,\; n = 6,\; 90\%;\; m = 5,\; 5\%$$

$$R = Ph,\; n = 4,\; 90\%;\; m = 5,\; 5\%$$

Ultrasonic waves also help in the cross-coupling reaction to make penta-silane [Eq. (4)]. This molecule is a key building block in the synthesis of strained polysilanes but is not feasible electrochemically.

$$SiCl_4 + 4Me_2SiClH + Li \xrightarrow{\;)))\;} (Me_2SiH)_4Si \;\; 60\% \qquad (4)$$

The formation of silicon–carbon bonds is also an important process that is accelerated by ultrasonic waves. There are two useful methods of generating organosilanes containing unusual and useful organic groups. They are unusual in that they differ from the simple alkoxy and alky groups normally used as precursors to polymer materials and useful in that there is residual functionality built into both the silicon portion and the organic moiety. In the first reaction [Eq. (5)] Li acts as a base, liberating hydrogen and creating the efficient nucleo-phile phenylacetylide which reacts readily with most chlorosilanes to give acetylene-substituted silanes. The triple bond is capable of a wide variety of reactions, including, after some modifications, polymerization. The second reaction [Eq. (6)] is novel in that Li causes reductive coupling of diphenyl-acetylene to give a unique dianion which

$$PhC\equiv CH + Li \xrightarrow{\;)))\;} PhC\equiv C:^- Li^+ \longrightarrow PhC\equiv CSiMe_3 \qquad (5)$$

$$PhC\equiv CPh + Li \xrightarrow{\;)))\;} \underset{\substack{PhC \ominus \;\; \ominus \;CPh \\ \\ Li^\oplus \qquad Li^\oplus}}{PhC{=}CPh} \longrightarrow \underset{\substack{PhC \qquad\quad CPh \\ \diagdown \;\;\; \diagup \\ Si \\ \diagup \;\; \diagdown \\ R \qquad R}}{PhC{=}CPh} \qquad (6)$$

forms cyclic silanes with unsaturation in the ring and with functionality on silicon if desired. Both of these syntheses can be executed at room temperature in a few hours with sonication compared to reaction conditions which require 10 to 24 hr at 60–70°C[14] without ultrasound.

The last reaction is one fundamental to organosilicon chemistry, hydro-silation. It is the most important method of preparing Si–C bonds after the direct process of Rochow. The most common procedure for executing this addition of Si–H bonds across olefins and acetylenes is to employ the very efficient homogeneous catalyst, chloroplatinic acid, also known as Speier's catalyst. Less well known is that the original patent called for the use of platinum on carbon and rigorous conditions of $T > 100°C$ at $P > 100$ psi. We have reexamined this original work by Wagner and Strother and found that with ultrasonic waves the hydrosilation reaction can be carried out in very high yield under very mild conditions [Eq. (7)].[15]

$$RCH{=}CH_2 + H{-}SiCl_3 + Pt/C \xrightarrow[0°]{)))} RCH_2CH_2SiCl_3 \quad 90\% \qquad (7)$$

hexene, 4-methylpentene, styrene, vinyltrimethylsilane

olefin:silane:Pt

0.05:0.10:0.005

ACKNOWLEDGMENT

This work was supported by the Air Force Office of Scientific Research through Grant No. 84-0008.

REFERENCES

1. For a review of this topic, see M. M. Baizer, *Organic Electrochemistry*, Dekker, New York, 1973.

2. R. Dessy, W. Kitching, and T. Chivers, Organometallic Electrochemistry. I. Derivatives of Group IV-B Elements, *J. Am. Chem. Soc.* **88**, 453–459 (1966).

3. E. Hengge and G. Litscher, Zur electrochemischen Bildung von Di-, Oligo- und Polysilanen, *Monatshefte fur Chemie* **109**, 1217–1225 (1978).

4. R. J. P. Corriu, G. Dabosi, and M. Martineau, *J. Organometal. Chem.* **188**, 63–72 (1980).

5. W. T. Richards and A. L. Loomis, The Chemical Effects of High Frequency Sound Waves. I. Preliminary Survey, *J. Am. Chem. Soc.* **49**, 2086–3100 (1927).

6. H. Freundlich, F. Rogowski, and K. Sollner, The Action of Ultrasonic Waves on Gels and Particularly Thixotropic Gels, *Kolloid-Beihefte* **37**, 223–241 (1933).

7. N. Moriguchi, Effects of Supersonic Waves on Chemical Phenomena. I., *J. Chem. Soc. (Japan)* **54**, 949–957 (1933); *Chem. Abstr.* **28**, 398 (1933).

8. A. Salzay, The Destruction of Highly Polymerized Molecules by Ultrasonic Waves, *Z. Phys. Chem.* **A164**, 234–240 (1933); *Chem. Abstr.* **27**, 3379 (1933).

9. T. Rummel, W. Esmarch, and K. Beuther, Degassing of Aluminum by Sound and Ultrasonic Waves, *Metallwirkschaft* **19**, 1029–1033 (1940); *Chem. Abstr.* **36**, 3760 (1942).

10. A. E. Thieman, The Tin Plating of Light Metals by Means of Ultrasound, *Automobiltech. Z.* **45** 668 (1942); *Chem. Zentr.* **II**, 367 (1943); *Chem. Abstr.* **38**, 4554 (1944).

11. A. Weissler, I. Pecht, and M. Anbar, Ultrasound Chemical Effects on Pure Organic Liquids, *Science* **150**, 1288–1289 (1965).

12. P. Boudjouk, B-H. Han, D. Thompson, and W. Ohrbom, unpublished studies. Preliminary report: P. Boudjouk and B-H. Han, Organic Sonochemistry. New Synthetic Applications of Ultrasonic Waves, Abstracts of the 183rd National Meeting of the American Chemical Society, (Abstract No. ORGN 190) March 28–April 2, 1982, Las Vegas, NV.

13. P. Boudjouk and B-H. Han, *Tetrahedron Lett.* **19**, 2757 (1981).

14. P. Boudjouk, B-H. Han, and R. Sooriyakumaran, unpublished studies.

15. B-H. Han and P. Boudjouk, Organic Sonochemistry, Ultrasonic Acceleration of the Hydrosilation Reaction, *Organometallics* **2**, 769–771 (1983).

40

SILICON DIOXIDE [O=Si=O]: NEW ROUTES TO CYCLIC SPIROSILOXANES; CHEMICAL VAPOR DEPOSITION OF SiO$_2$

WILLIAM P. WEBER, GEORGE K. HENRY, AND CLIFFORD D. JUENGST

Loker Hydrocarbon Research Institute
Department of Chemistry
University of Southern California
Los Angeles, California

There has been considerable interest in π-bonded reactive organosilicon intermediates in the last 15 years.[1] Transient species which possess silicon–carbon,[2-4] silicon–nitrogen,[5-10] silicon–oxygen,[11-14] silicon–sulfur,[15-17] silicon–phosphorous,[18-20] and silicon–silicon double bonds[21-23] have been generated. In this chapter we review the chemistry of silanones (R$_2$Si=O) and consider recent developments with silicon dioxide [O=Si=O].[12,24]

SILANONES

Silanones, transient reactive intermediates which possess a silicon–oxygen double bond, have been generated in several ways. Silenes [R$_2$Si=CH$_2$], produced by pyrolysis of silacyclobutanes, react with nonenolizable ketones to yield alkenes and cyclotrisiloxanes and cyclotetrasiloxanes.[12,14,25-32] These results were interpreted in terms of a [2 + 2] cycloaddition reaction between the silene and the carbon–oxygen double bond of the ketone to yield an unstable silaoxetane which undergoes a retro [2 + 2] cycloaddition reaction to yield an alkene and a silanone. Cyclooligomerization of the silanone gives the cyclotrisiloxane. Cyclotrimerization probably occurs in two steps. Initial head-to-tail dimerization of the polar silanone yields a cyclodisiloxane. Subsequent insertion of another reactive silanone into a Si—O single bond of the cyclodisiloxane gives the cyclotrisiloxane. Insertion of another silanone into a Si—O single bond of the cyclotrisiloxane will yield the cyclotetrasiloxane. Alternatively, dimerization of two cyclodisiloxanes may yield a cyclotetrasiloxane. Insertion of silicon multiple bonded intermediates into Si—O single bonds is a general reaction of these unsaturated species. The small Si—O—Si angle (86°) observed for tetramesitylcyclodisiloxane, the first stable example of this ring system, helps to account for the high reactivity of this system.[33,34] On the other hand, the reasons for the apparent instability of silaoxetanes remains an open question since no fully characterized example of this heterocyclic system has been reported.[35,36]

$$\boxed{}\!\!-\!Si(CH_3)_2 \xrightarrow{\Delta} H_2C{=}CH_2 + [H_2C{=}Si(CH_3)_2]$$

$$[H_2C{=}Si(CH_3)_2] + Ph_2C{=}O \rightarrow \left[\begin{array}{c} Ph \\ | \\ Ph{-}C{-}O \\ |\ \ \ | \\ H{-}C{-}Si(CH_3)_2 \\ | \\ H \end{array} \right] \rightarrow \begin{array}{c} Ph_2C{=}CH_2 \\ \\ [(CH_3)_2Si{=}O] \end{array} \tag{1}$$

$$2[(CH_3)_2Si{=}O] \rightarrow \left[(CH_3)_2Si \underset{O}{\overset{O}{\diamond}} Si(CH_3)_2 \right] \quad [(CH_3)_2Si{=}O]$$

$$\tag{2}$$

$$(D_3)$$

$$(D_4)$$

A second approach to the preparation of silanones is the high-temperature pyrolysis of cyclotetrasiloxanes which has been shown to yield cyclotrisiloxanes and cyclopentasiloxanes in equal amounts.[33,37-39] The first order dependence of the rate of this reaction on the concentration of cyclotetrasiloxane has been interpreted in terms of a rate-determining loss of silanone from the cyclotetrasiloxane with simultaneous formation of cyclotrisiloxane. The silanone thus formed rapidly inserts into a Si—O single bond of the cyclotetrasiloxane to yield the cyclopentasiloxane.

$$\xrightarrow{\Delta} \quad (D_3) \quad + [(CH_3)_2Si{=}O] \tag{3}$$

$$[(CH_3)_2Si{=}O] + D_4 \rightarrow (CH_3)_2Si \qquad (D_5) \tag{4}$$

A third approach to the preparation of silanones is the oxidation of silylenes [R_2Si:]. These silicon analogues of carbenes may be generated by pyrolysis of appropriate disilanes such as sym-dimethoxytetramethyldisilane[40,41] or by photolysis of polysilanes such as dodecamethylcyclohexasilane.[42] Both of these precursors yield dimethylsilylene which can be oxidized by sulfoxides to yield dimethylsilanone and sulfides.[43,44] Epoxides are also deoxygenated by silylenes to yield alkenes and silanones.[45] Tertiary amine oxides as well as N_2O also are effective oxidizing reagents for silylenes. In particular N_2O has been utilized to oxidize dimethylsilylene in an argon matrix at low temperature.[46]

$$[(CH_3)_2Si{:}] + DMSO \rightarrow [(CH_3)_2Si{=}O] + CH_3{-}S{-}CH_3 \tag{5}$$

$$+ [(CH_3)_2Si{:}] \longrightarrow + [(CH_3)_2Si{=}O] \tag{6}$$

A fourth route to silanones involves the intramolecular platinum catalyzed hydrosilation of α,α-dimethylallyloxydimethylsilane. This reaction proceeds by two competing pathways. The first yields 1-oxa-2-silacyclopentane, a stable compound, while the second gives an unstable silaoxetane which undergoes retro [2 + 2] cycloaddition to form an alkene and a reactive silanone. The most noteworthy feature of this reaction is that it proceeds in solution.[47]

$$(7)$$

Diels–Alder reaction of 1-oxa-2-silahexa-3,5-diene with hexafluoro-2-butyne provides a fifth route to silanones. This reaction probably initially yields 7-oxa-8-silabicyclo[2,2,2]-octadiene which is unstable and undergoes a retro-Diels–Alder reaction to give *ortho*-bis(trifluoromethyl)benzene and silanone:[48]

$$(8)$$

The reaction of 1,1-dimethyl-2,3-*bis*-(trimethylsilyl)-1-silirene with DMSO also yields dimethylsilanone.[64]

$$(9)$$

We have developed in collaboration with Drs. G. Bertrand and G. Manuel of the Laboratoire Organometallique, Universite Paul Sabatier, Toulouse,

France a versatile route to silanone intermediates. Thus, 6-oxa-3-silabicyclo-[3,1,0]hexanes undergo gas-phase pyrolysis to yield silanones and 1,3-dienes as the major products.[49] It appears that this reaction proceeds by initial homolytic cleavage of a carbon–oxygen bond of the epoxide to yield a diradical. The formation of 3-silacyclopent-4-en-1-ols in small amounts may result from intramolecular disproportionation of this initial diradical. The energy of activation for the reaction, which has been determined by gas-phase kinetics by Prof. I. Davidson, is found to be 227 kJ/mole.[50] This is consistent with the energy required for homolytic scission of a carbon–oxygen single bond assuming that the strain energy of the epoxide[51] is released in the transition state.

The major product forming reaction pathway involves beta scission of the carbon radical to yield a silyl radical which combines with the alkoxy radical to yield a 2-vinyl silaoxetane which undergoes a retro [2 + 2] cycloaddition to yield the silanone and 1,3-diene. These silanone intermediates have been trapped by insertion into Si—O single bonds of cyclic siloxanes such as 2,2,5,5-tetramethyl-1-oxa-2,5-disilacyclopentane as outlined in reactions 10–12.

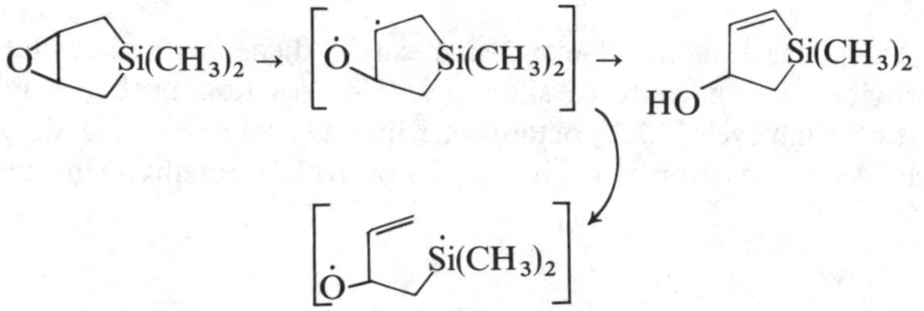

(10)

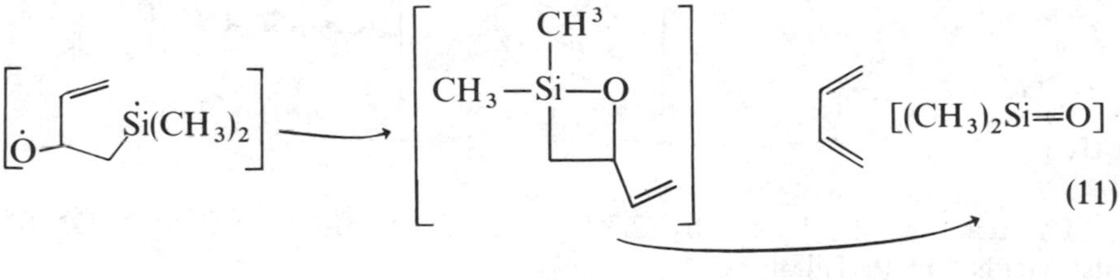

(11)

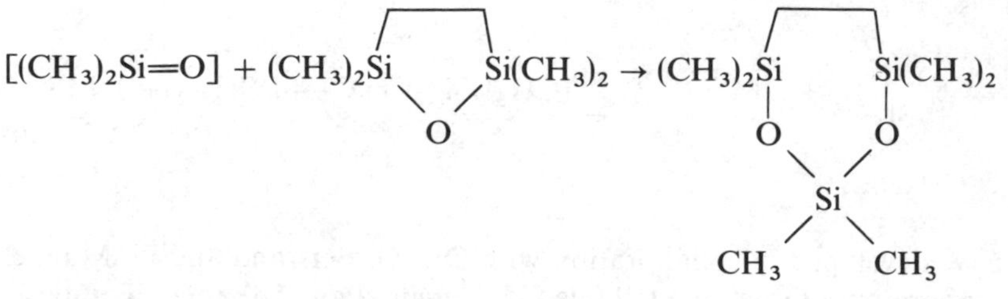

(12)

The silanone precursors, 6-oxa-3-silabicyclo[3,1,0]hexanes, are readily prepared in two steps. Dissolving metal reduction of 1,3-dienes in the presence of dichlorosilanes gives 1-silacylopent-3-enes.[52,53] Oxidation of these with *m*-chloroperbenzoic acid yields the desired 6-oxa-3-silabicyclo[3,1,0] hexanes in high overall yield. A wide range of substituents on the silyl center, such as vinyl, phenyl, as well as alkyl, can be tolerated.

SILICON DIOXIDE

While silicon dioxide [O=Si=O] is not as well characterized as silicon monoxide [Si=O], it is nevertheless an important species. For example, silicon dioxide has been detected in significant abundance in the stratosphere at an elevation between 50 and 80 km by a rocket borne negative ion mass spectrometer.[55] The infrared spectrum of matrix isolated [O=Si=O] has also been reported. A strong band at 1420 cm^{-1} has been assigned to the v_3 vibration of the [O=Si=O] molecule.[56] Silicon dioxide for this experiment was produced by reaction of [Si=O] evaporated from a Knudsen cell with atomic oxygen generated by microwave excitation.

Protective insulating films of silicon dioxide have been produced in a variety of ways by chemical vapor deposition reactions between silane (SiH$_4$) and a plasma of excited oxygen or N$_2$O gas. Electron impact has also been utilized to excite and dissociate gas-phase SiH$_4$ and N$_2$O molecules to form silicon dioxide films[57,58] These reactions may involve [O=Si=O] as a transient species.

Finally, fumed silica of high surface area, 200–300 m^2/g is prepared by the reaction of silicon tetrachloride with hydrogen and air in a flame at 2100°K. This gas-phase reaction may also involve [O=Si=O] as a primary intermediate.[59]

We have found that flash vacuum pyrolysis of 2,3,7,8-diepoxy-5-silaspiro-[4,4]nonanes yields [O=Si=O] and 1,3-dienes as major products.[49] The precursor to [O=Si=O] is easily prepared. Thus, dissolving metal reduction of isoprene with magnesium in THF in the presence of dichlorodiethoxysilane gives 5-silaspiro[4,4]nona-2,7-diene. Dichlorodiethoxysilane must be used since silicon tetrachloride cleaves THF at room temperature. While it has been reported that dichlorodiethoxysilane can be prepared by equilibration of tetraethoxysilane with silicon tetrachloride at 160°C in sealed glass tubes over 30 hr, we have found that addition of sodium ethoxide in THF to silicon tetrachloride in THF which has been cooled to −78°C gives excellent results in 2 hr and further has the advantage that it is not limited in scale. Finally, oxidation of 5-sila-spiro-[4,4]nona-2,7-diene to the diepoxide is accomplished with *m*-chloroperbenzoic acid.[54]

$$(EtO)_2SiCl_2 + \quad \underset{THF}{\overset{Mg}{\longrightarrow}} \qquad\qquad\qquad (13)$$

(14)

We have found that copyrolysis of 2,3,7,8-diepoxy-5-sila-spiro[4,4]nonanes with cyclic siloxanes such as hexamethylcyclotrisiloxane (D_3) leads to cyclic spirosiloxanes. Thus, the reaction of two molecules of D_3 with one reactive [O=Si=O] transient yields a 2,4,6,8,10,12,14-heptasila-1,3,5,7,9,11,13,15-octaoxaspiro[7,7]pentadecane system (I). This initial product will react with a second [O=Si=O] transient to yield a *bis*-spiro-siloxane (II). While the formation of I simply results from the sequential insertion of the silicon–oxygen double bonds of [O=Si=O] into silicon–oxygen single bonds of D_3, the formation of II is certainly more complicated and must involve a major reorganization of the silicon–oxygen bond framework. The structure of these products has been determined by a combination of mass spectrometry, infrared, ^{1}H, ^{13}C, and ^{29}Si NMR spectroscopy. In particular, ^{29}Si NMR is useful for the structure elucidation of these compounds. Thus the chemical shift of silicon in cyclic siloxanes depends significantly on ring size. The chemical shift of silicon in D_3 comes at -9.2 ppm whereas that in D_4 occurs at -20.0 ppm relative to the [29]Si signal of tetramethylsilane. While there is some variation in the [29]Si chemical shifts in specific compounds, the difference (~ 10 ppm) in chemical shift between six and eight membered cyclic siloxanes remains. There is significantly less difference in chemical shifts of the silicon resonances in larger siloxane rings. Thus the ^{29}Si NMR resonance of D_5 occurs at -22.8 ppm while that of D_6 is found at -23.0 ppm.[60] Finally, silicon atoms bonded to four oxygens in these spiro systems have unique chemical shifts which come in the region in which the [29]Si resonances of silicates are found, about -100 ppm upfield from the signal of TMS. This is a region in which the broad ^{29}Si resonances of glass (NMR tubes) are found. The structure of II was confirmed by X-ray crystallography in collaboration with Prof. R. Bau of USC. All ^{29}SiNMR chemical shifts are reported relative to TMS.

^{1}H NMR	0.1026 (S, 24H)
	0.0757 (S, 12H)
^{13}C NMR	0.761 (4C)
	0.599 (8C)
^{29}Si NMR	-17.570 (4Si)
	-18.768 (2 Si)
	-105.169 (1Si)

(I)

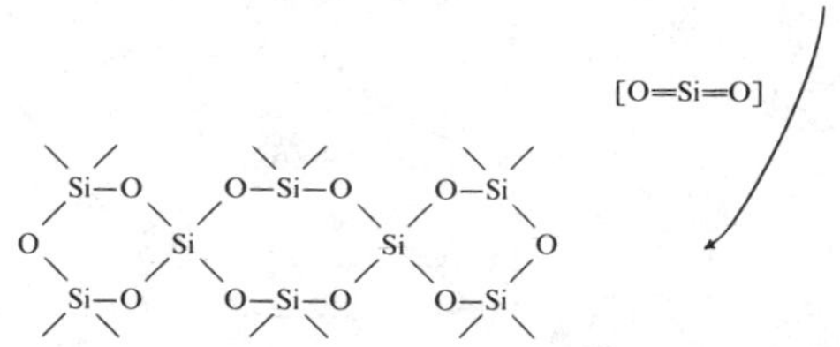

$$[O=Si=O]$$

(15)

(II)

^{1}H NMR 0.159 (S,24H) ^{13}C NMR 0.563 (8C) ^{29}Si NMR − 6.092 (4Si)
0.153 (S,12H) 0.075 (4C) − 15.629 (2Si)
− 100.761 (2Si)

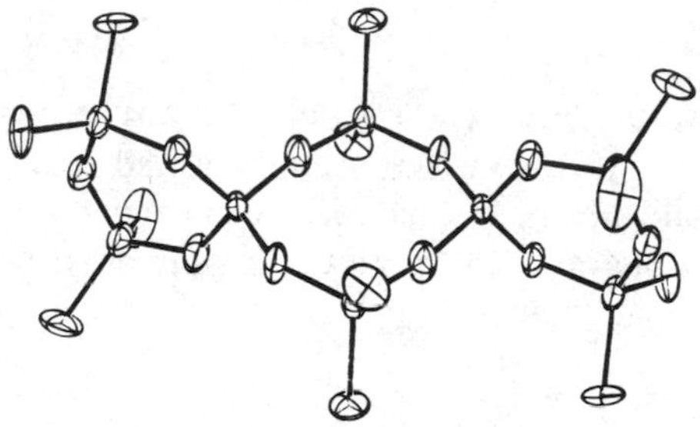

(16)

(II)

Likewise, flash vacuum copyrolysis of 2,2,5,5-tetramethyl-1-oxa-2,5-disila-cyclopentane with the precursor of $[O=Si=O]$ leads to a 2:1 adduct: 2,5,7,9,12-pentasila-1,6,8,13-tetraoxaspiro-[6,6]tridecane.

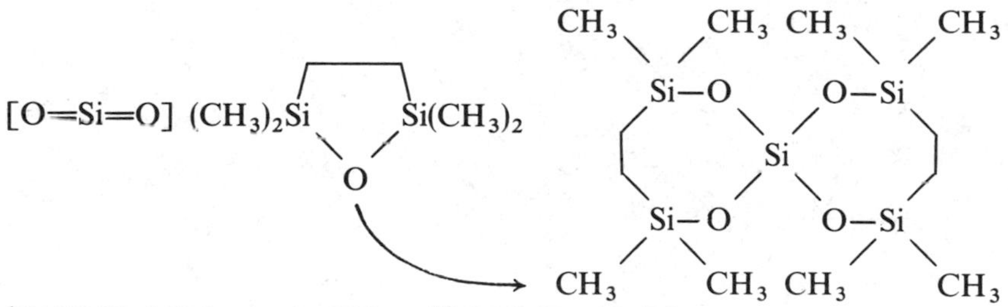

(17)

^{1}H NMR 0.733 ppm (s, 8H) ^{13}C NMR 10.422 ppm
0.085 ppm (s, 24H) −0.560 ppm

^{29}Si NMR 12.28 (4Si)
− 95.91 (1Si)

Other reaction patterns between $[O=Si=O]$ and cyclic siloxanes have also been observed. Thus $[O=Si=O]$ reacts with a single molecule of D_4 to yield 2,4,6,8,10-pentasila-1,3,5,7,9,11-hexaoxaspiro[5,5]undecane.[61,62] Obviously, transannular insertion reactions are not possible with smaller cyclic siloxanes. 2,4,6,8,10-Pentasila-1,3,5,7,9,11-hexaoxaspiro[5,5]-undecane has previously been prepared by cohydrolysis of silicon tetrachloride and dimethyldichloro-silane in very low yield (1–8%).[61−62]

$$
\begin{array}{c}
\text{CH}_3 \qquad \text{CH}_3 \\
\text{O–Si} \\
\text{(CH}_3)_2\text{Si} \qquad \text{O} \\
\text{O} \qquad \text{Si(CH}_3)_2 \\
\text{Si–O} \\
\text{CH}_3 \qquad \text{CH}_3
\end{array}
\xrightarrow{\ [\text{O}=\text{Si}=\text{O}]\ }
\begin{array}{c}
\text{CH}_3 \quad \text{CH}_3 \ \text{CH}_3 \quad \text{CH}_3 \\
\text{Si–O} \qquad \text{O–Si} \\
\text{O} \qquad \text{Si} \qquad \text{O} \\
\text{Si–O} \qquad \text{O–Si} \\
\text{CH}_3 \qquad \text{CH}_3 \ \text{CH}_3 \qquad \text{CH}_3
\end{array}
$$

(18)

^{1}H NMR 0.183 ^{13}C NMR 0.617 ^{29}Si NMR -6.314 (4Si)
-95.916 (1Si)

Cyclic siloxanes substituted with vinyl or chloromethyl groups also react with [O=Si=O] to yield spirosiloxanes which are substituted with vinyl or $-\text{CH}_2\text{Cl}$ functional groups. Specifically, reactions have been carried out between penta-methylvinylcyclotrisiloxane, and chloromethylheptamethyl-cyclotetrasiloxane and [O=Si=O] as outlined below.

$$
\begin{array}{c}
\text{CH}_3 \qquad \text{CH}=\text{CH}_2 \\
\text{Si} \\
\text{O} \qquad \text{O} \\
\text{(CH}_3)_2\text{Si} \qquad \text{Si(CH}_3)_2 \\
\text{O}
\end{array}
+ [\text{O}=\text{Si}=\text{O}] \longrightarrow
$$

$$
\begin{array}{c}
\text{CH}_3 \quad \text{CH}_3 \ \text{CH}_3 \quad \text{CH}_3 \\
\text{CH}_3 \qquad \text{O–Si–O} \qquad \text{O–Si–O} \\
\text{Si} \qquad \text{Si} \qquad \text{Si(CH}_3)_2 \\
\text{H}_2\text{C}=\text{CH} \qquad \text{O–Si–O} \qquad \text{O–Si–O} \\
\text{CH}_3 \qquad \text{CH}_3 \ \text{H}_2\text{C}=\text{CH} \ \text{CH}_3
\end{array}
$$

(19)

+ other isomers

$$
\begin{array}{c}
\text{CH}_3 \qquad \text{CH}_3 \\
\text{Si–O} \qquad \text{CH}_3 \\
\text{Cl–H}_2\text{C} \quad \text{O} \qquad \text{Si} \\
\text{Si} \qquad \text{O} \quad \text{CH}_3 \\
\text{CH}_3 \qquad \text{O–Si} \\
\text{CH}_3 \qquad \text{CH}_3
\end{array}
+ [\text{O}=\text{Si}=\text{O}] \longrightarrow
$$

$$
\begin{array}{c}
\text{CH}_3 \quad \text{CH}_3 \; \text{CH}_3 \quad \text{CH}_2\text{Cl} \\
\text{Si} - \text{O} \qquad \text{O} - \text{Si} \\
\text{O} \qquad \text{Si} \qquad \text{O} \\
\text{Si} - \text{O} \qquad \text{O} - \text{Si} \\
\text{CH}_3 \quad \text{CH}_3 \; \text{CH}_3 \quad \text{CH}_3
\end{array}
\qquad (20)
$$

Pyrolysis of 2,3,7,8-diepoxy-5-silaspiro[4,4]nonane has also been carried out in the absence of trapping reagents at 500°C and a pressure of 10^{-4} mm. Under these conditions a finely divided white matrix builds up inside the glass pyrolysis tube beginning in the hot zone and extending toward the cooled exit port (Fig. 40.1). This material adheres to the glass tube rather strongly, but can

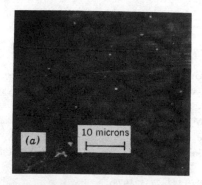

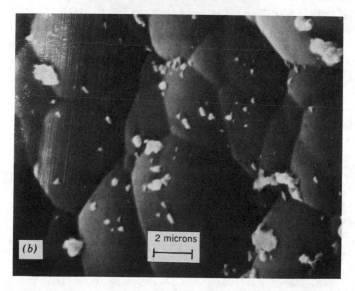

Figure 40.1 Scanning electron micrographs of chemically vapor deposited SiO_2. Resolution 100 Å.

be removed by scraping with a spattula. Silicon dioxide [O=Si=O] would be expected to insert into the Si—O single bonds found on the surface of the glass tube. Surface area measurements on this material were carried out by comparison of helium and nitrogen gas adsorption.[63] We have found surface areas for this material from 210 m²/g to over 300 m²/g from different preparations. These values are in the range reported for fumed silica.[59] We believe that this new method for the chemical vapor deposition of SiO_2 may have utility for support of heterogeneous catalysts as well as in other applications. We are currently studying the properties and structure of this interesting new material.

ACKNOWLEDGMENT

This work was supported by the Air Force Office of Scientific Research— 82-0333. W.P.W. thanks NATO for a travel grant 845/83.

REFERENCES

1. For a review see L. E. Gusel'nikov and N. S. Nametkin, *Chem. Rev.* **79**, 529 (1979).
2. N. S. Nametkin, V. M. Vdovin, L. E. Gusel'nikov, and V. I. Zav'yalov, *Izv. Akad. Nauk SSSR Ser. Khim.*, 584 (1966).
3. L. E. Gusel'nikov and M. C. Flowers, *J. Chem. Soc. Chem. Commun.*, 64 (1967).
4. M. C. Flowers and L. E. Gusel'nikov, *J. Chem. Soc. B*, 419 (1968).
5. M. Elseikh and L. H. Sommer, *J. Organometal. Chem.* **186**, 301 (1980).
6. D. R. Parker and L. H. Sommer, *J. Am. Chem. Soc.* **98**, 618 (1976).
7. U. Klingebiel, *Chem. Ber.* **111**, 2735 (1978).
8. S. A. Kazoura and W. P. Weber, *J. Organometal. Chem.* **268**, 19 (1984).
9. S. A. Kazoura and W. P. Weber, *J. Organometal. Chem.* **271**, 47 (1984).
10. W. Clegg, U. Klingebiel, G. M. Sheldrick, and D. Stalke, *J. Organometal. Chem.* **265**, 17 (1984).
11. D. N. Roark and L. H. Sommer, *J. Chem. Soc. Chem. Commun.*, 167 (1973).
12. C. M. Golino, R. D. Bush, and L. H. Sommer, *J. Am. Chem. Soc.* **97**, 7371 (1975).
13. I. M. T. Davidson and J. F. Thompson, *J. Chem. Soc. Chem. Commun,* 251 (1971).
14. T. J. Barton and J. A. Kilgour, *J. Am. Chem. Soc.* **96**, 2278 (1974).
15. L. H. Sommer and J. McLick, *J. Organometal. Chem.* **101**, 171 (1975).
16. H. S. D. Soysa and W. P. Weber, *J. Organometal. Chem.* **165**, C1 (1979).
17. C. W. Carlson and R. West, *Organometallics* **2**, 1798 (1983):
18. C. Couret, J. Escudie, J. Satge, J. D. Andriamizaka, and B. Saint-Roch, *J. Organometal. Chem.* **182**, 9 (1979).
19. A. H. Cowley and T. H. Newman, *Organometallics* **1**, 1412, (1982).
20. C. N. Smit, F. M. Lock, and F. Bickelhaupt, *Tetrahedron Lett.*, 3011 (1984).
21. D. N. Roark and G. J. D. Peddle, *J. Am. Chem. Soc.* **94**, 5837 (1972).
22. C. L. Smith and J. Pounds, *J. Chem. Soc. Chem. Commun.*, 910 (1975).
23. A. H. Cowley, *Polyhedron* **3**, 389 (1984).
24. G. Bertrand, G. Manuel, and P. Mazerolles, *Tetrahedron* **34**, 1951 (1978).
25. L. E. Gusel'nikov, M. S. Nametkin, and V. M. Vdovin, *Accts. Chem. Res.* **8**, 18 (1975).

26. T. J. Barton and E. Kline, Third International Symposium Organosilicon Chemistry, Madison, Wisconsin 1972.

27. M. Golino, R. D. Bush, P. On, and L. H. Sommer, *J. Am. Chem. Soc.* **97**, 1957 (1975).

28. R. D. Bush, C. M. Golino, G. D. Homer, and L. H. Sommer, *J. Organometal. Chem.* **80**, 37 (1974).

29. W. Ando and A. Sekiguchi, *J. Organometal. Chem.* **133**, 219 (1977).

30. P. B. Valkovich and W. P. Weber, *J. Organometal. Chem.* **99**, 231 (1975).

31. T. J. Barton, G. T. Burns, E. V. Arnold, and J. Clardy, *Tetrahedron Lett.*, 7 (1981).

32. Y. Nakadaira, H. Sakaba, and H. Sakurai, *Chem. Lett.*, 1071 (1980).

33. M. J. Fink, K. J. Haller, R. West, and J. Michl, *J. Am. Chem. Soc.* **106**, 822 (1984).

34. M. J. Michalczyk, R. West, and J. Michl, *J. Chem. Soc. Chem. Commun.*, 1525 (1984).

35. W. Ando, A. Sekiguchi, and T. Sato, *J. Am. Chem. Soc.* **104**, 6830 (1982).

36. T. J. Barton and G. P. Hussmann, *Organometallics* **2**, 692 (1983).

37. N. S. Nametkin, L. E. Gusel'nikov, T. H. Islamov, M. V. Shishkina, and V. M. Vdovin, *Dokl. Akad. Nauk. SSSR* **175**, 136 (1967).

38. N. S. Nametkin, T. H. Islamov, L. E. Gusel'nikov, A. A. Sobtsov, and V. M. Vdovin, *Izv. Akad. Nauk. SSSR Ser. Khim.*, 90 (1971).

39. L. E. Gusel'nikov, N. S. Nametkin, T. H. Islamov, A. A. Sobtsov, and V. M. Vdovin, *Izv. Akad. Nauk. SSSR. Ser. Khim.*, 84 (1971).

40. D. R. Weyenberg and W. H. Atwell, *Pure Appl. Chem.* **19**, 343 (1969).

41. W. H. Atwell and D. R. Weyenberg, *J. Am. Chem. Soc.* **90**, 3438 (1968).

42. M. Ishikawa and M. Kumada, *J. Organometal. Chem.* **42**, 325 (1972).

43. H. S. D. Soysa, H. Okinoshima, and W. P. Weber, *J. Organometal. Chem.* **133**, C17 (1977).

44. I. S. Alnaimi and W. P. Weber, *J. Organometal. Chem.* **241**, 171 (1983).

45. W. F. Goure and T. J. Barton, *J. Organometal. Chem.* **199**, 33 (1980).

46. C. A. Arrington, R. West, and J. Michl, *J. Am. Chem. Soc.* **105**, 6176 (1983).

47. T. H. Lane and C. L. Frye, *J. Organometal. Chem.* **172**, 213 (1979).

48. G. Hussmann, W. D. Wulff, and T. J. Barton, *J. Am. Chem. Soc.* **105**, 1263 (1983).

49. G. Manuel, G. Bertrand, W. P. Weber, and S. A. Kazoura, *Organometallics* **3**, 1340 (1984).

50. I. M. T. Davidson, A. Fenton, G. Manuel, and G. Bertrand, *Organometallics*, **4**, 1324 (1985).

51. H. K. Eiogenmann, D. M. Golden, and S. W. Benson, *J. Phys. Chem.* **77**, 1687 (1973).

52. G. Manuel, P. Mazerolles, and G. Cauquy, *Synth. React. Inorg. Met. Org. Chem.* **4**, 133 (1974).

53. D. R. Weyenberg, L. H. Toporcer, and L. E. Nelson, *J. Org. Chem.* **33**, 1975 (1968).

54. D. Terunuma, S. Hatta, T. Araki, T. Ueki, T. Okazaki, and Y. Suzuki, *Bull. Chem. Soc. Japan* **50**, 1545 (1977).

55. A. A. Viggiano, F. Arnold, D. W. Fahey, F. C. Fehsenfeld, and E. E. Fergurson, *Planet Space Sci.* **30**, 499 (1982).

56. H. Schnockel, *Angew. Chem. Int. Ed.* **17**, 616 (1978).

57. D. W. Hess, *ASTM Spec. Tech. Pub.* **804**, 218 (1983).

58. L. R. Thompson, J. J. Rocca, K. Emery, P. K. Boyer, and G. J. Collins, *Appl. Phys. Lett.* **43**, 777 (1983).

59. Cab-O-Sil Properties and Functions, Cabot Corp., 1983.

60. A. L. Smith, *Analysis of Silicones*, Wiley, New York, 1974, p. 319.

61. D. W. Scott, *J. Am. Chem. Soc.* **68**, 356 (1946).

62. F. R. Mayo, *J. Polym. Sci.* **55**, 59 (1961).

63. S. Brunauer, P. H. Emmett, and E. Teller, *J. Am. Chem. Soc.* **60**, 309 (1938).

64. D. Seyferth, T. F. O. Lim, and D. P. Duncan, *J. Am. Chem. Soc.*, **100**, 1626 (1978).

41

ELECTRONIC INTERACTIONS IN ORGANOMULTIMETALLIC SYSTEMS

T. M. PETTIJOHN AND J. J. LAGOWSKI
Department of Chemistry
The University of Texas
Austin, Texas

INTRODUCTION

Interest in polymers incorporating metallic moieties has grown over the years because of the premise that such materials could exhibit any one, or all, of a variety of interesting and useful properties. For example, "molecular metals" have attracted considerable attention in the recent past[1] because they have the potential of combining unique electrical and optical characteristics with structural elements, typified by organic compounds, which are attractive from the point of view of molecular engineering. In addition, the use of polymer-bound redox species as modified electrode surfaces has been a topic of intense interest in the last few years.[2-6] Current practice for the preparation of such materials involves either the treatment of preformed polymers with a suitable dopant or the synthesis of polymers that incorporate electronically appropriate units. Little is known of the properties of organometallic systems with metal atoms in the polymer backbone.

We have designed synthesis schemes which produce organometallic polymers with metal atoms in the polymer backbone. Oligomers have been made incorporating the structural units of interest in order to investigate the nature of the electronic interactions through organic moieties between multiple metal sites.

The target molecules are of the type shown by structure (I).

$$\square = 2e^-, CH_2, SiR_2, \ce{>C=C<}$$

EXPERIMENTAL

Bis(polyphenyl)chromium compounds of type I were prepared by the metal atoms synthesis technique using either a stationary reactor[7] or a rotating reactor.[8] The compounds prepared by this method include those shown as II. These compounds were purified by sublimation and characterized by their mass and ^{1}H NMR spectra.

The compounds of type II were "capped" with metal carbonyl moieties by reacting them with metal hexacarbonyls, for example,

$$RC_6H_5 + M(CO)_6 ----→ RC_6H_5M(CO)_3 + 3CO \qquad (1)$$

Using this reaction, we were able to prepare $\mu[\eta^6:\eta^6$-bis(biphenyl)-chromium]-bis(chromium tricarbonyl) (III). This compound was purified by recrystallization and characterized by mass, IR, and NMR spectra. Although a number of other possible capping reactions are available, we have concentrated on the metal carbonyl route exclusively.

Electrochemical oxidation and reduction processes were performed with conventional PAR equipment in tetrahydrofuran solution. The results of a

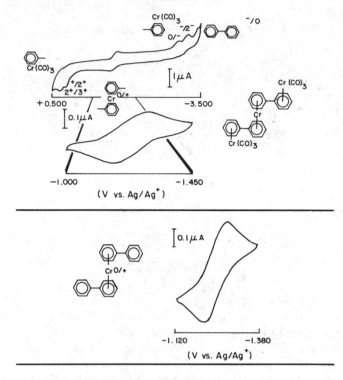

Figure 41.1 Cyclic voltammograms of μ-[η^6:η^6-bis(η^6-biphenyl)chromium]-(bischromiumtricarbonyl) and bis(η^6-biphenyl)chromium in THF with 0.2M tetrabutylammoniumhexafluorophosphate.

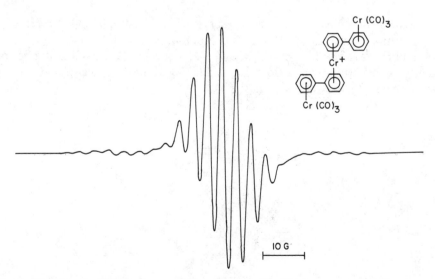

Figure 41.2 ESR spectrum of μ-[η^6:η^6-bis(η^6-biphenyl)chromium]-(bischromiumtricarbonyl) in trifluoroacetic acid.

typical electrochemical experiment are shown in Fig. 41.1. ESR spectroscopy was performed on oxidized samples of the multimetal compound III; a typical spectrum is shown in Fig. 41.2.

DISCUSSION

Type II compounds were first prepared in hopes that organometallic oligo-mers containing two to three metal centers could be produced by simple, direct metal atom reactions. Unfortunately, in each reaction except those in which biphenyl was the substrate, mononuclear species were the sole products.

The two phenyl rings that each type II complex possesses are still open to metal π-complexation. Therefore, we then turned to using them as precursors to multinuclear compounds. Compound III was synthesized from the II biphenyl analog by using the process described by Eq. (1); III was used as a model for electrochemical and ESR study of metal–metal electronic interactions.

We first performed cyclic voltammetric (CV) experiments on the type II complexes in order to gain an understanding of how the substituent on the π-complexed arene affects the metal species (Fig. 41.1). An electron-donating substituent increases the π-electron density at the metal causing its oxidation to proceed at a lower potential.[9] Therefore, from the redox half-wave potentials $(E_{1/2})$, we concluded that the donation of π-electron density into the vacant metal orbitals followed the substituent series:

$$\text{Substituent}-CH_2-Ph > -CH=CH-Ph > -Ph$$

$$E_{1/2}^* : \quad -1.229 \text{ V} \quad\quad -1.169 \text{ V} \quad\quad -1.099 \text{ V}$$

Each bis(arene)chromium complex (II) exhibited a one-electron reversible redox couple with cathodic–anodic peak separation of approximately 65 mV.

The CV of compound III did not resemble a one-electron process. Upon preliminary investigation, it appeared as though there was one broad wave with cathodic–anodic peak separation on the order of 126 mV. Figure 41.1 shows the CVs of both the uncapped (II) and the capped (III) bis(biphenyl)chromium complexes. Tafel plots of the two CVs clarified the voltammogram of III somewhat (Fig. 41.3).

The Tafel plot of III contains a definite change in slope indicating that two one-electron processes could be occurring.

The ESR spectrum of compound III was obtained in trifluoroacetic acid (TFA) at 5°C ($g = 1.988$) (Fig. 41.2). Only the central Cr atom is oxidized in TFA since the acid is a good oxidant for bis-arene chromium systems. The hyperfine splitting in the spectrum (Fig. 41.2) is attributed to (1) strong coupling of the nuclear spins with the hydrogen nuclei associated with the arene on the

*Versus ferrocene.

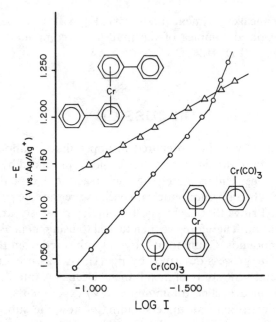

Figure 41.3 Tafel plots of μ-[η^6:η^6-bis(η^6-biphenyl)chromium]-(bischromiumtricarbonyl) and bis(η^6-biphenyl)chromium cyclic voltammograms.

central Cr atom [$a(^1H) = 3.56$] and (2) weak coupling with the ^{53}Cr isotope (9.5% natural abundance) which has a nuclear spin of 3/2. In earlier work,[10] the latter appeared as three to four very faint satellites at the extreme edges of the resonances derived from the ^{1}H coupling. In the spectrum of III, six lines can be seen on either side supporting possible through-space coupling between the center Cr$^+$ and a ^{53}Cr at an adjacent site. The validity of this conclusion will be tested by using a capping metal unit that has a nuclear spin other than 3/2.

CONCLUSION

1. We have shown that organometallic oligomeric species can be readily synthesized from simple mononuclear arene chromium complexes.
2. Electrochemical and ESR experiments on the oligomer containing three metal centers indicate that the further investigation of metal–metal interactions is merited. Future methods of analysis will include (a) metal (e.g.,^{53}Cr) NMR studies, (b) ESR studies of heterometallic systems, and (c) electrochemical convolution studies.
3. Information gained upon completion of these model studies should enable physicochemical predictions of the properties of higher organo-metallic polymers, and, perhaps, guide synthetic strategies for the

formation of organometallic polymers with optimal electronic characteristics.

REFERENCES

1. (a) J. T. Devreese, R. P. Evard, and V. E. van Doren, Eds., *Highly Conducting One-Dimensional Solids*, Plenum Press, New York, 1979. (b) W. E. Hatfield, Ed., *Molecular Metals*, Plenum Press, New York, 1979.

2. N. Oyama and F. C. Anson, *Anal. Chem.* **52**, 1192 (1980).

3. M. F. Daytartes and J. F. Evans, *J. Electroanal. Chem.* **109**, 301 (1980).

4. A. Bellelheim, R. J. Chan, and T. Kuwana, *J. Electroanal. Chem.* **110**, 93 (1980).

5. J. B. Kerr, L. L. Miller, and M. R. Van DeMark, *J. Amer. Chem. Soc.* **102**, 3383 (1980).

6. K. N. Kuo and R. W. Murray, *J. Electroanal. Chem.* **131**, 37 (1982).

7. V. Graves and J. J. Lagowski, *Inorg. Chem.* **15**, 577 (1976).

8. R. J. Markle, T. M. Pettijohn, and J. J. Lagowski, *Organometallics*, **4**, 1529 (1985).

9. T. Li and C. H. Brubaker, *J. Organomet. Chem.* **216**, 223 (1981).

10. R. D. Feltham, *J. Inorg. Nucl. Chem.* **16**, 197 (1961).

42

A THEORETICAL STUDY OF THE PYROLYSIS OF METHYLSILANE AND DISILANE

MARK S. GORDON, THANH N. TRUONG,
AND ELIZABETH K. BONDERSON
Department of Chemistry
North Dakota State University
Fargo, North Dakota

INTRODUCTION

In the present work, the potential energy surfaces for several alternative uni-molecular decomposition modes of methylsilane and disilane are analyzed. These calculations are of particular interest since there is some question regarding the relative activation energies of 1,1- versus 1,2-elimination of molecular hydrogen.[1,2]

COMPUTATIONAL METHODS

Several basis sets were used in this study. Because we are interested in extending the studies of the structures and conformational analyses to larger molecules, geometry predictions were performed using both 3-21G*[3] and 6-31G*.[4] This provides a measure of the ability of the smaller basis set to predict the properties of interest. For the analyses of the reaction surfaces, larger basis sets, including 6-31G**,[5] 6-311G**,[6] and MC-311G** were used. The latter combines the standard 6-311G** basis set for carbon and hydrogen with the extended basis set developed for silicon by McClean and Chandler.[7] The silicon basis is, of course, augmented by polarization functions.[4] Because calculations at the SCF molecular orbital level are not adequate for the prediction of reaction energetics or singlet–triplet splittings, correlation corrections have been added in the computation of these properties. The principal method used here for incorporating correlation effects is Moller–Plesset (MP) perturbation theory as formulated by Pople and coworkers.[8,9]

Because chemical reactions involving movement of several atoms can be rather complicated, determination of the transition state structure is frequently followed by an analysis of the minimum energy path leading from reactants through the transition state to products. The interpretation of this path, referred to as the intrinsic reaction coordinate (IRC),[10–14] is aided by generating localized molecular orbitals (LMOs)[15] at selected points on the path. This illustrates the metamorphosis of the reacting bonds and lone pairs as the reaction proceeds.

RESULTS AND DISCUSSION

Pyrolysis of Methylsilane

The results of the calculations on the various alternative pyrolysis pathways of methylsilane are summarized in Table 42.1. The energy differences have been obtained at the MP4/MC-311G** level of accuracy, using geometries optimized

TABLE 42.1 Energetics for Pyrolyses (kcal/mole)[a]

Reaction	ΔE	E_b
$CH_3SiH_3 \rightarrow CH_4 + SiH_2$	52.8	76.2
$CH_3SiH_3 \rightarrow CH_3SiH + H_2$	57.8	78.8
$CH_3SiH_3 \rightarrow CH_2{=}SiH_2 + H_2$	56.9	106.0
$CH_3SiH_3 \rightarrow CH_3 + SiH_3$	89.7	89.7
$CH_3SiH_3 \rightarrow CH_2 + SiH_4$	122.6	122.6
$CH_3SiH_3 \rightarrow SiH_3CH + H_2$	125.9	125.9
$Si_2H_6 \rightarrow SiH_4 + SiH_2$	54.0	54.0
$Si_2H_6 \rightarrow SiH_3SiH + H_2$	59.2	57.3
$Si_2H_6 \rightarrow SiH_2{=}SiH_2 + H_2$	52.2	88.7
$Si_2H_6 \rightarrow 2SiH_3$	74.3	74.3

[a]Thermodynamic ΔE's were calculated with MP4(SDTO) wavefunctions and barriers E_b with MP3, both with the MC-311G** basis set.

at the MP2/6-31G** level. According to these calculations, there are three thermodynamically competitive pathways: the extrusion of silylene to form methane and the 1,1- and 1,2-eliminations of molecular hydrogen. The endothermicities of these three reactions are within 5 kcal/mole of each other, with that for methane formation being the least endothermic. The homolytic cleavage of the Si—C bond to form methyl and silyl radicals is found to require an additional 30 kcal/mole, the predicted bond strength being in excellent agreement with experiment.[16] Elimination of methylene to form silane and elimination of H_2 to form silylmethylene are even more unlikely from a thermodynamic point of view, with endothermicities of 122.6 and 125.9 kcal/mole, respectively.

The competition among the three thermodynamically most facile reactions is unraveled to some degree by analysis of the transition states and associated energy barriers. The essential features of the transition state structures, obtained using MP2/6-31G** wavefunctions, are summarized in Fig. 42.1. For the elimination of silylene, it is interesting that the silicon–carbon bond breaks much more rapidly than Si—H. Similarly, for the 1,2-elimination of molecular hydrogen, the C—H bond breaks much sooner than Si—H. In fact, in the latter case both leaving hydrogens are closer to silicon than to carbon in the transition state. This is an example of the usefulness of having the ability to follow the

Methylsilane Transition States

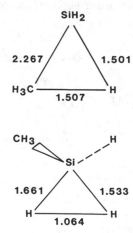

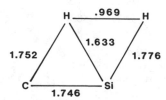

Figure 42.1   Transition state structures for methylsilane pyrolysis. (a) Extrusion of silylene; (b) 1,1-elimination of H_2; (c) 1,2-elimination of H_2. Bond lengths in Å, angles in degrees.

intrinsic reaction coordinate for a reaction, since the corresponding reactants and products are not always clear from the structure of the transition state. The normal mode corresponding to the reaction path for the 1,2-elimination is illustrated in Fig. 42.2. It appears from this figure that the atoms are moving in the appropriate directions (e.g., toward methylene). This is nicely verified in Fig. 42.3, where the structures are shown for several points along the IRC. Note that in the interest of conserving computer time the IRC was obtained at the 3-21G SCF level of computation.

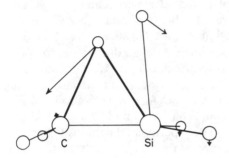

NORMAL MODES OF $H_2C SiH_2 + H_2$
        3-21G   TS

Figure 42.2 Reaction path normal mode for 1,2-H_2 elimination from methylsilane.

$$H_2 + CH_2 = SiH_2 \longrightarrow CH_3 - SiH_3$$

$H_2C \xrightarrow{1.69} SiH_2$
$2.00 / \quad \backslash$
$2.29 \backslash / \quad \backslash 2.20$
$H \xrightarrow{.73} H$

$H_2C \xrightarrow{1.72} SiH_2$
$1.74 / \quad \backslash$
$1.93 \backslash / \quad \backslash 1.88$
$H \xrightarrow{.81} H$

$H_2C \xrightarrow{1.74} SiH_2$
$1.68 / \quad \backslash$
$1.78 \backslash \quad \backslash 1.79$
$H \xrightarrow{.92} H$

$H_2C \xrightarrow{1.75} SiH_2$
$1.67 / \quad \backslash$
$1.75 \backslash / \quad \backslash 1.78$
$H \xrightarrow{.96} H$

$H_2C \xrightarrow{1.76} SiH_2$
$1.66 / \quad \backslash$
$1.62 \backslash / \quad \backslash 1.73$
$H \xrightarrow{1.11} H$

$H_2C \xrightarrow{1.79} SiH_2$
$1.72 / \quad \backslash$
$1.33 \backslash / \quad \backslash 1.64$
$H \xrightarrow{1.44} H$

$H_2C \xrightarrow{1.83} SiH_2$
$1.96 / \quad \backslash$
$1.09 / \quad \backslash 1.51$
$H ------ H$
$\quad 1.87$

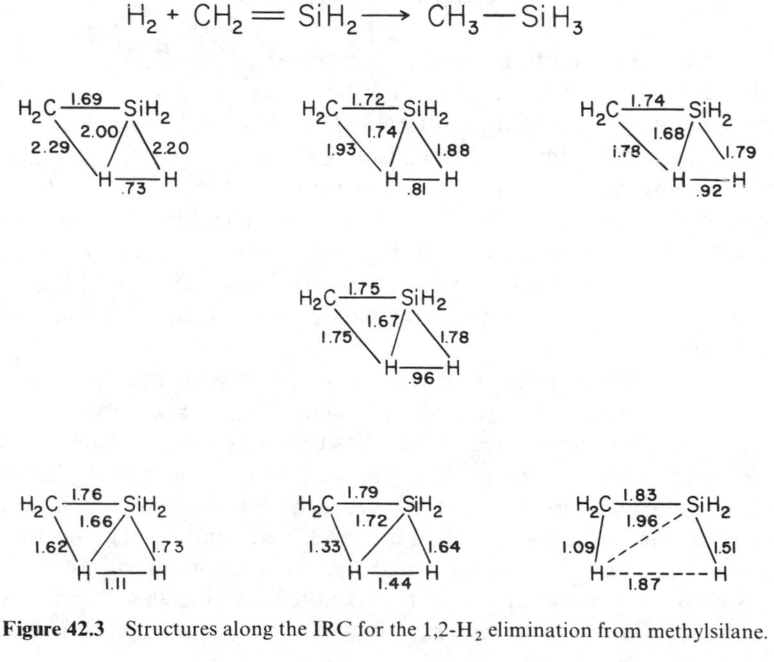

Figure 42.3 Structures along the IRC for the 1,2-H_2 elimination from methylsilane.

The calculated barriers for the methylsilane pyrolysis reactions are listed in Table 42.1. While the 1,2-elimination of H_2 is *thermodynamically* competitive with the 1,1-elimination and silylene extrusion, this is clearly not the case when the energy barriers are taken into account, since the barrier for the former reaction is predicted to be 30 kcal/mole greater than the latter two. The extrusion of silylene appears to be slightly favored, both thermodynamically and kinetically, over the 1,1-elimination of H_2; however, the energetics for these reactions are close enough that the relative values may reverse when the higher level calculations are completed.

The pyrolysis of methylsilane has been studied experimentally by Ring, O'Neal, and coworkers.[1] These authors were unable to obtain separate activation energies for the two different H_2 eliminations. However, they find the formation of methane to have a rather lower efficiency than that for the composite elimination of molecular hydrogen and speculate that the 1,2-elimination is less important than the 1,1-elimination. The latter is clearly in agreement with the current calculations. The experimental estimate for the activation energy for the extrusion of silylene to form methane is 66.7 kcal/mole, about 10 kcal/mole lower than the theoretical prediction. The experimental composite barrier for H_2 elimination is 64.7 kcal/mole. It is expected that the calculated activation barriers will decrease at higher levels of theory, but the qualitative prediction of a very high barrier for the 1,2-elimination is not likely to change. Indeed, we find that homolytic cleavage of the C—Si bond is kinetically favored over the 1,2-elimination.

Pyrolysis of Disilane

The energetics for alternative pyrolysis reactions of disilane are summarized in Table 2.1. As for the case of methylsilane, three reactions are found to be thermodynamically competitive at the MP4/MC-311G** level of accuracy: the elimination of silylene to form silane and the 1,1- and 1,2-eliminations of H_2 to form silylsilylene and disilene, respectively. The 1,2- and 1,1-hydrogen eliminations are found to be the least and most endothermic of the three, respectively. Homolytic cleavage of the Si—Si bond is found to require 74.3 kcal/mole, in good agreement with the experimental value of 74 kcal/mole,[16] and much higher than the thermodynamic requirements for the molecular eliminations.

As noted in an earlier paper, the elimination of silylene to form silane occurs with a monotonic energy increase. The same is expected for the homolytic cleavage. The essential features of the transition state structures for the remaining two reactions are summarized in Fig. 42.4. Most interesting is the fact that, as noted for the analogous methylsilane case, both leaving hydrogens are closer to one silicon in the transition state for the 1,2-elimination. The normal mode for the reaction coordinate is shown in Fig. 42.5, and the progress of the IRC is followed schematically in Fig. 42.6. It is apparent from this diagram that the initial attack of H_2 on disilene occurs in such a way as to form a loose complex

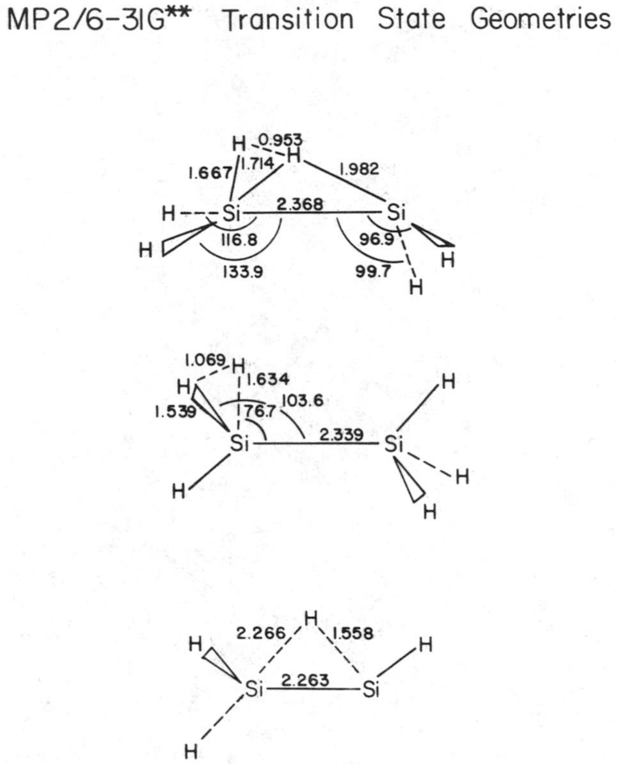

MP2/6-31G** Transition State Geometries

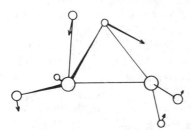

Figure 42.4 Transition state structures for disilane pyrolysis. (*a*) 1,2-H$_2$ elimination; (*b*) 1,1-H$_2$ elimination; (*c*) Disilene-Silyl silylene isomerization. Bondlengths in Å, angles in degrees.

Figure 42.5 Reaction path normal mode for the 1,2-H$_2$ elimination from disilane.

between silane and silylene. The lone pair on the latter moiety is oriented in such a way that one of the attacking hydrogens is able to move across the Si—Si bond and form a new Si—H bond.

The evolution of the reacting bonds and lone pairs for the 1,2-elimination reaction is illustrated in Figs. 42.7 and 42.8, using localized orbitals along the 3-21G* IRC. The transformation of the H—H bond into an Si—H bond on the SiH$_4$ end is apparent in Fig. 42.7. In Fig. 42.8 one can see one of two equivalent

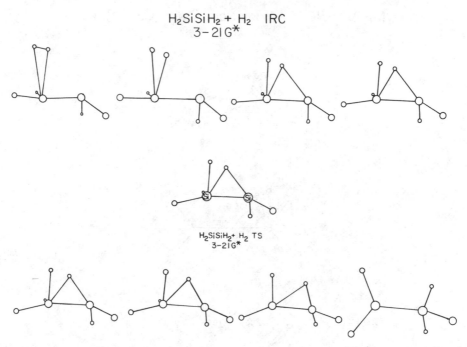

Figure 42.6   Structures along the IRC for the 1,2-H_2 elimination from disilane.

bent Si—Si bonds in disilene transform into a lone pair at intermediate geometries and finally into an Si—H bond. This is consistent with the interpretation given in the preceding paragraph.

The MP3/MC-311G** energy barriers are also listed in Table 42.1. Since the formation of silane and the Si—Si homolytic cleavage occur monotonically (the reverse barriers are zero), the barriers for these reactions are the same as the reaction endothermicities. As noted above for methylsilane, the barrier for the 1,2-elimination of H_2 is rather large (88.7 kcal/mole). So, despite the fact that this reaction is calculated to be most favored thermodynamically, it is clearly least favored kinetically. In contrast, the 1,1-elimination reaction is nearly competitive with the silane elimination. The latter is predicted to be the most likely process, in agreement with experimental predictions.[2] Note that the transition state for the 1,1-H_2 elimination is actually lower in energy than the separated products. Since all structures were obtained at the MP2/6-31G**, the saddle point is, in fact, real. This suggests that a long-range minimum similar to that found previously for the silane decomposition[17] exists on this surface.

Finally, the transition state for the isomerization from disilene to silyl silylene is also shown in Fig. 42.4. The MP4/MC-311G** energy difference is 7 kcal/mole, with the disilene lying lower, and the predicted barrier at the same level of theory is 17 kcal/mole.

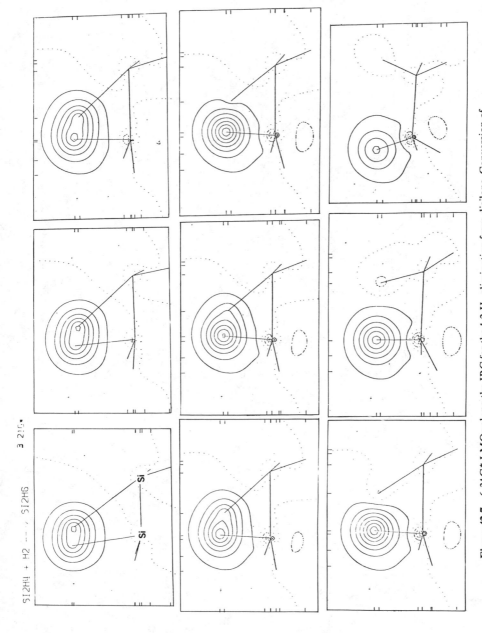

Figure 42.7 6-31G* LMOs along the IRC for the 1,2-H_2 elimination from disilane. Conversion of H—H to Si—H.

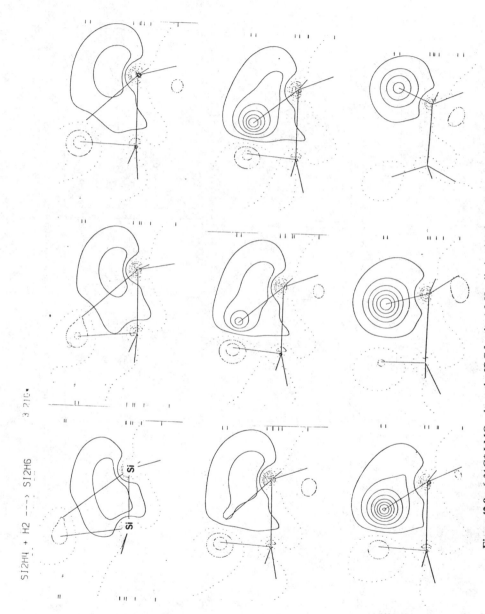

SI2H4 + H2 ---> SI2H6

Si—Si

Figure 42.8 6-31G* LMOs along the IRC for the 1,2-H_2 elimination from disilane. Conversion of Si—Si to Si—H.

CONCLUSION

The key result found in this work is that for both methylsilane and disilane the 1,2-molecular elimination of hydrogen is unfavorable relative to other pathways, because the barrier to the elimination is rather high in both instances. For methylsilane, the extrusion of silylene to form methane and the 1,1-elimination of H_2 to form methylsilylene have similar reaction energetics, with the former slightly favored at the current level of theory. For disilane the analogous processes are predicted to be the most favorable energetically, with the formation of silane being the lowest energy process, in agreement with experiment.

ACKNOWLEDGMENTS

This work was supported by grants from the Petroleum Research Fund (13025-AC6) and the National Science Foundation (CHE-8309948). The computer time made available by the North Dakota State University is greatly appreciated.

REFERENCES

1. B. A. Sawrey, H. E. O'Neal, M. A. Ring, and D. Coffey, Jr., *Int. J. Chem. Kinetics* **16**, 7, 23, 31 (1984).
2. G. Olbrich, P. Potzinger, B. Reimann, and R. Walsh, *Organomet.* **3**, 1267 (1984).
3. W. J. Pietro, M. M. Francl, W. J. Hehre, D. J. DeFrees, J. A. Pople, and J. S. Binkley, *J. Am. Chem. Soc.* **104**, 5039 (1982).
4. M. S. Gordon, *Chem. Phys. Lett.* **76**, 163 (1980).
5. P. C. Hariharan and J. A. Pople, *Theor. Chim. Acta* **28**, 213 (1973).
6. R. Krishnan, J. S. Binkley, R. Seeger, and J. A. Pople, *J. Chem. Phys.* **72**, 650 (1980).
7. A. D. McLean and G. S. Chandler, *J. Chem. Phys.* **72**, 5639 (1980).
8. J. S. Binkley and J. A. Pople, *Int. J. Quantum Chem.* **9**, 229 (1975); J. A. Pople, J. S. Binkley, and R. Krishnan, *Int. J. Quantum Chem. Symp.* **10**, 1 (1976).
9. R. Krishnan, M. J. Frisch, and J. A. Pople, *J. Chem. Phys.* **72**, 4244 (1980).
10. D. G. Truhlar and A. Kupperman, *J. Am. Chem. Soc.* **93**, 1840 (1971).
11. C. P. Baskin, C. F. Bender, C. W. Bauschlicher, Jr., and H. F. Schaefer III, *J. Am. Chem. Soc.* **96**, 2709 (1974).
12. K. Ishida, K. Morokuma, and A. Komornicki, *J. Chem. Phys.* **66p**, 2153 (1977).
13. K. Fukui, *Accts. Chem. Res.* **14**, 363 (1981).
14. M. W. Schmidt, M. S. Gordon, and M. Dupuis, *J. Am. Chem. Soc.*, in press.
15. J. M. Foster and S. F. Boys, *Proc. Roy. Soc. Ser A* **293**, 543 (1966).
16. R. Walsh, *Accts. Chem. Res.* **14**, 246 (1981).
17. M. S. Gordon, J. S. Binkley, M. J. Frisch, and D. R. Gano, *The Ground State Dissociation of Disilane*, in preparation.

43

APPLICATION OF MNDO TO
SILICON CHEMISTRY

LARRY P. DAVIS AND LARRY W. BURGGRAF
Department of Chemistry
United States Air Force Academy
Colorado Springs, Colorado

INTRODUCTION

Silicon chemistry is increasingly important to many products of materials science including catalysts, semiconductors, organosilicon polymers, ceramics, glasses, and composites. In all of these applications an efficient, accurate, theoretical model for predicting silicon chemistry could produce a great economy of effort. In particular, a detailed model of the mechanisms and products for silanol polymerization would be of great help in tailoring silica materials for specific purposes. To this end, we began a combined theoretical and experimental program to study silicon reactions and silica surfaces several years ago.

The goal is to perform theoretical calculations for rather large molecular systems without resorting to theoretical models which are expensive, time consuming, and, thus, self-defeating for our purposes. At the same time, we needed a theoretical model that could produce approximate geometries for transition states and products and yield good enthalpies of reaction and enthalpies of activation (to within about 10 kcal/mole) for these processes.

Recently, theoretical methods have become available to study reactions of silicon-containing compounds. Dewar's MNDO method[1] has been parameterized for silicon,[2] and his MINDO/3 program has included silicon parameters for a number of years.[3] Recently, Dewar's group has reparameterized silicon using a wider range of experimental data,[4] and the results have been vastly improved. Unfortunately, only a few MNDO studies of silicon reactions have appeared,[5,6] probably due to the poorer agreement of the calculated results using the older set of silicon parameters with experimental data for silicon-containing compounds as compared to MNDO results for second row elements.[7] Recently we completed a study of reactions of fluorine atom and fluoride ion with methane and silane, and compared the MNDO results with high-level *ab initio* calculations.[8] Even though that set of calculations was done with the old MNDO parameters, the results were quite encouraging for the use of MNDO to study larger silicon-containing systems. Other high-level *ab initio* calculations on simple silicon systems have also recently appeared.[9-12] The state-of-the-art now seems to be adequate to begin applying theoretical chemical calculations to large silicon-containing molecules of interest to experimental chemists.

The initial goal in these calculations was to evaluate silanol polymerization mechanisms, both in acidic and basic media. That goal remains, but first we must address the question of the applicability of MNDO to silicon-containing molecules. We have studied this problem both from the points of view of comparing our calculated results with experimental data, when available, or with high-level *ab initio* calculations, when experimental data are not available. We also have done some preliminary work on the anionic silanol polymerization mechanism. Each of these topics is addressed in this chapter.

CALCULATIONS

All calculations were performed with the MNDO method developed by Dewar and coworkers.[1] The MNDO method is a semiempirical molecular orbital method based on a neglect of diatomic differential overlap (NDDO) scheme. It is parameterized by comparisons with experimental data in the form of heats of formation, molecular geometries, ionization potentials, and dipole moments for a small, "basis set" of molecules. The method, in the form of a computer program called MOPAC,[13] is capable of optimizing geometries of stable molecules or transition states, or carrying out reactions along selected reaction coordinates. Options are also available to carry out force constant and thermodynamic calculations on selected geometries. MNDO is now parameterized for all second row elements except lithium and neon, and some third row elements. The silicon parameters were obtained from a recent reparameterization.[4]

For each reaction, we picked some geometric parameter (generally a bond distance or bond angle that changed appreciably during the reaction) and ran a reaction path calculation by holding that parameter (called the reaction coordinate) fixed at a number of selected values while allowing the rest of the molecule to optimize completely. The geometry along this reaction path that appeared closest to the highest enthalpy point was then used to begin an optimization of the transition state geometry. All stationary points on the potential surface (reactants, products, and transition states) were optimized using procedures supplied with the MOPAC package of programs. Force constant calculations were carried out on all suspected transition states. In each case, diagonalization of the force constant matrix yielded one, and only one, negative eigenvalue, thereby verifying the transition states.

RESULTS AND DISCUSSION

To achieve our objectives of first testing the applicability of MNDO and then applying it to possible silanol polymerization reactions, we organize this chapter into three topics. First, we discuss reactions of fluorine atoms with methane and silane, and compare the MNDO results to experimental and *ab initio* results for these reactions. This comparison is interesting not only because of the chemistry but also because this severe case reflects the magnitude of error in MNDO calculations for tetra- and pentavalent silicon species. Next, of possible major importance to silanol polymerization mechanisms, we describe results on decomposition modes of the five-coordinate silicon anion $H_3SiCH_3OH^-$. We compare these results to experimental data on the decomposition of similar silicon anion complexes in the gas phase. Finally, we discuss heats of reaction for pentacoordinate silicon species which may be important in silanol polymerization.

Fluorine Atom Reactions with Methane and Silane

The first system used to test the applicability of MNDO in the general area of silicon chemistry is fluorine atom attack on methane and silane.[8] There is some experimental data on this system, as well as high-level *ab initio* calculations. In addition, it gives us a chance to compare a common carbon reaction with the analogous silicon reaction. We consider both fluorine substitution, resulting in $H + SiH_3F$ (or CH_3F), and fluorine abstraction, resulting in $HF + SiH_3$ (or CH_3).

As a first step, we consider how well MNDO predicts heats of formation for reactants and products. These results are given in Table 43.1. The agreement between the calculated heats of formation and the experimental ones are reasonable (better than 10 kcal/mole) for all species. There is some uncertainty in the experimental data for some of the silicon compounds, most notably SiH_3. The calculated results support the lower of the two experimental values cited, the result of 46.4 kcal/mole determined by Walsh. Still, the good agreement for all of the species displayed in the table supports the conclusion that MNDO should be useful for these types of reactions.

We next consider reactions of the type $F + AH_4 \rightarrow AH_3F + H$, in which A stands for either C or Si. The results of these calculations and experimental results, where available, are presented in Table 43.2. Both the MNDO and

TABLE 43.1   Heats of Formation of Reactants and Products

Molecule	$\Delta H_f^\circ(25°C)$ (MNDO) (kcal/mole)	$\Delta H_f^\circ(25°C)$ (Experimental)[a] (kcal/mole)
H	52.1	52.103
F	18.9	18.86
H_2	0.7	0.0
HF	− 59.7	− 65.14
CH_3	25.8	34.82
CH_4	− 11.9	− 17.895
CH_3F	− 60.9	− 56
SiH_3	36.9	54.25[b], 46.4[c]
SiH_4	1.2	8.2
SiH_3F	− 96.4	− 90

[a]All experimental values, unless otherwise indicated, are obtained from the *JANAF Thermochemical Tables*, Dow Chemical Company, Midland, MI, 1965, and updates.

[b]J. B. Pedley, B. S. Iseard, A. Kirk, S. Seilman, and L. G. Heath, *Computer Analysis of Thermochemical Data: Silicon Compounds*, Carton and Co., Southwick, England, 1972.

[c]R. Walsh, Bond Dissociation Energy Values in Silicon-Containing Compounds and Some of Their Implications, *Accts. Chem. Res.* **14**, 246–252 (1981).

TABLE 43.2   Results for Fluorine Atom Reactions

	$\Delta H°$ (reaction) (kcal/mole)			$\Delta H°^+$ (kcal/mole)	
Reaction	MNDO	ab initio[b]	Exp[a]	MNDO	ab initio[b]
$F + CH_4 \rightarrow CH_3F + H$	− 15.8	2.4	− 4.9	57.7	48.8
$F + SiH_4 \rightarrow SiH_3F + H$	− 64.4	− 50.4	− 65.0	30.4	5.2
$F + CH_4 \rightarrow CH_3 + HF$	− 40.9	− 24.3	− 31.1	27.3	6.5
$F + SiH_4 \rightarrow SiH_3 + HF$	− 42.9	− 35.6	− 45.8[c]	19.6	0
			− 37.9[d]		

[a]All experimental values calculated from the *JANAF Tables*, footnote 1 to Table 42.1, unless otherwise noted.

[b]From Ref. 8.

[c]Value calculated from the Walsh heat of formation for SiH_3, footnote 3 to Table 42.1.

[d]Value calculated from the Pedley et al. heat of formation for SiH_3, footnote 2 to Table 42.1.

ab initio results for $\Delta H°$ of the reactions are reasonably close to the experimental values, with MNDO doing somewhat better for the silicon case and the *ab initio* results slightly better for the carbon case. The *ab initio* results are too endothermic, while the MNDO results are too exothermic in the carbon case, but nearly correct in the silicon case.

The transition state is predicted to be a five-coordinate trigonal bipyramid for both reactions. MNDO predicts the $CH_4 + F$ transition state to be slightly distorted from the ideal trigonal bipyramid. The prediction for the $SiH_4 + F$ transition state is more distorted from the ideal trigonal bipyramid. In this transition state, there is very little Si—H bond breaking. In fact, the fluorine is so far away from the silane molecule that this calculated transition state appears very much like a tetrahedral silane only slightly perturbed by the fluorine atom in the vicinity. It appears that in this case MNDO is predicting a much-too-long bond length because of spuriously large core—core repulsions, as have been observed in previous calculations as well.[7] This effect for species in which there will be longer than usual bond distances (such as transition states) is probably due to the fact that MNDO was parameterized for "normal" ground state molecules with "normal" bond distances. The core—core repulsion function in MNDO is thus appropriate to "normal" bond distances, but decreases much too slowly as the bond distance is increased.[7]

The activation enthalpies (Table 43.2) are relatively high for the carbon substitution reaction. The *ab initio* value is about 10 kcal/mole lower than the MNDO value, and is probably more accurate given the tendency of MNDO to overestimate activation enthalpies. The latter occurs particularly for cases in which abnormally high repulsions result from transition state geometries in which there is a distance(s) of approximately van der Waals length. This effect is even more pronounced for the silicon substitution reaction. For silicon substitution the *ab initio* result indicates a very small activation barrier, but MNDO predicts a barrier of 30 kcal/mole. Note that the overprediction of the barrier is more pronounced in the silicon case, because a larger proportion of the MNDO-calculated activation enthalpy is due to these spuriously large core—core repulsions. In the carbon case, the barrier is large anyway due to steric crowding of the required five atoms in the transition state around the smaller carbon atom. Still, the trend in both methods indicates a much easier substitution reaction at silicon than at carbon.

Next we consider reactions of the type $AH_4 + F \rightarrow AH_3 + HF$, in which A is either carbon or silicon. The results for these reactions are also given in Table 43.2. In the carbon case, MNDO predicts the $\Delta H°$ about as well as the *ab initio* procedure, but the errors are in opposite directions. In the silicon case, both methods do reasonably well, although the experimental value is not certain due to the disagreement of values in the literature for the heat of formation of SiH_3.

The calculations predict a transition state of C_{3v} symmetry for the carbon abstraction reaction. The MNDO-predicted transition state occurs much earlier in the reaction than the *ab initio* result,[8] leading us to suspect that spuriously large repulsions are distorting the transition state geometry and producing an inordinately large activation enthalpy. The activation enthalpies are about 20 kcal/mole higher for MNDO than the *ab initio* results shown in the table. Sana et al.[14] calculated values ranging from 2.2 to 3.1 kcal/mole, depending on the *ab initio* method they used. They also give an experimental value of 1.15

kcal/mole. Thus it does appear that MNDO overpredicts the activation enthalpy by about 20 kcal/mole. Note again that this overprediction occurs for a case in which there is a bond being formed in the transition state that MNDO predicts to be far too long, no doubt due to the spuriously large core–core repulsions.

The MNDO-calculated geometry of the transition state for the silicon analog was qualitatively similar to the carbon case (C_{3v} symmetry). The process is predicted to occur without activation according to the *ab initio* results, but MNDO predicts a barrier of 20 kcal/mole, again about 20 kcal/mole greater than the *ab initio* results. Once again, MNDO has overpredicted the activation enthalpy in a case that it also predicts the bond being formed in the transition state to be too long. Thus, in situations in which there is neither a very small anion nor an overestimated core–core repulsion, we expect the calculated activation enthalpy to be within 10 kcal/mole of the true value.

Decompositions of the Hydroxide Adducts of Silanes

We have calculated a number of pentacoordinate silicon anions, and, as Dewar and Healy found in an earlier paper,[5] we have found a majority of them to be stable species with regards to decomposition to an anion of any of the constituent ligands and the four-coordinate species that would remain. The geometry of these species almost invariably is a trigonal bipyramid, with two axial substituents and three equatorial substituents. (The single exception is H_4SiF^-, which may also exist in a tetragonal pyramidal form.) There usually is some distortion of the ideal bipyramid if the molecule is asymmetric because of different constituent ligands. Because of the stability of these five-coordinate species, we propose that they may play an important role in silicon reactions involving anions, and, in particular, anionic silanol polymerization.

We chose the simple five-coordinate silicon anion $H_3Si(CH_3)OH^-$ to use in a study of various possible decomposition modes of these types of anions. There are four possible isomers of our trigonal bipyramidal structure: (1) both methyl and hydroxide axial, (2) both methyl and hydroxide equatorial, (3) methyl axial and hydroxide equatorial, and (4) methyl equatorial and hydroxide axial. Isomer 2 (both equatorial) is the most stable by about 10 kcal/mole compared with isomer 1 (both axial). The other two isomers are not stationary points according to MNDO, but transform to the most stable isomer (both equatorial).

The $H_3Si(CH_3)OH^-$ anion (called the anion complex or adduct) can decompose via various routes to give anions. In particular, the anions H^-, OH^-, and CH_3^- can be produced simply by stretching the appropriate bonds in the complex and allowing the rest of the structure to optimize at each point along the reaction path. The results indicate that in each case, decomposition by production of these anions results in a monotonic increase in enthalpy to the enthalpy of the final products. These results are summarized in Fig. 43.1. MNDO has a known deficiency in the prediction of too-high heats of formation for

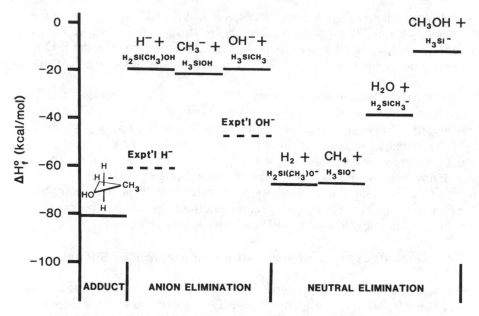

Figure 43.1   Heats of formation of products of decomposition of the anion complex.

small anions, such as OH^- and H^-, and the results in the figure show corrections for these cases in addition to the actual calculated MNDO values. For the elimination of CH_3^-, we carried out the reaction on both stable isomers. In the case of the most stable isomer, stretching the Si—C bond eventually resulted in elimination of methane instead of the CH_3^- anion, after the Si—C bond had been almost broken. This reaction will be discussed later. For the less stable isomer, stretching the Si—C bond results in the expected CH_3^- product, since the departing anion is on the opposite side of the silicon from the hydroxide group. In any case, the anion complex is significantly more stable than any product anion and four-coordinate silicon product.

There are other routes for the anion complex to decompose, however. In particular, the neutral molecules CH_4, H_2, CH_3OH, and H_2O may be formed by appropriate rearrangements of the complex anions. Indeed, DePuy et al.[15] have done a gas-phase study in which they reacted various silanes with OH^- and then studied the products. The products were generally those which would be formed by a transfer of the hydroxide proton to a departing anion, resulting in products like methane, hydrogen, ethane, and so on. Their kinetics suggested the formation of a reasonably stable five-coordinate adduct which then rapidly decomposed into the stated products. They were also able to correlate product ratios to gas-phase acidities of the products, suggesting a transition state which is composed of an almost-dissociated anion abstracting a proton from the hydroxide group. This experimental work by DePuy's group is extremely important to our work for two reasons: (1) as a test of our theoretical work in its ability to explain his experimental results, and (2) in showing the stability of five-coordinate silicon anions and their possible decomposition routes.

To compare with DuPuy's results and to assess the decomposition modes which result in neutral product elimination and a silane anion, we searched for the transition states for CH_4, H_2, CH_3OH, and H_2O elimination. In the methane case, the reaction coordinate was chosen to be the distance between the methyl carbon and the hydroxide proton. The transition state was located and found to be one in which the Si—C bond had essentially broken, and the O—H bond only slightly stretched. The portion of the molecule to become the tetravalent silicon anion upon completion of the reaction already looks very much like the product; that is, a slightly distorted tetrahedron. This anionic dissociative proton transfer transition state appears to be a characteristic feature of low-energy neutral species elimination from pentacoordinate silicon anions. This structure, shown in Fig. 43.2, is qualitatively in agreement with our interpretation of DuPuy's experimental work.[15] Thus, since the transition state looks very much like an anion abstracting a proton from the hydroxyl group, we would expect the relative reaction rates of different abstraction products to correlate with the gas-phase acidities of the conjugate acids of these anions. In the hydrogen case, the reaction coordinate was chosen to be the distance between the hydroxide proton and the hydrogen atom being removed from the silicon. Again, the transition state structure (Fig. 43.2) shows a Si—H bond which is

TRANSITION STATE GEOMETRIES

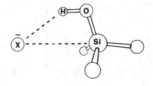

X	SPECIES	BOND LENGTHS (Å)		
		X – Si	X – H	O – H
H	ADDUCT	1.49	2.29	0.93
H	T. S.	3.23	1.04	1.12
CH$_3$	ADDUCT	1.87	3.52	0.93
CH$_3$	T. S.	3.32	1.86	0.97

Figure 43.2 Structures of the transition states for neutral product elimination from the anion complex.

already essentially broken, an O—H bond somewhat stretched, and an only slightly distorted silicon tetrahedral species left behind. For the water elimination, the reaction coordinate was chosen to be the distance between the hydroxide oxygen and the hydrogen atom being removed from the silicon. The transition state occurs only after the distance between these two gets fairly small, about 1.2 Å. For the methanol elimination, the reaction coordinate was chosen to be the distance between the hydroxide oxygen and the methyl carbon. The results are quite similar to the water elimination, with methanol being formed only when the distance between the carbon and the oxygen is very close to that in methanol.

Figure 43.3 shows a profile of the enthalpy changes as the elimination reactions proceed. We also show the enthalpy of the reactants (hydroxide ion and methyl silane). The five-coordinate intermediate is formed without activation from the reactants. Each elimination reaction to a neutral product molecule and a silicon-containing anion has an activation barrier to final product formation. In each case, the barrier height is above the enthalpy of the initial reactants if we use the experimental value for an hydroxide ion. Thus, for this species, we would not expect an activation-less process as DuPuy observes for related systems.[15] It is interesting to note, however, that the barrier to hydrogen elimination is lower than that for methane elimination, in agreement with his experimental results on relative ease of product formation. This correlates

REACTION PROFILES

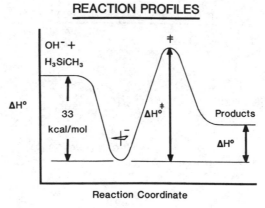

NEUTRAL PRODUCT	$\Delta H^{\ddagger}$ (kcal/mol)	ΔH° (kcal/mol)
H_2	42	13
CH_4	45	15
H_2O	96	43
CH_3OH	92	70

Figure 43.3 Profile of enthalpy changes as the anion complex is formed and then dissociated.

well with the larger gas-phase acidity of H_2 relative to that of methane. In addition, our preliminary results on his experimental silicon anion systems (for example, replacing all of the hydrogens attached to the silicon with methyls) indicate that the MNDO-predicted barrier to these eliminations decreases substantially and may disappear entirely.

In the cases of water and methanol elimination, the heats of formation increase drastically before water or methanol is actually formed and removed from the molecule. This is due to the fact that we are attempting to form these neutral molecules from two species that are both negatively charged. Thus, until the two are very close, the negative charge on the hydroxide repels that of the hydride (or the methide group) and the enthalpy gets very large before the reaction occurs. Thus both of these processes have very high activation enthalpies, over 90 kcal/mole. Our overall conclusion is that MNDO shows great promise in understanding the chemistry of large silicon-containing anions.

Pentacoordinate Silicon in Silanol Polymerization

We now turn our attention to a preliminary assessment of the importance of pentacoordinate silicon in silanol polymerization. Starting with silicic acid, silica can be prepared by silanol polymerization using either acidic or basic catalysis, although the basic catalysis route predominates at pH's greater than 2.[16] In this chapter we will consider only the first few steps of this polymerization mechanism, and that only for anionic conditions. In addition, we report here only heats of reaction and not activation enthalpies.

Figure 43.4 shows the set of reactions that we considered along with the heats

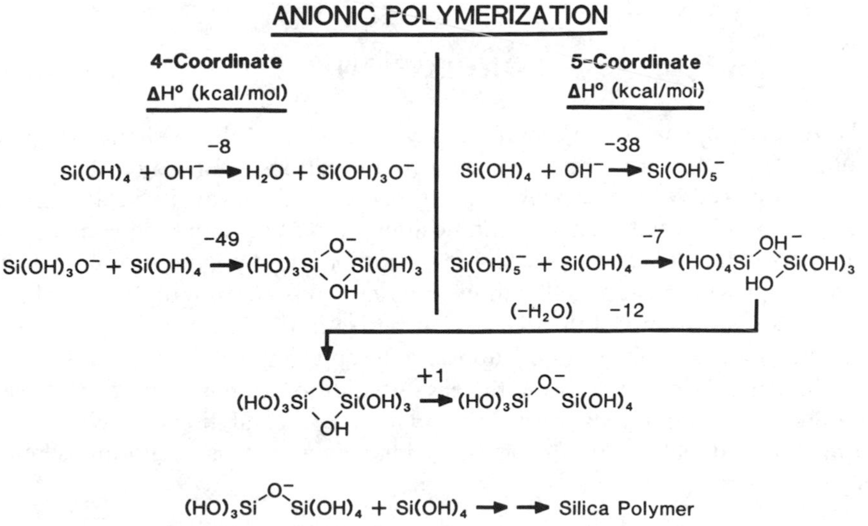

Figure 43.4 Reactions involved in first steps of anionic silanol polymerization and their heats of reaction.

of reaction. Silicic acid in basic media can either act as an acid, transferring a proton to hydroxide to form water and a four-coordinate anion, or it can accept a hydroxide to form the five-coordinate silicon anion $Si(OH)_5^-$. Because of the stability of this five-coordinate anion, it appears likely to be the dominant form of monomeric silanol under basic conditions. Either anion can attack a neutral silicic acid, resulting in silicon anions each containing two silicons. In the simpler four-coordinate anion case, the resulting anion can eliminate hydroxide regenerating the catalyst and forming the partially polymerized structure $(OH)_3Si-O-Si(OH)_3$. This process can continue to eliminate the remaining hydroxides on this product and eventually form silica. The heats of reaction for this process are all negative except for the last step of removing the hydroxide; however, the heat of reaction for the overall process $2Si(OH)_4 \rightarrow H_2O + (OH)_3Si-O-Si(OH)_3$ is negative ($\Delta H° = -15$ kcal/mole).

Attack of the five-coordinate anion $Si(OH)_5^-$ onto silicic acid is more complicated. Several anions containing two silicons and nine hydroxides are possible; our figure shows only the one predicted to be the most stable. In addition, to complete the reaction to the partially polymerized $(OH)_3Si-O-Si(OH)_3$ requires loss of both water and hydroxide. It may be possible to continue the polymerization indefinitely, however, by simply building larger and larger anions which can then continue to attack silicic acid molecules. Thus, it may not be necessary to eliminate hydroxide during the bulk polymerization (with either the four- or five-coordinate anion) provided we can eliminate water easily (in the case of the five-coordinate anion). We are now actively pursuing calculation of activation enthalpies for these processes to determine which are kinetically feasible. It does appear, however, that five-coordinate silicon anions can and probably do play a key role in basic catalysis of silanol polymerization.

CONCLUSION

MNDO does appear to be a useful study for studying silicon chemistry, particularly if we recognize and deal with its limitations. It gives results which agree quite well with reasonable interpretations of the available gas-phase kinetic data on formation and decomposition of five-coordinate silicon anions. It appears ideal for studying the first few steps of silanol polymerization, which cannot at present be attacked with *ab initio* methods because of the size of the species required to model the silanol polymerization.

Pentacoordinate silicon anions appear to be quite important in systems which contain small anions because of the ease of addition of these anions to tetrahedral silicon compounds. Thus, our preliminary calculations support these pentacoordinate silicon anions as the prime chain-carriers in anionic silanol polymerization.

ACKNOWLEDGMENTS

We wish to thank Professors Michael J. S. Dewar of the University of Texas, Austin, Robert Damrauer of the University of Colorado, Denver, and Mark S. Gordon of North Dakota State University for many helpful discussions.

REFERENCES

1. M. J. S. Dewar and W. Thiel, Ground States of Molecules. 38. The MNDO Method. Approximations and Parameters, *J. Am. Chem. Soc.* **99**, 4899–4907 (1977).

2. M. J. S. Dewar, M. L. McKee, and H. S. Rzepa, MNDO Parameters for Third Period Elements, *J. Am. Chem. Soc.* **100**, 3607 (1977).

3. M. J. S. Dewar, D. H. Lo, and C. A. Ramsden, Ground States of Molecules. XXIX. MINDO/3 Calculations of Compounds Containing Third Row Elements, *J. Am. Chem. Soc.* **97**, 1311–1318 (1975).

4. M. J. S. Dewar, G. L. Grady, E. F. Healy, and J. J. P. Stewart, Revised MNDO Parameters for Silicon, in preparation for publication.

5. M. J. S. Dewar and E. Healy, Why Life Exists, *Organometallics* **1**, 1705–1708 (1982).

6. W. S. Verwoerd, MNDO Calculations of Silicon-Containing Molecules, *J. Comput. Chem.* **3**, 445–450 (1982).

7. M. J. S. Dewar, private communication.

8. L. P. Davis, L. W. Burggraf, M. S. Gordon, and K. K. Baldridge, A Theoretical Study of Fluorine Atom and Fluoride Ion Attack on Methane and Silane, submitted for publication.

9. M. S. Gordon and C. George, Theoretical Study of Methylsilanone and Five of Its Isomers, *J. Am. Chem. Soc.* **106**, 609–611 (1984).

10. M. S. Gordon, Hydrogen Abstraction by Triplet Methylene and Silylene, *J. Am. Chem. Soc.* **106**, 4054–4055 (1984).

11. M. O'Keeffe and G. V. Gibbs, Defects in Amorphous Silica; Ab Initio MO Calculations, *J. Chem. Phys.* **81**, 876–879 (1984).

12. U. Brandemark and P. E. M. Siegbahn, The Reactions Between Negative Hydrogen Ions and Silane, *Theo. Chim. Acta (Berlin)* **66**, 233–243 (1984).

13. Quantum Chemistry Program Exchange Program Numbers 455 and 464, Department of Chemistry, Indiana University, Bloomington, Indiana 47405.

14. M. Sana, G. Leroy, and J. L. Villaveces, A Theoretical Study of Hydrogen Abstraction Reactions $CH_4 + R \rightarrow CH_3 + HR$, Geometrical, Energetical and Kinetic Aspects, *Theo. Chim. Acta (Berlin)* **65**, 109–125 (1984).

15. C. H. DePuy, V. M. Bierbaum, and R. Damrauer, Relative Gas-Phase Acidities of the Alkanes, *J. Am. Chem. Soc.* **106**, 4051–4053 (1984).

16. R. K. Iler, *The Chemistry of Silica*, Wiley, New York, 1979, Chapter 3.

PART 4

Ultrastructure in Macromolecular Materials

44

NETWORK THEORY AND GELATION

PAUL J. FLORY
Department of Chemistry
Stanford University
Stanford, California

Polymerization of silicic acid and of its derivatives dissolved in water or other suitable solvent leads readily to the formation of a gel, that is, an elastic solid having (initially) a very low modulus of elasticity. The manifestation of high deformability with full recovery is a feature that is an exclusive characteristic of structures consisting of mobile molecular chains capable of assuming numerous spatial configurations. The polymerization of the silicate moiety resembles polycondensation of polyfunctional organic compounds,[1] ramified molecules of increasing complexity being generated through combination of one species with another.[2] These molecules are randomly configured and copiously interspersed, each within the region of space pervaded by numerous other homologous molecules. This process stands in sharp contrast to the aggregation of a dispersion of globular particulates each of which preempts a compact region of space.

The gel point observed in a polycondensation process may be identified with the incipience of a network of a macroscopic size.[2-4] The occurrence of gelation requires that the branching probability α, defined as the probability that a molecular chain extending from one branch point in the structure leads to another branch, must exceed its critical value given by

$$\alpha_c = \frac{1}{f-1}$$

where f is the functionality of the branch point (e.g., for polymeric silicates $f = 4$). The network, or gel, makes its appearance abruptly in the course of the polymerization which, however, continues without interruption at and beyond the gel point.[3] The proportion of gel increases continuously at the expense of the sol-comprising species of finite size. Simultaneously, the density of reticulation in the network increases, as is reflected by the increase in the elastic modulus.[3,4]

This description of the gelling process is abundantly confirmed by comparisons of theory with experimental observations on branched and crosslinked polymers.[5-7] The theory also describes the distribution of the various molecular species and the apportioning of the polymerizing material between sol and gel. Experimental results on molten silicates appear to be consistent with the theory briefly outlined.[8]

The condensation of silicic acid and its derivatives involves processes as follows:

$$(-O \rightarrow)_m Si(-OR)_{4-m} + (RO \rightarrow)_{4-n} Si(-O-)_n \longrightarrow$$

$$(-O \rightarrow) - Si - O - Si(-O-)_n + R_2O$$
$$\qquad\qquad | \qquad\qquad | $$
$$\qquad (OR)_{3-m} \quad (OR)_{3-n}$$

where R may be an akyl group, a hydrogen atom, a complementary cation or a mixture of these, and $m, n = 1-4$; a bridging oxygen between two silicon atoms is denoted by $(-O-)$. Experimental determination of the relative reactivities of the hydroxyl group for diverse values of m would be required for quantitative application of the theory briefly outlined. The incidence of branching relative to linear chain growth depends on the reactivity for $m > 1$ compared with that for $m = 1$. Such information is essential to the evaluation of α as a function of the overall degree of condensation. Only if the reactivity for $m = 0$ or 1 greatly exceeds that for $m > 1$ is the formation of particulates possible under ordinary conditions of polycondensation. If, however, the polycondensation is carried out at high dilution, intramolecular processes may dominate and thereby lead to the formation of discrete particles.[1]

The extensive investigations of Hurd[1] show convincingly that the process whereby silicic acid gels are formed proceeds through chemical reactions between functional groups, that is, that it proceeds by a polycondensation, and not by flocculation of particles. The elastic characteristics of the gel (before drying and heating) are uniquely diagnostic for a molecular network.

REFERENCES

1. C. B. Hurd, *Chem. Revs.* **22**, 403 (1938); *J. Phys. Chem.* **57**, 678 (1953).
2. P. J. Flory, *J. Phys. Chem.* **46**, 132 (1942).

3. P. J. Flory, *Principles of Polymer Chemistry*, Chapter IX. Cornell University Press, Ithaca, NY, 1053,

4. P. J. Flory, Introductory Lecture, General Discussion on Gels and Gelling Processes, *Faraday Discussions of the Chem. Soc.* **57**, 7–18 (1974).

5. W. H. Stockmayer and L. L. Weil, in *Advancing Fronts in Chemistry*, S. B. Twiss, Ed., Reinhold, New York, 1945, Chapter 6.

6. C. A. L. Peniche-Covas, S. B. Dev, M. Gordon, M. Judd, and K. Kkkajiwara, *Discuss. Faraday Soc.* **57**, 165 (1974).

7. R. F. T. Stepto, *Discuss Faraday Soc.* **57**, 69 (1974).

8. P. C. Hess, *Geochim. Cosmochim. Acta* **35**, 289 (1971).

45

MORPHOLOGY OF ORIENTED POLY(*p*-PHENYLENEVINYLENE): A CONDUCTING POLYMER FROM A SOLUBLE PRECURSOR POLYMER

FRANK E. KARASZ, ROBERT W. LENZ,
DAVID R. GAGNON, AND THIERRY GRANIER
Department of Polymer Science and Engineering
University of Massachusetts
Amherst, Massachusetts

INTRODUCTION

The effect of polymer microstructure on mechanical properties has long been known to be of the utmost importance. In the field of conducting polymers, structure/property relationships are not so easily studied due to the intractable and insoluble nature of the rigid conjugated chemical structure necessary for charge carrier delocalization. To overcome this problem a relatively new technique of conducting polymer synthesis has been employed in which a soluble precursor polymer is prepared and processed prior to its conversion to the conjugated polymer, poly(*p*-phenylene vinylene). This technique allows the systematic variation of the polymer morphology and the resultant electrically conducting properties.

Conjugated polymers consist of long chains of alternating single and double bonds in which charged species are delocalized over many repeat units. Two of the simplest structures fulfilling this requirement are polyacetylene (**1**) and poly(*p*-phenylene) (**2**).

(**1**)   (**2**)

The reactions of these polymers with electron-donating or electron-accepting reactants are called *n*- and *p*-doping and result in the injection of either negative or positive charges, respectively, into the conjugated matrix. Conductivity is a bulk measure of the ability of these charges to flow through the matrix when an electrical potential is applied across the sample. The magnitude of conductivity (σ) is defined by $\sigma = ne\mu$, where n is the number of charge carriers of charge e, and μ is the net mobility of these carriers in the direction of the applied field. The number of carriers is related to the concentration of dopant. The mobility of these resultant charges is determined primarily by the structure of the matrix and the degree to which the charge is delocalized.

Poly(*p*-phenylene vinylene), PPV, can be described as an alternating co-polymer of structures I and II. Due to solubility limitations, it had only been prepared as low molecular weight oligomeric powders[1] prior to the work of Wessling and Zimmerman,[2,3] and Kanbe[4] on the precursor method of PPV synthesis which is used in this work. The conductivity of the PPV oligomers after *p*-doping with AsF$_5$ was ~ 1 S/cm,[5] whereas isotropic films of PPV pre-pared by us using the precursor method attained conductivities ~ 10 S/cm.[6] The increase is probably due primarily to the coherent nature of a film as com-pared to grains of a pressed powder and not to the increase in molecular weight. The real advantage to the precursor route to PPV is that films of the precursor polymer can be stretched to give an ultrastructure having high degree of uniform molecular orientation. After doping the conductivity in the direction parallel to the orientation is increased by over two orders of magnitude to a value of ~ 2500 S/cm.[7,8]

The first step toward understanding the reason for such a marked increase in electrical conductivity is to characterize fully the PPV ultrastructure. In this work, wide angle X-ray diffraction (WAXD) and electron diffraction (ED) were used to characterize both unoriented and oriented films of PPV.

EXPERIMENTAL

Figure 45.1 outlines the synthetic method used to obtain the PPV samples for this study.[9,10] Synthesis of the monomer, *p*-phenylenedimethylene-*bis*(dimethyl-sulfonium chloride) (**3**), was carried out by the reaction of α,α'-dichloro-*p*-xylene with excess dimethylsulfide at 50°C in a methanol:water (80:20) solution for 20 hr. This material was purified by concentration of the reaction solution, pre-cipitation in cold acetone, filtration, and vacuum drying. The product obtained was a white crystalline powder. The polymerization to produce the poly(*p*-

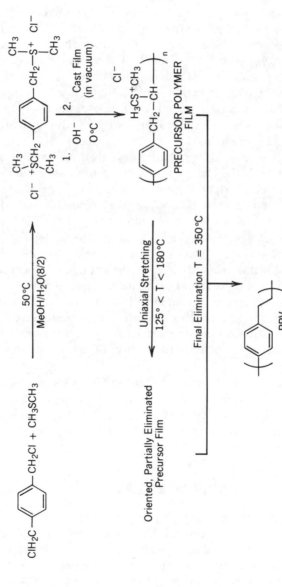

Figure 45.1 Synthesis of poly(p-xylene-α-dimethylsulfonium chloride), the precursor polymer; and the processing steps to obtain unoriented and oriented PPV films.

xylyene-α-dimethylsulfonium chloride) precursor polymer (**4**) was carried out by the aqueous reaction of (**3**) with an equimolar amount of sodium hydroxide at 0°C for 1 hr under conditions which vigorously excluded oxygen. The reaction was quenched by neutralization with HCl to a slightly acidic pH. The resultant solution was highly viscous and showed UV fluorescence. The polyelectrolyte was separated from residual monomer and NaOH by dialysis against H_2O for 3 days. The polyelectrolyte was wet cast into film form by evaporation of the H_2O *in vacuo* at 25°C. The polymer film was clear and could be redissolved in polar solvents or processed into PPV films or oriented films as follows. Simple heating of the polyelectrolyte film would result in the loss of $(CH_3)_2S$ and HCl, yielding a free standing, yellow film of PPV. The duration of time and the temperature at which the film was annealed determined the amount of unsaturation induced in the polymer. To obtain maximum elimination, the films were annealed at 350°C for more than 2 h in vacuum.

Uniaxially oriented films were obtained by clamping the two ends of a piece of precursor film (typically measuring about 1 cm × 4 cm) and stretching while heating between 125 and 180°C in an apparatus comprised of two glass slides separated by 3 mm thick Teflon®* spacers and wrapped with Nichrome heating wire. In this way the oriented film was stretched with necking in the heat zone. The film was passed through the heat zone under stress until orientation to the desired draw ratio was uniform. Even with this rather crude technique uniformity of orientation was satisfactory. The thermal elimination to oriented PPV was completed by heating to 350°C under vacuum for 2 hr.

WAXD patterns were obtained with Cu $K\alpha$ radiation on flat X-ray films in the Statton configuration. ED patterns were taken with a JEOL 100CX transmission electron microscope on thin samples prepared by mechanical fibrillation of the oriented samples.

RESULTS AND DISCUSSION

The stretching process is novel in that it occurs only during the thermal elimination of the precursor polyelectrolyte film to PPV. The extent of elongation and hence the final orientation is determined by the instantaneous concentration of the volatile elimination products which act as a diluent in the matrix to lubricate or plasticize the material. The concentration of Me_2S and HCl depend upon the initial degree of elimination to PPV and upon the elimination temperature. Films which are less than 20% converted to PPV and temperatures between 125° and 180°C have been found to be optimal to obtain highly oriented films without the bubbling or foaming effect seen above 180°C.

WAXD patterns of unoriented PPV films show three diffuse rings at *d*-spacings of 4.39, 3.20, and 2.10 Å. (Fig. 45.2*a*). These spacings represent most probable interchain close packing distances and do not represent a well-defined

*Registered trademark of E. I. DuPont de Nemours, Inc., Wilmington, Delaware.

crystallographic lattice as shown by the diffusivity of the reflections and by the lack of higher order reflections. The drawn samples, on the other hand, show distinct layer lines and sharp equatorial reflections resembling classic fiber patterns of polymers (Fig. 45.2b). The layer line spacings correspond to a chain repeat unit (00l) spacing of 6.58 Å. High orientation is evidenced by the fact that up to fourth order layer lines are seen and that the angle made by the

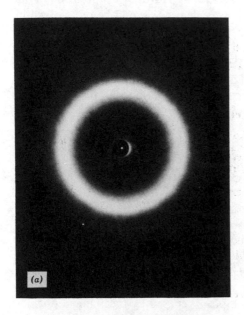

(a)

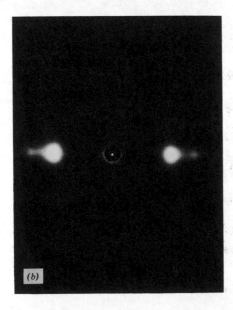

(b)

Figure 45.2 Wide-angle X-ray diffraction pattern of (a) unoriented PPV film and (b) highly oriented PPV film (draw ratio $l/l_0 = 10$).

equatorial reflections with the origin is $<8°$. The repeat unit distance measured corresponds exactly to the dimensions calculated for *trans*-PPV using the model of Roberson[11] based on *trans*-stilbene. This *trans*-configuration about the double bond and the high molecular orientation is also confirmed by polarized infrared spectroscopy.

Uniaxial structures of this type can have various degrees of order ranging from an ideal 3-D extended chain single crystal, to a loose packed bundle of highly oriented chains with no interchain ($hk0$) correlations or interactions at all.

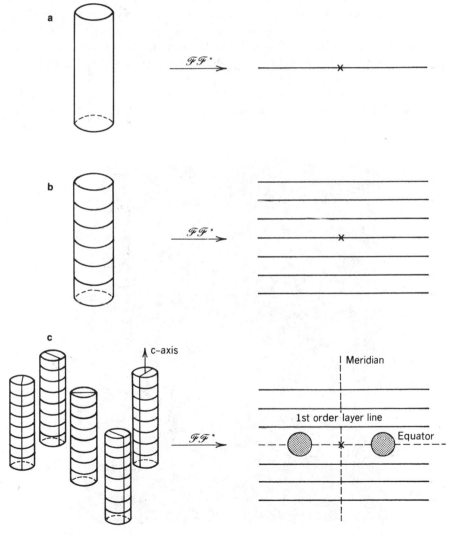

Figure 45.3 Diffraction pattern (*a*) from a rod, (*b*) from a rod made of repeat units, and (*c*) from a uniaxially oriented bundle of rods of repeat units with no 3-D registry.

In the latter case the diffraction pattern expected would consist of equatorial spots corresponding to most probable interchain packing distances, and a number of orders of meridianal layer lines with distances corresponding to the *c*-axis chain repeat unit length. This type of pattern can be understood in the context of Figure 45.3 which shows the diffraction pattern of (*a*) a single infinitely long rod, (*b*) a single rod made up of discrete repeat units, and (*c*) a bundle of these rods. When the repeat unit is molecular the intensity of each layer line will vary as one scans along it perpendicular to the meridianal direction.

Using cylindrically averaged molecular transform calculations[12] which assume free translation along the chain axis and are based upon the structure factor of the repeat unit, the layer line intensity profile for this first approximation of our PPV model can be calculated and compared to that of the experimental patterns. Electron diffraction is most useful for this purpose because it gives local structure information with short exposure times and without the need for Ewald sphere corrections.

The comparison of the experimental layer line intensity profile to that of the calculated profile shows good qualitative agreement for the fourth and higher

Figure 45.4   Electron diffraction pattern of highly oriented PPV ($l/l_0 = 14$).

order layer lines. This demonstrates that there is no long-range interchain registry. On the other hand, lower order layer line intensity profiles do not match very well with calculations and this shows the beginning of 3-D ordering which affects lower order layer lines most strongly.

PPV morphological structure is thus paracrystalline in nature and can be analyzed using lattice distribution functions based upon lattice distortions of the second kind. That is, the position of each unit is statistically distributed from the position of the nearest neighbor and the registry becomes poorer as one moves away from the starting unit. The modeling of this packing requires the calculation of this 3-D distribution function, the variables of which can be extracted semiempirically from the position, width, and relative intensity of the measured equatorial reflections and the layer line intensity profiles. We have recently been able to obtain high enough quality electron diffraction patterns from which to perform these calculations (Fig. 45.4). These patterns show five equatorial reflections and up to nine orders of layer lines which should give enough information. The analysis is now in progress.

CONCLUSIONS

There is a strong correlation between the molecular order and the final conducting properties of AsF_5 doped PPV. Electronic charge carriers propagate through a conjugated system along the chain via delocalization and/or by hopping to a neighboring chain. The relative importance of these two mechanisms has not been clearly understood mainly because of the lack of a clearly defined morphology in most conducting polymers. The definition of undoped morphology is only the first step. The location of dopant molecules in the matrix must also be known to understand conduction mechanisms.

The undoped morphology of PPV is nearly understood at this time, and its strong effect on final electronic properties has been clearly shown. Future study of doped PPV in order to determine the location of the dopant in the matrix should lead to a more thorough understanding of the conduction mechanism, and hence the criteria for designing a better polymeric conductor.

ACKNOWLEDGMENT

The authors are grateful for the support provided by the AFOSR Grant # AF84-0033. DRG thanks the Plastics Institute of America for partial support in the form of a supplemental fellowship award.

REFERENCES

1. R. N. McDonald and T. W. Campbell, The Wittig Reaction as a Polymerization Method, J. Am. Chem. Soc. **82**, 4669 (1960).

2.   R. A. Wessling and R. G. Zimmerman, Polyelectrolytes from *Bis*-Sulfonium Salts, U.S. Pat. # 3,401,152 (1980).

3.   See Ref. 2, Polyxylyidene Articles, U.S. Pat. # 3,706,677 (1972).

4.   M. Kanbe and M. Okawara, Synthesis of Poly-*p*-Xylidene from *p*-xylene-*bis*(dimethylsulfonium tetrafluoroborate), *J. Polym. Sci.* (*A*-1) **6**, 1058 (1968).

5.   G. E. Wnek, J. C. W. Chien, F. E. Karasz, and C. P. Lillya, Electrically Conducting Derivative of Poly(*p*-phenylene-vinylene), *Polym.* **20**, 1441 (1979).

6.   D. R. Gagnon, J. D. Capistran, F. E. Karasz, and R. W. Lenz, Conductivity Anisotropy in Oriented Poly(*p*-phenylene vinylene), *Polym. Bull.* **12**, 293 (1984).

7.   I. Murase, O. Toshihiro, N. Takanobu, and H. Masaaki, Highly Conducting Poly(*p*-phenylene vinylene) Prepared from a Sulfonium Salt, *Polymer Comm.* **25**, 327 (1984).

8.   T. Granier, D. R. Gagnon, R. W. Lenz, and F. E. Karasz, in press.

9.   J. D. Capistran, D. R. Gagnon, R. W. Lenz, and F. E. Karasz, Synthesis and Electrical Conductivity of High Molecular Weight Poly(arylene vinylenes), *ACS Polym. Prepr.* **25**(2), 282 (1984).

10.   D. R. Gagnon, J. D. Capistran, F. E. Karasz, and R. W. Lenz, Conductivity Anisotropy of Doped Poly(*p*-phenylene vinylene) films, *ACS Polym. Prepr.* **25**(2), 284 (1984).

11.   J. M. Roberson and I. Woodward, X-Ray Analysis of Dibenzyl Series IV-Detailed Structure of Stilbene, *Proc. R. Soc. London* (*Ser.A*) **162**, 568 (1937).

12.   K. Suchiro, Y. Chatani, and H. Tadakoro, Structural Studies of Polyesters. VI. Disordered Crystal Structure (Form II) of Poly(*β*-propiolactone), *Polym. J.* **7**, 352 (1975).

46

DESIGN, ULTRASTRUCTURE, AND DYNAMICS OF POLYMERIC THIN FILMS

PARAS N. PRASAD
Department of Chemistry
State University of New York at Buffalo
Buffalo, New York

INTRODUCTION

Organic macromolecular systems rich in π-electrons have received considerable attention during recent years because of their promises as electroactive polymers.[1] Also, these polymers show strong nonlinear optical effects because of the large π-electron contribution to the electrical susceptibilities.[2] For the latter reason, an organic polymer can play an important role in any future development of optical computers.

Our research program focuses on the design, ultrastructure, dynamics, and possible device application of micrometer to monomolecular thin polymeric films. The design of polymeric thin films involves chemistry and controlled fabrication of polymers of desired chemical structure, physical properties, and thickness. In determining the ultrastructure, our approach involves the study of molecular structure and domain structure in thin films. Investigation of the dynamics of polymer films deals with the study of physics of electronic and optical effects, and nonlinear interactions. Our interest in device fabrication centers on such applications as Schottky diodes, nonlinear optical devices using polymers as optical waveguides, and batteries using polymer-coated electrodes.

Only design and ultrastructure analysis will be discussed in this chapter.

DESIGN OF POLYMERIC THIN FILMS

Our emphasis on design of polymeric thin films is on the following aspects: (1) synthesis of novel electroactive polymers; (2) control of thickness to the ultrasubmicrometer level; (3) control of physical properties, principally electrical conductivity and optical properties; (4) fabrication of polymer composites and copolymers as a means to control the physical properties; (5) formation of layered structures for device fabrication such as Schottky diodes and optical multiplexers.

For the design of electroactive polymeric films, interfacial processes are found to be highly suitable. Specifically, the following three methods have been used:

Langmuir–Blodgett Method

This method is used for producing polymer films from monomers, which in solid state, can be thermally or photochemically polymerized. Typically, a film of reactive monomer is coated, at a constant surface pressure, on a substrate by dipping the substrate in a Langmuir–Blodgett trough in which the monomer solution is dispersed and compressed to form a condensed film. The monomer film deposited on the substrate is then photopolymerized. The monomer film can also be polymerized on the water surface and then transferred to the substrate.[3] The advantage of the Langmuir–Blodgett method is production of highly oriented monolayer or successively built multilayer (by repetitive dipping) polymer films. Such films produced can vary in thickness from 20 to 1000 Å. We have used this method to deposit one class of polymers, polydiacetylene. These polymers are suggested to be a 1-D semiconductor and have shown indications of large third-order susceptibilities $(\chi)^3$ necessary for application in nonlinear optical devices. The monomer diacetylene $R-C\equiv C-C\equiv C-R'$ was polymerized by UV light. Two different monomers were used:

$$R=R'=-CH_2-OSO_2C_6H_4CH_3 \tag{1}$$

$$R=CH_3-(CH_2)_{11}, \ R'=(CH_2)_8-\overset{\displaystyle O}{\overset{\displaystyle \|}{C}}-OH \tag{2}$$

The first monomer is symmetric having polar groups at both ends. Therefore, it was not expected to form a traditional Langmuir–Blodgett film. However, we were successful in obtaining multimolecular oriented films of ~ 100 Å thickness.[3] The second monomer has both hydrophobic and hydrophilic end groups and, therefore, forms a traditional Langmuir–Blodgett film. We were able to form a condensed oriented monomolecular polymer film in this case.

Solid–Gas Interface Reaction

Many organic solids, in monomeric state, react with a strong Lewis acid, such as AsF_5, to produce a well-oriented polymeric film. These reactions are diffusion controlled. We have made two new polymers by this method. A thin film of furil (I) was vacuum deposited on a glass substrate and, then, exposed for a period of ~ 24 hr, whereby a polymer film of polyfuril, metallic in appearance, was obtained.[4] We have also polymerized azulene (II) to produce polyazulene.

I   II

In the oxidized form, the polymer backbone has a positive charge which is compensated by the inclusion of an anion (AsF_6^-). The oxidized forms generally have high electrical conductivity. When reduced, these films reject the anion and become insulators. The polymeric films obtained by this method are crystalline, but contain a considerable amount of lattice disorder.

Electrochemical Polymerization

Thin films can be produced by electrochemical oxidation of the monomer solution at the anode surface.[1] We have used this method in collaboration with Professor S. Bruckenstein of our Department to produce polymers of azulene by electrochemical oxidation of azulene in acetonitrile solution containing either ClO_4^- or BF_4^- as the supporting electrolyte anion. In the oxidized state, the polymer obtained has a positive charge and includes the anion. In this state the polymer has high electrical conductivity, but becomes an insulator in the reduced form. Electrochemical polymerization produces amorphous polymeric films whose thickness can easily be controlled to $100\,\text{Å}$. The films are electrochromic, going through a distinct color change between oxidized and reduced states. We have used this method to produce a random copolymer of azulene and pyrrole.

These polymers have the drawback that once formed they cannot be processed. They have a fairly high thermal stability (infusable, decomposition temperature $\geqslant 500°C$) and tend to be brittle. For this reason, effort has also been made to form polymer composites. Azulene has been polymerized in the matrix of PMMA and polystyrene.

Another advantage of this method is that one can form layered structures containing several polymers. This feature is useful to study electronic processes at junctions. We have fabricated sandwich structures containing polyazulene and polypyrrole.

ULTRASTRUCTURE

We use laser Raman spectroscopy to investigate the vibrational spectra of the polymeric films from 10 to 3500 cm^{-1}. This range covers both phonon motions and intramolecular motions. The phonon spectra, 10 to 200 cm^{-1}, corresponds to intermolecular motions. The longitudinal accordian motions (LAM) of polymers also show up in this region. In determining the ultrastructure with phonon spectroscopy, the underlying principle is that these motions are governed by short-range interactions and, therefore the molecular organization at $\leqslant 100$ Å level is probed.[5] Polarized Raman spectroscopy in conjunction with laser scanning (microprobing) can be used to investigate local anisotropy and homogeneity of domain structures. Crystalline polymers exhibit phonon spectra which consist of sharp structures. The presence of disorder broadens these phonon peaks. In amorphous polymers, a broad peak due to local phonon density of states appears.

We use two types of geometries for obtaining laser Raman spectra of polymeric thin films: (1) low-angle laser scattering, with the laser incident at a small angle with the plane of the polymer film; (2) an optical waveguide technique using a 1–5 μm thick polymer film on a glass substrate. Under appropriate matching of refractive indices of the prism, polymer film, and the glass substrate, this film acts as a waveguide through which the light propagates.[6] In order to obtain the Raman spectra of polymer films of $\leqslant 100$ Å thickness, we coat the thin films on the surface of an optical waveguide (which may be a different polymer). By this method one can also study any interfacial interaction between the two polymer films.

The first example of laser Raman spectroscopy in ultrastructure determination is the study of the molecular mechanism of photopolymerization of p-toluene sulfonate (PTS) diacetylene in the bulk. Figure 46.1 shows the phonon spectra of the monomer, a partially converted sample, and the polymer. The phonon spectra consist of a well-defined resonance enhanced Raman peak of the polymer. This peak gradually shifts. The polymerization is a one-phase (homogeneous blend of the monomer and the polymer) single crystal-to-single crystal transition.[5]

Low-angle laser Raman scattering was used to investigate a 100 Å thick PTS polydiacetylene film deposited by a modified Langmuir–Blodgett technique.[3] The spectra in the phonon regions are the same as in the bulk, indicating that the molecular organization in these films are the same and that the phonon motions are indeed governed by short-range interactions that cover an interaction radius of $\leqslant 100$ Å. The intramolecular vibrations show that the conformations of the polymer in the thin film and in the bulk are the same.

Polarized Raman spectra were used to probe orientational effects resulting

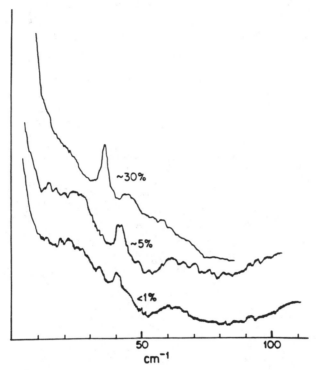

Figure 46.1 The phonon spectra of the *p*-toluene sulfonate diacetylene monomer, a partially polymerized sample and the polymer at room temperature are compared.

from preferred physical orientations within thin films formed at surface pressure of 10 dyne/cm. Typical spectra of the 1400–1800 cm^{-1} region were obtained with orthogonal polarizations. The spectrum obtained by a vertically polarized light, as shown in Fig. 46.2, is enhanced with respect to the spectrum obtained by horizontal polarization. The polarization effect indicates that orientational anisotropic effects are present. However, the anisotropic effects appear to depend on the local environment since the relative intensities of the polarized spectra change from one location to another within the same film.

Diacetylene was dispersed in polystyrene and investigated by the optical waveguide technique in order to: (1) examine the nature of polymer composite; (2) investigate the use of polymer composites as optical waveguides. Solutions of polystyrene and diacetylene were thoroughly mixed and spread on a glass slide to obtain a ~ 3 μm film. The film was exposed to UV light and the formation of polydiacetylene was detected by the appearance of its characteristic vibrational bands in integrated optics laser Raman scattering. The phonon spectral region of the polymer composite is compared with that of the pure polystyrene and the polydiacetylene bulk in Fig. 46.3. The polymer composite does not contain the sharp phonon feature of polydiacetylene bulk which suggests that the polydiacetylene is not in the form of crystallites in the polystyrene

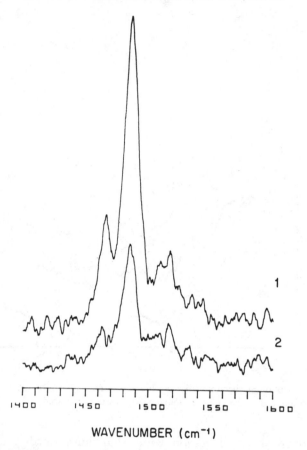

Figure 46.2 Polarized Raman spectra of a 100 Å thick p-toluene sulfonate diacetylene polymer film at room temperature in the internal vibration region 1400–1600 cm^{-1}. Spectrum 1 was obtained by vertically polarized laser light and spectrum 2 by horizontally polarized laser light.

matrix but more homogeneously mixed at the molecular level to form an amorphous polymer composite.

We have also successfully studied a submicrometer thick polymer composite of polystyrene and polydiacetylene by depositing it on a glass of ~ 100-μm thickness which acts as a waveguide.

To test the feasibility of studying a monolayer polymer film by laser Raman scattering, we investigated the diacetylene-containing carboxylic group (second type of diacetylene monomer discussed above) which readily forms a monolayer by the Langmuir–Blodgett technique. This film was deposited on the top of an optical waveguide of PMMA polymer and then polymerized. The Raman spectra reveal a different packing arrangement of the polymer than that in the bulk form.

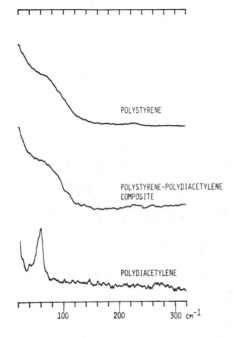

Figure 46.3 The phonon spectral region of the (a) polystyrene; (b) polystyrene and p-toluene sulfonate polydiacetylene composite; and (c) the p-toluene sulfonate polydiacetylene bulk at room temperature are compared.

ACKNOWLEDGMENT

This research was supported by the Air Force Office of Scientific Research. I am grateful to Dr. Donald R. Ulrich for his encouragement and support.

REFERENCES

1. W. J. Allbury and A. R. Hillman, *Ann. Rept R. Soc. Chem.* London, 377 (1981).
2. D. J. Williams, ed., *Nonlinear Optical Properties of Organic and Polymeric Materials*, American Chemical Society, Washington, D.C., 1983.
3. R. R. McCaffrey, P. N. Prasad, M. Fornalik, and R. Baier, *J. Polym. Sci. Polym. Phys. Ed.* (in press).
4. R. Burzynski and P. N. Prasad, accepted for publication in the *J. Polym. Sci. Polym. Phys. Ed.*
5. P. N. Prasad, J. Swiatkiewicz, and G. Eisenhardt, *App. Spectrosc. Rev.* **18**, 59 (1982).
6. J. D. Swalen , M. Tacke, R. Santo, K. E. Rieckhoff, and J. Fischer, *Helv. Chim. Acta* **61**, 960 (1978).

47

CONFORMATIONAL ANALYSIS OF SOME POLYSILANES, AND THE PRECIPITATION OF REINFORCING SILICA INTO ELASTOMERIC POLY(DIMETHYLSILOXANE) NETWORKS

J. E. MARK
Department of Chemistry
University of Cincinnati
Cincinnati, Ohio

INTRODUCTION

The polysilanes [—SiRR'—] are a new class of semi-inorganic polymers with fascinating properties and considerable promise in a variety of applications. For example, some members of this series can be cast into transparent films, spun into fibers, and converted into silicon carbide at high temperatures.[1,2] They can also be used as photoinitiators,[3] resists in UV lithography,[2,4] p-type semiconductors when properly doped,[2] and as reinforcing media in ceramics when converted *in situ* into β-SiC fibers.[5]

Relatively little is known about the conformational characteristics of the polysilanes from either an experimental or theoretical point of view, although some work is in progress.[6,7] For this reason, conformational energies were calculated for two of the simpler polysilanes. Information thus obtained can

434

be used to predict the regular conformations in which the polymers should crystallize,[8] and the equilibrium flexibility of the chains in the undiluted amorphous state and in solution.[8]

The structurally related polysiloxanes [—SiRR'O—] have long been known and extensively studied.[8] The dimethyl polymer has been of particular interest because of its unusual flexibility. This property is exploited, for example, in the use of the crosslinked polymer as an elastomeric material[9,10] in low-temperature applications. These elastomers, however, unlike their competitors such as natural rubber and butyl rubber, cannot undergo strain-induced crystallization.[11,12] They are therefore inherently weak and require reinforcement with a high surface area filler in practically all applications.[9] Blending such fillers into (highly viscous) polymers prior to crosslinking can be very difficult and filler agglomeration is almost impossible to avoid. For these reasons it could be highly advantageous to generate such filler *in situ*, for example, by the hydrolysis of silicates sufficiently nonpolar to dissolve in typical elastomers such as poly-(dimethylsiloxane) (PDMS). Such techniques have now been developed, and the results obtained should transcend the area of elastomer reinforcement, giving information useful as well in the area of sol–gel–ceramics technology.[13]

CONFORMATIONAL ANALYSIS OF SOME POLYSILANES

Computational Details

The first polymer of interest was polysilane (PSL) itself, [—SiH$_2$—], and the specific sequence investigated is shown in Fig. 47.1.[7] The length l of the Si—Si skeletal bonds is 0.234 nm, which is considerably larger than the 0.153-nm length of the C—C bonds in the hydrocarbon analogue, polyethylene (PE) [—CH$_2$CH$_2$—].[8] This should reduce repulsive interactions in polysilanes, but could be partially offset by the increased length of the Si—H bond relative to the C—H bond (0.148 versus 0.110 nm). Skeletal bond angles in PSL are approximately tetrahedral, as they are in PE.[8] Rotational states are *trans* (T), *gauche*

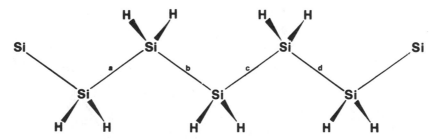

Figure 47.1 Sketch of the polysilane chain.[7] Conformational energies are calculated as a function of the rotation angles about skeletal bonds *b* and *c*.

positive (G^+), and *gauche* negative (G^-), and are expected to occur in the vicinity of the symmetric locations specified by the rotational angles $\phi = 0°$, $120°$, and $-120°$, respectively. The second polymer was poly(dimethylsilylene) (PDMSL) [$-Si(CH_3)_2-$], shown in Fig. 47.2[7] The model for it was similar to that for PSL, but with the rotational angles of the methyl side groups representing additional variables.

Distances between all pairs of atoms were calculated in the usual manner,[7,8] as a function of the skeletal and side-chain rotational angles. Conformational energies were then calculated from these distances using empirical potential energy functions, and a torsional contribution corresponding to a barrier height of 0.4 kcal/mole (which is considerably smaller than that for PE, 2.8 kcal/mole).[8] Entire contour maps of the energy against rotational angles were obtained, and then used to calculate configurational partition functions and to average the energies and rotational angles about the minima. In this way, configurational statistical weights were refined to include a preexponential or entropy factor. The statistical weights were then used in a matrix multiplication scheme to calculate values of the *characteristic ratio* $\langle r^2 \rangle_0 / nl^2$, where $\langle r^2 \rangle_0$ is the chain dimension as unperturbed by excluded volume effects[14] and n is the number of skeletal bonds. This ratio is much used as an inverse measure of equilibrium chain flexibility.[8]

Results for Polysilane

Polysilane was found to show a preference for pairs of *gauche* states of the same sign ($G^\pm G^\pm$) over the corresponding *trans* states (TT) by *ca.* 0.5 kcal/mole, in contrast to the analogous *n*-alkanes which prefer TT over $G^\pm G^\pm$ by *ca.* 1.0 kcal/mole.[8] Even $G^\pm G^\mp$ states, commonly found to be prohibitively repulsive for most polymers, were preferred over the TT states by 0.4 kcal/mole.[7] The predicted crystalline state conformation could thus be described as helical, of a pitch similar to that shown by polyoxymethylene [$-CH_2O-$]. It is thus quite different from the PE preferred form, which is the planar, all-*trans*, zig-zag

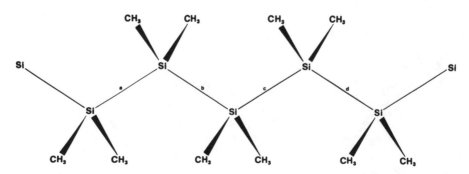

Figure 47.2   Sketch of the poly(dimethylsilylene) chain.[7]

conformation. The same conclusion was reached in an earlier theoretical study of this chain[6] which focused exclusively on discrete minima. As can be seen from Fig. 47.3, nearly all regions of configurational space were within 2 kcal/mole of the minima, indicating considerable chain flexibility. This was confirmed by the unusually low value, 4.0, calculated for the characteristic ratio. The value for PE is approximately 7.5.

Results for Poly(Dimethylsilylene)

Previous calculations[6] on this polymer indicated the $G^{\pm}G^{\pm}$ conformation again preferred. The present results,[7] however, indicate $G^{\pm}G^{\pm}$ and $TG^{\pm}$ (or $G^{\pm}T$) conformations to have essentially the same energies (0.08 versus 0.00 kcal/mole). If the energy difference is indeed this small, the conformation actually adopted by the chain upon crystallization would probably be determined by differences in chain packing energies.

Location of $G^{\pm}$ states at angles that minimize the energy would place them at $\pm 95°$. This revision in the direction of the T state, and the diminished number

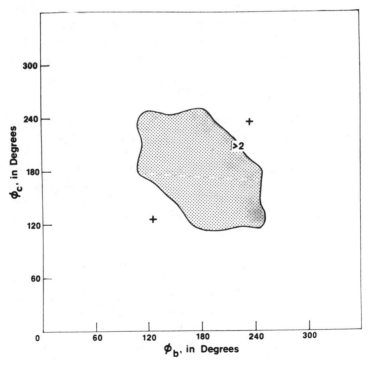

Figure 47.3 Conformational energy map for the polysilane segment giving the energy, in kcal/mole relative to the conformational energy minima designated by "+" on the map. The energy contours are shown as a function of the rotational angles ϕ_b and ϕ_c.[7] The shaded region corresponds to energies greater than 2 kcal/mole above the minima.

of compact $G^{\pm}$ states should increase the characteristic ratio to the vicinity of 15. This would make the PDMSL chain considerably less flexible than both PSL and PE.

PRECIPITATION OF REINFORCING SILICA INTO ELASTOMERIC NETWORKS

Some Experimental Details

Silica may be prepared by the hydrolysis

$$Si(OC_2H_5)_4 + 2H_2O \rightarrow SiO_2 + 4C_2H_5OH \tag{1}$$

of tetraethylorthosilicate (TEOS), in the presence of any of a variety of catalysts. There are three techniques by which silica thus precipitated can be used to reinforce an elastomeric material. First, an already cured network, for example, prepared from PDMS, may be swollen in TEOS and the TEOS hydrolyzed *in situ*.[15-20] Alternatively, hydroxyl-terminated PDMS may be mixed with TEOS, which then serves simultaneously to tetrafunctionally end link the PDMS into a network structure and to act as a source of SiO_2 upon hydrolysis.[21-23] Finally, TEOS mixed with vinyl-terminated PDMS can be hydrolyzed to give a SiO_2-filled polymer capable of subsequent end linking by means of a multifunctional silane.[24]

Precipitation Rates

The rates of the precipitation reaction were studied through plots of weight percent filler against time. Typical results for the $C_2H_5NH_2$-catalyzed system within an already cross linked PDMS elastomer are shown in Fig. 47.4.[19] Although the rates increase with catalyst concentration,[20] as expected, they are seen to vary in a complex manner. One complication is the deswelling of the network due to migration of TEOS and the by-product ethanol to the surrounding aqueous solution. The loss of TEOS should be smaller in the case of the more dilute $C_2H_5NH_2$ solution (since it is more hydrophilic), and this would explain the relatively simple monotonic form of the corresponding precipitation curves. In the case of the more concentrated $C_2H_5NH_2$ solutions the curves level off, because of the TEOS migration, and then turn downward, presumably because of loss of colloidal silica. At constant time, less filler is precipitated in the case of the networks having the larger value of the molecular weight between crosslinks, and this is probably due to larger losses of TEOS and silica from the larger "pores" in these networks in the highly swollen state.[19]

Mechanical Properties of the Filled Elastomers

The elastomeric properties of primary interest here are the nominal stress $f^* \equiv f/A^*$ (where f is the equilibrium elastic force and A^* the undeformed

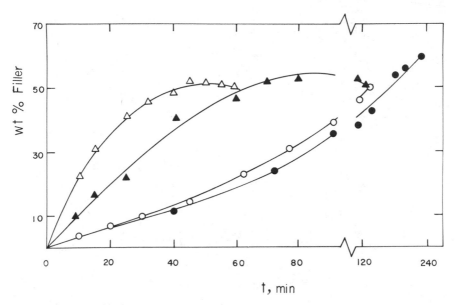

Figure 47.4   Weight percent filler precipitated as a function of time.[19] The circles locate the results for 2.0 wt% ethylamine and the triangles the results for 25.0 wt%; the open symbols are for networks having a molecular weight between crosslinks of 8.0×10^3 g/mole, and the filled symbols 21.3×10^3 g/mole.

cross-sectional area), and the reduced stress or modulus[25] $[f^*] \equiv f^*/(\alpha - \alpha^{-2})$ (where $\alpha = L/L_i$ is the elongation or relative length of the sample).

Typical stress–strain isotherms obtained on the *in situ* filled PDMS networks are given in Fig. 47.5. The data show[25] the dependence of the reduced stress on reciprocal elongation. The presence and efficacy of the filler are demonstrated by the large increases in modulus, with marked upturns at the higher elongations. Figure 47.6 shows the data of Fig. 47.5 plotted in such a way that the area under each stress–strain curve corresponds to the energy E_r of rupture, which is the standard measure of elastomer toughness. Increase in % filler decreases the maximum extensibility α_r but increases the ultimate strength f^*_r. The latter effect predominates and E_r increases accordingly. In some cases, extremely large levels of reinforcement are obtained. Such networks behave nearly as thermosets, with some brittleness (small α_r), but with extraordinarily large values of the modulus $[f^*]$.[19]

Characterization of the Filler Particles

Transmission electron microscopy,[16] and light scattering and neutron scattering measurements[26] are being used to study the filler particles. As illustration, an electron micrograph for a PDMS elastomer in which TEOS has been hydrolyzed is shown in Fig. 47.7.[16] The existence of filler particles in the network, originally hypothesized on the basis of mechanical properties,[15] is clearly confirmed. The particles have average diameters of approximately 250 Å, which is in the range

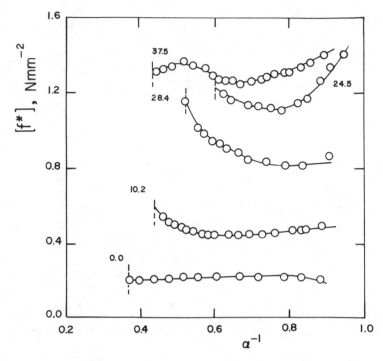

Figure 47.5   The reduced stress as a function of reciprocal elongation, at 25°C, for PDMS networks filled in 2.0 wt% $C_2H_5NH_2$ solution.[19] Each curve is labeled with the wt% filler in the network including results for the unfilled elastomer.

of particle sizes of fillers typically introduced into polymers in the usual blending techniques. The distribution of sizes is relatively narrow, with most values of the diameter falling in the range 200–300 Å.[16]

Strikingly, particles aggregation invariably present in the usual types of filled elastomers is absent. These materials should be useful in characterizing effects of aggregation, and could be of practical importance as well.[16]

Other Novel Filling Techniques

In typical filled systems, anisotropy of mechanical properties can arise only if the filler particles or their agglomerates are asymmetric, since they are then oriented as a result of the flow of the un-crosslinked mix during processing operations. In fact, fibrous fillers are often used for the express purpose of introducing mechanical anisotropy. Recent studies, however, show that even when the particles are spherical, if they are magnetic and the filled elastomer is cured in a magnetic field, then highly anisotropic thermal[27] and mechanical[28] properties can be obtained.

The filler used in one study[28] was an extremely fine commercial magnetic powder (MG-410 Magnaglo) in which the particles are very nearly spherical,

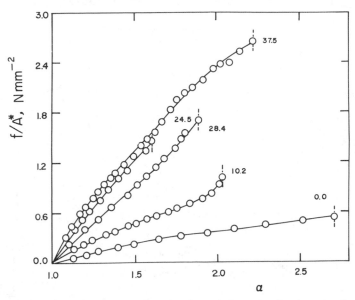

Figure 47.6 Reduced stress as a function of reciprocal elongation for the networks filled in the 50.0 wt% $C_2H_5NH_2$ solution.

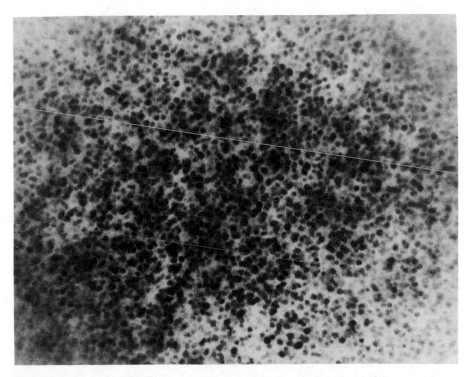

Figure 47.7 Transmission electron micrograph (118,800X) for *in-situ* filled **PDMS** network containing 34.4 wt% filler.[16] The average particle diameter is 250 Å.

441

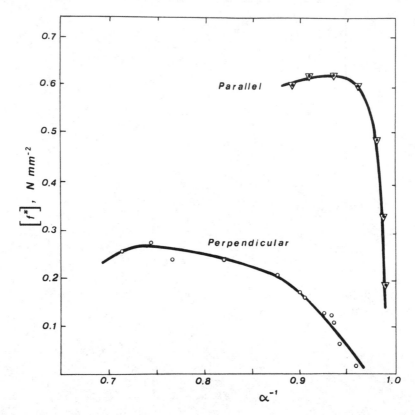

Figure 47.8 Reduced stress shown as a function of reciprocal elongation for magnetic-particle filled PDMS strips cut parallel and perpendicular to the magnetic field imposed during the curing procedure.[28]

with an average diameter of approximately 10 μm. The particles and benzoyl peroxide were mixed into high molecular weight PDMS, and the mixture was cured in a magnetic field provided by a 580-gauss permanent magnet. (The final product contained about 31.5% by weight of magnetic filler, which corresponds to roughly 6% by volume.) The cured sheet was cut into rectangular strips parallel to and perpendicular to the vector of the magnetic field.

The stress–strain isotherms obtained are shown in Fig. 47.8.[28] They are seen to be very different from those usually obtained, which almost invariably have a constant, positive slope in the region of low to moderate elongation.[25] The isotherms also show the highly anisotropic nature of the reinforcement obtained in the presence of the magnetic field. For example, the maximum value of [f^*] for the strip cut parallel to the field exceeds the corresponding value for the perpendicular strip by a factor of nearly 3.

Another novel technique is the generation of magnetic filler particles by the *in situ* thermal or photochemical decomposition[29,30] of carbonyl compounds of iron, nickel, or cobalt, with elastomer curing being carried out in a magnetic field.

ACKNOWLEDGMENTS

It is a pleasure to acknowledge the financial support provided by the Air Force Office of Scientific Research through Grant AFOSR 83-0027, and by the National Science Foundation through Grant DMR 84-15082.

REFERENCES

1. J. P. Wesson and T. C. Williams, *J. Polym. Sci., Polym. Chem. Ed.* **18**, 959 (1980); **19**, 65 (1981).

2. R. West, L. D. David, P. I. Djurovich, K. L. Stearley, K. S. V. Srinivasan, and H. Yu, *J. Am. Chem. Soc.* **103**, 7352 (1981); P. Trefonas III, P. I. Djurovich, X.-H. Zhang, R. West, R. D. Miller, and D. Hofer, *J. Polym. Sci. Polym. Lett. Ed.* **21**, 819 (1983); P. Trefonas III, R. West, R. D. Miller, and D. Hofer, *J. Polym. Sci. Polym. Lett. Ed.* **21**, 823 (1983).

3. A. R. Wolfe, R. West, and D. G. Peterson, Eighteenth Organosilicon Symposium, April 1984.

4. D. C. Hofer, R. D. Miller, and C. G. Willson, *Proc. SPIE* **469**, 16 (1984).

5. R. West, L. D. David, P. I. Djurovich, H. Yu, and R. Sinclair, *Ceram. Bull.* **62**, 899 (1983).

6. J. R. Damewood, Jr. and R. West, *Macromolecules* **18**, 159 (1985).

7. W. J. Welsh, L. C. DeBolt, J. E. Mark, J. R. Damewood, Jr., and R. West, manuscript in preparation.

8. P. J. Flory, *Statistical Mechanics of Chain Molecules*, Wiley-Interscience, New York, 1969.

9. E. L. Warrick, O. R. Pierce, K. E. Polmanteer, and J. C. Saam, *Rubber Chem. Technol.* **52**, 437 (1979).

10. J. E. Mark, *Adv. Polym. Sci.* **44**, 1 (1982).

11. J. E. Mark, *Polym. Eng. Sci.* **19**, 254 (1979).

12. J. E. Mark, *Polym. Eng. Sci.* **19**, 409 (1979).

13. L. L. Hench and D. R. Ulrich, Eds., *Ultrastructure Processing of Ceramics, Glasses, and Composites*, J. Wiley, New York, 1984.

14. P. J. Flory, *Principles of Polymer Chemistry*, Cornell University Press, Ithaca, NY, 1953.

15. J. E. Mark and S.-J. Pan, *Makromol. Chemie, Rapid Comm.* **3**, 681 (1982).

16. Y.-P. Ning, M.-Y. Tang, C.-Y. Jiang, J. E. Mark, and W. C. Roth, *J. Appl. Polym. Sci.* **29**, 3209 (1984).

17. C.-Y. Jiang and J. E. Mark, *Makromol. Chemie* **185**, 2609 (1984).

18. C.-Y. Jiang and J. E. Mark, *Colloid Polym. Sci.* **262**, 758 (1984).

19. Y.-P. Ning and J. E. Mark, *Polym. Eng. Sci.* **25**, 000 (1985).

20. J. E. Mark and Y.-P. Ning, *Polym. Bulletin* **12**, 413 (1984).

21. M.-Y. Tang and J. E. Mark, *Polym. Eng. Sci.* **25**, 29 (1985).

22. J. E. Mark, C.-Y. Jiang, and M.-Y. Tang, *Macromolecules* **17**, 2613 (1984).

23. M.-Y. Tang, A. Letton, and J. E. Mark, *Coll. Polym. Sci.* **262**, 990 (1984).

24. Y.-P. Ning and J. E. Mark, *J. Appl. Polym. Sci.* **30**, 3519 (1985).

25. J. E. Mark, *Rubber Chem. Technol.* **48**, 495 (1975).

26. D. W. Schaefer, Y.-P. Ning, and J. E. Mark, unpublished results.

27. G. J. L. Griffin, in *Advances in Polymer Friction and Wear*, Vol. 5B, L.-H. Lee, Ed., Plenum Press, New York, 1972.

28. Z. Rigbi and J. E. Mark, *J. Polym. Sci., Polym. Phys. Ed.* **23**, 1267 (1985).

29. T. W. Smith and D. Wychick, *J. Phys. Chem.* **84**, 1621 (1980).

30. S. Reich and E. P. Goldberg, *J. Polym. Sci., Polym. Phys. Ed.* **21**, 869 (1983).

48

ORDERED POLYMERS AND MOLECULAR COMPOSITES

W. WADE ADAMS AND THADDEUS E. HELMINIAK
Polymer Branch, Materials Laboratory
Air Force Wright Aeronautical Laboratories
Wright-Patterson Air Force Base, Ohio

INTRODUCTION

Current efforts in ultrastructure processing of ceramic materials to enhance physical properties consider the organic and physical chemistry of the starting materials in conjunction with processing science to tailor final forms having desired properties. An analogous situation exists in the area of high-performance polymeric materials. By innovative synthetic polymer chemistry approaches, the U.S. Air Force has developed a new class of synthetic macromolecules which possess unprecedented physical properties, but require special processing in order to control their morphology. These polymers, dubbed "Ordered Polymers," have been the subject of an extensive research program sponsored by the Air Force Wright Aeronautical Laboratories, Materials Laboratory, and the Air Force Office of Scientific Research. The goal of the program is to attain from organic polymers mechanical properties and environmental resistance similar to materials now used for fiber-reinforced composites, but without the use of fiber reinforcement.

This research has evolved polymeric materials in the form of fiber, uniaxial, and biaxial film, and "molecular composites." This chapter reviews the brief history of the program, identifying the key concepts that prompted the polymer chemists to develop rigid rod polymers. Since the polymer chains must be organized into an ordered morphology to maximize properties, emphasis will be placed on the morphology of these systems. Current and future work using rigid rod molecules to reinforce conventional polymers at or near molecular levels are also described.

LADDER POLYMERS

The Nonmetallic Materials Division of the Materials Laboratory has long been interested in aromatic–heterocyclic polymers as potential structural materials because of their high thermooxidative stability and environmental resistance. Following work on the polybenzimidazoles in the 1950s, two ladder-like polymers were synthesized in the 1960s: BBB,[1] a semiladder containing a flexible joint, and BBL,[2] a true ladder polymer containing at least two parallel chemical bonds at all points along the chain. Both easily formed films by casting

Benzimidazo Benzophenanthroline (BBB)

Benzimidazoisoquinoline (BBL)

from acid solutions, but BBL exhibited an unusual precipitation/consolidation phenomenon. Solid suspensions of the polymer, obtained from acid reprecipitation with subsequent washing, were collected upon a fritted glass filter. After drying, the polymer was removed in the form of a tough, durable film having an intense golden luster giving the appearance of a metal foil. Films prepared by this technique possessed tensile strengths of approximately 10 ksi. Thus, instead of obtaining the polymer in the form of a disorganized solid or powder, a smooth film was obtained which compared favorably with the 16 ksi tensile strengths for films prepared by casting from methanesulfonic acid. Obtaining any measurable strength from precipitated films was very surprising since the precipitated film was formed by the coalescence of *discrete* particles of *solid* matter without passing through a melt phase.

  Further research exploring the relationship between polymer chain structures and this film-forming tendency of certain aromatic heterocyclic polymers showed that the molecular geometry of the polymer backbone is a critical factor in determining film forming tendencies.[3] Even though the intermolecular inter-action due to the high degree of π-bonding was present in all the polymers considered, not all formed precipitated films. It was concluded that if geometric configurations create kinks in the polymer chain backbone, the tendency to form precipitated films can be reduced or eliminated. Furthermore, it was proposed that the high strength of these unique precipitated films stemmed from their "composite" character.[4,5] They were considered "composites" because the precipitated films were formed by the aggregation and coalescence of individual microscopic sheets of precipitated polymer. The high strength was attributed to the inherent strength of microscopic sheets due to their high degree of molecular order.

RIGID ROD POLYMERS

Prior to this work on ladder polymer films, approaches to "neat" composites had been under consideration in the late 1960s to eliminate problems associated with adhesion and thermal expansion in fiber-reinforced composites. "Neat" composites refers to a composite where the reinforcing fiber material and resin matrix material are of similar molecular composition, differing principally in morphology. Research began with attempts to improve molecular orientation and packing by chain extension of flexible polymers into rodlike conformations but eventually led to the synthesis of rigid rod extended chain polymers.

This work was predicated on the assumption that if a polymer molecule existed in a rigid rod, extended chain form it could, like stiff uncooked spaghetti in its box, be coaxed into extraordinary alignment which would provide a high-strength, high-modulus, anisotropic structural material. This approach hinged on being able to obtain and process a *para*-configured PBI instead of the *meta*-configured molecular structure that had been previously used. This approach was followed using a different synthetic route and three *p*-ordered polymers subsequently were synthesized consisting of benzimidazole (PDIAB)[6] benz-oxazole (PBO),[4,7,8] and benzothiazole (PBT)[9,10] heterocyclic units.

Poly (p-phenylenebenzodimidazole) (PDIAB)

Poly (p-phenylenebenzobisoxazole) (PBO)

Poly (p-phenylenebenzobisthiazole) (PBT)

The earliest examination of these *para* polymers showed the tendency to precipitate from dilute solution in a sheet form previously observed *only* for ladder polymers. The fact that a *nonladder* polymer possessed these unusual properties prompted reconsideration of the precipitated film formation phenomena. These observations led to the hypothesis that a nonreinforced composite with useful mechanical properties might be achieved with appropriate polymer chain geometry and suitable processing. The key factor for success would be the ability to achieve a high degree of molecular order.

Shortly thereafter, disclosures from DuPont described the attainment of high-modulus fibers from anisotropic dopes of aromatic polyamides.[11] These results were unique for fibers but with respect to our goals for "neat" composites, the DuPont work reinforces the decision to pursue the more desirable, but more difficult, rigid rod approach. Further, like the all-*para* aromatic amides, the tendencies of our rigid rod polymers to form liquid crystalline strong acid solutions suggested the possibility of highly anisotropic fiber and ribbon formation. The Materials Laboratory with the support of AFOSR undertook the task of exploring the potential of this "ordered polymer" approach to structural materials for aerospace applications. The initial efforts of this program have been principally concerned with the choice of the macromolecular chemical structure while concurrent studies, theoretical and experimental, have been considering the problems of characterization, evaluation, and processing into high-strength fibers, ribbons and "molecular composites."

Although these polymeric materials have been successfully synthesized, they have presented special processing problems because of the extended chain, rigid rod structural character of the molecules. Present processing requires strong mineral or organic acid solvents such as methanesulfonic acid or polyphosphoric acid. This problem of solubility has manifested itself in several ways during the course of the research: the limited number and type of solvents, relatively low solubility, and difficulty in polymerization to high molecular weight due to apparent insolubility. This problem has been addressed with three approaches. The first was an extensive solubility study to discover noncorrosive organic solvents or to determine organic chemical structures that could be expected to dissolve these polymers. The results did not provide any single organic or mixed solvent.[12] The second approach consisted of polymer chain structural modifications either in the form of pendant additions[7,10] or the introduction of swivel joints into the main chain forming an articulated molecular structure.[13] This approach has also not yet succeeded, although work on pendant additions is still underway.

The third approach resulted in a major breakthrough in the polymerization process. Polymerization at sufficiently high rod–polymer concentration in the solvent, polyphosphoric acid, could be carried out in a liquid crystalline phase, which resulted in a significantly higher ultimate degree of polymerization, or molecular weight.[14] This technique also worked for stiff chain (but not rigid rod) polymers of ABPBO and ABPBT, resulting in increases in the molecular weight and dope concentrations, thereby improving processibility for fiber or film formation.[14]

Poly(2,5-benzimidazole) (ABPBI)

Poly(2,5-benzoxazole) (ABPBO)

Poly(2,5-benzothiazole) (ABPBT)

MOLECULAR COMPOSITES

Concurrent to these efforts to process the neat rigid rod polymers into fiber and film, the rigid rod molecules were used as reinforcement in flexible aromatic–heterocyclic polymer and thermoplastic polymer matrices to provide composites at the molecular level which are analogous to chopped fiber composites.[15] The object was to obtain reinforcement by dispersion of the rigid rod polymer as individual molecules. This molecular composite concept having its origins in the earliest "neat composite" approach had been deferred in favor of the film and fiber studies of the rigid rod polymer. Initially investigations of cast films from dilute solution-blended rods and coils revealed that these blends exhibited increases in tensile strength and modulus, simply due to the reinforcement effect of an undesirable domain structure. Solution studies of the rod–coil blends determined the existence of a phase transition due to variation in component concentrations.

Subsequently Flory's theory for ternary mixtures of rod and coil molecules in solution[16] provided a rationale for solution processing at concentrations below a critical concentration to avoid phase separation. In the isotropic phase, the rod and coil polymer molecules are not phase separated, and this molecular level mixing is retained in the bulk form when the solvent is removed quickly enough. Kinetics is used to prevent the inevitable thermodynamic phase separation, resulting in molecular or near molecular dispersion of rod molecules in a coil matrix. When fibers were formed by this molecular composite process, the modulus obeyed a volume fraction rule of mixtures, indicating that the reinforcement effect of the rigid rod molecule was as efficient as that of a conventional fiber-reinforced composite.[17]

RIGID ROD POLYMER CHARACTERIZATION

The ultimate modulus of a polymeric material is determined both by the continuity of chemical bonds along the backbone of the molecule (chain rigidity) and by the arrangement of the chains on a supermolecular scale (ordered molecular packing). Chain rigidity is achieved in nature by intramolecular hydrogen bonds or by helical conformations, and in the synthetic aromatic polyamides by semiflexible bonds. For the all-*para* aromatic heterocyclic rigid rod polymers, chain rigidity is assured by the absence of bond rotational freedom except axially. If stiff molecules can be ordered laterally, and if the lateral interactions between structural units are sufficiently strong, then high strength can also be achieved, both in tension and in compression.

The motivation for study of the nature of rigid rod molecules is clear. The ability to influence the order by chemical structure modification or by processing is critical to successful utilization of these materials. Only by understanding the molecular and supermolecular structure can one hope to achieve the full performance potential of this new class of polymers.

Extensive characterization of PBT and PBO in solution has been performed, establishing that the polymers exhibit rodlike behavior, with inherent viscosity up to about 47 dL/g in methanesulfonic acid (molecular weight of about 45,000, degree of polymerization about 170). The molecular weight distribution may be somewhat narrower than the most probable distribution,[18] while the Mark Houwink exponent is 1.8, as expected for a rod. The persistence length in solution has been measured to be longer than the contour length of the molecule. The molecule is protonated in solution, and is a rigid rod.

The aromatic heterocyclic rigid rod polymers have excellent thermal and oxidation stability. For example, PBO suffers a weight loss of less than 10% after 200-hr exposure to air at 370°C. None of these materials exhibit a melting point or a glass transition temperature before they degrade (above 600°C in nitrogen).[19] When spun into fibers from anisotropic solutions, the rigid rods have remarkable mechanical properties. Table 48.1 lists typical heat-treated fiber mechanical properties for PBT, PBO, and ABPBO

By extruding the anisotropic dope in the form of a ribbon, uniaxial films of PBT have been formed with mechanical properties close to those of fiber. By film-blowing methods, biaxial (more properly, nonuniaxial) films have been

TABLE 48.1 Mechanical Properties of High-Performance Fibers

Material	Modulus (Msi)	Tensile Strength (ksi)	Elongation (%)
PBT	50	600	1
PBO	60	600	1
ABPBO	30	500	2

produced with excellent lateral strength.[22] These material forms offer the composite designer the potential to fabricate a composite part with higher reinforcement loading and planar isotropic properties using simpler processing methods.

RIGID ROD POLYMER MORPHOLOGY

The morphology of the rigid rod polymers has been studied primarily at the AF Materials Laboratory and at the University of Massachusetts, and the material most studied has been as-spun and heat-treated fiber. Structure–property correlation studies have related processing conditions, morphology, and mechanical properties.[21,23,24] High tensile modulus is achieved by high spin–draw ratio processing followed by heat treatment under tension at high temperatures. Tensile strength is dependent upon flaws, and it increases significantly with heat treatment. Compressive strength of PBT is similar to that of the aromatic polyamides, and reflects the tendency of the fibers to internally buckle, which is manifested by kink-band formation.[25]

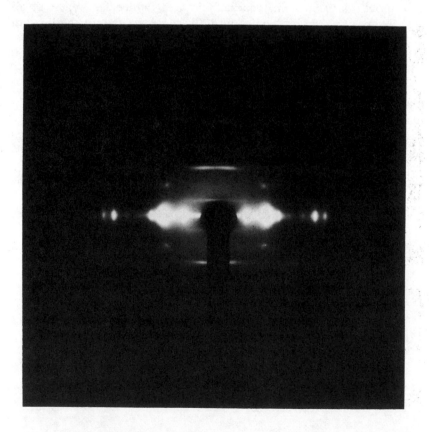

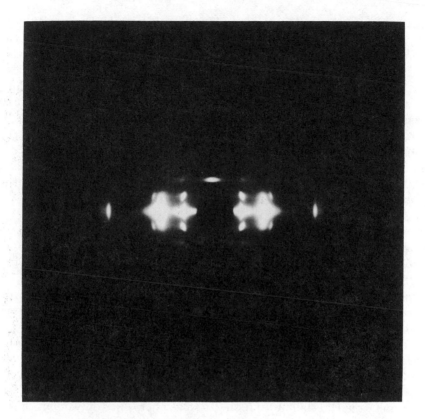

Figure 48.1 WAXD patterns for PBO (left) and ABPBO (above), chain axis vertical.

The morphology of rigid rod polymers reflects the anisotropy of the molecule. Oriented material forms are fibrillar in nature, whether they are fibers, films, or thicker sheets. The existence of an elemental fibril has not been confirmed, but microtomy of as-spun fibers by special techniques reveals a network of microfibers of about 10-nm width.[26] These fibrils then pack into dense oriented arrays to form micrometer-sized fibers. Upon breaking of a fiber, bundles of fibrils are found, which are about 0.5 μm in diameter, similar to those found in poly(paraphenylene terephthalamide), PPTA.[27] Unlike PPTA, however, no banding or pleated texture has been observed for PBT, PBO, or PDIAB. The skin-core phenomenon reported for PPTA[27] has also been observed for PBT,[28] indicating inhomogeneity in packing order of the fibrils across the cross section. The intense equatorial small-angle X-ray scattering (SAXS) observed for PBT[23] arises from elongated voids between fibrils or fibrillar bundles, probably in the core of the fiber.

The molecular chains are highly ordered laterally in the fibrils, as shown by wide-angle X-ray diffraction (WAXD). No amorphous halo is seen and the suggested single-phase structure for PBT consists of lateral 2-D order with axial translateral disorder.[29] Only a small fraction of regions of 3-D order in PBT have been observed by electron microscopic lattice imaging.[30] Defects due to chain ends or chemical imperfections (but not conformational defects), must exist along the microfibrils, and disrupt the high degree of order, resulting in lower modulus and strength than theoretically possible. Small-angle X-ray scattering reveals a periodic defect structure for other high modulus systems,[31] and a very weak periodic meridional SAXS pattern has been observed for PBT.[32] The chain extension through any defect zones, however, is extensive enabling effective load distribution throughout the fibrils which results in extraordinary levels of mechanical properties. True 3-D crystallinity is not required for such properties, but the extremely high orientation is very important.

PBO fibers exhibit the same high lateral order, but also have more axial chain registry, or 3-D crystallinity than PBT. (See Fig. 48.1). The SAXS pattern is also considerably more intense. Apparently the increased crystalline order more than compensates for the increased number or severity of defects along the fibrils. ABPBO, an extended chain but not a rigid rod, exhibits clear 3-D order[33] and the most intense periodic SAXS pattern of all.[21,32] In this case the introduction of a conformation angle into the chain lowers the chain modulus, but increases the ability of the molecules to crystallize, resulting in good mechanical properties. The increased defect structure in the laboratory spun ABPBO fibers offers hope for improvements in properties with better processing methods.

All the morphology studies to date have focused on the final fiber state, yet it is during the coagulation process that the rod molecules undergo a transformation from a liquid crystalline solution to a solid, after which only minor changes in morphology occur. The morphological nature of the initial acid solution changes in the solution during various processing methods, and the state of the polymer in the coagulation and postspinning process are all areas of current investigation.

MOLECULAR COMPOSITE MORPHOLOGY

When rigid rod polymer molecules are dispersed at the molecular level in a flexible polymer matrix, the morphology study becomes more difficult. The morphology of the two individual components (initially PBT and ABPBI) in fiber form is more or less known. Studies of case films[15] revealed clear phase separation in the form of rod-rich oblate ellipsoids (1–5 μm in size), formed as the two components partially phase separated upon evaporation of the acid solvent. By rapid coagulation of a solution with total polymer concentration below the critical concentration,[16] phase separation was prevented, and no large-scale heterogeneity was observed. Extensive morphological studies by TEM showed no crystalline phase separation larger than 3 nm, while WAXD indicated, by breadth and intensity of the prominent 0.59-nm reflection characteristic of

PBT, that the rods were in very small (or highly disordered) bundles.[34] Preliminary solid-state NMR studies have also shown the absence of significant phase separation, although more work will be required to quantify the local molecular environment of a rod molecule.[35] Regardless of the degree of molecular dispersity of the rod molecules in the host matrix, the anisotropy of the rod molecule provides an aspect ratio for reinforcement that greatly improves mechanical properties of the matrix.[17] Current work in this area centers on thermoplastic matrices which offer processing options not currently available with aromatic–hetrocyclic matrix polymers.

REFERENCES

1. R. L. Van Deusen, *J. Polym. Sci.* **B4**, 211 (1966).

2. F. E. Arnold and R. L. Van Deusen, *J. Appl. Polym. Sci.* **15**, 2035 (1971).

3. A. J. Sicree, F. E. Arnold, and R. L. Van Deusen, *J. Polym. Sci.* **12**, 265 (1974).

4. T. E. Helminiak, F. E. Arnold, and C. L. Benner, *Polym. Prepr. Am. Chem. Soc., Div. Polym. Chem.* **16**(2), 659 (1975).

5. T. E. Helminiak et al., U.S. Patent 4,051,108, Sept. 27, 1977.

6. R. F. Kovar and F. E. Arnold, *J. Polym. Sci., Polym. Chem. Ed.* **14**, 2807 (1976).

7. J. F. Wolfe and F. E. Arnold, *Macromolecules* **14**, 909 (1981).

8. E. W. Choe and S. N. Kim, *Macromolecules* **14**, 920 (1981).

9. J. F. Wolfe, B. H. Loo, and F. E. Arnold, *Polym. Prepr. Am. Chem. Soc., Div. Polym. Chem.* **19**(2), 1 (1978).

10. J. F. Wolfe, B. H. Loo, and F. E. Arnold, *Macromolecules* **14**, 915 (1981).

11. S. L. Kwolek, US Patent 3,600,350, Aug. 17, 1971.

12. D. C. Bonner, AFML-TR-77-73, 1977.

13. R. C. Evers, F. E. Arnold, and T. E. Helminiak, *Macromolecules* **14**, 925 (1981).

14. J. F. Wolfe and P. D. Sybert, AFWAL-TR-82-4191, 1983.

15. G. Husman, T. Helminiak, M. Wellman, W. Adams, D. Wiff, and C. Benner, AFWAL-TR-80-4034, 1980.

16. P. J. Flory, *Macromolecules* **11**, 1138 (1978).

17. W-F. Hwang, D. R. Wiff, C. L. Benner, and T. E. Helminiak, *J. Macromol. Sci.* (*Phys.*) **B22**, 231 (1983).

18. D. B. Cotts and G. C. Berry, *Macromolecules* **14**, 930 (1981).

19. E. Soloski et al., AFWAL-TR-84-4019 (1984).

20. W-F. Hwang, private communication.

21. W-F. Hwang et al., AFWAL-TR-85-XXX, in press (1985).

22. E. Chenevey, AFWAL-TR-85-XXX, in press (1985).

23. J. R. Minter, AFWAL-TR-82-4097 (1982).

24. S. R. Allen, AFWAL-TR-83-4065 (1983).

25. S. J. DeTeresa, AFWAL-TR-84-XXX (1984).

26. Y. Cohen and E. L. Thomas, *Polym. Engr. Sci.*, submitted Nov. 1984.

27. M. Panar, P. Avakian, R. C. Blume, K. H. Gardner, T. D. Gierke, and H. H. Yang, *J. Polym. Sci., Polym. Phys. Ed.* **21**, 1955 (1983).

28. L. Drzal, private communication.

29. E. J. Roche, T. Tokahashi, and E. L. Thomas, *Am. Chem. Soc. Symp. Fiber Diffraction Methods* **141**, 303 (1980).

30. K. Shimamura, J. R. Minter, and E. L. Thomas, *J. Mater. Sci. (Lett.)* **2**, 54 (1983).

31. L. I. Slutsker, L. E. Utevskii, Z. Yu. Chereiskii, and K. E. Perepelkin, *J. Polym. Sci., Polym. Symp.* **58**, 339 (1977).

32. S. J. Bai, private communication.

33. A. V. Fratini, E. M. Cross, J. F. O'Brien, and W. W. Adams, *J. Macromol. Sci (Phys.)*, in press (1985).

34. S. J. Krause, J. F. O'Brien, and W. W. Adams, AFWAL-TR-85-XXX, in Press (1985).

35. D. L. VanderHart, AFWAL-TR-85-XXX, in press (1985).

49

MORPHOLOGY AND PROPERTIES OF PARTICLE/POLYMER SUSPENSIONS

R. F. STEWART AND D. SUTTON
ICI Corporate Colloid Science Group
The Heath
Runcorn, Cheshire, England

INTRODUCTION

The ability to control the structure and properties of particle/polymer suspensions is of great importance owing to the widespread occurrence of these systems in areas such as ceramic processing. The behavior of such suspensions is poorly understood relative to the state of knowledge for electrolyte-coagulated dispersions, at least partly a consequence of the complexity of the phenomena observed. Using well-characterized colloidal dispersions, the effects of varying polymer type and concentration on suspension structure and properties have been investigated. The polymers have been classified as follows:

1. Medium molecular weight polymer of opposite charge to particles.
2. High molecular weight polymer of same charge as particles or uncharged.
3. High molecular weight polymer of opposite charge to particles.

MATERIALS AND METHODS

The principal dispersions used in the work were surfactant-free, (anionic) charge-stabilized polystyrene lattices prepared following the methods of Goodwin et al.[1] Monodisperse suspensions with diameters ranging from 0.2–4.0 μm were produced by appropriate modification of reaction conditions.

The polymers used included a cationic polyelectrolyte, polyethyleneimine (PEI) of medium molecular weight (MW ~ 50,000) and a variety of high MW (> 10⁶), polyacrylamide-based flocculants.

The polymers used included a cationic polyelectrolyte, polyethyleneimine (PEI) of medium molecular weight (MW $\sim 50{,}000$) and a variety of high MW ($> 10^6$), polyacrylamide-based flocculants.

Optical microscopy was used to determine the structure in dilute systems. In concentrated samples morphology was characterised by freeze-etch microscopy.[2] The technique involves sample crash freezing, fracture, etching by sublimation at reduced temperature and pressure, surface replication with C and Pt, and examination of the replica by TEM.

Mechanical property measurements made on the suspensions included continuous shear rheological behavior (Deer PDR 81 rheometer or Haake RV3 rotoviscometer), shear modulus (Rank Brothers Limited Pulse-Shearometer), and centrifugation characteristics. A compressive modulus was determined from the centrifuge data using the methods of Buscall.[3]

ELECTROLYTE-COAGULATED SYSTEMS

The properties of electrolyte-coagulated materials provide a baseline from which the properties of particle/polymer suspensions may be assessed. It has been

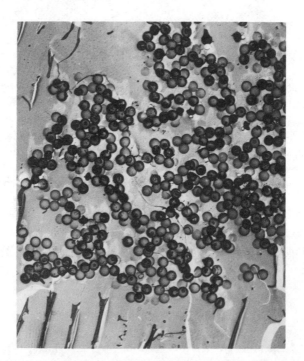

Figure 49.1 Freeze fracture of salt-coagulated 0.5-μm latex. Note the high porosity of the aggregates.

reported[4,5] that the addition of salt to a charge-stabilized colloidal suspension results in the formation of highly porous flocs such as those shown in Fig. 49.1. At some specific solids loading these aggregates begin to interact strongly to form a space-filling network. This loading is observed to be a function of particle size. As the size decreases there is a convergence to a limiting gel point of 8% v/v solids, in harmony with the simulation work of Vold.[6] Above the gel point properties such as shear moduli and continuous shear viscosities rise rapidly with further increase in solids loading as low coordination linkages in the network are eliminated.

PARTICLE/POWER SUSPENSIONS

Medium Molecular Weight Polymer of Opposite Charge to Particles

The suspension structures observed in systems containing medium molecular weight cationic polymer (PEI) are only slightly different to those seen on electrolyte coagulation of the same material. We believe that this polymer induces aggregation by a "charge-patch" mechanism.[7] The particles, which are observed to be in contact, are nearly randomly aggregated into small floccules which strongly engage as the concentration rises. Analysis reveals that the individual flocs tend to be more dense than in the salt-coagulated system, reflecting the differing flocculation mechanism. Not surprisingly, suspension properties differ only in degree rather than in kind for the two aggregation mechanisms. Though the use of PEI slightly increases network strength compared to salt-coagulated material, solids content and particle size remain the dominant influences on the material properties.

The continuous shear viscosity is observed to be more influenced by the flocculation mechanism, being greater in the PEI-flocculated system than in the salt-coagulated material. We believe this is due to long-range, attractive Coulombic interactions between the charged particles. This hypothesis is corroborated by the observation that the viscosity decreases upon the addition of salt, presumably due to the screening of the Coulombic interactions by the additional ions.

High Molecular Weight Nonionic or Anionic Polymers

The addition of high molecular weight nonionic or anionic polymers to our latices results in the formation of aggregates in which the particles are connected by polymer linkages (Fig. 49.2). The properties of the material reflect this morphology: the flocs settle rapidly to high solids contents ($\sim$ 50% v/v); network moduli are negligible except as the particles approach close packing; viscosities, though higher than in the absence of polymer, are much lower than those measured in systems in which strong, open networks are formed. The properties of

Figure 49.2 Micrograph showing structure of 0.5-μm latex flocculated by high molecular weight anionic polymer.

these systems, however, do exhibit sensitivity to flocculant charge or the presence of electrolyte. With no added salt the particles remain widely spaced as in Fig. 49.2. The addition of small amounts of electrolyte results in the formation of more compact structures (cf, the observations of Hachisu[8]) as shown in Fig. 49.3. Low levels of addition of salt with the polymer increase the flocculation rate. As the concentration of ions is increased the suspension properties revert to those of the salt-coagulated material.

High Molecular Weight Polymers of Opposite Charge to the Particles

Systems to which high molecular weight polymer of opposite charge to the particles has been added exhibit more complex behavior than the other systems studied. With these agents the flocs frequently have a "stringy" appearance and the mechanical properties of the system show exceptional sensitivity to shear conditions during flocculation. Freeze-etch examination of the diluted flocculant solution, and of flocculated suspensions, reveals why this strong dependence on flocculation history occurs: when the flocculant solution is diluted, even in the absence of particles, dispersion of the gel-like concentrated solution is not immediate. During the normal procedure of adding polymer, in which the solution is added to the particulate dispersion, coating of these dropletlike species occurs (Fig. 49.4). Subsequent distortion and rupture of these particle/polymer

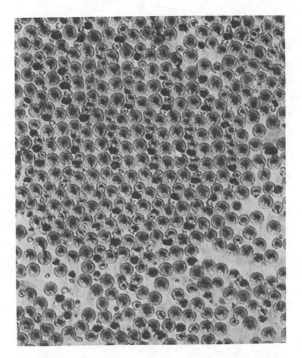

Figure 49.3 Dense aggregates produced by addition of high molecular weight nonionic polymer or high molecular weight anionic polymer with low levels of electrolyte.

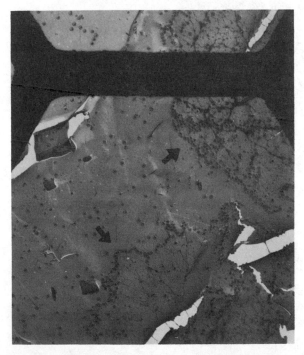

Figure 49.4 Anionic latex aggregated by addition of high molecular weight cationic polymer. Structures resulting from coating of droplets of polymer solution with particles are arrowed.

solution entities, under the prevailing shear conditions, determines the final floc form. The mechanism of particle–particle binding is thought to be a combination of double-layer compression, "charge-patch" attraction and bridging, the two former effects being dominant. Despite the more complicated aggregation mechanism the modulus behavior is not very different to that observed for electrolyte-coagulated material.

CONCLUSION

Efficient handling of colloidal particle/polymer dispersions requires careful control of suspension properties. From consideration of the effects of changes in basic variables on suspension structure, using techniques such as freeze-etch microscopy, unifying principles concerning material characteristics can be determined. This should enable more reliable operation of processes involving suspensions of this sort.

REFERENCES

1. J. W. Goodwin, J. Hearn, C. C. Ho, and R. H. Ottewill, *Colloid Polym. Sci.* **252**, 464 (1974).
2. P. F. Luckham, B. Vincent, J. McMahon, and Th. F. Tadros, *Colloids Surf.* **6**, 83 (1983).
3. R. Buscall, *Colloids Surf.* **5**, 269 (1982).
4. R. F. Stewart and D. Sutton, Control of Structure in Particulate Solids Suspensions, in *Solid-Liquid Separation*, J. Gregory, Ed., Ellis Harwood, Chichester, 1984, pp. 111–128.
5. R. F. Stewart and D. Sutton, *Chemistry and Industry*, 373 (1984).
6. M. J. Vold, J. Colloid Sci. **14**, 168 (1959).
7. J. Gregory, *J. Colloid Interface Sci.* **42**, 448 (1973).
8. A. Kose and S. Hachisu, *J. Colloid Interface Sci.* **55**, 487 (1976).

PART 5

Micromorphology

50

COLLOID SCIENCE OF
COMPOSITE SYSTEMS

EGON MATIJEVIĆ

Department of Chemistry and Institute
of Colloid and Surface Science,
Clarkson University, Potsdam, NY 13676

INTRODUCTION

Until recently, most studies of colloidal systems have dealt with one kind of particles, preferably of simple composition. At first the essential requirement was to produce reasonably stable dispersions in order to assess their longevity as a function of various experimental parameters (sol concentration, solvent composition, nature of additives, surface charge, etc.). The next stage consisted of analogous investigations using suspended uniform colloids, preferably spherical in shape. The latter condition is desirable as most theoretical models apply to spherical geometry. For the same reason polymer latexes have been the subject of extensive investigations. In recent years a variety of inorganic colloids have been prepared with well-defined particles of different shapes, including spheres.[1] This development made it possible to extend the knowledge of interfacial and stability properties of materials of different chemical and surface characteristics that are of interest in numerous applications, as corrosion, catalysis, flotation, and so on.

In contrast to single systems, in nature and in a great many applications one deals with composite materials, either with respect to chemical composition or in terms of particle heterogeneity. Such mixtures are considerably more difficult to treat, both experimentally and theoretically. Because of their importance, complex colloids are receiving increasing attention. In the experimental part of this chapter the preparation of some composite, yet uniform particles is described. In addition, it is shown that mixed dispersions of different kinds can be obtained if one metal oxide is precipitated in the presence of another pre-formed colloidal material. In the theoretical part electrostatic interactions of unlike particles, with consequences related to the stability of multicomponent dispersions, are discussed.

All these aspects of colloids are of essence to the science of ceramics proces-sing, particularly if we are to develop a better understanding of the formation and characteristics of high-performance materials.

PREPARATION OF COMPOSITE SYSTEMS

It is well known that properties of materials may change drastically if another component is incorporated, to yield particles of mixed composition. The amount of the additive can vary over a broad range and so can the nature of the composite particles. In some cases the latter solids have a stoichiometric com-position while in some other cases the admixed matter is either adsorbed (as ions, molecules, or coatings) or included in the bulk in arbitrary ratios. Magnetic, optical, catalytic, and surface charge characteristics, as well as density, hardness, solubility, particle size, and shape, and so on, can be affected by introducing a second (or additional components) into a given solid.

In this chapter composite systems will be illustrated consisting either of particles of mixed composition or of dispersions of different kinds of materials.

Composite Particles

Adsorption

Adsorption of a small amount of Co^{2+} on magnetite or maghemite particles increases significantly the coercive force of such recording materials. The same ion in the form of the radioactive isotope (Co^{60}), when attached to corrosion products (namely, iron oxides), causes a major contamination problem in water-cooled nuclear reactors. Doping of various solids with metal ions can greatly affect their catalytic activities. Thus, adsorbed ruthenium sensitizes titania in catalytic decomposition of water. These are but a few examples of property changes when particle surfaces are modified by adsorption of often analytically insignificant quantities of another component.

Figure 50.1 shows that the adsorption density of Co^{2+} on colloidal hematite particles is strongly pH dependent.[2] This effect is caused by the change in sur-face charge of the adsorbent as the proton concentration in solution is varied. It should be noted that conditions in these experiments preclude precipitation of cobalt hydroxide.

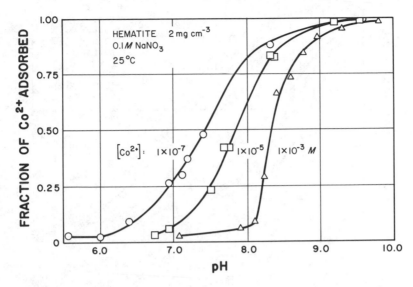

Figure 50.1   Fraction of Co^{2+} adsorbed on spherical hematite particles (125 $\pm$ 10 nm in diameter) in a dispersion containing 2 mg α-Fe$_2$O$_3$/cm^3 as a function of pH at three different initial concentrations of Co(NO$_3$)$_2$ in the presence of 0.1 M NaNO$_3$.[2]

Several mechanisms can account for the uptake of ionic species on metal oxide surfaces, including ion exchange, surface complexation or condensation, ion pairing, and so on. For example, in the case of Co^{2+} adsorption on magnetite, the ion exchange process seems to best explain the experimental data.[3]

Metal ions can adsorb on any other kind of material, such as on polymer latexes, or on a variety of inorganic compounds different from metal oxides. It has been amply demonstrated that the uptake of these cations is dramatically enhanced once they become hydrolyzed.[4,5] A thermodynamic model was offered to account for these effects.[6]

Solid surfaces can be even more substantially altered by adsorption of charged organic species, particularly if the latter form complexes with constituent ions of the substrate. Figure 50.2 gives the electrokinetic mobilities of finely dispersed uniform spherical hematite particles as a function of pH in the absence and in the presence of ethylenediamine tetraacetate ion (EDTA). The addition of this complexing anion in micromole quantity already causes a shift in the isoelectric point (i.e.p.) from ~7.2 to 4.3.[7] Such a change in surface charge characteristics has a profound effect on various interactions with adsorbent particles.

Rather efficient adhesion of colloidal hematite on steel can be achieved in acidic media (pH 3–4) if the particles are pretreated with EDTA (Fig. 50.3). Very little uptake is observed under the same conditions if iron oxide is not reacted with this chelating agent.[8]

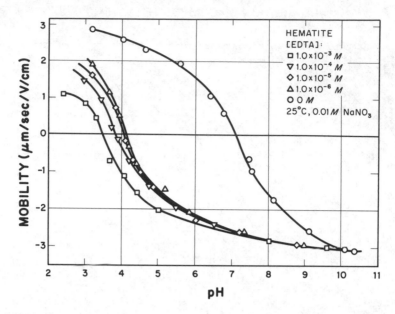

Figure 50.2 Electrophoretic mobilities of spherical α-Fe$_2$O$_3$ particles (100 nm in diameter) in the absence and in the presence of varying concentrations of EDTA as a function of pH at 25°C. Ionic strength: 0.01M.[7]

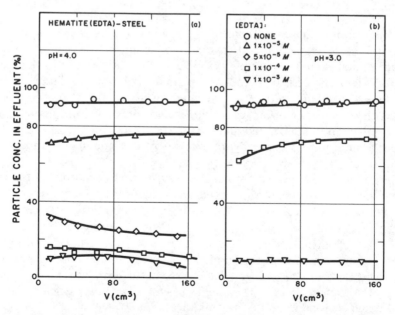

Figure 50.3 The percentage of the number of hematite particles (120 $\pm$ 10 nm in diameter) remaining in the effluent as a function of the volume of the sols pretreated with EDTA solutions of different concentrations passed through a bed of steel beads at pH 4 (*a*) and at pH 3 (*b*). Circles give data in the absence of EDTA. Concentration of the sol: $\sim$2 $\times$ 10^8 particles/cm^3; flow rate: $\sim$2 cm^3/min; steel used: 5 g.[8]

The examples offered above clearly illustrate the importance of the surface composition on the properties and interactions of colloidal particles. They also show that the state of the surface can be readily modified by addition of very small amounts of properly chosen adsorbate species. Since the properties of composite materials may be controlled by appropriate conditioning of powder components, the consequences for ceramics are obvious.

Particles of Mixed Composition

In "doping" a solid by adsorption of ions or molecules only a small amount of a new component is attached to the surface of the particles. In many instances it is desirable to have the bulk of the solids consist of mixed constituents. While such materials are commonly found in nature and have been frequently synthesized, procedures for preparation of uniform colloidal particles of mixed composition is a recent development. In this chapter ferrites, metal sulfides, and polymer/inorganic particles will be described as examples of well-defined colloids of mixed chemical composition.

Ferrites

Ferrites are known minerals, as well as products of various precipitation processes (such as in corrosion), which can be considered as adducts of magnetite, Fe_3O_4 obtained by substitution of either Fe(II) or Fe(III) constituent ions by other metals [Co(II), Ni(II), Cr(III), etc.]. Because of their magnetic, catalytic, and other properties, such spinels represent a family of metal oxides of considerable practical interest.

A procedure has been developed which makes it possible to generate magnetite and other ferrites as spherical colloidal particles of narrow size distribution and of varying modal diameters. In principle, a ferrous hydroxide gel is first precipitated and then partially oxidized by a mild oxidizing agent (e.g., KNO_3) in weakly acidic media and aged at elevated temperatures ($\sim 100°C$) for several hours. Under certain experimental conditions the resulting magnetite particles are spherical, the average size of which can be altered by the excess of $FeSO_4$.[9] Co-, Ni-, or Co-Ni-ferrites are obtained either by coprecipitating the corresponding metal hydroxides with ferrous hydroxide and proceeding as described above, or by exchanging a certain amount of ferrous ions for other divalent metals before the oxidation and aging processes take place. The exchange is rapid and it is accomplished by contacting the gel with solutions of Co(II) or Ni(II) or both salts.[10-12]

Figure 50.4 illustrates two such ferrites. The left SEM (a) shows Co-Ni-ferrite particles crystallized from a gel containing a mixture of $Fe(OH)_2$, Co-$(OH)_2$, and $Ni(OH)_2$ in the presence of KNO_3, under conditions given in the legend.[12] It is apparent that rather uniform spheroidal particles are generated by the described technique. The composition of such particles can be varied by using different ratios of metal hydroxides in the original gel.

Mixed ferrites can also be prepared by ion exchange of iron in $Fe(OH)_2$ gel for other ions prior to aging.[10,11] Such a method was used in the synthesis of Ni– or Co–ferrites resulting in spherical colloidal particles which resembled closely those illustrated in Fig. 50.4a and will not be reproduced here. Figure 50.5 shows, that in the case if Ni^{2+}, the exchange for Fe^{2+} in the gel is indeed a rather fast process which, on aging, results in the incorporation of nickel into the solid spinel.

Electrophoretic mobilities of several ferrites, consisting of particles with different contents in nickel, as a function of dispersion pH are given in Fig. 50.6. The same curve can be drawn through all points, yielding an i.e.p. of 6.7.[10]

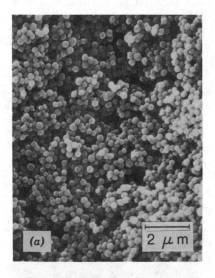

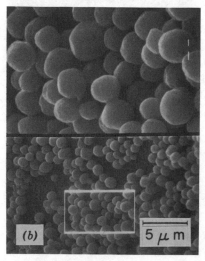

Figure 50.4 (a) Scanning electron micrograph (SEM) of cobalt–nickel ferrite particles crystallized from a gel of mixed hydroxides with the ratios of $([Co^{2+}] + [Ni^{2+}])/[Fe^{2+}] = 0.20$, and $[Ni^{2+}]/[Co^{2+}] = 3.7$, with $[Fe(OH)_2] = 0.025$ mol/dm³ and $[KNO_3] = 0.20M$. The concentration of OH^- was equivalent to the total concentration of divalent metal ions. The gel was aged for 4 hr at 90°C.[12] (b) SEM of chromium ferrite particles obtained from a gel initially containing $[Cr^{3+}]/[Fe^{3+}] = 0.25$ which was subsequently aged at 150°C for 16 hr. The solid particles were then reduced with hydrogen at $\sim 370°C$. In ferrite particles $[Cr]/[Fe] = 0.18$.

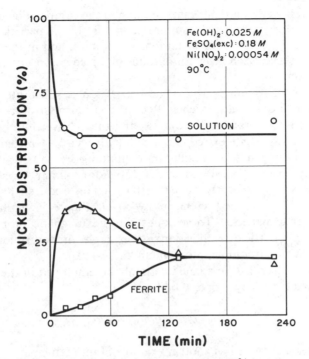

Figure 50.5 Distribution of nickel during the incorporation of Ni^{2+} ions into magnetite crystallizing from a ferrous hydroxide gel at 90°C. Initial composition: $Fe(OH)_2 = 0.025$ mol/cm³, $[Fe^{2+}] = 0.18M$, $[NO_3] = 0.2M$, and $[Ni^{2+}] = 5.4 \times 10^{-4}M$. Different cruves show the change in the content of nickel in solution in contact with the gel ($\bigcirc$), in the gel ($\triangle$), and in ferrite particles ($\square$).[10]

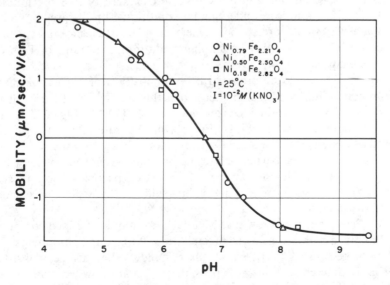

Figure 50.6 Electrophoretic mobilities of nickel ferrite particles of different chemical composition as a function of the dispersion pH.[10]

These results show that a significant variation in the composition of the studied nickel ferrites does not alter the surface properties of the particles; obviously, the content of the constituent metal ions in the bulk and in the surface layer must be at variance. Such behavior of solids of mixed composition seems to be rather common.

Preparation of ferrites in which the Fe(III) ion is substituted by another metal required a different approach. Recently, uniform Cr–ferrites have been obtained by first generating $Cr_xFe_{2-x}O_3$ and then converting the solid oxide to a corresponding ferrite by reduction with hydrogen.[13] A mixed gel was precipitated by adding base to a solution containing $Fe(NO_3)_3$ and $Cr(NO_3)_3$ and after repeated washing it was aged at 150°C for ~15 hr. Crystalline mixed oxide particles were separated and Fe(III) partially reduced by heating the solids at 340–400°C in the presence of hydrogen. Figure 50.4b illustrates such chromium ferrite particles. These solids showed magnetic properties which depended on the particle composition. For example, the saturation magnetization decreased as the content in chromium increased.

Techniques described here could certainly be employed in the preparation of other mixed oxides by proper modifications.

Mixed Metal Sulfides

Metal sulfides are offered as another example of inorganic particles consisting of mixed components. It was demonstrated that exceedingly uniform spherical particles of cadmium sulfide[14] or zinc sulfide[15] could be prepared by slow decomposition of mildly acidic solutions of thioacetamide (TAA) in the presence of dissolved respective metal nitrates. Prismatic crystals of PbS precipitated in lead nitrate–TAA solutions under similar conditions. In all cases a "seed" sol consisting of very small corresponding metal sulfide particles was first generated and then further grown either by admixing additional amounts of TAA or by raising the temperature.

Mixed sulfides could be prepared if decomposition of TAA took place in solutions containing two kinds of metal ions, that is, Zn^{2+} and Cd^{2+}, or Pb^{2+} and Cd^{2+}.[16] Transmission electron micrograph (TEM, Fig. 50.7a) shows ZnS·xCdS particles and the SEM (Fig. 50.7b) illustrates PbS·xCdS particles prepared as described in the legends. X-ray data indicate that the solids consist of mixed metal sulfides, particularly if the original systems are aged at higher temperatures (70°C). The morphology of PbS·xCdS particles also represents a combination of the geometries of individual sulfides, that is, of CdS (spherical) and PbS (prismatic).

Again, these examples show that well-defined colloids of mixed composition can be obtained by homogeneous precipitation of reacting components.

While in both cases (ferrites and mixed sulfides) the particles consisted of two or more different cations, it is possible to prepare powders with the same cation but different anions. Uniform cadmium and lead selenides were generated by decomposition of selenourea in solutions of the corresponding metal

Figure 50.7 (a) Transmission electron micrograph (TEM) of ZnS·xCdS particles prepared by aging zinc sulfide "seeds" and the original supernatant solution in the presence of 1×10^{-3} mol/ dm^3 cadmium ion at 70°C for 100 min. (b) SEM of PbS·xCdS particles obtained by the addition of 0.50 cm^3 of $4.3 \times 10^{-2} M$ Cd(NO$_3$)$_2$ and 1.00 cm^3 of $2.1 \times 10^{-2} M$ Pb(NO$_3$)$_2$ to 20 cm^3 of the PbS "seed" sol and further aging at 80°C for 30 min.

salts. Mixed sulfide–selenide systems should form in solutions containing both TAA and selenourea under properly selected conditions.

Mixed Organic Polymer/Inorganic Systems

A variety of polymer colloids (generally known as latexes) have been produced by different techniques among which emulsion and dispersion polymerizations are the most common. Considerable interest has developed in polymer particles that include inorganic components in order to achieve special magnetic, conductive, density, strength, and other properties. The usual preparation techniques do not lend themselves readily to the production of such organic/inorganic mixed systems.

Recently, polymer colloids have been obtained by an entirely different experimental approach. First aerosol droplets, uniform in size, of a monomer liquid are generated by evaporation/condensation and then these aerosols are brought into contact either with an initiator in the vapor phase, or with a fast reacting comonomer. In both cases polymerization takes place yielding solid spherical particles. The advantage of the aerosol technique is in the predictability of the colloid size (which depends on the size of the precursor droplets) and in the purity of the products (since no solvents or surfactants are required). The technique can also be employed to prepare polymer colloids that cannot be generated by the conventional procedures. Polystyrene particles[18] and copolymers of divinylbenzene/ethylvinylbenzene[19] have been obtained if trifluoromethanesulfonic acid vapor initiator is added to the aerosol consisting of respective monomer droplets. Polyurea powders were prepared by direct interaction of toluene diisocyanate (TDI) or hexamethylene diisocyanate (HDI) droplets with ethylenediamine (EDA) vapor.[20]

The aerosol technique seemed suitable, in principle, for achieving the objective of generating mixed organic/inorganic colloidal particles. The probability of success was substantiated by the fact that inorganic oxides (titania, alumina, and mixed titania/alumina) were obtained by hydrolysis of metal alkoxide droplets in the presence of water vapor.[21–23] Indeed, if polyurea particles in gaseous media are contacted with metal alkoxide vapors and subsequently reacted with water vapor, a certain amount of metal (hydrous) oxide is incorporated in the polymer beads.[20] Figure 50.8a shows polyurea particles obtained from HDI droplets that reacted with EDA vapor under conditions given in the legend. Figure 50.8b shows the same particles after they had been exposed to titanium(IV) isopropoxide and subsequently reacted with water vapor. Atomic absorption analysis proved that these particles contained the inorganic component ($\sim 3\%$ by weight of pure titanium metal). The metal oxides are distributed throughout the bulk of the polymer particles rather than as a surface layer. This was concluded from electrophoretic measurements which gave identical mobilities as a function of pH for dispersions of polyurea, polyurea/titania, and polyurea/alumina particles. In all cases the i.e.p. was at pH ~ 4.5; if alumina formed a coating, the i.e.p. would have been at pH ~ 9. The incorporation of the metal oxide is substantiated by the increase in particle size of the mixed particles relative to pure polyurea used as the precursor. Alkoxide

Figure 50.8 (a) SEM of polyurea particles prepared by reacting hexamethylene diisocyanate (HDI) droplets (obtained at a boiler temperature of 60°C at a helium flow rate of 1.6 dm³/min) with ethylenediamine (EDA) vapor carried by helium at a flow rate of 65 cm³/min. (b) SEM of polyurea/TiO₂ particles prepared by contacting the polymer aerosol shown in (a) with titanium (IV) isopropoxide vapor at 40°C carried by helium at a flow rate of 0.9 dm³/cm and subsequently exposing the system to water vapor at 20°C.

vapor diffuses into the polymer beads causing them to swell, after which the hydrolysis with water vapor takes place.

It is evident that a large number of such mixed organic/inorganic particles could be prepared by the aerosol technique.

Mixed Dispersions

Dispersions consisting of more than one kind of solids can be achieved in principle by[24]:

1. Homogeneous precipitation from solutions of mixed chemical composition.
2. Precipitation of solid(s) in a dispersion already containing one or more kinds of preformed particles.
3. Mixing dispersions of different particles.

In the experimental part of this chapter only phenomena observed in case (2) are discussed. Several different situations may be encountered when one kind of solid matter is precipitated in the presence of already existing particles: (1) the newly precipitated solid can exist in the form of independent stable particles; (2) the precipitate can adhere to the surface of preformed particles; (3) the precipitated matter can form heteroflocs with the existing particles; (4) the nucleation and crystal growth can take place on the surface of preformed particles; and (5) a film of the precipitated material can envelope the existing solids. Cases (2) and (5) could be classified as particles of mixed composition discussed in the previous section.

As illustration recent results are described, obtained by precipitation of iron oxide (hematite, α-Fe_2O_3) in aqueous dispersions containing uniform spherical titania particles. The latter were prepared by the aerosol technique[21] and had a modal diameter of 0.44 ± 0.09 μm. The titania powder was readily suspended in aqueous solutions containing varying amounts of $FeCl_3$ and HCl, which were then aged for desired periods of time (a few hours to a week) at 75 or 100°C. The nature of the resulting dispersions was sensitive to the conditions, specifically to the concentrations of $FeCl_3$, HCl, and to the amount of suspended titania.

Transmission electron micrographs in Fig. 50.9 illustrate two extreme cases. Figure 50.9(a) shows that independent hematite particles (the small ones) are precipitated in situ in the presence of preformed titania (the larger spheres), whereas in Fig. 50.9(b), the conditions described yield titania particles coated with hematite while no separately dispersed iron oxide could be detected. In the latter case the titania particle surface was first chemically modified by the addition of a very small amount of tungstophosphoric acid ($H_7PW_{12}O_{42}$). Heteropoly anions are known to interact strongly with hematite particles in acidic media.[25] It appears that hematite nucleates on the surface-modified titania and continues to grow, forming a film around the core particles.

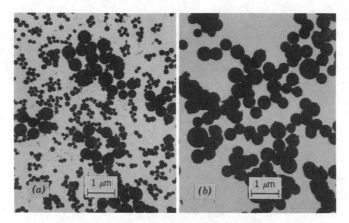

Figure 50.9   (a) Transmission electron micrograph (TEM) of a dispersion obtained by aging for 72 hr at 100°C a suspension containing 35 mg/dm^3 TiO$_2$ particles (larger spheres 0.44 μm in diameter) in the presence of 2 × 10^{-2} mol/dm^3 FeCl$_3$. (b) TEM of a dispersion obtained by aging for 24 hr at 100°C a suspension containing 35 mg/dm^3 TiO$_2$ particles in the presence of 2 × 10^{-4} mol/dm^3 FeCl$_3$, 5 × 10^{-3} HCl, and 1 × 10^{-5} mol/dm^3 tungstophosphoric acid (H$_7$PW$_{12}$O$_{42}$).

Figure 50.10 shows two cases of partial interactions of hematite, precipitated *in situ*, with the preformed titania particles. For comparison purposes electron micrographs of the iron oxide obtained under identical conditions in the absence of titania are also given. In Fig. 50.10a, the TEM illustrates an exceedingly fine hematite sol obtained in the absence of TiO$_2$, yet α-Fe$_2$O$_3$ particles turned out to be even smaller when aging was carried out with titania in the system (Fig. 50.10b). Furthermore, there is evidence that in the latter case some of the dispersed iron oxide adheres to TiO$_2$ particles. In Fig. 50.10c is illustrated the effect of a much higher concentration of the preformed particles. The micrograph (50.10c) shows that under conditions of this experiment rather large aggregates of α-Fe$_2$O$_3$ are generated, while in the presence of a considerable amount of TiO$_2$ (Fig. 50.10d) essentially the entire amount of precipitated hematite is associated with titania.

The examples clearly show the complexity of phenomena involved when precipitation of a solid occurs in the presence of preformed particles. However, a systematic control of the experimental conditions can result in composite materials with a variety of properties useful in different applications. It is of particular interest that, depending on conditions, either surface modifications of preformed particles may take place or entirely new mixed dispersions may be achieved.

In some cases the reasons for such diverse behavior can be inferred from the interaction energies of unlike particles as well as from the nucleation effects of preformed particles on the precipitation of the new phase.

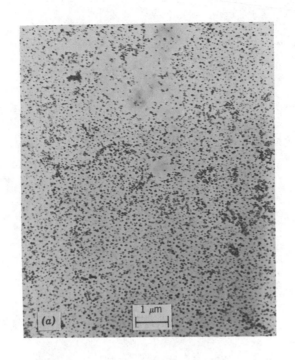

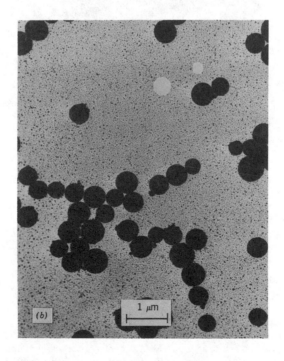

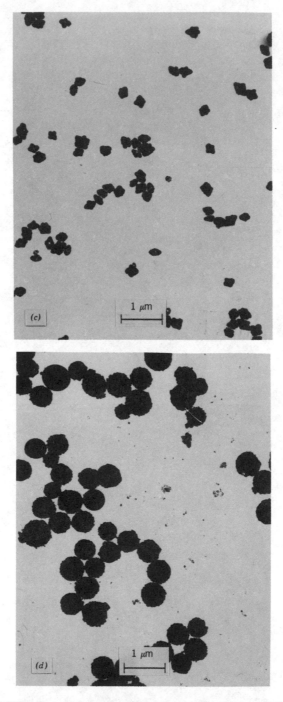

Figure 50.10 (*a*) TEM of hematite particles obtained by aging for 72 hr at 100°C a solution 2×10^{-4} mol/dm^3 in FeCl$_3$ and 1×10^{-4} mol/dm^3 in HCl. (*b*) The same system as (*a*) to which were added 35 mg/dm^3 TiO$_2$ particles before aging. (*c*) TEM of hematite particles obtained by aging for 72 hr at 100°C a solution 2×10^{-3} mol/dm^3 in FeCl$_3$ and 1.7×10^{-2} mol/dm^3 in HCl. (*d*) The same system as (*c*) to which were added 500 mg/dm^3 TiO$_2$ particles before aging.

THEORETICAL

The evaluation of the interaction energy between two charged surfaces is the classical problem of colloid science, which has to be resolved if the phenomena of dispersion stability, particle adhesion, and homo-, hetero, or selective coagulation are to be understood. The generally accepted approach is to add the repulsion energy, due to the electrical double layers and the attractive energy, resulting from London–van der Waals attractive forces, as developed by Derjaguin, Landau, Verwey, and Overbeek (DLVO theory).

While the original treatment is reasonably easily applied to interacting parallel plates or identical spheres of equal potential, the double layer theory for a system of unequal spheres with different magnitude of potential is considerably more complicated. Since an explicit solution of the Poisson–Boltzman (P.-B.) equation, which describes the potential of the double layer, is difficult to obtain, different approximations have been developed to cope with the problem. The most consistently applied approximate expression for the interaction energy of two unequal spheres (E_{HHF}) was derived by Hogg, Healy, and Fuerstenau[26] by generalizing Derjaguin's method for equal spheres.[27] Those authors[26] clearly stated the limitations due to the approximations, but these were too frequently disregarded by various investigators who applied their expression to experimental systems which did not conform with the model. For this reason considerable discrepancies between calculated and experimentally determined interaction energies have been repeatedly noted. Another reason for such disagreement seemed to be the method by which the symmetry of the system was treated.[26]

More recently an expression was derived, based on the nonlinear P.-B. equation, which is applicable to two parallel plates or to two spheres of any size and potential.[28,29] A comparison of calculated electrostatic energies using this approach (E_{BM}) with E_{HHF} has clearly delineated conditions of applicability of the HHF model and also situations when the HHF approximations fail.[30,31]

Probably the most intriguing feature of the P.-B. equation in its simplest form is the prediction that two surfaces of like sign but different magnitude of potential may *electrostatically* attract each other below a certain critical separation. This effect was discussed for the configuration of two parallel plates by Bierman[32] and Parsegian and Gingell.[33]

When applied to spheres, the consequences of this phenomenon are even more interesting. Owing to the curvature, portions of the surfaces may be attractive even though the rest of the surfaces may be repulsive. As a result the overall repulsion energy decreases or even changes its sign to net attraction. The magnitude of the effect at a given separation depends on the sizes of the spheres and on the difference in potentials of like sign. The quantitative calculation of the effect was made possible by the derivation of the P.-B. equation in its 2-D form.[28] The contribution of the electrostatic attraction to the interaction energy of unequal spheres is one of the major reasons for the deviation between E_{HHF} and E_{BM}. The lower calculated energies, using the latter expression, often bring the experimental and calculated results into better agreement.[34,35]

Several examples of computed electrostatic energies for unlike spheres are offered to illustrate the above phenomena and indicate the consequences. Figure 50.11 is a plot of electrostatic energies as a function of separation as calculated using the HHF and BM equations at constant ionic strength (expressed as inverse Debye distance κ). The difference in the size of the spheres is large enough for this case to correspond to a plate/sphere configuration which is, therefore, characteristic of the process of particle adhesion on a plane surface. It is quite obvious that E_{BM} changes from net repulsion to net attraction at a finite separation and that at all separations at which repulsion prevails $E_{HHF} > E_{BM}$.

Figure 50.12 shows the ratios E_{BM}/E_{HHF} for the system described in Fig. 50.11 as well as for a case of two spheres similar in size but of larger potential difference. Again the HHF overestimates the electrostatic interaction energy while the change in net attraction to repulsion takes place at an even larger separation.

The total interaction energies that include the van der Waals attraction for two unlike spheres of fixed sizes and potentials as a function of separation at four different concentrations of 1-1 electrolyte (expressed in terms of κ) are given in Fig. 50.13. The result reveals that in this system, at reasonably small distances, the repulsion barrier rises with increasing electrolyte concentration, which is never the case with identical spheres. In consequence, under such conditions, the likelihood of the formation of heteroflocs decreases and that of the homoflocs

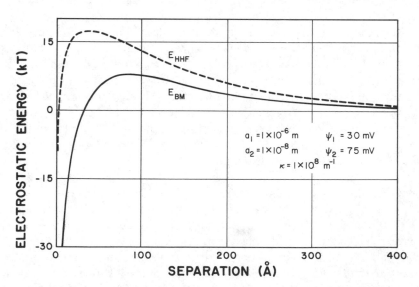

Figure 50.11 Electrostatic energy in kT units as a function of separation (in Å) for two spheres of radii $a_1 = 1 \times 10^{-6}$ m and $a_2 = 1 \times 10^{-8}$ m with surface potentials $\psi_1 = 30$ mV and $\psi_2 = 75$ mV for $\kappa = 1 \times 10^8$ m^{-1} calculated using models of Hogg, Healy, and Fuerstenau (E_{HHF}, Ref. 26) and Barouch and Matijević (E_{BM}, Ref. 29).

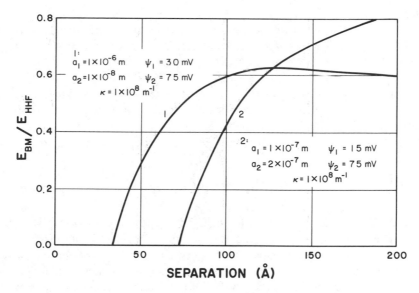

Figure 50.12 The ratio E_{BM}/E_{HHF} as a function of separation (in Å) for the system shown in Fig. 50.11 (curve 1) and for two spheres of radii $a_1 = 1 \times 10^{-7}$ m and $a_2 = 2 \times 10^{-7}$ m with surface potentials $\psi_1 = 15$ mV and $\psi_2 = 75$ mV, for $\kappa = 1 \times 10^8$ m^{-1} (curve 2).

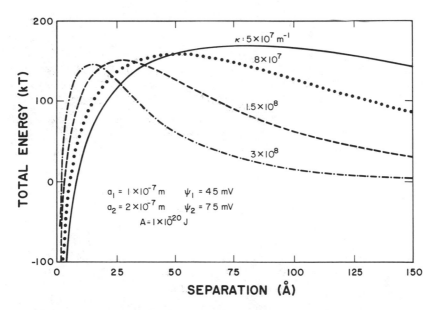

Figure 50.13 Total interaction energy as a function of separation (in Å) for two spheres of radii $a_1 = 1 \times 10^{-7}$ m and $a_2 = 2 \times 10^{-7}$ m with surface potentials $\psi_1 = 45$ mV and $\psi_2 = 75$ mV for $\kappa = 5 \times 10^7$ m^{-1} (——); 8×10^7 m^{-1} ($\cdots\cdots$); 1.5×10^8 m^{-1} (-----); and 3×10^8 m^{-1} (—·—·—). Hamaker constant used to calculate the London–van der Waals energy $A = 1 \times 10^{-20}$ J.[30]

479

increases as the ionic strength becomes larger. In practice such behavior, termed selective flocculation, is well known and is significant in the formation of composite materials. The example offered in Fig. 50.13 is not rare; indeed computations of other combinations of particle sizes and potentials have given the same trends.[30]

The recent refinement in the treatment of interactions of unlike particles has made it possible to better explain some experimental observations of mixed dispersions. The theory also allows for predictions of the behaviors of such complex systems which are of essence in ceramics.

CONCLUSION

In this chapter some insight into the experimental and theoretical problems related to mixed systems is offered. The latter may consist of composite particles or of dispersions of different materials. In either case considerable progress has been made in recent years, although the science of such mixed systems is still in its infancy, particularly if well-defined materials are required. From the theoretical point of view avenues are indicated for the treatment of the energy of two unlike particles, which now must be extended to multiple particle interactions.

ACKNOWLEDGMENTS

This work was supported by NSF grants CHE-8318196 and CPE-8111612. The author is indebted to his associates and students, whose names appear in the cited references, for their contributions to this program. The aerosol research is based on a cooperative project with Professor R. E. Partch and the theoretical studies result from a joint program with Professor E. Barouch, both of Clarkson University. The computations carried out by Mr. T. H. Wright (Clarkson University) are gratefully acknowledged.

REFERENCES

1. E. Matijević, *Acc. Chem. Res.* **14**, 22 (1981).

2. I. Kobal and E. Matijević, unpublished results.

3. H. Tamura, E. Matijević, and L. Meites, *J. Colloid Interface Sci.* **92**, 303 (1983).

4. E. Matijević, *J. Colloid Interface Sci.* **43**, 217 (1973).

5. E. Matijević and D. W. White, *Colloids Surf.* **9**, 355 (1984).

6. R. O. James and T. W. Healy, *J. Colloid Interface Sci.* **40**, 65 (1972).

7. H.-C. Chang, T. W. Healy, and E. Matijević, *J. Colloid Interface Sci.* **92**, 469 (1983).

8. C.-C. Lo, E. Matijević, and N. Kallay, *J. Phys. Chem.* **88**, 420 (1984).

9. T. Sugimoto and E. Matijević, *J. Colloid Interface Sci.* **74**, 227 (1980).

10. A. E. Regazzoni and E. Matijević, *Corrosion* **38**, 212 (1982).

11. H. Tamura and E. Matijević, *J. Colloid Interface Sci.* **90**, 100 (1982).

12. A. E. Regazzoni and E. Matijević, *Colloids Surf.* **6**, 189 (1983).

13. C. M. Simpson and E. Matijević, unpublished results.

14. E. Matijević and D. Murphy Wilhelmy, *J. Colloid Interface Sci.* **86**, 476 (1982).

15. D. Murphy Wilhelmy and E. Matijević, *J. Chem. Soc., Faraday Trans. I* **80**, 563 (1984).

16. D. Murphy Wilhelmy and E. Matijević, *Colloids Surf.*, (1985) in press.

17. J. Gobet and E. Matijević, *J. Colloid Interface Sci.* **100**, 555 (1984).

18. R. Partch, E. Matijević, A. W. Hodgson, and B. E. Aiken, *J. Polymer Sci., Polym. Chem. Ed.* **21**, 961 (1983).

19. K. Nakamura, R. E. Partch, and E. Matijević, *J. Colloid Interface Sci.* **99**, 118 (1984).

20. R. E. Partch, K. Nakamura, K. J. Wolfe, and E. Matijević, *J. Colloid Interface Sci.*, **105**, 560 (1985).

21. M. Visca and E. Matijević, *J. Colloid Interface Sci.* **68**, 308 (1979).

22. B. J. Ingebrethsen and E. Matijević, *J. Aerosol Sci.* **11**, 271 (1980).

23. B. J. Ingebrethsen, E. Matijević, and R. E. Partch, *J. Colloid Interface Sci.* **95**, 228 (1983).

24. E. Matijević, *Pure Appl. Chem.* **53**, 2167 (1981).

25. E. Matijević, R. J. Kuo, and H. Kolny, *J. Colloid Interface Sci.* **80**, 94 (1981).

26. R. Hogg, T. W. Healy, and D. W. Fuerstenau, *Trans. Faraday Soc.* **62**, 1638 (1966).

27. B. V. Derjaguin, *Disc. Faraday Soc.* **18**, 85 (1954); *Kolloid-Z.* **69**, 155 (1934); *Physicochim. Acta USSR* **10**, 333 (1939).

28. E. Barouch, E. Matijević, T. A. Ring, and J. M. Finland, *J. Colloid Interface Sci.* **67**, 1 (1978); **70**, 400 (1979).

29. E. Barouch and E. Matijević, *J. Chem. Soc., Faraday Trans. I*, **81**, 1797 (1985).

30. E. Barouch and E. Matijević, and T. H. Wright, *J. Chem. Soc., Faraday Trans. I*, **81**, 1819 (1985).

31. E. Barouch and E. Matijević, *J. Colloid Interface Sci.*, **105**, 552 (1985).

32. A. Bierman, *J. Colloid Interface Sci.* **10**, 231 (1955).

33. A. V. Parsegian and D. Gingell, *Biophys. J.* **12**, 1192 (1972).

34. R. J. Kuo and E. Matijević, *J. Colloid Interface Sci.* **78**, 407 (1980).

35. F. K. Hansen and E. Matijević, *J. Chem. Soc., Faraday Trans. I* **76**, 1240 (1980).

51

SYNTHESIS AND PROCESSING OF SUBMICROMETER CERAMIC POWDERS

R. H. HEISTAND II, Y. OGURI, H. OKAMURA,
W. C. MOFFATT, B. NOVICH, E. A. BARRINGER,
AND H. K. BOWEN
Ceramics Processing Research Laboratory
Materials Processing Center
Massachusetts Institute of Technology
Cambridge, Massachusetts

INTRODUCTION

The physical and chemical characteristics of ceramic powders are receiving new research emphasis because of their recognized relationship with green microstructure uniformity.[1] The size distribution of voids and particles in green microstructures, as well as particle size and coordination number, affect microstructure evolution during sintering, especially in systems without a liquid phase present during densification.[2-5] Progress has also been made in dispersion and surface sciences, and in the development of binders and our understanding of how these affect forming operations.[4-6]

Because of this research activity and the need for better ceramics, new laboratory and pilot plant processes for forming ceramic powders have been developed. Some involve precipitation from a gas or liquid phase, or reactions with solids. New opportunities now exist to study the interrelationships between powder characteristics and forming and sintering behavior.

This chapter presents results of synthesizing and processing four ceramic systems: ZnO, TiO_2, Al_2O_3–TiO_2, and Al_2O_3–ZrO_2. The synthesis routes involve different chemistries and demonstrate new procedures for making "ideal" ceramic powders. Control of the dispersion process and particle packing have yielded uniform green microstructures. The reported sintering studies indicate the very different microstructures that can be created through various sintering and heat-treating schemes.

SYNTHESIS OF NARROW SIZE DISTRIBUTION ZnO POWDER

Commercial synthesis of zinc oxide powder is principally by the direct oxidation of zinc vapor.[7] In varistor applications, a submicrometer powder with a narrower particle size distribution than is now commercially available could produce higher voltage devices with higher alpha values.[8] Attempts to extend the alkoxide hydrolysis process used in producing submicrometer, monosized SiO_2, TiO_2, and ZrO_2[9] to ZnO have encountered many obstacles; such alkoxide precursors as $Zn(OEt)_2$, $Na_2Zn(OMe)_4$, and $LiZn_2(OMe)_5$ have proved unsatisfactory.[10–11]

To meet the necessary hydrolysis conditions, a soluble, low molecular weight zinc compound is required. Preferably no other metal or halide species should be present that could lead to contamination. A literature search revealed that a series of alkylzinc alkoxide complexes meet these requirements and are readily synthesized from dialkylzinc and alcohols.[12–15] In the solid state and in moderately concentrated solutions, these compounds tetramerize to form cubic structures (Fig. 51.1),[14–15] the highest state of alkylzinc alkoxide complexation. The cubic precursor structure may encourage nucleation and growth of the hexagonal close-packed zinc oxide crystal lattice produced by the following synthesis.

In an inert atmosphere of nitrogen, 3.25 g of ethylzinc-t-butoxide[14] is dissolved in 30 mL of anhydrous toluene. The solution is filtered and admixed quickly with an anaerobic solution of 3 mL deionized water and 67 mL of anhydrous ethanol. Stirring is maintained for 4 hr. Zinc oxide is formed quantitatively, collected by centrifugation, redispersed ultrasonically in absolute ethanol, and then recollected by centrifugation. The last two steps are repeated once, followed by ultrasonic dispersion of the centrifugate in absolute ethanol. This method produced a dispersion of 0.17–0.20 μm spherical zinc oxide particles, as characterized by scanning and transmission electron microscopy (TEM), Stokes settling, and photon correlation spectroscopy.

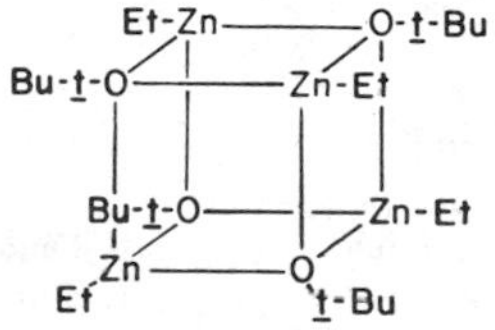

Figure 51.1   Crystal structure of ethylzinc-t-butoxide.[14]

Figures 51.2*a* and 51.2*b* show bright and dark field transmission electron micrographs, respectively, of ZnO powder particles formed this way. The spheroidal particle has a diameter of approximately 0.2 μm and consists of zincite crystallites on the order of 200 Å, which is confirmed by X-ray diffraction line broadening. Figure 51.2*c* shows a scanning electron micrograph (SEM) of a thin layer produced from the powder, which again demonstrates the packing advantage of a narrow size distribution powder. Photon correlation spectroscopy (PCS) reveals a particle size distribution less than 20% of the mean value, which ranges from 0.15 to 0.30 μm, depending on processing conditions. Stokes settling measurements corroborate PCS and electron microscopy.

By the proper selection of the organometallic precursor, therefore, polycrystalline, submicrometer, spheroidal, nonagglomerated, narrow size distribution zincite is formed by hydrolysis and can be processed into fine ultrastructures.

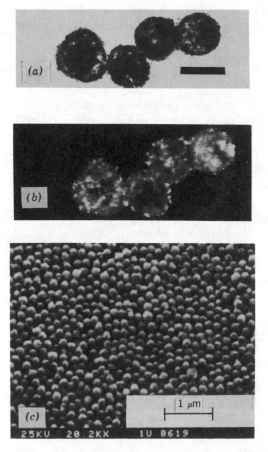

Figure 51.2 Transmission electron micrographs of ZnO powder, (*a*) bright field, and (*b*) dark field (bar = 0.2 μm). (*c*) Scanning electron micrograph of the top surface of a ZnO film.

PROCESSING OF HYDROTHERMALLY TREATED TiO$_2$ POWDERS

Significant advances have recently been made on the low-temperature sintering of electronic ceramics, of which TiO$_2$ is often a major component. Submicrometer, spheroidal, nonagglomerated, amorphous TiO$_2$ particles with a narrow size distribution were synthesized in this study by controlled hydrolysis of titanium tetraethoxide in ethanolic solution. Hydrolysis is difficult to complete, however, and residual alkoxide groups cause problems during subsequent processing. Amorphous TiO$_2$ is easily converted to anatase and rutile by calcination, but the powder is difficult to redisperse in liquid, and particle agglomeration during calcination seems unavoidable.[16] Hydrothermal treatment can complete the hydrolysis reaction and produce crystalline oxides directly in aqueous dispersion, a convenient form for subsequent processing steps.

Spheroidal TiO$_2$ particles containing 9–10 wt% organic and water were produced from titanium tetraethoxide by the continuous flow reaction method,[17] which gives a slightly different powder from the batch process.[5] A molar ratio [H$_2$O]/[Ti(OC$_2$H$_5$)$_4$] of 0.98/0.26, and a residence time of 11.9 s were selected. The precipitate was washed with deionized water three times, then dispersed in basic (pH 10) aqueous solution. After the resulting dispersion's solid content was adjusted to 5 wt%, the dispersion was heated in an autoclave at 200–300°C (1.56–8.60 MPa pressure) for 5 hr, then cooled. A typical hydrothermal treatment is illustrated in Fig. 51.3.

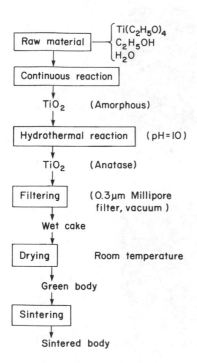

Figure 51.3   Hydrothermal treatment of TiO$_2$ powder.

The original powder particles ranged from 0.3–0.8 μm in diameter and had smooth surfaces before autoclaving (Fig. 51.4a). These became soft agglomerates of particles consisting of 100–500 Å crystals, having a raspberry-like appearance (Fig. 51.4b, c). The agglomerates broke down into individual crystals during ultrasonication. X-ray diffraction showed the amorphous TiO_2 had changed to the anatase crystal form, and BET measurement showed a reduction in surface area, from 320 m^2/g (amorphous hydrolyzed spheres) to 90 m^2/g (anatase crystals). Differential thermal analysis (DTA) indicated that heating the amorphous TiO_2 in air produced a large exothermic peak near 400°C that

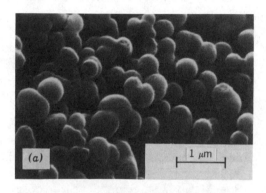

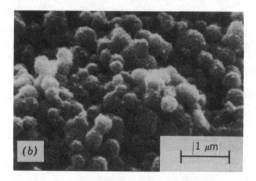

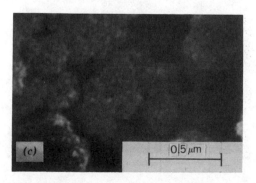

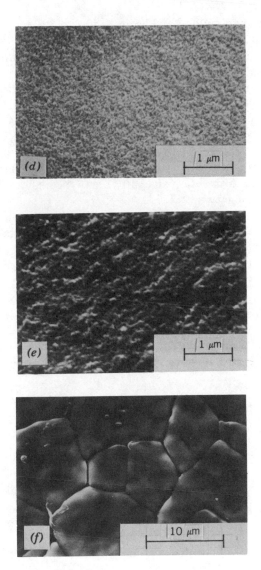

Figure 51.4 Scanning electron micrographs of (a) TiO$_2$ powder before and (b, c) after hydrothermal treatment; (d) anatase green compact; (e) compact after firing at a rate of 10°C/min to 900°C, and (f) after additional sintering at 1200°C for 5 min.

was not observed for the autoclaved powder. These results suggest that the following two reactions occur during hydrothermal treatment:

$$\text{Ti—OH} + \text{HO—Ti} \rightarrow \text{Ti—O—Ti} + \text{H}_2\text{O} \qquad (1)$$

$$\text{Ti—OR} + \text{HO—Ti} \rightarrow \text{Ti—O—Ti} + \text{ROH} \qquad (2)$$

Compacts were made from hydrothermally treated powder by filter casting.

These had a lustrous surface and were ~50% of theoretical density (TD) (Fig. 51.4d). They did not crack or warp during subsequent drying and sintering, and underwent homogeneous shrinkage. Compacts heated to 900°C at a rate of 10°C/min and then cooled had no trace of rutile; they showed only well-crystallized anatase, and had densities >97% TD. SEM confirmed the removal of pores, and indicated negligible grain growth (Fig. 51.4e).

Usually a high packing density cannot be obtained with 100–500 Å alkoxide-derived powders because of agglomeration. In this study, however, the crystallized particles were readily dispersed and cast into uniform green bodies. The uniformity in void distribution, high particle–particle coordination number, and narrow particle size distribution all contribute to the high fired density (97.7%) with negligible grain growth.[5,18]

In another sintering cycle, the anatase pellet was placed in a preheated furnace at 1200°C for 5 min. The TiO_2 transformed completely to rutile, and sintered to >99% TD. The grain size for this heat treatment cycle ranged from 5 to 15 μm (Fig. 51.4f). Further study is necessary to investigate the high grain boundary mobility apparently due to the phase transformation from anatase to rutile.

In conclusion, hydrothermal treatment converted alkoxide-derived TiO_2 particles to small anatase crystal aggregates that were readily dispersed and cast. The green bodies could be sintered to high densities as anatase, anatase–rutile mixtures, or rutile, depending on the firing conditions.

SYNTHESIS AND PROCESSING OF Al_2O_3 —TiO_2 COMPOSITE POWDER

Aluminum titanate (Al_2TiO_5) is well known as a low thermal expansion material.[19] It is ordinarily prepared from a stoichiometric mixture of TiO_2 and Al_2O_3 powders by conventional processing methods such as ball-milling and isostatic pressing.[20] Such methods, however, often do not yield powders mixed homogeneously on a microscopic scale, and do not allow control of particle size distribution and shape. To obtain an improved, reliable composite ceramic, a green composite powder with a specific chemical composition and particle size is desired.

A coating process was investigated (Fig. 51.5). Hydrolysis of the coating, titanium alkoxide, in a colloidal dispersion of the substrate component, Al_2O_3, proved effective in producing a composite powder for the Al_2O_3—TiO_2 system; stepwise hydrolysis of alkoxides was found to be effective in producing unagglomerated particles. A typical composite powder synthesis procedure is shown in Fig. 51.6. The starting material used was size-classified, high-purity alumina. Growth of the composite particles during the TiO_2-coating process investigated is illustrated in Fig. 51.7.

By controlling the total amount of alkoxide hydrolyzed, the amount of TiO_2 coating the Al_2O_3 powder can be controlled. By selecting the starting powder's particle size, the final particle size can also be controlled.

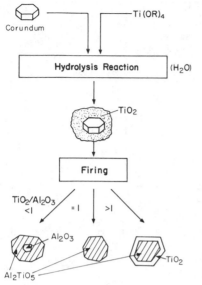

Figure   51.5   Al$_2$O$_3$—TiO$_2$ composite   powder preparation.

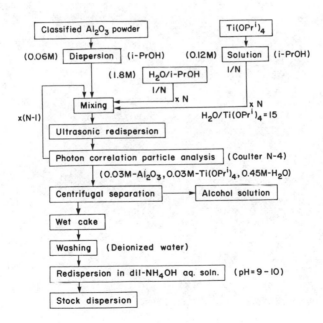

Figure 51.6   Typical TiO$_2$ coating procedure.

Figure 51.8 shows TiO$_2$-coated Al$_2$O$_3$ (TiO$_2$/Al$_2$O$_3$ = 1) powder made by stepwise hydrolysis of Ti(OPr^i)$_4$ in an alumina dispersion. The TiO$_2$-coated powders are spheroidal, and have a narrow particle size distribution and large surface area (190 m^2/g).

The crystal phase of TiO$_2$ changed from amorphous to anatase, then to rutile, and finally to aluminum titanate with increasing firing temperature.

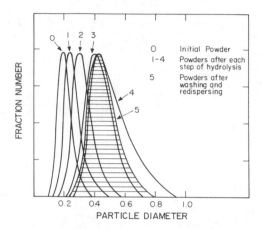

Figure 51.7 Composite particle growth (in μm) during TiO_2 coating procedure (determined by photon correlation spectroscopy).

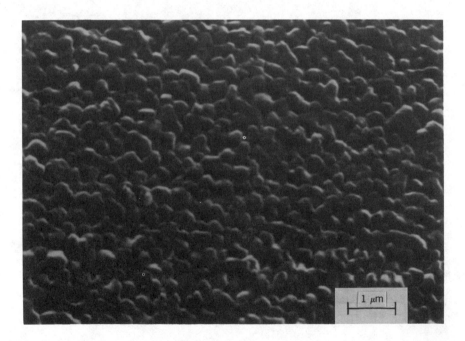

Figure 51.8 Scanning electron micrograph of TiO_2-coated Al_2O_3 powder compact (50 mol% TiO_2).

Compacts of the coated powder were prepared by slip casting or centrifugal casting from a basic (pH 10) NH_4OH aqueous dispersion.

Compacts fired above 1300°C formed aluminum titanate (100%) and yielded high relative densities (95–99%) (Fig. 51.9a), but this firing schedule resulted in crystallographic domain formation. Cracks between domains formed due to the

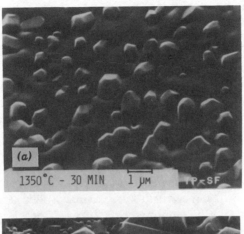

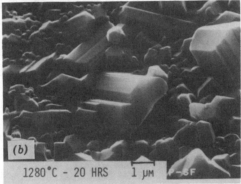

Figure 51.9 Scanning electron micrographs of TiO_2-coated Al_2O_3 compacts fired (a) at 1350°C for 30 min, and (b) at 1280°C for 20 hr.

anisotropy of thermal contraction and the large domain size. Compacts fired below 1300°C resulted in a two-phase dense structure of rutile and corundum. Prolonged heat treatment (1280°C, 20 hr) caused rutile crystal growth, as shown in Fig. 51.9b. Domain structure was not observed in this low-temperature sintered specimen.

In conclusion, Al_2O_3–TiO_2 composite powder can be sintered to aluminum titanate or rutile–corundum composite ceramics by controlling sintering conditions. The eutectoid decomposition of Al_2TiO_5 to TiO_2 and Al_2O_3 also allows for the formation of a three-phase structure. More detailed studies including variation of the TiO_2/Al_2O_3 ratio are being performed.

SYNTHESIS AND PROCESSING OF Al_2O_3 —ZrO_2 COMPOSITE POWDER

Alumina–zirconia composite powders show promise for ceramic applications requiring high strength and toughness. Synthesis techniques for this system have included milling of powders in air, water, or solvents,[21] oxidation or decomposition of soluble, often organic, ceramic precursors,[22] and hetero-coagulation of sols[23] or high shear mixing of flocced sols.[24]

Nominally 8 wt% Al_2O_3—ZrO_2 composite powder was produced by a continuous flow reactor method by mixing a water-free stream of alumina dispersed in a solution of zirconium tetra-n-propoxide in dry ethanol with a "wet" stream, ethanol in water, under plug flow conditions. The wet stream hydrolyzed alkoxide located both interstitially and on alumina particle surfaces, forming a dispersion of the composite powder.

The resulting powder was washed once in deionized water, then half the sample was autoclaved in deionized water for 5 hr at ~250°C. The composite powder samples were consolidated at 5000 psi in a colloid press (Fig. 51.10), and air dried at low temperature. Autoclaving produced a macroscopically homogeneous powder sample more satisfactory for consolidation in the fritted press than unautoclaved powder.

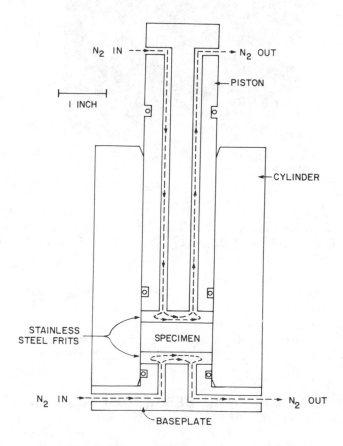

Figure 51.10   Cross section of cylindrical brass colloid press.

A TEM micrograph of the autoclaved powder (Fig. 51.11a) shows that most of the zirconia precursor is attached to alumina particle surfaces. Considering zirconium tetra-n-propoxide's stabilizing effect on dispersions of alumina in ethanol, this is likely due to hydrolysis of alkoxide adsorbed on the dispersed particles. Thermogravimetric analysis (TGA) of the autoclaved-powder compact indicated an abrupt weight loss between 400 and 500°C, presumably from dehydration of zirconium hydroxide. In contrast, the unautoclaved powder compact showed a gradual weight loss between room temperature and 500°C, suggesting organic material burnout.

The relative green densities of compacts of the autoclaved and unautoclaved composite powder were 59.5 and 51.4%, respectively. An SEM photograph of the green body made from autoclaved powder is shown in Fig. 51.11b. Dilatometry results showed significant shrinkage beginning at 1000°C for the autoclaved powder compact, and at 1100°C for the unautoclaved compact. This

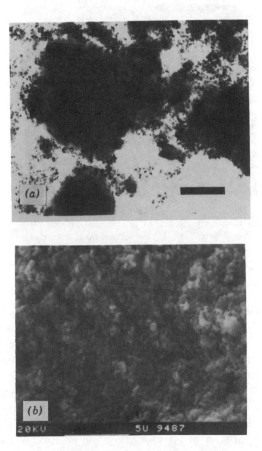

Figure 51.11 (a) transmission electron micrograph of autoclaved Al$_2$O$_3$—ZrO$_2$ composite powder (bar = 0.2 μm), and (b) scanning electron micrograph of unfired compact fracture surface.

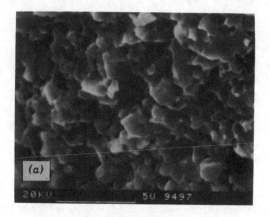

Figure 51.12 Scanning electron micrographs of fracture surfaces of (*a*) unautoclaved-powder compact fired at 1500°C for 100 min, and (*b*) autoclaved-powder compact fired at 1500°C for 100 min.

difference may be largely due to the density difference between green bodies. Figure 51.12*a* shows the fracture surface of a sintered sample of the unauto-claved powder specimen. SEM investigation of the fracture surface shows great variability in structure. Figure 51.12*b* shows the fine-grained, high-density microstructure formed by sintering the autoclaved powder specimen at the same temperature. The uniformity of the green body produced from the autoclaved composite powder, and the absence of observable macrodefects on the fired ceramic's fracture surface suggest that a uniform high-strength composite ceramic was formed.

In conclusion, synthesis of Al_2O_3–ZrO_2 composites by hydrolysis of zirconium alkoxides on alumina particle surfaces in a continuous flow reactor can yield powder that sinters to a fine-grained, uniform, high-density ceramic.

CONCLUSIONS

Ceramic powders of controlled composition, crystallinity, size, and size distribution have been formed. New variations on the chemical precursor to the precipitation process and the use of classified powders as substrates for coating have been demonstrated. The use of hydrothermal treatment of dispersed powders to crystallize precipitates and remove unreacted precursors has also been shown to be advantageous in powder formation techniques. In all cases studied (ZnO, TiO_2, Al_2O_3–TiO_2, Al_2O_3–ZrO_2), processes for forming unagglomerated, narrow size distribution powders were shown to be feasible.

These powders, when properly dispersed in aqueous or nonaqueous liquids using techniques such as colloid pressing, slip casting, filter casting, or centrifugal casting, produced homogeneous green bodies. Sintering of the packed powders resulted in high densities, yet widely varying microstructures, depending on firing conditions. The example systems studied clearly demonstrate new possibilities for the control of microstructure, phase, and phase distribution in sintered bodies, and thus new opportunities for discovering performance–property relationships of polycrystalline ceramics.

ACKNOWLEDGMENTS

Three authors have been visiting scientists at MIT, on leave from their companies: Dr. R. H. Heistand II (Dow Chemical), Dr. Y. Oguri (Mitsubishi Chemical), and Dr. H. Okamura (Nippon Soda). This support is gratefully acknowledged. All of the authors acknowledge funding for this reasearch from the Air Force Office of Scientific Research and the Massachusetts Institute of Technology Ceramics Processing Industrial Consortium.

REFERENCES

1. H. K. Bowen et al., Basic Needs in High-Temperature Ceramics, *J. Mater. Sci. Eng.* **44**, 1–56 (1980).

2. W. H. Rhodes, Agglomeration and Particle Size Effects on Sintering Yttria-Stabilized Zirconia, *J. Am. Ceram. Soc.* **64**, 19–22 (1981).

3. E. A. Barringer et al., Processing Monosized Powders, in *Ultrastructure Processing of Ceramics, Glasses, and Composites*, L. L. Hench and D. R. Ulrich, Eds., Wiley, New York, 1984, pp. 315–333.

4. F. F. Lange et al., Processing Related Fracture Origins: I, II, and III, *J. Ceram. Soc.* **65**, 396–408 (1983).

5. E. A. Barringer and H. K. Bowen, Formation, Packing, and Sintering of Monodispersed TiO_2 Powders, *J. Am. Ceram. Soc.* **65**(12), C199–C201 (1982); Effects of Particle Packing on the Sintered Microstructure, *Physics of Sintering* **16** (1985).

6. M. Sacks et al., Preparation of SiO_2 Glass Model Powder Compacts, *J. Am. Ceram. Soc.* **67**(8), 526–537 (1984); Milling and Suspension Behavior of Al_2O_3 in Methanol and Methyl Isobutyl Ketone, *J. Am. Ceram. Soc.* **66**(7), 488–494 (1983).

7. H. E. Brown, *Zinc Oxide Properties and Applications*, International Lead Zinc Research Organization, Inc., New York, 1976, pp. 7–9.

8. R. J. Lauf and W. D. Bond, Fabrication of High-Field Zinc Oxide Varistors by Sol-Gel Processing, *Ceram. Bull.* **63**(2), 278–281 (1984).

9. B. Fegley and E. A. Barringer, Synthesis, Characterization, and Processing of Monosized Ceramic Powders, in *Better Ceramics Through Chemistry*, C. J. Brinker, Ed., Elsevier, New York, 1984, pp. 187–197.

10. B. Fegley, H. Okamura, L. Rigione, and C. Sobon, *Zinc Alkoxide Hydrolysis*, Report #Q2, MIT Ceramics Processing Research Lab. (1984).

11. C. Sobon, *Alkoxide Precursors*, Report #Q1, MIT Ceramics Processing Research Lab. (1984).

12. G. Allen, J. M. Bruce, D. W. Farren, and F. G. Hutchinson, Organozinc Compounds. Part II. The Methanolysis of Dimethyl-and Diphenyl-zinc, *J. Chem. Soc.* (*B*), 799–803 (1966).

13. J. M. Bruce, B. C. Cutsforth, D. W. Farren, F. M. Rabagliati, and D. R. Reed, Organozinc Compounds. Part III. Further Alcoholyses of Dimethyl- and Diphenyl-zinc, *J. Chem. Soc.* (*B*), 1020–1024 (1966).

14. Y. Matsui, K. Kamiya, M. Nishikawa, and Y. Tomiie, The Crystal Structure of Ethylzinc *t*-Butoxide, *Bull. Chem. Soc. Jap.* **39**(8), 1828 (1966).

15. H. M. M. Shearer and C. B. Spencer, The Crystal Structure of Tetrameric Methylzinc Methoxide, *Chem. Comm.*, 194 (1966).

16. M. F. Yan and W. W. Rhodes, Low-Temperature Sintering of TiO_2, *Mater. Sci. Eng.* **61**, 59–66 (1983).

17. B. Novich, *Continuous Processing of Composite Ceramics*, Report #43, MIT Ceramics Processing Research Lab. (1985).

18. H. K. Bowen, Ceramics as Engineering Materials: Structure-Property-Processing, *Material Research Society Symposium*, Vol. 24, Elsevier, New York, 1984, pp. 1–11.

19. W. R. Bussem, N. R. Thielke, and R. V. Sarakauskas, The Expansion Hysteresis of Aluminum Titanate, *Ceram. Age* **60**, 38–40 (1952).

20. K. Hamano, Y. Ohya, and Z. Nakagawa, Microstructure and Mechanical Strength of Aluminum Titanate Ceramic Prepared from a Mixture of Alumina and Titania, *J. Ceram. Soc. Japan* **91**(2), 94–101 (1983).

21. I. A. Aksay, F. F. Lange, and B. I. Davis, Uniformity of Al_2O_3-ZrO_2 Composites by Colloidal Filtration, *J. Am. Ceram. Soc.* **66**(10), C190–C193 (1983).

22. B. Fegley, Jr., P. White, and H. K. Bowen, Preparation of Zirconia-Alumina Powders by Zirconium Alkoxide Hydrolysis, accepted for publication in *J. Am. Ceram. Soc. Comm.*

23. F. F. Lange and M. M. Hirlinger, Hindrance of Grain Growth in Al_2O_3 by ZrO_2 Inclusions, *J. Am. Ceram. Soc.* **67**(3), 164–168 (1984).

24. E. Carlstrom and F. F. Lange, Mixing of Flocced Suspensions, *J. Am. Ceram. Soc.* **67**(8), C169–C170 (1984).

52

LOW-TEMPERATURE SYNTHESIS OF CERAMIC POWDERS FOR STRUCTURAL AND ELECTRONIC APPLICATIONS

J. J. RITTER AND K. G. FRASE

Inorganic Materials Division
National Bureau of Standards
Gaithersburg, Maryland

INTRODUCTION

Most of the recent research effort in the development of low-temperature synthetic routes to produce ceramic powders has focused on the hydrolysis of metal alkoxides. Alkoxides are the product of the reaction of metals or metal chlorides with alcohols. Most metal alkoxides react readily with water to give hydrous metal oxides which can be calcined to the metal oxide. Some 63 elements are available as alkoxides.[1] When two or more metal alkoxides are reacted to give a mixed-metal alkoxide system, binary or ternary oxides are obtained upon hydrolysis.

Metal alkoxide chemistry, although widely applicable to oxide ceramic systems, is of course, inappropriate for the synthesis of nonoxide materials. Except for the polycarbosilane and aminosilane methods for SiC and Si_3N_4, the low-temperature synthesis of nonoxide materials is largely unexplored. Other chemistries are clearly needed to synthesize fine powders of such materials as TiB_2 and B_4C. We have developed a novel low-temperature synthesis of these nonoxide compounds using a modified Wurtz–Fittig reaction, common in organic chemistry.

METHOD

Metal chlorides are reacted with sodium in a nonpolar solvent to produce an amorphous precursor powder and NaCl. The NaCl can be distilled off, and the precursor crystallized at temperatures as low as 700°C:

$$10\text{Na} + \text{TiCl}_4 + 2\text{BCl}_3 \xrightarrow[\substack{\text{heptane} \\ 25-160°C}]{} \text{"TiB}_2\text{"} + 10\text{NaCl} \xrightarrow[\substack{\text{vacuum} \\ 700°C}]{} \text{TiB}_2$$

$$+$$

$$10\text{NaCl(g)} \qquad (1)$$

It is proposed that the initial reaction involves halogen abstraction from each of the metal halides and a coupling of the resultant products:

$$\text{BCl}_3 + \text{TiCl}_4 + 2\text{Na} \rightarrow \text{Cl}_2\text{B}-\text{TiCl}_3 + 2\text{NaCl} \qquad (2)$$

Further halogen abstraction and coupling reactions generate a 3-D B—Ti precursor matrix. We have generated silicon carbide and B_4C from $SiCl_4$, BCl_3 and CCl_4 in similar type reactions. These nonoxide ceramic materials, SiC, TiB_2, and B_4C have applications as microchip substrates, aluminum reduction cathodes, and high-temperature refractories, respectively. Additional exploration of this and other methods for the low-temperature synthesis of nonoxide powders is desirable.

DISCUSSION

There are a number of advantages to the low-temperature synthesis of precursor powders. Reagents can be highly purified and thus enhanced product purity can be achieved. Dispersing the reagents in a solvent provides mixing on the molecular level, rather than at the particulate level as in solid-state reactions. Molecular mixing has a number of ramifications, including product homogeneity and crystal symmetry. Finally, by using intimately mixed components at low temperatures, metastable compounds not accessible by solid-state reaction can be formed, some of which have particularly interesting properties. Each of these advantages will be examined in more detail below.

The high purity and homogeneity available via low-temperature synthesis is useful in the production of standard reference materials for instrument calibration. Submicrometer particles of TiO_2 have been produced with a homogenous doping of 1 to 3 wt% heavy metal (e.g., Zr, U, Pb) as proposed calibration standards for analytical electron microscopes. Particle size distribution can be controlled in a flow reactor through the manipulation of the hydrolysis reaction parameters, as shown in Fig. 52.1.

The molecular level mixing available through low-temperature synthesis also affects the crystallization of the final product. In two ionic conductor systems, the sodium β/β''-alumina and Y_3TaO_7 systems, the final crystalline materials produced by alkoxide synthesis differ from the products of the solid-state reaction.

In the sodium β/β''-alumina case, solid-state reaction at temperatures above 1500°C is required to reach the equilibrium phases, the binary β-alumina and the ternary sodium β''-alumina (Mg-stabilized). In contrast, alkoxide synthesis of materials with the β''-alumina composition yield crystalline products at 1100°C which have a distorted "solid solution" β-type structure (Fig. 52.2). The primary difference between the β'' and solid solution β structures is the Mg-substitution in the aluminum–oxygen layers. In the β'' structure, this substitution is ordered ,[2] while in the solid solution β structure it is random.

Preliminary Experiments: Model System TiO_2, 3%ZrO_2
Alkoxides in ethanol, 0.2M $Ti(OR)_4$, 0.003M $Zr(OR)_4$

Run	Hydrolysis Medium	H_2O injection rate (mmoles/s)
A	H_2O, NH_4OH, C_2H_5OH	0.173
B	H_2O, C_2H_5OH	0.162
C	H_2O, C_2H_5OH	4.7
D	H_2O, NH_4OH, C_2H_5OH	5.7

Particle Size Estimation by Sedigraph

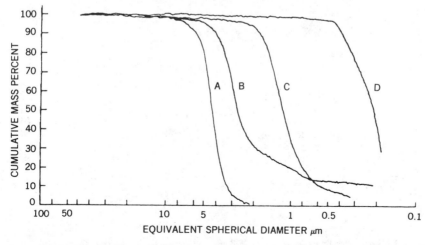

Figure 52.1   Reaction parameter effects on particle size and size distribution.

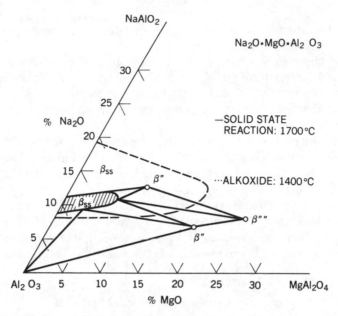

Figure 52.2 Expansion of solid solution areas in beta-alumina through the use of alkoxide synthesis.

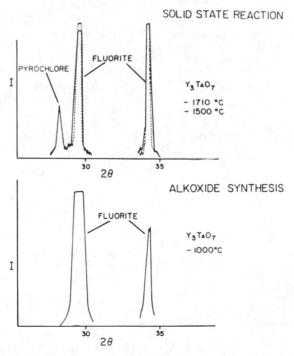

Figure 52.3 X-ray diffraction patterns of Y_3TaO_7 prepared by (a) solid-state reaction and (b) alkoxide hydrolysis.

Thus the molecular mixing inherent in alkoxide syntheses appears to favor the disordered structure.

This preference for higher symmetry through alkoxide synthesis is also observed in the Y_3TaO_7 case. When prepared by solid-state reaction, Y_3TaO_7 forms as a distorted pyrochlore at temperatures between 1500 and 1700°C.

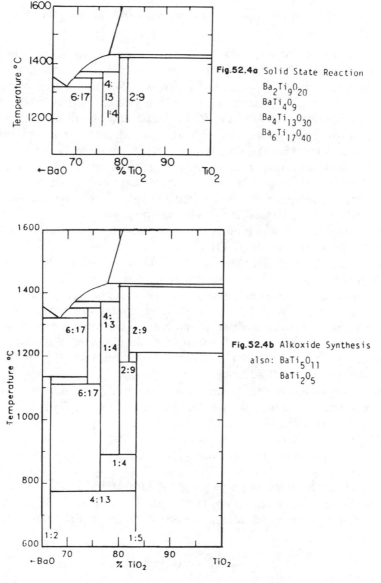

Figure 52.4 Phase stability diagrams of BaO—TiO$_2$ system derived from (a) solid-state reaction and (b) alkoxide hydrolysis.

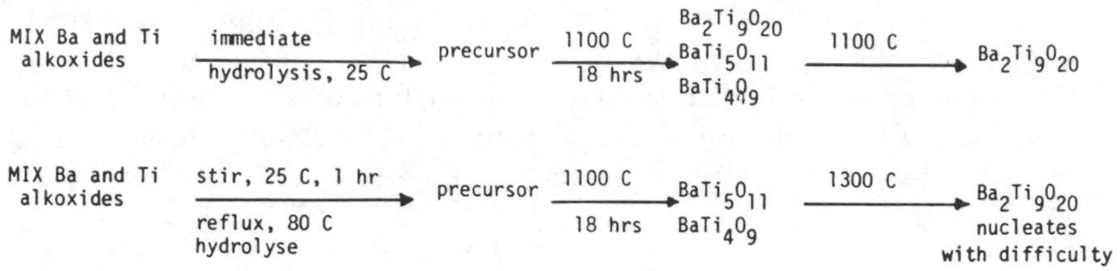

MIX Ba and Ti alkoxides → immediate hydrolysis, 25 C → precursor → 1100 C, 18 hrs → $Ba_2Ti_9O_{20}$, $BaTi_5O_{11}$, $BaTi_4O_9$ → 1100 C → $Ba_2Ti_9O_{20}$

MIX Ba and Ti alkoxides → stir, 25 C, 1 hr / reflux, 80 C / hydrolyse → precursor → 1100 C, 18 hrs → $BaTi_5O_{11}$, $BaTi_4O_9$ → 1300 C → $Ba_2Ti_9O_{20}$ nucleates with difficulty

Figure 52.5   Influence of prehydrolysis intermediates on final crystalline phases.

Above 1700°C a cubic fluorite structure forms which can be quenched in to lower temperatures. This cubic form is isostructural with cubic zirconia and is therefore a potential solid oxide ion electrolyte. When prepared by alkoxide hydrolysis, the Y_3TaO_7 material begins to crystallize as the cubic phase at temperatures as low as 1000°C (Fig. 52.3). The tendency for disordered phases to form from low-temperature syntheses may introduce a whole new group of oxide ion conductors based on "high-temperature" disordered crystal structures.

Finally, molecular mixing effectively shortens the necessary diffusion distance required for crystallization to occur. Crystallization can therefore occur at lower temperatures than those required for solid-state reaction. Metastable compounds which are unstable as higher temperatures can be formed through this combination of lower synthesis and crystallization temperatures. A case in point is the compound $BaTi_5O_{11}$.

The phase stability diagram for the TiO_2-rich end of the $BaO-TiO_2$ pseudobinary, based on solid state reactions above 1200°C, is shown in Fig. 52.4a. A number of compounds appear, of which two, $BaTi_4O_9$ and $Ba_2Ti_9O_{20}$, are used as microwave dielectric materials. A crystal structure for $BaTi_5O_{11}$ had been reported in the literature,[3] and the compound was expected to have good dielectric loss properties,[4] but it could not be made by solid-state reaction. However, by low-temperature alkoxide synthesis, both $BaTi_5O_{11}$ and $BaTi_2O_5$ can be made routinely (Fig. 52.4b).

The $BaO-TiO_2$ system is particularly interesting because it develops a series of intermediates in solution before hydrolysis, and performing the hydrolysis at these different stages produces different final products. For example, Fig. 52.5 shows the case for a composition of Ba:Ti ratio 2:9, where hydrolysis immediately upon mixing the two alkoxides yields a three-phase product which rapidly converts at 1100°C to single-phase $Ba_2Ti_9O_{20}$. However, hydrolysis at the final precipitated stage yields a mixture of $BaTi_4O_9$ and $BaTi_5O_{11}$, which only slowly converts to $Ba_2Ti_9O_{20}$ at 1300°C. Clearly the intermediate species play an important role in determining the crystallization reaction.

CONCLUSION

A clear understanding of the relationship between species in solution, the amorphous precursor and the final crystalline product could allow us to design phases with particular properties, such as ionic conductivity or dielectric strength. The exploration of this relationship, as well as expansion in the area of nonoxide materials, is necessary to exploit fully the advantages of low-temperature synthesis of ceramic powders: the control of particle size, purity, stability and ordering, all through the manipulation of the reaction parameters.

REFERENCES

1. D. C. Bradley, R. C. Mehrotra, and D. P. Gauer, *Metal Alkoxides*, Academic Press, London, 1978, p. 1.
2. K. G. Frase, PhD. Thesis, University of Pennsylvania, 1983.
3. E. Tillmanns, *Acta Crystallogr.* **B25**, 1444 (1969).
4. J. J. Ritter, R. S. Roth, and J. Blendell, Alkoxide Precursor Synthesis and Characterization of Phases in the Barium–Titanium Oxide System, *J. Amer. Ceram. Soc.* (1986).

53

TWO-DIMENSIONAL PARTICULATE STRUCTURES

GEORGE Y. ONODA
IBM Thomas J. Watson Research Center
Yorktown Heights, New York

INTRODUCTION

In recent years, interest in colloidal systems has spread into many disciplines as a result of several important advances. The first was the development of methods for producing spherical, uniform latex particles in the 1950s by Dow Chemical Company and later, similar colloids from hydrous oxides.[1,2] The next important observation was that concentrated dispersions of these colloids can form 3-D structures that undergo order–disorder transitions by changing the interparticle forces. This attracted the interest of physicists working with theoretical and experimental aspects of multibody problems. These and other advances have recently been reviewed by Pieranski[3] and Hirtzel and Rajagopalan.[4]

Most of the studies to date have dealt with 3-D colloidal structures; in contrast, relatively few have involved studies of 2-D structures. In most respects, the 2-D structures are equally of interest as the 3-D. Along with many analogies between 3-D atomic and colloidal systems, there are likewise analogies between 2-D atomic and colloidal systems. For example, the deposition of colloids onto solid surfaces is similar to the adsorption of molecules on solid surface. Experimentally, the 2-D structures are particularly attractive because of the greater ease by which the structures can be observed and characterized. Localization of particles in a single plane facilitates the focusing of various probes for observing a complete assemblage of particles over the viewed area.

This chapter reviews recent work by others and by the author and his collaborators in the area of 2-D colloidal structures. A 2-D colloidal structure is defined as a planar assemblage of particles partially or totally immersed in a liquid phase. The liquid phase is essential for forming a 2-D structure and is a means by which thermal energy is imparted to the particles (via Brownian motion). Since Brownian motion acts to diffuse particles away from a planar arrangement, there must exist constraining forces sufficiently greater than kT that hold the particles to the plane. Usually, these constraining forces are imposed by an adjacent or nearby interface. In this review, we discuss structures resulting from the constraints imposed by gas–liquid interfaces and liquid–solid interfaces.

STRUCTURES AT GAS–LIQUID INTERFACES

Pieranski has demonstrated[5] that 2450 Å polystyrene spheres can be trapped at a water–air interface. The particles are held by the surface tension of water, which provides an energy which is a factor of 10^6 greater than kT. Therefore, once captured at the interface, the particle cannot diffuse away by Brownian motion. The particles at the interface were found to order in a 2-D, trianglular array. Ordering occurred with particles spaced apart at distances that were 4

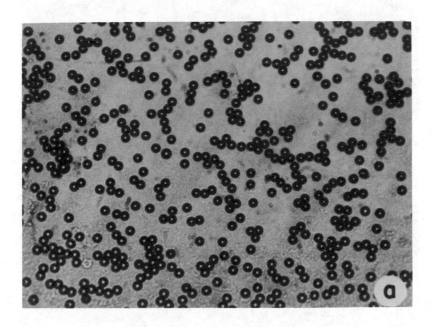

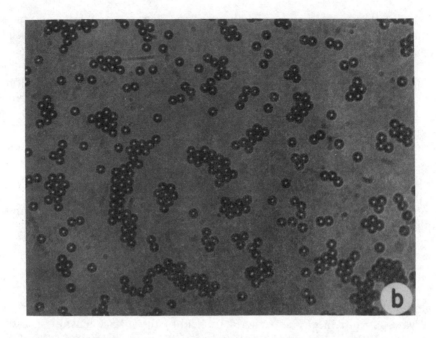

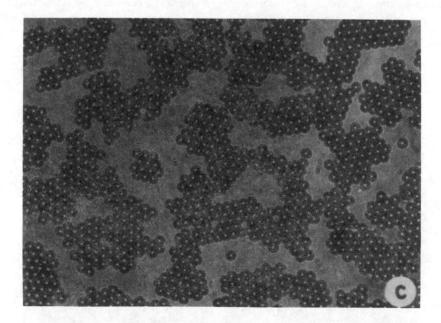

Figure 53.1 Clustering and growth of ordered regions of 5-μm latex spheres at the air–water interface. After (a) 5 min, (b) 10 min, and (c) 20 min. 400 × magnification.

to 40 times their diameter, indicating that the order was caused by longe range, electrostatic repulsive forces resulting from the double layers around particles.

Kalia and Vashishta[6] carried out a molecular dynamics study of colloids trapped at water–air interfaces. They showed that the system undergoes a first-order melting transition between a disordered state and a triangular lattice arrangement. This occurs when the ratio of the electrostatic repulsive energy between particles to the energy kT is between 59 and 65. The transition entropy per particle was found to be equal to 0.3 k.

We are studying the clustering and growth of 2–10 μm latex particles at the air–water interface. At this interface, there exists a capillary and a van der Waal attractive energy between neighboring particles. As a result, particles arriving at the air–water interface can diffuse laterally, form clusters, and grow to larger ordered clusters as shown with 5-μm particles (Fig. 53.1). When the particles are around 2 μm, the bonding and thermal energies are of a ratio that allows dynamic behavior, with particles joining and leaving clusters continuously and with small clusters having "liquidlike" structure rather than ordered structures.

STRUCTURES IN PLANAR SANDWICH ARRANGEMENTS

Two-dimensional colloidal structures can be formed by confining them in a narrow space between two solid planar surfaces, such as two closely spaced glass slides. If the charges on the walls and on the particles are of the same sign, the particles are repelled from both walls and find an equilibrium position midway between the walls. Using latices in water, it has been shown[7] that the colloids form a 2-D, triangular lattice. The spacings between particles is considerably greater than the size of the particles, indicating that the ordering is occurring from long-range electrostatic repulsion.

Stjeltorp[8,9] demonstrated that a similar 2-D structure can be created between two glass slides by using magnetic repulsive forces between particles instead of electrostatic. Instead of using magnetic particles, he immersed latex particles in a paramagnetic fluid (a mixture of very fine magnetite dispersed in kerosene). In the presence of a magnetic field, the latex particles function as "holes" in the magnetic liquid which have an apparent magnetic dipole associated with them. By orienting the field normal to the layer, magnetic repulsive forces are created between the particles. This promotes repulsive ordering in a triangular lattice, similar to what occurs with electrostatic forces. Without the field, the particles are disordered; when a critical field is applied, ordering occurs. Order is destroyed when the field is removed because of thermal disruption. He also observed that when the field was oriented parallel to the layer, the particles formed parallel chains. Under these conditions, there are attractive forces between particles in the direction of the chains and repulsive forces normal to the chains.

Recently, work was reported[10] in which conductive coatings were placed on two glass slides so that an electric field could be applied across the gap. The

originally random suspension of latex particles between the slides responded
to an ac field of $\sim$ 2 V at 1 kHz which caused a sudden attraction of the particles
to the surfaces of the slides. A patchy, fractal-like structure was formed. Close
inspection revealed ordered regions within the patchy areas. This phenomenon
did not occur at higher ($>$ 5 kHz) or lower frequencies. Small variations in
frequency around 1 kHz produced noticeable changes in the optical diffraction
characteristics of the coatings. The effect of frequency was thought to be related
to the mobility of the diffuse double layer around individual particles.

STRUCTURES AT LIQUID–SOLID INTERFACES

The deposition of colloids at solid–liquid interfaces has been an active research
area (see Chapter 5 in Ref. 4 for a recent review). Most studies have focused on the
mechanism and kinetics of deposition on flat substrates and in packed beds and
on the bonding forces as they relate to the chemistry of the system. In contrast,
few have been concerned with the types of structures developed in the deposited
layer. The limited works that deal with single layers and their structure are
reviewed below.

Disordered 2-D Layers

In 1966, Iler[11] described a technique for depositing single layers and multilayers
of colloidal particles on smooth surfaces from aqueous suspension. The basic
idea involves creating an opposite electric charge between the smooth surface
and the particle. In this way, the particle becomes bonded to the surface electrostatically while the particles do not bond together in the suspension.

To bond silica particles to a glass plate (both of which are normally negatively
charged in water), Iler first exposed the plate to a suspension of colloidal
boehmite alumina consisting of fibrils of A100H about 5 or 6 nm in diameter.
The excess sol was rinsed off and the surface air dried, leaving an invisible layer
of boehmite. The surface was then wetted with a suspension of colloidal silica
(100-nm diameter) at pH 3. At this pH, the silica is slightly negatively charged,
while the boehmite at the surface is positively charged. The excess suspension
was rinsed off with running distilled water, which removed all silica particles
that were not in direct contact with the surface. After rinsing, the surface was
dried, leaving a single layer of silica particles on the surface. If desired, the process
can be repeated a number of times, building a controlled multilayer structure
(Fig. 53.2).

Several surfactants to control surface charge were also explored by Iler,
such as cationic species that impart a positive charge to glass and bond the silica
colloid, and anionic species to reverse the charge of a positively charged surface.
In a somewhat different vein, Deckman and Dunsmuir[12] used thin films of
aluminum (presumed to have oxidized surfaces) to impart a positive charge on
substrates and attract negatively charged polystyrene latex particles.

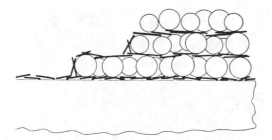

Figure 53.2   Schematic cross section of building layers of silica with alternating layers of boehmite fibrils (from Iler, Ref. 11).

Recently, we have shown that high-molecular weight polyacrylamides are particularly effective in bonding silica colloids to a wide range of surfaces.[13] Polyacrylamides are well-known flocculating agents for settling out particulate matter from aqueous suspensions. Our method is similar to Iler's boehmite method except that the boehmite is replaced by the cationic polyacrlylamide. Surfaces are coated with this polymer by immersion in a dilute solutions (100– 1000 ppm) for around 10 min, followed by rinsing and drying. Subsequent exposure to a colloidal suspension, followed by rinsing and drying leads to the formation of a single layer of particles.

The particulate coatings produced with the aid of the polyacrylamide were shown to have a random arrangement. When particles come in contact with the polymer-coated surface, they attach firmly and neither detach nor migrate along the surface. As a result, this deposition represents a good experimental model for what is known as the "random parking problem in two dimensions." In this problem, circles are placed one after another on a surface at random. If an added circle overlaps an existing circle, the new circle is not allowed. After enough circles are added, a point is reached where no more circles can be added. This point is called the random parking limit. To date, no analytical solution exists for this limit. However, several computer simulations have been carried out, yielding values between 0.50 and 0.55 for the limit, depending on the algorithm used.[14,15] Using silica spheres onto glass surfaces with the flocculant, we were able to provide the first experimental confirmation that the limit lies within the range predicted by the computer simulations.[13] The arrangement of silica spheres near the parking limit is shown in Fig. 53.3.

Ordered 2-D Layers

Deckman and Dunsmuir[12] produced ordered arrays of polystyrene latex particles on smooth substrates using a spin coating process. The layer was "polycrystalline," with grains having 50–1000 spheres. It was thought that ordering occurred because the sol flows across the substrate at high shear rates while the excess coating material is being dispelled to produce densely packed ordered arrays on the surface. It was observed that the ordering occurred when

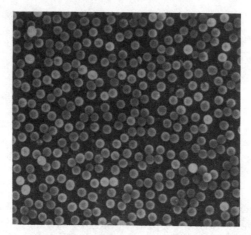

Figure 53.3   Deposited silica particles (0.6 μm) at the glass surface near the 2-D parking limit.

the particles and the substrate were of opposite charge. However, no systematic study was carried out to understand the process in detail or to determine the effects of different variables such as sol rheology, particle concentration, and so on.

We have studied 2-D ordering using silica and latex spheres. Figure 53.4 illustrates the type of ordering possible with silica on a glass slide. These were formed when a dilute suspension (0.1%) was allowed to dry on the glass slide. These ordered regions are found in areas near the center of where the drop existed; along the perimeter, thicker layers with varying degrees of order existed. For submicrometer particles, ordering is induced by the growth of ordered rafts at the air–water interface and its subsequent deposition during drying.

Figure 53.4   Ordered silica spheres (0.6 μm) produced during drying.

For large latex particles (5 μm and greater), ordering occurs at the liquid–solid interface through the crowding brought about by the diminishing area of contact of the drop during drying and the dragging effects of the triple point interface as the liquid recedes. Close inspection of these layers reveals a wide range of defects, including various types of vacancies, dislocations, stacking faults, and grain boundaries. Also, the disturbances in order caused by occasional larger particles are evident.

CONCLUSIONS

The studies on the structures of 2-D colloidal layers reveal many interesting phenomena, many of which are not well understood. Reversible order–disorder transitions are of particular interest because the possibility of directly observing (under suitable conditions) continuous transitions with an intervening "hexatic" phase of the type described by Kosterlitz and Thouless[16] and Halperin and Nelson.[17] In disordered systems, there is growing interest in random parking structures, random dense packing (if it exits), and dynamic clustering. In ordered structures, there remain many unanswered questions concerning the formation of defects (vacancies, dislocations, etc.), the size of individual "grains," as well as the virtually unexplored areas of 2-D structures resulting from the mixing of particles of different sizes. Once can only anticipate a rapid growth of interest in 2-D structures because of the inviting new possibilities for future research.

ACKNOWLEDGMENTS

The author acknowledges the contributions of coauthors for papers on which this article is based, including E. Liniger and P. Somasundaran. Thanks also to A. Skjeltorp, T. Shaw and D. Clarke for useful suggestions.

REFERENCES

1. W. Stoeber, A. Fink, and E. Bohn, *J. Colloid Interface Sci.* **26**, 62 (1968).

2. E. Matijevic, *Prog. Colloid Polym. Sci.* **61**, 24 (1968); E. Barringer, N. Jubb, B. Fegley, R. L. Pober, and H. K. Bowen, in *Ultrastructural Processing of Ceramics, Glasses, and Composites*, L. L. Hench and D. R. Ulrich, Eds., Wiley, New York, 1984, Chapter 16.

3. P. Pieranski, *Contemp. Phys.* **24**, 25 (1983).

4. C. S. Hirtzel and R. Rajagopalan, *Colloidal Phenomena: Advanced Topics*, Noyes, New Jersey, 1984.

5. P. Pieranski, *Phys. Rev. Lett.* **45**, 569 (1980).

6. R. K. Kalia and P. Vashishta, *J. Phys. C: Solid State Phys.* **14**, L643 (1981).

7. P. Pieranski, L. Strzelecki, and B. Pansu, *Phys. Rev. Lett.* **50**, 900 (1983).

8. A. T. Skjeltorp, *Phys. Rev. Lett.* **51**, 2306 (1983).

 9. A. T. Skjeltorp, *J. App. Phys.* **55**, 2587 (1984).

10. P. Richetti, J. Prost, and P. Barois, *J. Physique Lett.* **45**, L-1137 (1984).

11. R. K. Iler, *J. Colloid Interface Sci.* **21**, 569 (1966).

12. H. Deckman and J. H. Dunsmuir, *Appl. Phys. Lett.* **41**, 377 (1982).

13. G. Onoda and E. Liniger, *Phys. Rev. A* (1985).

14. M. Tanemura, *Ann. Inst. Statist. Math.* **31**, Part B, 351 (1979).

15. J. Feder, *J. Theor. Biol.* **87**, 237 (1980).

16. J. M. Kosterlitz and D. J. Thouless. *J. Phys. C* **6**, 1181 (1973).

17. D. Nelson and B. I. Halperin, *Phys. Rev. B* **19**, 2456 (1979).

54

STRUCTURES OF COLLOIDAL SOLIDS

ILHAN A. AKSAY AND RYOICHI KIKUCHI
Department of Materials Science and Engineering
College of Engineering
University of Washington
Seattle, Washington

INTRODUCTION

Colloidal systems containing spherical particles have been shown to display disorder–order transitions.[1,2] Since these transitions resemble the liquid to crystal transitions of atomic systems, many attempts have been made to outline the stability regions of the colloidal liquid and crystalline structures in phase diagram forms.[3] The structural characterization of these phases has also been the subject of numerous studies.[2] In the crystalline form, these structures are often easily recognized by their bright iridescent colors. In the most commonly accepted theoretical treatments, the existence of highly repulsive interparticle interactions are considered to be the essential requirement for the formation of colloidal crystals.[3]

In contrast to this prevailing view, in this chapter we point out our observation on the formation of colloidal crystals through spontaneous gas to condensed phase transitions which can only be realized when net attractive interactions exist. Including our new interpretations, we now present an all-inclusive treatment of phase stability in colloidal systems combining gas, liquid, and crystal phases and transitions among them. Further, we provide experimental data on the unifying hierarchical features of the colloidal solids and their interpretations.

513

BASIC PHASE DIAGRAM OF THREE AGGREGATES

In discussing the phase diagram of colloidal systems the main requirement is that the interparticle potential must be known as a function of the system parameters. In our experimental studies we specifically worked with systems where the particles interacted electrostatically. Therefore, the discussion in this paper will be limited to such interaction.

In electrostatic systems, the repulsion energy between two colloidal particles can be expressed by the DLVO theory.[4] The combination of this repulsive potential with the energy of attraction due to the van der Waals forces results in a pairwise potential that, in general, has a maximum (either in the repulsive or attractive range) separating two minima (both in the attractive range).[4] A complete particle–particle contact is prevented at a cutoff point due to the presence of a semiincompressible fluid envelope surrounding the particles.[4,5] Therefore, for the phase transitions with which we are concerned, the particles do not come to the primary minimum which is at a much smaller interparticle separation distance than permitted by the fluid envelope. In this work, the key property of the potential energy curves is that the binding energy ε at the cutoff point is approximately proportional to ζ^{-2} where ζ is the zeta potential.[4] When we use the lattice model in calculating the phase diagram of the system, as will be explained below, we can use the ε as the representative interaction potential, and then make an approximation that ε is proportional to ζ^{-2}:

$$\varepsilon \propto \zeta^{-2} \tag{1}$$

It is not the purpose of this chapter to emphasize the functional dependence ζ^{-2} but to make use of the general qualitative dependence of ε on the inverse power of ζ.

In understanding experimental observations of phase transitions in colloidal systems, we distinguish the essential features and additional features. The essential features can be provided by equilibrium phase diagrams of atomic systems. Since we are interested in qualitative properties of the system, 2-D phase diagrams, rather than three, are sufficient for understanding the basic phase diagram properties. Further, in order to make the theoretical treatment of phase diagrams easier for a variety of cases, we work with lattice-gas model rather than with the more rigorous continuum space computation which is time consuming. An example of the 2-D lattice gas model calculations is that of Kikuchi and Cahn,[6] which treated the grain boundary melting phenomenon, and is accepted as predicting essential features of the processes occurring in 3-D grain boundaries. In this chapter we also base our discussions on results due to the lattice gas model which was used in Ref[6].

The method of formulation leading to the theoretical phase diagram of Fig. 54.1a is sketched here. The calculation was done using the cluster variation method (CVM). The system contains one kind of particles and vacancies and

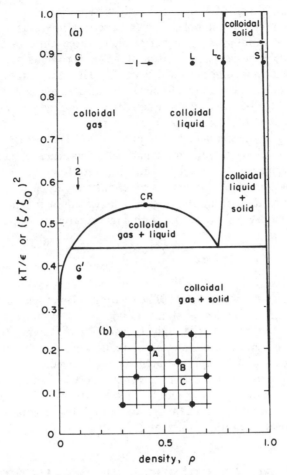

Figure 54.1   (*a*) Phase diagram for a one component colloidal system of monosize spherical part-icles calculated by the cluster variation method based on the two dimensional square lattice model of (*b*). Density scale is normalized with respect to a fully occupied lattice structure.

is described using the square lattice as the underlying structure Fig. 54.1*b*. The interparticle potentials are chosen in the same way as by Orban et al.[7] When a particle exists at *A*, the nearest-, second-, and third-neighbor lattice points are excluded, and a second particle can approach *A* up to point *B*. In the treatment, we used two interaction potentials ε_{AB} and ε_{AC}. Particle pairs farther away are not considered as contributing to the potential energy. The equilibrium state is derived by writing the Helmholtz free-energy *F* and calculating the minimum of *F* as outlined in Ref. 6. In the phase diagram of Fig. 54.1*a*, the temperature scale is normalized by $\varepsilon = |\varepsilon_{AC}|$. Similarly, the density scale is normalized by the density of a fully occupied lattice structure. We use

the phase diagram in Fig. 54.1 in *qualitatively* interpreting experimental observations. As many previous examples show, theoretical diagrams of 2-D systems
can safely be used in interpreting experiments of three dimensions. Naturally
we avoid accurate *numerical* comparison of 3-D experiments and 2-D theory.
We can also use the phase diagrams which are calculated for equilibrium systems
in interpreting nonequilibrium phenomena. In this case again, the equilibrium
calculations are to be used as a qualitative guideline; when numerical
comparison is to be attempted, special care is needed.

   In atomic systems, the reduced temperature scale kT/ε of the phase diagram
is only proportional to the thermodynamic temperature T since the pair
potential in a given system is usually treated as fixed. However, when we work
with colloidal systems, the interaction potential can be easily varied. Thus,
in the case of electrostatically interacting particle systems, using the approximate
property expressed in (1), we can work with a reduced temperature:

$$T_R = \frac{kT}{\varepsilon} = \left(\frac{\zeta}{\zeta_0}\right)^2 \tag{2}$$

where ζ_0 is a normalization constant. The relation (2) is crucial in interpreting
the phase diagram of colloidal systems as analogous to the three aggregate
phases of atomic systems. At the high T_R end of the system, our treatment
approaches that of repulsive interaction systems, while for the low T_R region
the attractive interaction is appropriately taken into account. The essential
point of our proposed interpretation (2) is that even when the thermodynamic
temperature T is kept fixed, the reduced temperature T_R can be varied by
changing the ζ-potential, and thus the "temperature" versus composition phase
diagram can be used even for a constant T.

HIERARCHICAL STRUCTURE OF CLUSTERS

We can now use the phase diagram in Fig. 54.1 in interpreting experimental
observations. In a dilute suspension, for example at the point G in Fig. 54.1a,
colloidal systems display gaslike behavior. As the number density ρ of the
colloidal particles is increased (along path 1), this gaslike state changes continuously to a liquidlike state at L. Finally, for ρ larger than a certain density
L_c, the liquidlike state transforms to a solidlike state S through a first order
transition.[1-3] In the phase diagram of Fig. 54.1a, the region where such
transitions are observed is above the critical point, CR.

   On the other hand, when this same suspension displaying gaslike behavior
is shifted (along path 2) from its dispersed state G, to a new state G' of a lower
kT/ε or ζ-potential value, we observe a distinct first order transition of gas to
condensed phase. Our work with nearly monosize (0.7 μm) SiO_2 particle
systems showed that, when a colloidal system is within the miscibility gap,
some of the particles in the suspension start forming permanent multiparticle

clusters, that is, colloidal solids.[8] In the gravitational field, due to their higher effective mass, these clusters are separated from the remaining portion of the primary particles as a result of differential settling. The primary particles, which stay dispersed until they eventually sediment due to gravitational force, result in the formation of a cloudy supernate, that is, colloidal gas.

This phenomenon of cloudy supernate formation has long been observed by various researchers in flocculating, that is, condensing, systems.[9,10] Experiments by Siano[10] suggested the possibility of spinodal-decomposition like fluid–fluid phase segregation in colloidal systems. This interpretation is in agreement with the phase diagram presented here. Similarly, the work of Vincent and coworkers[9] illustrated the coexistence of singlet particles as the gas phase with clusters of particles in weakly interacting particle systems in accordance with our interpretation. In recent theoretical treatments, the coexistence of colloidal gas and liquid phases below a critical point has also been predicted.[11] Thus, in view of our experiments and supportive evidence in the literature, we propose that the stability regions of colloidal phases are outlined in the generalized phase diagram of Fig. 54.1a.

Our theoretical treatment of the phase diagram (Fig. 54.1) has been done for an idealizing condition that the colloidal solid can be assumed to form a single crystal and its structure approaches a perfect state (i.e., no vacancies) as kT/ε approaches zero. However, experiments are usually far from ideal and the formation of polycrystalline structures is the rule rather than the exception.[12,13] Microstructural variations in these polycrystalline colloidal solids formed at three different ζ-potential levels of the phase diagram are illustrated in the scanning electron micrographs of Fig. 54.2. Particle clusters are formed during colloidal solidification. Note the special arrangement of these particle clusters. The first generation of clusters begin as domains and are formed by close packing of primary particles. The domain size decreases with increasing interparticle binding energy or decreasing T_R. The collection of a number of these domains results in the formation of second generation clusters. The structure of these second generation clusters displays continuous variations in the interdomain void space. At the high kT/ε end of the spectrum where the interparticle binding energy is low, tight domain interlocking results in polydomain structures which closely resemble polycrystalline atomic structures and are easily recognized as colloidal crystals due to their iridescent characteristics.[1,2]

With increasing interparticle binding energies, domain interlocking efficiency decreases since domains become increasingly rigid and thus behave as hard sphere packing units themselves. The combined effect of the decreasing domain size and the increasing interdomain void space is the loss of the iridescent property of the colloidal solids. At the low T_R end, an additional contribution to the low packing efficiency may be the formation of third generation clusters and the associated void space with the grouping of second generation clusters, Fig. 54.3. The most important concept illustrated by our observations is that a hierarchy of microstructures can be obtained by varying the degree of interaction between particles as described by T_R.

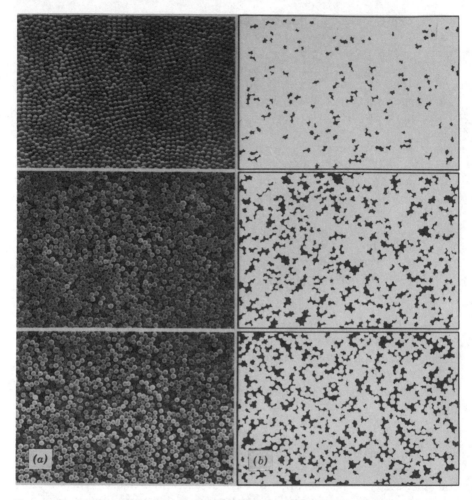

Figure 54.2 (a) The arrangement of particle domains formed by centrifugal sedimentation of SiO_2 microspheres (average diameter = 0.7 μm) in H_2O at $\zeta = 110$ mV (top), $\zeta = 68$ mV (middle), and $\zeta = 0$ mV (bottom);[12] (b) images where only the second generation voids are highlighted as dark regions in order to illustrate the continuous variations in the domain size with ζ-potential.

This experimental evidence on the structure of colloidal solids suggests the modification needed in interpreting the theoretical phase diagram of Fig. 54.1a when compared with experiments. Remember that the density of the solid phase in the theory is only with respect to the intradomain regions. In Fig. 54.3a, we provide experimental data on the density of colloidal solids, obtained by sedimentation volume measurements, as a function of kT/ε. As the interparticle binding energy is increased (going to smaller kT/ε values), a significant decrease in the density of colloidal solids is observed below the critical point. The prime cause of this density decrease, in the subcritical region,

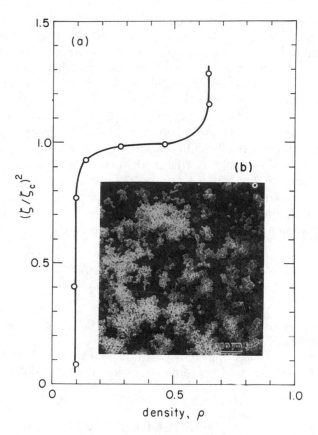

Figure 54.3 (*a*) Variations in the sedimentation density (relative to the total volume) of colloidal solids as a function of $(\zeta/\zeta_{cr})^2$ where ζ_{cr} is the zeta potential at the critical point; (*b*) scanning electron micrograph of particle clusters formed at low T_R values.

is the retention of third generation void space in the colloidal solids formed by gravitational settling. However, when such a system is consolidated further by centrifugation, all the third generation voids can be eliminated completely and the density becomes greater than that shown in Fig. 54.3*a*.[12]

CONCLUSION

In colloidal systems the interparticle energy ε can be controlled through chemical adjustments to the fluid matrix surrounding the colloidal particles. An important implication of this property is that even when the thermodynamic temperature is kept fixed, changing the potential energy by other means than T has the same effect as changing T of phase diagrams for atomic systems. The stability regions of colloidal gas, liquid, and solid phases are outlined with a theroetically calculated kT/ε versus density phase diagram. In experiments, it is noted that the colloidal solids display (continuous) variations in the size and packing of particle clusters as a function of kT/ε. The first, second, and third generations of clusters and voids are clearly distinguished. This observation on the hierarchical nature of the colloidal solids plays a key role in our generalized treatment of the colloidal phase transitions.

The types of phase transitions discussed in this paper are of interest in diverse fields.[2] For instance, colloids play an important role in the processing of high-technology ceramics.[14] Colloidal fluid-to-solid transitions have been widely observed in biological systems.[2] Furthermore, these colloids are ideal models for simulating atomic systems. Although the theoretical phase diagram presented here is based on idealized colloidal phase structures, it serves[12,14,15] the purpose of outlining colloidal phase stability regions in a generalized form with sufficient accuracy when we accept the new hierarchical interpretation of clusters and voids observed in the solid structure.

ACKNOWLEDGMENTS

The major part of this work was sponsored by the Office of Naval Research under Contract No. N00014-82-K-0336 and in part by the Advanced Research Projects Agency of the Department of Defense and was monitored by the Air Force Office of Scientific Research under Grant No. AFOSR-83-0375.

REFERENCES

1. T. Alfrey, Jr., E. B. Bradford, and J. W. Vanderhoff, *J. Opt. Soc. Amer.* **44**, 603 (1954); W. Luck, M. Klier, and H. Wesslau, *Ber. Bunsenges. Phys. Chem.* 67, 75, 84 (1963); P. A. Hiltner and I. M. Krieger, *J. Phys. Chem.* **73**, 2386 (1969); A. Kose, M. Ozaki, K. Takano, Y. Kobayashi, and S. Hachisu, *J. Colloid Interface Sci.* **44**, 330 (1973).

2. I. F. Efremov, in *Surface and Colloid Science*, E. Matijević, Ed., Wiley, New York, 1976, Vol. 8, pp. 85–192; and P. Pieranski, *Contemp. Phys.* **24**, 25 (1983) provide extensive reviews on ordered colloidal systems.

3. J. A. Beunen and L. R. White, *Colloids and Surfaces* **3**, 371 (1981); P. M. Chaikin, P. Pincus, S. Alexander, and D. Hone, *J. Colloid Interface Sci.* **89**, 555 (1982); and references cited in these to earlier literature.

4. J. Th. G. Overbeek, *J. Colloid Interface Sci.* **58**, 408 (1977).

5. G. Frens and J. Th. G. Overbeek, *J. Colloid Interface Sci.* **38**, 376 (1972).

6. R. Kikuchi and J. W. Cahn, *Phys. Rev.* **B21**, 1893 (1980).

7. J. Orban, J. van Crean, and A. Bellmans, *J. Chem. Phys.* **49**, 1778 (1968).

8. A detailed account of this work which was performed with an X-ray absorption unit will be published elsewhere.

9. J. A. Long, D. W. J. Osmond, and B. Vincent, *J. Colloid Interface Sci.* **42**, 545 (1973); and C. Cowell and B. Vincent, ibid **87**, 518 (1982).

10. D. B. Siano, *J. Colloid Interface Sci.* **68**, 111 (1979).

11. M. J. Grimson, *J. Chem. Soc., Faraday Trans.* **2**, 79, 817 (1983); and J. M. Victor and J. P. Hansen, *J. Phys. Lett.* **45**, L-307 (1984).

12. R. M. Allman, III and I. A. Aksay, to be published.

13. P. Pieranski, in *Physics of Defects*, R. Balian et al., Eds., North-Holland, Amsterdam, 1981, pp. 183–200.

14. I. A. Aksay, in *Advances in Ceramics*, J. A. Mangels and G. L. Messing. Eds., American Ceramic Society, Columbus, OH, 1984, Vol. 9, pp. 94–104.

15. M. Yasrebi, Kinetics of Flocculation in Aqueous Alpha Alumina Suspensions, M.Sc. Thesis, UCLA, 1984.

55

RHEOLOGICAL SCIENCE IN
CERAMIC PROCESSING

MICHAEL D. SACKS
Department of Materials Science and Engineering
University of Florida
Gainesville, Florida

INTRODUCTION

Rheology is important in numerous ceramic processing operations, including (1) wet mixing/milling, (2) shape forming (e.g., slip casting, tape casting, extrustion, injection molding, fiber drawing, fluidized bed spray granulation, etc.), and (3) coating/deposition (e.g., dipping, spinning, spraying, etc.). Proper control over rheological properties is essential, not only for efficiency in processing, but also for the development of optimum physical properties in the processed material (powder, green body, surface coating, etc.). In addition, rheological measurements are extremely useful in the structural characterization of particle–liquid systems (i.e., suspensions, injection molding mixes, etc.). In this chapter illustrative examples of the role of rheological science in ceramic processing are provided.

EXPERIMENTAL

Rheological properties were determined by several methods. Shear stress, τ, versus shear rate, $\dot{\gamma}$, flow curves were generated using concentric cylinder*

*Model RV-100/CV-100, Haake, Inc., Saddle Brook, NJ.

and cone-plate* viscometers. Viscosity values, η, were obtained from the flow curves using the relationship:

$$\eta = \frac{\tau}{\dot{\gamma}}$$

(1)

Intrinsic viscosity measurements were made using a capillary viscometer.† A microelectrophoresis apparatus‡ was used to measure the electrokinetic mobility of particles. Zeta potentials were determined from the electrophoretic mobilities using the Helmholtz–Smoluchowski equation.[1] Polymer adsorption measurements were made by the gravimetric solution depletion method. Further details on experimental methods and materials are provided elsewhere.[2-6]

RESULTS AND DISCUSSION

Characterization of the State of Dispersion

Effect of pH

Rheological measurements are very useful in assessing the state of particulate dispersion in powder–liquid suspensions. Figure 55.1a shows the rheological flow curves (i.e., shear stress versus shear rate behavior) for aqueous suspensions containing 20 vol% silica. The flow behavior for the pH = 3 suspension indicates that extensive flocculation has occurred. The viscosity values are high and flow is highly shear thinning, i.e., the viscosity decreases with increasing shear rate (Fig. 55.1b). In flocculated suspensions, liquid is immobilized in the interparticulate void space of the flocs and floc networks. This increases the "effective" solids loading (relative to a well-dispersed suspension), resulting in higher viscosities. If the flocs are broken down by shearing the suspension, occluded liquid is released and the viscosity decreases, shear thinning flow is observed. In contrast to this behavior, the pH = 10 suspension shows Newtonian flow, that is, the viscosity is independent of shear rate. This is characteristic of a well-dispersed suspension.

The effect of pH on the state of dispersion is explained by acid–base reactions at the silica surface which alter the particle surface charge:

$$SiOH_{(s)} + H^+_{(aq)} = SiOH^+_{2(s)}$$

(2)

$$SiOH_{(s)} + OH^-_{(aq)} = SiO^-_{(s)} + H_2O_{(l)}$$

(3)

*Wells-Brookfield Engineering Laboratories, Inc., Stoughton, MA.
†ASTM Size 0B Ubblelohde, Industrial Research Glassware, Ltd., Union, NJ.
‡Mark II, Rank Brothers, Cambridge, England.

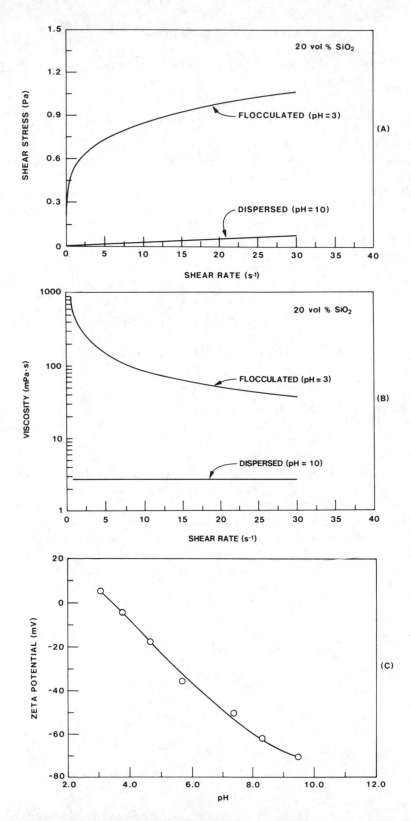

Electrokinetic measurements (Fig. 55.1c) show that zeta potentials vary by ~ 75 mV (absolute value) over the pH range ~ 3–10. With other factors constant, improved dispersion is expected as the zeta potential (absolute value) increases since this results in larger electrostatic repulsion forces between particles.[1]

The state of dispersion in suspension has an important influence on the green microstructure which develops during powder consolidation (and liquid removal). Well-dispersed suspensions tend to produce high-density, homogeneous compacts. Sedimentation of monosized, spherical silica particles from a well-dispersed suspension (e.g., pH = 10) produces highly ordered compacts (Fig. 55.2a). In contrast, compacts formed from flocculated suspensions (e.g., pH = 3) tend to have larger total porosity and median pore radius.[4] Figure 55.2b shows that the compact contains tightly packed particle clusters (flocs) with large amounts of intercluster (interfloc) porosity. These differences in green microstructure have a dramatic effect on the subsequent sintering behavior.[7] The highly ordered compact sinters to near theoretical density at 1000°C (Fig. 55.3a). In contrast, the sample prepared from the flocculated suspension remains highly porous after sintering under the same conditions

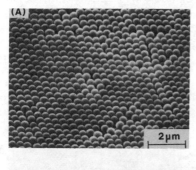

(A)

2 μm

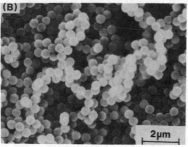

(B)

2μm

Figure 55.2 Scanning electron micrographs of compacts prepared from (A) well-dispersed (pH = 10) and (B) flocculated (pH = 3) suspensions.

Figure 55.1 (A) Plots of shear stress versus shear rate for well-dispersed (pH = 10) and flocculated (pH = 3) suspensions. (B) Plots of suspension viscosity versus shear rate for well-dispersed (pH = 10) and flocculated (pH = 3) suspensions. (C) Plot of zeta potential versus pH for silica.

(Fig. 55.3*b*). Only local densification (of tightly packed particle clusters) occurs. Enhanced sintering of the ordered compact is due to the large number of nearest-neighbor contacts per particle and uniform small pore size. Rearrangement and differential microdensification processes, which produce regions with larger pores and lower shrinkage rates, are minimized in ordered compacts.

Effect of Adsorbed Polymer

Rheological measurements are also useful in assessing the effect of polymeric additives (e.g., binders, plasticizers, etc.) on particulate dispersion. Figures 55.4 and 55.5 show the effect of polyvinyl butyral resin (0.5 vol% PVB-1 addition) on the rheological behavior of Al_2O_3 and silicate glass suspensions (30 vol% solids) prepared with two nonaqueous liquids, methanol (MEOH) and methyl isobutyl ketone (MIBK). All suspensions without polymer were flocculated, as indicated by highly shear thinning behavior (Figs. 55.4 and 55.5). PVB additions (0.5 vol%) had essentially no effect on the flow behavior of Al_2O_3 and silicate glass suspensions prepared with MEOH (Figs. 55.4*a* and 55.5*a*, respectively). However, in suspensions prepared with 75 vol% MIBK in the liquid phase (3:1, MIBK/MEOH), the polymer addition produced well-dispersed suspensions with approximately Newtonian behavior (Figs. 55.4*b* and 55.5*b*, respectively). The available evidence indicates that these suspensions are

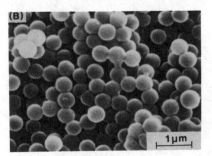

Figure 55.3 Scanning electron micrographs of sintered (1000°C, 24 hr) compacts prepared from (A) well-dispersed and (B) flocculated suspensions.

sterically stabilized. There are three important requirements for effective steric stabilization[8]:

1. The particle surface should be well covered by adsorbed polymer. Figure 55.6 shows that the amount of polymer adsorbed on Al_2O_3 and silicate glass powders increases with the MIBK content in the suspension liquid. The higher surface coverage of the 3:1, MIBK/MEOH suspensions (~ 2.1 mg PVB/g of powder) relative to the MEOH suspensions (~ 0.7–0.9 PVB mg/g of powder) is consistent with the good dispersion of the former suspensions.

2. The stabilizing moieties (i.e., the polymer "loops" and "tails" which extend into the liquid solution phase) should be well solvated by the suspension liquid medium. (The liquid should be better than a theta solvent.) In this study, the solvent quality was evaluated for various liquid (MIBK/MEOH) ratios by intrinsic viscosity measurements. The intrinsic viscosity provides information on the hydrodynamic size of the polymer molecules in solution. In poor quality solvents, those worse than theta solvents, the polymer molecules are tightly coiled and low intrinsic viscosities are measured. As the quality of the solvent improves, the polymer molecule becomes more expanded and higher intrinsic viscosities are measured. The results in Fig. 55.7 indicate that 3:1, MIBK/MEOH is the best solvent for the PVB, while the MEOH is the worst solvent. Again, this is consistent with the good stability against flocculation of the 3:1, MIBK/MEOH suspensions.

3. The polymer should be strongly anchored to the particle surface. In this study, good anchoring may be attributed to the hydroxyl groups in the PVB resin. These groups can form hydrogen bonds with hydroxyl groups and/or adsorbed water on the oxide surfaces. The importance of hydroxyl groups was demonstrated by comparing the adsorption behavior of PVB resins with different hydroxyl contents (see Table 55.1). In each case, lower adsorption was observed with the lower hydroxyl content polymer (PVB-4).

Effect of Solids Loading

The state of particulate dispersion often is affected by the suspension solids loading (i.e., the solid/liquid ratio). Rheological characterization is particularly useful, since measurements can be made over a wide range of loadings. Figure 55.8 shows viscosity versus shear rate curves for aqueous and nonaqueous silicon suspensions prepared with various solids loading. The aqueous suspension with 30 vol% solids has a low viscosity and Newtonian flow behavior. The measured zeta potential is ~ 50 mV, indicating that the suspension is

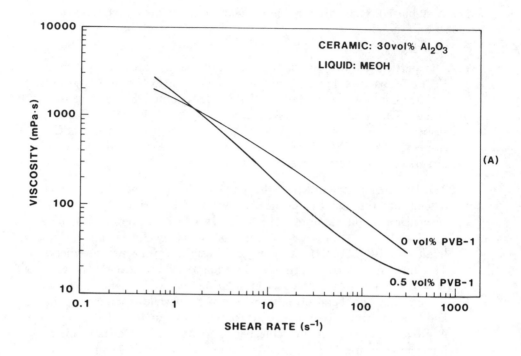

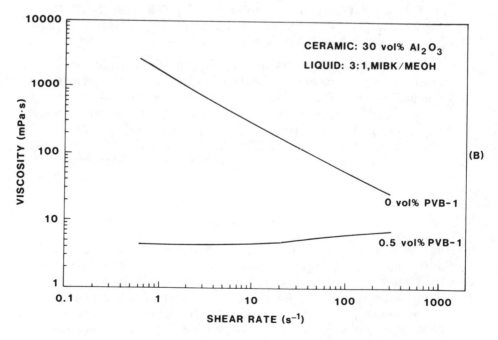

Figure 55.4 Plots of viscosity versus shear rate for 30 vol% Al_2O_3 suspensions prepared using (A) MEOH and (B) 3:1, MIBK/MEOH with indicated polymer (PVB-1) concentrations.

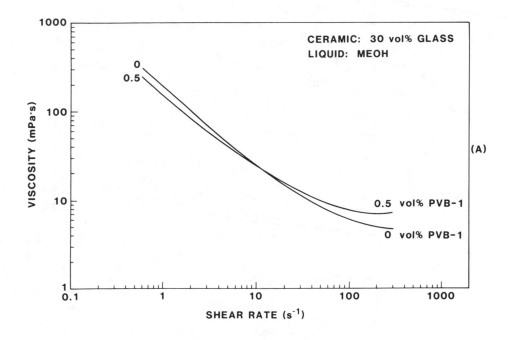

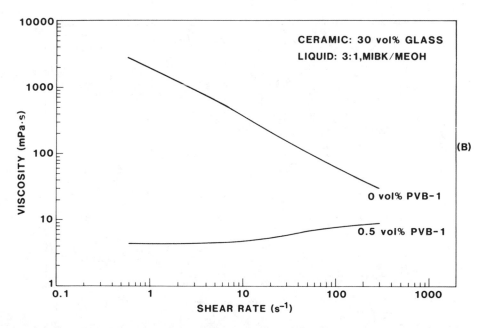

Figure 55.5 Plots of viscosity versus shear rate for 30 vol% silicate glass suspensions prepared using (A) MEOH and (B) 3:1, MIBK/MEOH with indicated polymer (PVB-1) concentrations.

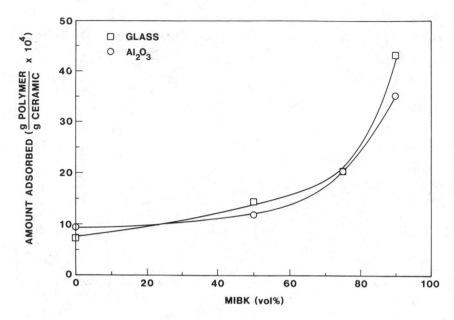

Figure 55.6 Plots of amount of adsorbed polymer (PVB-1) versus MIBK content of liquid phase for Al$_2$O$_3$ and silicate glass suspensions.

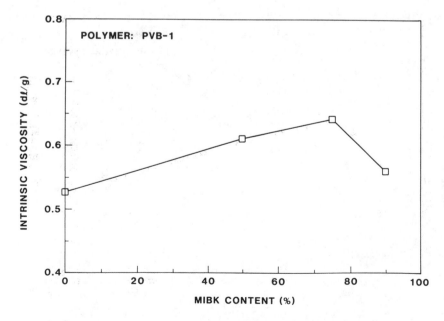

Figure 55.7 Plot of intrinsic viscosity for PVB-1 solutions prepared with various MIBK/MEOH ratios.

TABLE 55.1 Effect of Hydroxyl Content in PVB Resin on Adsorption Behavior[a]

Ceramic	Polymer	Amount Adsorbed (g polymer/g ceramic) $\times 10^4$
Al_2O_3	PVB-1[b]	25
Al_2O_3	PVB-4[c]	6
Glass	PVB-1[b]	25
Glass	PVB-4[c]	3

[a]Suspension composition was 30 vol% ceramic, 69 vol% 3:1, MIBK/MEOH, and 1.0 vol% PVB.
[b]B-98, Monsanto, Co., St. Louis, MO. Hydroxyl content 18–20%.
[c]B-79, Monsanto, Co., St. Louis, MO. Hydroxyl content 9–13%.

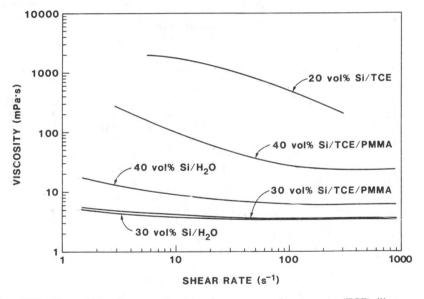

Figure 55.8 Plots of viscosity versus shear rate for aqueous and nonaqueous (TCE) silicon suspensions prepared with indicated solids loadings. 30 and 40 vol% silicon suspensions prepared with trichloroethylene (TCE) contained 1.0 wt% polymethyl methacrylate (PMMA).

electrostatically stabilized. Good dispersion is also obtained in a 30 vol% silicon/trichloroethylene (TCE) suspension containing 1.0 wt% polymethyl methacrylate (PMMA). A suspension prepared with only 20 vol% silicon in TCE, but with no PMMA, is highly flocculated. Several observations indicate that the 30 vol% silicon/TCE/PMMA suspension is sterically stabilized. First, strong adsorption (~ 3 mg PMMA/g silicon) was observed. Silicon particle

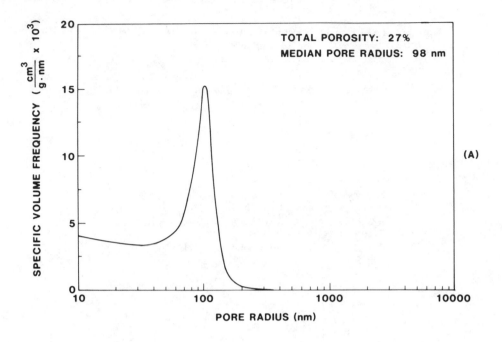

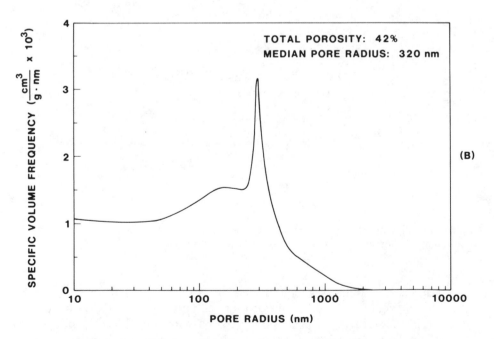

Figure 55.9 Plots of specific pore volume frequency function versus pore radius for compacts prepared from well-dispersed (A) aqueous and (B) nonaqueous (TCE/1.0 wt% PMMA) 30 vol% silicon suspensions.

surfaces were substantially oxidized, resulting in a "silica-like" surface.[5] A number of previous studies have also shown that PMMA in certain chlorinated solvents adsorbs strongly on silica surfaces.[9-11] Second, TCE is known to be a good quality solvent for PMMA.[10]

Despite almost identical behavior at 30 vol% solids loading, the aqueous and nonaqueous silicon suspensions with 40 vol% solids have extremely different flow characteristics (Fig. 55.8). The aqueous suspension, with only a moderate increase in viscosity, remains relatively well dispersed, while the TCE suspension, with high viscosities and highly shear thinning behavior, is extensively flocculated. In the TCE suspension, the repulsive energy barrier, which prevents flocculation in the suspension with only 30 vol% solids, is evidently surmounted when the average particle–particle separation distance is reduced, due to the increased particle concentration. The same effect occurs in electrostatically stabilized aqueous suspensions when the interparticle repulsion forces are lower, that is, due to lower zeta potentials.[5] This effect may have serious consequences in forming operations, such as slip casting, since the solids loading increases as the liquid phase is removed. This problem was illustrated by slip casting the two (aqueous and nonaqueous) well-dispersed (30 vol% solids) suspensions. Pore size distributions for the consolidated samples were determined by mercury porosimetry* (Fig. 55.9). The aqueous suspension, with large interparticle repulsive forces, forms a powder compact with high density ($\sim 27\%$ total porosity) and small pore size (~ 98 nm median pore radius), Fig. 55.9a. In contrast, the TCE suspension flocculates during casting and forms a compact with much lower density ($\sim 42\%$ total porosity) and larger pore size (~ 320 nm median pore radius), Fig. 55.9b.

Process Optimization

Coating/Fiber Drawing

Certain ceramic processing operations are dependent upon achieving a particular type of rheological flow behavior. Shear thinning flow is important in some coating operations. At high shear rates, as the coating is applied, the low viscosity allows rapid, uniform coverage of the coating surface. After the applied shear stress is removed, the coating will not "run" due to the high viscosity of the shear thinning material.

The type of flow behavior is also important in drawing fibers from solutions. Sols suitable for fiber drawing can be prepared by hydrolysis/condensation of tetraethylorthosilicate (TEOS) under acid-catalyzed, low water content conditions.[12,13] Figure 55.10 shows shear stress versus shear rate curves at various aging times for a sol prepared with a water/TEOS mole ratio of 2.0 and a nitric acid/TEOS mole ratio of 0.1. "Spinnability" was tested by dipping a glass rod into the sol and attempting to draw a fiber. Fibers could not be drawn

*Model SP-100, Quantachrome Corporation, Syosett, NY.

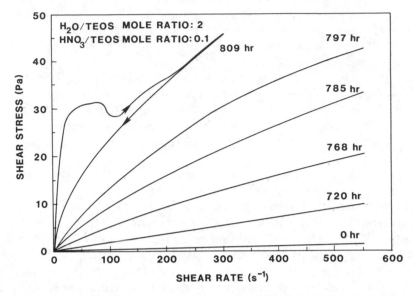

Figure 55.10 Plots of shear stress versus shear rate for a "silica" sol at indicated aging times.

during the period of Newtonian flow or during the early stages of shear thinning flow. As the shear thinning flow behavior became more pronounced (785 hr), the sol developed "spinnability." Optimum "spinnability" was observed at 797 hr, that is, just prior to the development of hysteresis in the flow curve (thixotropic behavior). Sols were no longer "spinnable" when the flow curves showed large yield stresses and extensive hysteresis (i.e., highly thixotropic behavior).[6]

Mixing/Milling

The efficiency of wet milling operations depends on the suspension rheology. In particular, suspension viscosity affects the velocity of the grinding media, and, therefore, affects the energy and frequency of impact events in the mill. The importance of viscosity is illustrated in Fig. 55.11 for the vibratory milling* of aluminum oxide. Milling was carried out for 2 hr using various solids loadings (14, 20, and 25 vol% Al_2O_3) and deflocculant concentrations (0–0.2 mole/L sodium citrate). For each solids loading, the median particle size decreases sharply (i.e., grinding is enhanced) with the initial addition of sodium citrate (Fig. 55.11a). This is directly related to the large initial decrease in suspension viscosity (Fig. 55.11b). Sodium citrate additions electrostatically stabilize the

*M-18-5, Sweco, Inc., Florence, KY.

Figure 55.11 Plots of (A) median particle size, (B) suspension viscosity, and (C) zeta potential versus sodium citrate concentration for 14, 20, and 25 vol% Al_2O_3 suspensions milled 2 hr.

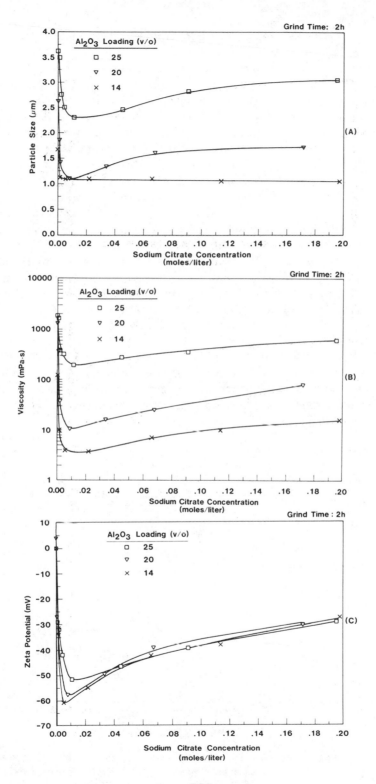

suspension, as indicated by the large increase (absolute value) in the zeta potential, Fig. 55.11c. At sodium citrate concentrations $\gtrsim 10^{-2}$ mole/L, the median particle size after grinding gradually increases for the 20 and 25 vol% Al_2O_3 suspensions. This increase is directly related to the increase in viscosity and the decrease in zeta potential (absolute value). In contrast, the particle size for the 14 vol% Al_2O_3 suspensions remains approximately constant at the higher sodium citrate concentrations, despite the increase in viscosity and decrease in the zeta potential (absolute value). These results indicate that grinding efficiency is dependent on viscosity except when viscosity values are very low, as in this case, below $\sim$ 10–15 mPa·s at a shear rate of 375 s^{-1}. To maximize material throughput in a milling operation, well-dispersed suspensions should be used since this allows for higher solids loading at a given viscosity value.

The efficiency of deagglomeration processes can be assessed by rheological measurements. A study was done in which several suspension preparation procedures were compared: (1) mixing the components for 15 min using a paint shaker, (2) mixing the components for 15 min using a paint shaker, followed by 15 min of ultrasonication, and (3) "washing" the powder with methanol, followed by mixing the components for 15 min using a paint shaker. Suspensions were prepared with agglomerated Al_2O_3 powder (30 vol%), 3:1, MIBK/MEOH liquid (69.5 vol%), and PVB-1 (0.5 vol%). "Soft" agglomerates of Al_2O_3 were intentionally formed by tumbling the as-received powder in a polyethylene container on a roller. Each preparation method produced suspensions with

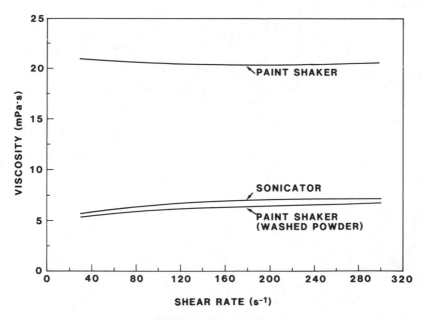

Figure 55.12 Effect of suspension preparation procedure on the viscosity versus shear rate behavior of a 30 vol% Al_2O_3 suspension prepared using 3:1, MIBK/MEOH and 0.5 vol% PVB-1.

approximately Newtonian flow behavior (Fig. 55.12). Based on the results shown in Figs. 55.4-.7, this is due to steric stabilization of the "particles," that is, primary particles and/or agglomerates. However, the suspension prepared by the first method (paint shaker only) had a higher viscosity, indicating that less agglomerate breakdown had occurred. As in suspensions which contain flocs, liquid is immobilized in the interparticulate void space of the agglomerates. This results in a higher "effective" solids loading, and higher viscosity, compared to a suspension of well-dispersed primary particles.

The results in Fig. 55.12 illustrate that suspensions with Newtonian flow behavior are not necessarily well dispersed, that is, dispersed as *primary* particles. To some extent, the state of dispersion can be evaluated from the viscosity value, measured at a known shear rate on a suspension of known solids loading. However, the viscosity depends on many factors in addition to shear rate and solids loading, such as particle size distribution, particle shape, nature of the interparticle forces, and so on. For an informative review, see the paper by Goodwin.[14] Although qualitative trends are well established, most quantitative treatments are restricted to suspensions of spherical, monosized particles with low solids loadings and minimal particle–particle interactions.

CONCLUSION

Certain processing operations, such as fiber drawing from TEOS-derived sols, require a particular type of rheological flow behavior. In other cases, processing efficiency is improved with proper control over rheological characteristics. Rheological measurements are particularly useful in characterizing the state of dispersion in concentrated particle–liquid suspensions. In turn, the state of dispersion has an important effect on shape-forming operations, green micro-structural development, and subsequent sintering behavior.

ACKNOWLEDGMENTS

This paper is based on the work of many people at the University of Florida, including C. S. Khadilkar, H. W. Lee, O. Rojas, G. W. Scheiffele, R.-S. Sheu, and T.-Y. Tseng. Their contributions are gratefully acknowledged, as is the financial support provided by Dresser Industries, Garrett Turbine Engine Co. (through DOE-NASA contract DEN3-167), and the National Science Foundation (DMR-8024320).

REFERENCES

1. J. Th. G. Overbeek, The Interaction Between Colloid Particles, in *Colloid Science*, H. R. Kruyt, Ed., Elsevier, Amsterdam 1952, pp. 245–277.

2. M. D. Sacks and T.-Y. Tseng, Role of Sodium Citrate in Aqueous Milling of Aluminum Oxide, *J. Am. Ceram. Soc.* **66**(4), 242–247 (1983).

3. M. D. Sacks and C. S. Khadilkar, Milling and Suspension Behavior of Al_2O_3 in Methanol and Methyl Isobutyl Ketone, *J. Am. Ceram. Soc.* **66**(7), 488–494 (1983).

4. M. D. Sacks and T.-Y. Tseng, Preparation of SiO_2 Glass from Model Powder Compacts: I, Formation and Characterization of Powders, Suspensions, and Green Compacts, *J. Am. Ceram. Soc.* **67**(8), 526–532 (1984).

5. M. D. Sacks, Properties of Silicon Suspensions and Cast Bodies, *Am. Ceram. Soc. Bull.* **63**(12), 1510–1515 (1984).

6. M. D. Sacks and R. S. Sheu, Rheological Characterization During the Sol-Gel Transition, Chapter 10, this volume.

7. M. D. Sacks and T.-Y. Tseng, Preparation of SiO_2 Glass from Model Powder Compacts: II, Sintering, *J. Am. Ceram. Soc.* **67**(8), 532–537 (1984).

8. T. Sato and R. Ruch, *Stabilization of Colloidal Dispersion by Polymer Adsorption*, Marcel Dekker, New York, 1984.

9. E. Hamori, W. C. Forsman, and R. E. Hughes, Adsorption of Poly (Methyl Methacrylate) from Dilute Solution by Silica and Silicic Acid, *Macromolecules* **4**(2), 193–198 (1981).

10. J. M. Herd, A. J. Hopkins, and G. J. Howard, Adsorption of Polymers at the Solution-Solid Interface, IV. Styrene-Methyl Methacrylate Copolymers on Silica, *J. Polym. Sci.: Part C,* No. 34, 221–226 (1971).

11. F. M. Fowkes and M. A. Mostafa, Acid-Base Interactions in Polymer Adsorption, *Ind. Eng. Chem. Prod. Res. Dev.* **17**(1), 3–7 (1978).

12. S. Sakka, Gel Method for Making Glass, in *Treatise on Materials Science and Technology, Vol. 22, Glass III.* M. Tomozawa and R. Doremus, Eds., Academic Press, New York, 1982, pp. 129–167.

13. S. Sakka and K. Kamiya, Preparation of Shaped Glasses Through the Sol-Gel Method, in *Materials Science Research*, Vol. 17, *Emergent Methods for High-Technology Ceramics*, R. F. Davis, H. Palmour III, and R. L. Porter, Eds., Plenum Press, New York, 1984, pp. 83–94.

14. J. W. Goodwin, The Rheology of Dispersions," in *Colloid Science*, Vol. 2, D. H. Everett (Senior Reporter), The Chemical Society, London, 1975, pp. 246–293.

56

COLLOIDAL BEHAVIOR OF SILICON CARBIDE AND SILICON NITRIDE

M. J. CRIMP
Department of Metallurgy and Materials Science
Case Western Reserve University
Cleveland, Ohio

R. E. JOHNSON, Jr.
Central Research and Development Department
Experiment Station
E. I. DuPont de Nemours and Company
Wilmington, Delaware

J. W. HALLORAN
Department of Metallurgy and Materials Science
Case Western Reserve University
Cleveland, Ohio

D. L. FEKE
Department of Chemical Engineering
Case Western Reserve University
Cleveland, Ohio

INTRODUCTION

The control of the packing structure of particles is the key to successful processing of high-reliability ceramics. Agglomeration and suspension rheology can be controlled by manipulating particle charge and double layer interactions. In

539

aqueous systems, particle charge can arise from adsorption of hydrogen or other ions at certain sites along the surface. For hydrous oxides, these surface functional groups are usually amphoteric, with an acidity described by an equilibrium constant pK_A and a basicity described by pK_B. The colloidal behavior depends strongly on the type and number of the surface functional groups and their relative acidity, expressed by $\Delta pK = pK_A - pK_B$. These factors, with the appropriate equilibrium constants for ion adsorption, allow the prediction of surface charge and surface potential, and can be used to interpret experimentally accessible properties such as hydrogen ion adsorption and zeta potential.[1,2]

Much information is available on the nature and behavior of surface functional groups on the hydrous oxides,[3] but, with the exception of carbon blacks,[4] relatively little is known about nonoxide covalent inorganics such as silicon carbide and silicon nitride. It is not clear if the surfaces of the nitride or carbide itself can be examined, since the particle surfaces may be oxidized to some extent by an aqueous environment. If the surface were significantly oxidized, the powder would have some of the character of silica which has interfacial chemistry dominated by surface hydroxide silanol groups.[5] Silanols on silica show a large degree of acidity, having a ΔpK of about 10.[6] The behavior of silicas is known to be strongly dependent on the relative population of hydrophilic silanols and hydrophobic siloxanes, even to the point of changing the character of the material from a hydrophobic to a hydrophilic solid.[5,7] The silanol/siloxane ratio depends upon the temperature–humidity history of the powder.[8] By analogy it might be expected that the behavior of silicon nitride and carbide powders depend on the state of oxidation and hydroxylation of the surface.

Infrared spectroscopy by Crimp[9] has demonstrated the existence of silanols on the surfaces of silicon carbide and nitride, which indicates that these powders should show acidic silanol behavior. Silicon nitride additionally shows surface amine groups by infrared spectroscopy, which suggests that basic surface sites may exist on these powders.

In this chapter, we present data on the zeta potential of aqueous suspensions of these powders and early data on hydrogen ion adsorption.

MATERIALS AND EXPERIMENTAL PROCEDURES

The silicon carbide powder* was an Acheson-type α-SiC with a reported Stokes diameter of 1.7 μm and a BET surface area of 15.5 m^2/g. The chemical analysis supplied by the vendor was < 0.2 wt% free Si, < 0.4 wt% free C, and < 0.7 wt% O. The powder dispersed readily in deionized water to make a stable suspension with pH of 5.0 $\pm$ 0.5 at 43 vol% solids. Addition of HNO_3 to adjust the pH to 2.7 caused rapid flocculation and subsequent rapid settling, leaving a

*LONZA UF-10, Lonza Inc., Fairlawn, NJ.

voluminous sediment and slightly turbid supernate. Suspensions at pH 5 and pH 10.3 settled very slowly forming a compact sediment and clear supernate.

The silicon nitride powder was an α-Si_3N_4 prepared by the liquid ammonia process* with a reported mean Stokes diameter of 0.5 microns and a BET surface area of 12.7 m^2/g. The powder was 97.5% alpha phase with a composition of 38 wt% N and 1.5 wt%0. When mixed with deionized water at 42 vol% solids, the suspension reached a pH of 8–9. A portion of the powder, probably very large hard agglomerates, settled immediately. The fines remaining in suspension settled very slowly, eventually forming a compact sediment and a clear supernate. When the pH was adjusted to 6, the suspension flocculated immediately and settled within minutes to leave a voluminous sediment and a clear supernate.

To create a controlled surface, powders were "washed" by ultrasonically dispersing the powder in high-purity water, centrifuging, and oven drying the centrifuge cake. These washed powders redispersed readily. Atomic absorption spectroscopy detected 50 ppm Si in the supernate from a 20 wt% Si_3N_4 suspension and less than 5 ppm Si in the supernate from a similar SiC suspension. Titration of the supernate revealed a substantial quantity of an acidic species in the SiC supernate and a basic species in the Si_3N_4 supernate.

To determine if the powder behavior was affected by mild oxidation or dehydroxylation, samples of washed powders were heated at 500°C for 24 hr in dry oxygen or dry argon. The argon treatment was intended to reduce the population of surface hydroxyls while the oxygen treatment was intended to fully oxidize the particle surfaces.

Electrophoresis was performed at the E. I. DuPont laboratories using the PEN-KEM System 3000. Zeta potentials were calculated using the method of O'Brien and White.[10] Dilute suspensions of the washed powders in high purity water (0.5 MΩ or better) were prepared at various ionic strengths using KNO_3. The pH was adjusted with HNO_3 and KOH. For each data point, at least two separate samples were prepared from the powder and were equilibrated at pH 2.0 before attempting the measurement. Zeta potentials were determined for at least four separate aliquots of the suspensions and are reported as the average. Typical reproducibility† was ±1 mV. Samples were kept under a blanket of N_2 to prevent CO_2 contamination.

Titrations were performed on a Radiometer automatic titration system. Suspensions of 1 wt% solids were examined under CO_2-free conditions. A titration experiment during which the pH was changed from pH 3 to pH 10.5 typically required 20 min.

A MATEC System 8000 was used to determine the electrokinetic sonic amplitude (ESA) as a function of pH and ionic strength for the same samples used in the titrations. The ESA is the measured sonic signal (in microvolts) generated by the suspension as it is excited by a 1-Mhz electrical signal.

*UBE SNE-10, Ube Industries, New York, NY.
†Error limits reflect the larger of either the range of observed values or the calibration of the pH probe.

ESA and electrophoresis measurements were performed on the identical suspensions on which the hydrogen ion-adsorption titrations were done. With the experimental apparatus on hand, no dilutions or special preparations were necessary and artifacts of sample-to-sample variations were eliminated.

The rheological behavior of a 42 vol% SiC suspension was determined with a HAAKE RV-100 viscometer at 20°C.

RESULTS

Electrokinetic Measurements

Electrophoresis demonstrated that the SiC powder has an acidic character similar to silica, possessing a negative zeta potential in all but the lowest pH range. Figure 56.1 shows the zeta potential of once-washed SiC in 0.001 and $0.01N$ KNO_3. Both curves interect zero at the same point, yielding an isoelectric point (IEP) of 2.5 ± 0.2. The gradient in the zeta potential at the IEP is about 20 mV/pH unit, well below the Nernst value of 59 mV/pH. Heating at 500°C in dry oxygen or argon did not change the IEP as shown in Figure 56.2. Note that the magnitude of the zeta potential is increased by heat treatment, especially for the oxygen treated powders, which show about 60% higher zeta potentials at pH 4.

Silicon nitride is quite different, possessing a much more basic character than SiC. Figure 56.3 shows the zeta potential of once-washed Si_3N_4 in 0.001 and $0.01N$ KNO_3. Both curves cross zero at the same point yielding an IEP of 6.5 ± 0.5. The gradient in the zeta-potential curve at the IEP is about 40 mV/pH in $0.001 N$ electrolyte, much higher than the value displayed by SiC. Heating in

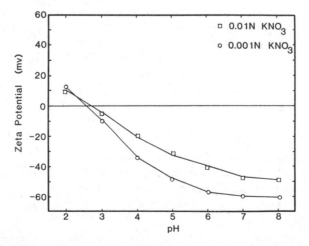

Figure 56.1   Zeta potential of SiC suspensions at varying ionic strength.

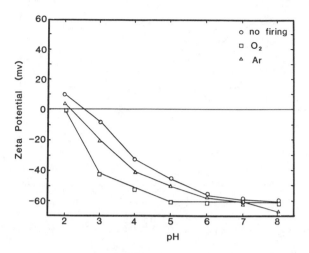

Figure 56.2   Variation of the zeta potential of SiC in $0.01N$ KNO_3 with type of heat treatment.

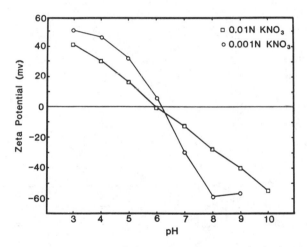

Figure 56.3   Zeta potential of Si_3N_4 suspensions at varying ionic strength.

oxygen or argon does not appear to have a significant effect on the zeta potential for the silicon nitride powders. Figure 56.4 shows the curves at $0.01N$ KNO_3 to be similar, intersecting zero between pH 6 and 6.3, which are indistinguishable within the accuracy of the measurement.

For the powders studied, the ESA and the zeta potential were closely related. In particular, the pH at which the ESA is zero correlates well with the IEP. In our experiments, the agreement was within one-half of a pH unit. Table 56.1 summarizes the ESA zero points and the IEP for our suspensions. Figure 56.5 shows the variation of ESA with pH and ionic strength for 1 wt% suspensions of SiC and Si_3N_4.

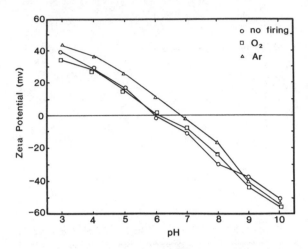

Figure 56.4 Variation of the zeta potential of Si_3N_4 in $0.01N$ KNO_3 with type of heat treatment.

TABLE 56.1 Isoelectric Points and Zero Points in the Electrokinetic Sonic Amplitude

	SiC	IEP, ± 0.3	pH (ESA = 0), ± 0.3	Si_3N_4	IEP, ± 0.3	pH (ESA = 0), ± 0.3
	$0.01N$	2.5	2.8	$0.01N$	6.2	6.4
	$0.001N$	2.5	1.7	$0.001N$	5.9	6.1
Ar, 500°C	$0.01N$	2.0	2.1	$0.01N$	6.5	5.7
	$0.001N$	1.7	1.9	$0.001N$	6.9	6.1
O_2, 500°C	$0.01N$	2.1	2.3	$0.01N$	6.4	6.3
	$0.001N$	2.2	1.8	$0.001N$	7.2	6.1

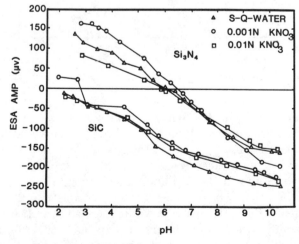

Figure 56.5 Electrokinetic sonic amplitude measured as a function of pH and ionic strength for 1% suspensions of SiC and Si_3N_4.

544

Rheology of SiC Suspensions

Figure 56.6 shows the flow curves for a 42 vol% SiC suspension at pH 2.3, 2.8, 5.0, and 10.3 (all $\pm$ 0.4). At the higher pH the suspensions are relatively fluid and slightly dilatant, consistent with a well-dispersed system. At pH 2.3 and 2.8, near the IEP, the suspensions are viscous and pseudoplastic with an apparent yield stress around 0.35 Pa.

Titration Results

It was necessary to wash the powders to obtain titration data which could be reliably related to hydrogen ion adsorption on the particle surfaces. Suspensions prepared from as-received unwashed powders contained a substantial amount of titratable species in their supernates. Figure 56.7 compares the titration behavior of the supernate obtained from unwashed Si_3N_4 powder and the washed powder. Shown is the difference in the hydrogen ion consumption between the supernate or powder and a $0.10N$ KNO_3 blank. Both the SiC and Si_3N_4 supernates consumed more titrant than its parent powder, introducing a large artifact in the apparent hydrogen ion adsorption of the unwashed powders.

The nature of the titratable supernatant species differed between the two powders. The silicon carbide supernate contained a complex acidic species, while the silicon nitride supernate contained a basic species. Most of this species could be removed by washing the powder in water and discarding the supernate. These titratable species may represent a contaminant in the as-received powder, or may indicate that the powders react with water. The silicon carbide could be more successfully cleaned. Titration of the supernates taken

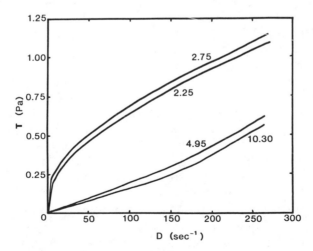

Figure 56.6 Stress (T)–strain rate (D) measurements for 42 vol% SiC suspensions held at various pH.

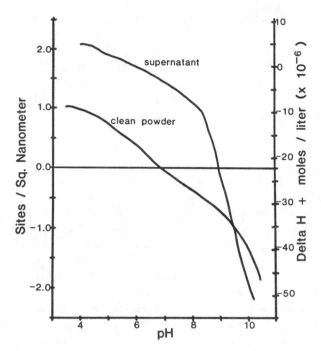

Figure 56.7 Calculated hydrogen ion adsorption for washed Si_3N_4 suspensions and the supernates from the first wash in $0.01\,N\,KNO_3$.

after a second, third, fourth or fifth wash showed only a small amount of the titratable species. The silicon nitride supernate from a second wash contained appreciable titratable material, although subsequent washings showed reduced amounts.

The question of hydrolytic stability of these powders is still unresolved. For short ($\sim$ 20 min) periods, comparable to the time required to complete a titration experiment, pH-Stats indicate that both powders are stable. However, Whitman[11] has shown that the silicon nitride powders are reactive over long ($\sim$ 2 hr) periods.

Hydrogen ion adsorption on multiply washed silicon carbide is shown in Fig. 56.8. Repeated washing caused little change in the hydrogen ion adsorption, indicating that one washing is probably sufficient to produce a controlled surface for silicon carbide. The data display a shallow plateau with relatively little hydrogen ion adsorption. A point of zero charge (PZC) occurs at pH 5.1 $\pm$ 0.3. The adsorption behavior was not affected by either heating treatment. Figure 56.9 shows similar adsorption curves for powders heated in oxygen and argon. Points of zero charge between pH 4.5 and 5.5 are indicated.

We have not yet obtained reliable hydrogen adsorption data for silicon nitride. Preliminary indications suggest an extremely shallow adsorption curve with a wide plateau. A PZC appears to be at pH 7 $\pm$ 1, although it is difficult

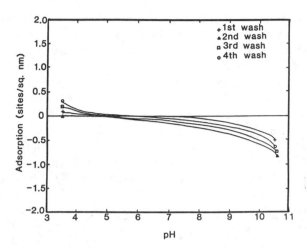

Figure 56.8 Titration of SiC in 0.01N KNO₃ showing the effects of repeated washings on the titration behavior.

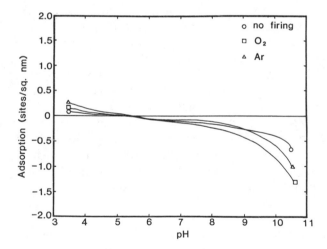

Figure 56.9 Hydrogen ion adsorption for SiC suspensions showing the effect of varying heat treatment.

to discern due to the very shallow slope. The behavior of the silicon nitride at high pH suggests that the powder may be reacting with the solution.

DISCUSSION

The silicon carbide and silicon nitride both behave as hydrophobic colloids. With an IEP of 2.5, the silicon carbide is quite acidic and similar to silica, so silanol is probably the major functional group on SiC powder surfaces. The silicon nitride, however, with its IEP of 6.5, is too basic to be consistent with a silanol-dominated surface. Our IEP measurement for silicon nitride is similar to that reported by Clarke and Shaw.[12]

The as-received silicon carbide powder differs from washed powder since it is contaminated with some acidic material. The washed SiC is reproducible and appears to be hydrolytically stable. The washed powder is not affected by an oxidative heat treatment, suggesting that the surfaces are appreciably oxidized. Heating in dry argon to partially rehydroxylate the surface had no effect. Apparently rehydroxylation is rapid. In practical terms for powder processing, these results indicate that the behavior is insensitive to mild heat treatment such as drying.

Relatively little hydrogen ion adsorption occurs on the silicon carbide. For most of the pH range studied, less than one ion adsorbs or desorbs per nm^2, as compared to the 4–6 sites/nm^2 for complete charging of typical silica surface site populations. The adsorption curves are extraordinarily shallow. To explain such a low gradient through the PZC would require a large ΔpK. The observed gradient in the zeta-potential curve at the IEP is also indicative of a large ΔpK.

The silicon carbide suspensions displayed a large difference between the IEP (pH 2.5 ± 0.5) and the hydrogen ion PZC (pH 5.1 ± 0.3). Since the PZC is more basic than the IEP, adsorption of an anionic species must be occurring. It is not likely that adsorption of a simple nitrate anion could cause such a large discrepency. Rather, adsorption of a complex anion is indicated. This complex anion is probably associated with the acidic species detected in the titration of the supernate. One possibility is a silicate complex.

The as-received silicon nitride also behaved differently from the washed silicon nitride powders. The presence of (or generation of) a complex basic species in the supernate of the as-received powder is indicated. Again, heating in oxygen or argon appeared to have little effect, suggesting that the powder surface is either fully oxidized or else rapidly reacts with the water. The powder clearly does not behave as if its surface were covered by impermeable silica layers. The IEP is too basic for a silanol dominated surface. To interpret the zeta potential results, it may be necessary to invoke a two-site model, featuring a very acidic site and a very basic site. Silanol and amine are likely candidates. Titration data suggest a relatively low hydrogen ion adsorption, with a PZC equivalent to the IEP. However, these results are inconclusive since the chemical stability of silicon nitride powders in water is questionable at present.

CONCLUSIONS

Silicon carbide is similar to silica, with an IEP at pH 2.5 ± 0.5, consistent with a silanol dominated surface. Silicon nitride, which exhibits an IEP at pH 6.5 ± 0.5, cannot possess only silanol surface functional groups.

Hydrogen ion adsorption locates the PZC of silicon carbide at pH 5.1 ± 0.3.

As-received powders produce supernates with acidic species for SiC and basic species for Si_3N_4. Water washing removes these species.

Both powders show ESA data that are similar to the zeta potential data. The zero point of the ESA data and the measured IEP correspond.

ACKNOWLEDGMENT

The authors acknowledge P. K. Whitman for her discussions and permission to cite the results of work in progress. The invaluable assistance of Jerry Hughes and Edvins Krams of DuPont is gratefully acknowledged.

REFERENCES

1. R. E. Johnson, Jr., *J. Colloid Interface Sci.* **100**(2), 540 (1984).

2. J. A. Davis, R. O. James, and J. O. Leckie, *J. Colloid Interface Sci.* **74**, 32 (1980).

3. M. Anderson and A. J. Rubin, Eds., *Adsorption of Inorganics at Solid–Liquid Interfaces*, Ann Arbor Science, Ann Arbor, 1981.

4. A. I. Medalia and D. Rivin, in *Characterization of Powder Surfaces*, G. D. Parfitt and K. S. W. Sing, Eds., Academic Press, New York, 1976, Chapter 7.

5. R. K. Iler, *The Chemistry of Silica*, Wiley, New York, 1979.

6. T. W. Healy and L. R. White, *Adv. Colloid Interface Sci.* **9**, 303 (1978).

7. J. Eisenlauer and E. Killmann, *J. Colloid Interface Sci.* **74**, 108 (1980).

8. D. Barby, in *Characterization of Powder Surfaces*, G. D. Parfitt and K. S. W. Sing, Eds., Academic Press, New York, 1976, Chapter 8.

9. M. J. Crimp, M. S. Thesis, Case Western Reserve University (1985).

10. R. W. O'Brien and L. R. White, *J. Chem. Soc. Faraday II* **74**, 1607 (1978).

11. P. K. Whitman, Ph. D. Dissertation, Case Western Reserve University (in progress 1985).

12. D. R. Clark and T. M. Shaw, Technical Report SC5178, TR4, Rockwell International Science Center, 1000 Oaks California, for DOE Contract DE-ACO3-78ERO1885.

57

EFFECT OF SURFACE HYDRATION ON POLYMER ADSORPTION

BRIJ M. MOUDGIL AND YEU-CHYI CHENG
Department of Materials Science and Engineering
University of Florida
Gainesville, Florida

INTRODUCTION

Ultrastructure processing of materials involves the manipulation and control of surfaces and interfaces to produce high-performance materials. Manipulation of fine particles in controlled assemblages can be adversely affected by aggregation, which however, can be minimized either by increasing Coulombic repulsion, or by steric hindrance of adsorbed polymers. For preparing a stable dispersion of desired rheological behavior, it is essential to understand the adsorption mechanism of polymers on the substrate. The amount of polymer adsorbed and the conformation of the adsorbed species is known to have a significant effect on the stability of the dispersed particles. Forces responsible for polymer adsorption result mainly from hydrogen, electrostatic, and covalent bonding. It has been observed that hydration of the substrates can significantly affect the adsorption of polymer on surfaces such as SiO_2, Fe_2O_3 which adsorb predominantly through hydrogen bonding.[1-4]

The objective of this chapter is to establish the role of surface hydration in the adsorption of a nonionic polymer polyethylene oxide (PEO) on silica.

550

EXPERIMENTAL

Materials

Polyethylene oxide (PEO) with a molecular weight of 5 million as stated by the manufacturer (Polysciences, Inc., PA) was used as received.

The silica particles used in this study were synthesized through controlled hydrolysis and condensation reactions of tetraethylorthosilicate (TEOS) in a mixture of ethanol and ammonium hydroxide.[5] By using a reactant ratio of 3 mole NH_4OH/liter ethanol and 0.2 mole TEOS/liter ethanol, spherical silica particles with a narrow size distribution were produced.

Techniques

Adsorption measurements were carried out, after equilibrating the powder with the desired amount of polymer for 12 hr followed by centrifuging at 900 g to achieve the solid–liquid separation. The residual PEO concentration in the supernatant was determined by a turbidity method developed by Attia and Rubio.[6]

Flocculation tests were conducted by adding the polymer solution slowly to a well-dispersed silica suspension which was then left undisturbed for 1 hr. The optical absorbance of the sample supernatant was then measured and compared with that of a blank to which no flocculant was added. Results are reported as "relative absorbance" (Ar)[7]:

$$Ar = \frac{\text{absorbance of sample}}{\text{absorbance of blank}}$$

Thus a value of Ar equal to 1 indicates that suspension stability is not affected by the polymer addition, while a value less than 1 indicates destabilization of the suspension.

RESULTS AND DISCUSSION

Effect of Calcination Temperature on Surface Hydration

The FTIR spectra in Fig. 57.1 show the effect of calcination temperature on the degree of surface hydration of the silica particles. It is observed that as calcination temperature increases, the broad reflectance band at $\sim$ 3100–3650 cm^{-1} narrows, which is associated with the removal of hydrogen bonded water and silanol groups. The peak intensity at $\sim$ 3747 cm^{-1} decreases at higher calcination temperatures indicating the removal of "free" or isolated silanol groups.

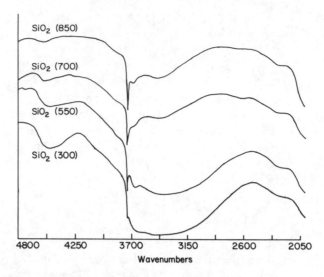

Figure 57.1   FTIR spectra of silica powders calcined at various temperatures.

PEO Adsorption on SiO₂

It is observed from the adsorption isotherms of PEO on calcined silica powder plotted in Fig. 57.2 that the adsorption density increases with an increase in the calcination temperature from 300 to 700°C, and it decreases at 850°C.

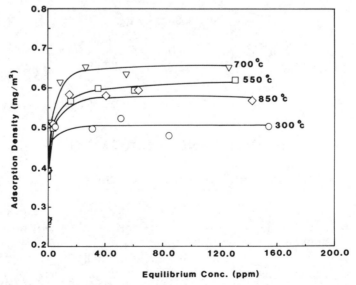

Figure 57.2   Adsorption isotherms of PEO onto silica as a function of calcination temperature at pH = 5.6.

In other words, moderate dehydration increases the adsorption of PEO on silica and further dehydration reduces it.

Aggregation of PEO-Coated SiO$_2$ Particles

Flocculation results of different silica suspensions as a function of polymer dosage are presented in Fig. 57.3. It is clear that destabilization of the suspension occurs at a PEO dosage of about 5 mg/kg. At higher polymer dosage the particles are restabilized, possibly due to a steric stabilization process. An optimum in flocculation is observed for powders calcined at 700°C. It was determined that at a PEO dosage of 5 mg/kg all the polymer is depleted from the suspensions indicating that polymer adsorption had occurred on all the samples to the same extent. The differences in flocculation, therefore, cannot be attributed to the amount of polymer adsorbed. It is suggested that conformation of the adsorbed polymer molecules is responsible for the variations in the flocculation behavior of silica particles calcined at different temperatures.

Heating is known to cause dehydration of the silica surface.[8-10] Generally, a precipitated silica surface is covered with a dense layer of silanol groups which are mostly hydrogen bonded. Heating above 400°C has been determined to cause irreversible dehydration and also yield a surface with isolated silanol

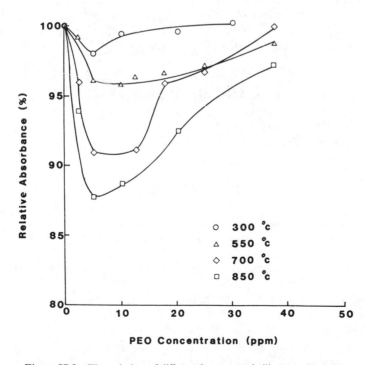

Figure 57.3 Flocculation of different heat treated silicas at pH = 5.6.

groups. Removal of these isolated silanol groups is believed to start at a calcination temperature of 700°C. However, complete dehydration of the surface does not take place until above 1000°C, after which the surface is covered with siloxane groups. In the above discussion it was presented that the adsorption density is maximum for silica calcined at 700°C. Thus, an explanation of the observed adsorption behavior is that the ether oxygens of the PEO molecules bond to the isolated surface silanol groups. However, one cannot rule out the possibility of water molecules competing with PEO segments for surface sites, since a change in the surface state of silica alters the solvent–substrate and polymer-substrate interactions.

CONCLUSION

Monospherical silica particles of 0.5 μm size were produced by controlled hydrolysis and condensation reactions of tetraethylorthosilicate. Calcination of silica at 300, 550, 700, and 850°C yielded powders with different degrees of surface hydration. No significant rehydration of the dehydrated powders was observed within the adsorption time used in this study. Maximum adsorption of PEO occured on the powder calcined at 700°C. This was attributed to a combination of removal of water and isolated silanol groups and competition of water molecules for the surface sites.

At a given polymer adsorption density, maximum aggregation occurred for powders heated at 700°C. It has been suggested that conformation of the adsorbed species is responsible for the observed aggregation behavior.

ACKNOWLEDGMENT

Financial support of this work by Air Force Office of Scientific Research (Contract #F49620-83-C-0072) is acknowledged.

REFERENCES

1. O. Griot, and J. A. Kitchener, *Trans. Faraday Soc.* **61**, 1032 (1965).
2. Greenland quoted in *Formation and Properties of Clay-Polymer Complexes*, Elsevier, New York, 1979, p. 6.
3. B. M. Moudgil, The Role of Polymer-Surfactant Interactions in Interfacial Processes, Eng. Sc. D. Thesis, Columbia University, New York, 1981.
4. J. Rubio, and J. A. Kitchener, *J. Colloid Interface Sci.* **57**, 132 (1976).
5. W. Stober, A. Fink, and G. Bohr, *J. Colloid Interface Sci.* **26**, 62 (1968).
6. Y. A. Attia and J. Rubio, *Br. Polym. J.* **7** 135 (1975).
7. G. J. Fleer and J. Lyklema, *J. Colloid Interface Sci.* **46**, 1 (1974).
9. B. A. Morrow and J. A. Cody, *J. Phys. Chem.* **77**, 1465 (1973).
10. G. J. Young, *J. Colloid Interface Sci.* **13**, 67 (1958).

58

AC FIELD-INDUCED ORDER IN DENSE COLLOIDAL SYSTEMS

ALAN J. HURD
Sandia National Laboratories
Albuquerque, New Mexico

SETH FRADEN AND ROBERT B. MEYER
Martin Fisher School of Physics
Brandeis University
Waltham, Massachusetts

INTRODUCTION

The effects of electric fields on colloidal suspensions have been studied for many years, with much attention paid to the various aspects and applications of electrophoresis in static fields. More recently, suspensions in alternating fields have been studied,[1,2] particularly in order to understand the electrical properties of suspensions such as the conductivity and dielectric constant. For the most part, these theoretical investigations have been restricted to the simplest case of a very dilute suspension with no spatial correlations among particles.

In this chapter, we demonstrate experimentally that dipole moments, induced by a strong, alternating external field, can create strong spatial correlations: ordering occurs first as chains of particles, then, at higher fields, as close-packed structures. In addition to being of fundamental interest, these correlations are important for understanding the electrical behavior of dense suspensions.

Our objective is to point out the potential application of electric field effects in ceramics. Since large voids in the particle packing are generally known to decrease the strength of the sintered ceramic, the ideal packing would be a close-packed colloidal crystal. Pressure is commonly applied to densify "green bodies," but it seldom produces maximal packing density: there is little free energy to gain (in the PV term) in going from random close packing ($f = 0.64$) to close packing ($f = 0.74$). The difference in defect distribution may be critical, however. An additional interaction energy would provide a deeper free-energy minimum for ordered packing and a stronger driving force to unravel topological defects. In view of our results, an alternating electric field seems to be a good candidate for this since it has the additional advantage of being adjustable in strength. In the following sections, we will discuss the basic physics of the ordering phenomenon as well as problems and questions that it raises concerning ceramic powder compaction.

EXPERIMENTAL DETAILS

The experiment was designed to demonstrate the effects of alternating current on a dense colloid in a controlled manner. Uniform, polystyrene latex particles, with a diameter of 1.053 μm and suspended in distilled water at a volume fraction of 7%,[3] were constrained to a thin layer (5–20 μm) between glass microscope slides. Transparent, indium–tin oxide electrodes were on one slide, etched by photolithography from precoated glass slides,[4] thereby leaving conductive strips on the glass. A 200-μm gap between electrodes was used in this study. Alternating voltages were applied to the electrodes using a standard function generator capable of producing 35 V peak to peak at a range of frequencies extending to 10 MHz. Typically, the sample was sealed by an O-ring and a metal clamp, although rapid results could be obtained using a cover slip as the top slide and sealing it to the electrode slide with immersion oil. Finally, the region between electrodes was observed through an optical microscope (Fig. 58.1).

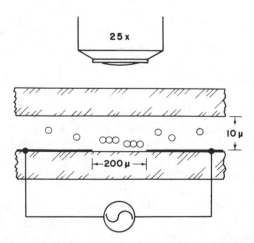

Figure 58.1   Schematic of thin film colloidal sample and electrode geometry.

Figure 58.2 Structures formed at various voltages. (a) 5 V; (b) 15 V; (c) 25 V. The frequency was 1000 Hz and the electrode gap was 200 μm.

Observations were made at various frequencies and voltages, spanning the capabilities of the function generator. Low frequencies (< 20 Hz) and dc offsets were avoided because the electrodes degraded rapidly under these conditions.

Interesting effects begin to occur when the voltage drop across a particle becomes comparable to the surface potential of the particle (ca. 100 mV). Short, transient strings of particles form at 5 V, constantly breaking up and reforming with Brownian motion (Fig. 58.2). As the voltage is increased, the dipole moments become stronger and the strings grow to include as many as 100 particles at 15V. At this stage, the strings repel side-to-side, and the attraction is strong enough to prevent the strings from breaking up, although Brownian motion continues to distort them. Eventually, as the voltage is increased, the strings attract side to side; adjacent strings usually stick together in perfect registry for close packing. The field strengths for these stages of ordered-packing growth are somewhat dependent on the thickness of the sample, especially with respect to the development of electrohydrodynamic instabilities that destroy the order. These instabilities will be discussed in a later section.

INTERPRETATION

The two mechanisms of string formation in ac fields have been understood since 1970[1]: a dipole moment is induced either by distorting the counterionic cloud around the particle (Fig. 58.3a) or by directly polarizing the dielectric solvent and the dielectric particle. For long chains to form, the frequency must be high enough that the particles cannot respond to it by electrophoretic motion yet low enough that the counterions can diffuse from one side of the particle to the other on each half-cycle. At higher frequencies when the counterions can no longer respond, only dielectric polarization occurs. For polystyrene particles in salt-free water, both effects give rise to a dipole moment that opposes the field[2]:

$$\mu = C_0 a^3 \mathbf{E}, \qquad C_0 = -\frac{1}{2} - \frac{3}{4} \frac{\varepsilon_p/\varepsilon}{1 - (4\pi K/\omega\varepsilon)^2} \tag{1}$$

where ε_p and ε are the dielectric constants of the particle and water, respectively, a is the particle radius, $\mathbf{E}$ is the electric field, ω is the frequency, and K is the conductivity of the water. The dipole attraction between two particles competes with the monopole repulsion (attributable to the surface charges) and with Brownian motion. The potential at position $\mathbf{r}$ from a particle is given by the sum of a screened Coulomb potential and a screened dipole,

$$\psi(\mathbf{r}) = \frac{Q}{\varepsilon} \frac{e^{-\kappa r}}{r} + \frac{\mu}{\varepsilon} e^{-\kappa r} \left(\frac{\kappa}{r} + \frac{1}{r^2} \right) \cos \theta \tag{2}$$

where Q is the charge and κ^{-1} is the Debye screening length. The interaction energy of two particles separated by $\mathbf{r} \simeq a\hat{z}$ is

$$W = Q\psi(\mathbf{r}) + \mu \cdot \nabla\psi \mathbf{r} \simeq \frac{e^{-\kappa r}}{\varepsilon a} \left(Q^2 - \frac{2\mu^2}{a^2} \right) \tag{3}$$

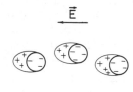

(a)

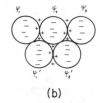

(b)

Figure 58.3   Origin of attractive interactions in external field. (a) Distorted double layers give rise to dipoles that oscillate with the field, causing the particles to attract along field lines. (b) At higher field strengths, pockets of charge may allow chains to attract side-to-side in close-packed arrangement.

Since $Q \propto a^2$ and $\mu \propto a^3$, the ratio of the dipole attraction to the monopole repulsion goes as $a^2 E^2$; for small particles the field must be large to form chains.

The presence of long chains indicates that the energy of attraction is much greater than the thermal energy, $W \gg kT$, but the process has been found to be reversible: the chains fly apart when the field is turned off. In long chains, the interaction of each dipole is "saturated" when it is in contact with two other dipoles end-to-end. There is no reason for two chains to attract side to side; in fact, they tend to repel owing to the fringing fields at the ends. The situation is analogous to chains formed by ferromagnetic grains.[5]

The dipole model breaks down when the field is large enough to cause the strings to attract side-to-side, for then the chain cannot be considered to consist of saturated dipoles. Instead, an arrangement of alternating charges, similar to an ionic crystal, must arise (Fig. 58.3b). It is not clear whether the counterions collect in compact pockets between spheres or remain diffuse; nor is it clear whether the surface charge on the spheres remains constant or varies as the proximity of other charged surfaces forces a new ionization equilibrium. For particles with constant surface potential (such as metal particles), the voltage drop along a chain would cause it to attract another because surfaces of dissimilar potentials can attract by induced image charges.[6] Thus, if $\psi_1 < \psi_1' < \psi_2 < \psi_2' \ldots$ in Fig. 58.3b, then a close packing arrangement (in 2-D, at least) would naturally arise. Whatever the mechanism of side-to-side chain attraction, the important point is that 3-D close packings can be achieved.

PRACTICAL CONSIDERATIONS

While the experiments described above demonstrate the possibility of electric-field-aided compaction, a number of important questions must be answered before the technique can be applied. The most serious drawback is the occurrence of a convective electrohydrodynamic instability associated with certain voltage–frequency relationships. The instability is more prevalent in thicker samples, and since it destroys any order in the colloid, it is important to avoid it. Another important question is how ceramic powders will respond to ac fields, particularly uncharged particles. It may be important to find the optimum ionic strength and dielectric constant of the solvent to get a large effect. Since large voltages will be required for macroscopic molds, it may be difficult to achieve the relatively high frequencies required. Finally, it would be convenient to find a way to hold the particles irreversibly in their crystalline positions with the field off.

ACKNOWLEDGMENTS

Alan J. Hurd would like to thank Professor Carlos Bustamante of the University of New Mexico for his hospitality during part of this work. The research was supported by the U.S. Department of Energy under Contracts No. DE-ACO4-76DP00789 (Alan J. Hurd) and DE-FGO2-84ER45084 (Seth Fraden and Robert Meyer).

REFERENCES

1. S. S. Dukhin and V. N. Shilov, Kinetic Aspects of Electrochemistry of Disperse Systems. Part II. Induced Dipole Moment and the Non-Equilibrium Double Layer of a Colloidal Particle, *Adv. Colloid Interface Sci.* **13**, 153 (1980).

2. R. W. O'Brien, The Response of a Colloidal Suspension to an Alternating Electric Field, *Adv. Colloid Interface Sci.* **16**, 281 (1982).

3. Interfacial Dynamics Corporation, P.O. Box 40306, Portland, OR 97240.

4. Continental Industries Corporation, 2698 Marine Way, Mountain View, CA 94043.

5. P. G. de Gennes and P. A. Pincus, Pair Correlations in a Ferromagnetic Colloid, *Phys. Kondens. Mater.* **11**, 189 (1970).

6. E. Matijević, this volume, Chapter 50.

59

INFLUENCE OF PARTICLE ARRANGEMENT ON SINTERING

F. F. LANGE
Rockwell International Science Center
Thousand Oaks, California

BRUCE KELLETT
Department of Materials Science and Engineering
University of California
Los Angeles, California

INTRODUCTION

Past theories concerning sintering have dealt with the transport of mass to the position of contact between two particles and the kinetics associated with different transport phenomena.[1] Although these theories are limited to the early stage of the contact area growth (due to approximations made to simplify geometical parameters), they appear to be consistent with experimental observations concerning linear arrays of identical spheres. Use of these theories to explain the sintering behavior of powder compacts has been tenuous. For example, the theories suggest that densification kinetics should scale with inverse particle size. However, ultrafine particle size powders can be more difficult to sinter than coarser powders with nearly identical chemistry. Other phenomena such as rearrangement (particle movement during contact area growth) and grain growth during densification are neither predicted nor included in earlier theories. The tenuous relation between past sintering theories

561

and the observed sintering behavior of powder compacts is due, in part, to how particles are arranged in space. Particle arrangement will be shown to influence the condition for pore closure* and forces acting on particles that produce rearrangement. Grain growth in powder compacts will also be discussed and will be shown to be beneficial during most stages of densification.

EQUILIBRIUM CONFIGURATIONS FOR SIMPLE PARTICLE ARRAYS

Mass transport during sintering is driven by the decrease in free energy associated with the decrease in external (vapor–solid) surface area of particles undergoing contact area (neck) growth. For crystalline particles, the contact area is a grain boundary; thus, the free energy change must account not only for the decrease in external surface, but also for the increase in grain boundary area during neck growth. Thus, for identical particles (e.g., spheres) which form an array in which each particle can undergo identical changes in shape during contact area growth, the free energy per particle can be written as the sum of the external surface energy, the grain boundary energy and the internal energy:

$$E = A_s \gamma_s + A_b \gamma_b \tag{1}$$

where A_s = the particle's external surface area, A_b = the grain boundary area, γ_s = external surface energy per unit area, and γ_b = the grain boundary energy per unit area (surface energies are assumed to be independent of crystal orientation and grain pair misorientation). The ratio of the surface energies can be related to an equilibrium dihedral angle (ϕ_e) using Young's relation:

$$\cos \phi_e/2 = \frac{\gamma_b}{2\gamma_s} \tag{2}$$

allowing us to rewrite Eq. (1) as

$$E = \gamma_s \left(A_s + 2 \cos \frac{\phi_e}{2} A_b \right) \tag{3}$$

Differentiating Eq. (3), one can conclude that the contact area will only grow ($dE < 0$) when the external surface area decreases ($dA_s < 0$), and it will cease to grow when $dA_s/dA_b = 2 \cos(\phi_e/2)$; that is, when $dE = 0$. The configuration of the particle array when the contact area ceases to grow is defined as the equilibrium configuration of the array. If any one particle within the array has a lower free energy than the rest, it will grow at the expense of the others by longer range transparticle mass transport and the equilibrium configuration will be a single

---

*Past theories concerning either two particles or linear arrays did not, by definition, include void space encompassed by particles, and therefore were unable to define conditions for pore closure.

sphere. If, on the other hand, the particles maintain identical free energies requiring identical shape changes and flat grain boundaries, contact area growth will occur by short-range, interparticle mass transport; and the particles will maintain their identity and will conserve their mass to produce a multiple-particle equilibrium configuration. Once the equilibrium configuration is reached, the orthogonal surface curvature of each particle must be independent of location, which requires that each particle be a sphere subtended by its neighbors.

To determine an array's free energy change during contact area growth, one needs to define an initial particle shape, how the particle changes its shape during mass transport, and the relation between the external surface area and grain boundary area during mass transport. Theory and experiments have shown that mass is transported to the particle contact position due to the negative surface curvature at this position relative to other positions on the particle surface. This results in a nonuniform external surface curvature and makes definition of the particle shape change analytically intractable. For this reason, we have assumed that the external surface of the particle remains spherical during contact area growth, but the particle's radius increases in order to maintain constant mass. Since the equilibrium configuration of the array is defined by particle surface and grain boundary configurations that minimize free energy, it will be independent of the configurational change of the particles prior to reaching equilibrium. Thus although the assumed particle configurational change used to examine the free energy change is different from the complex shape changes that take place in real systems, their endpoint configuration will be identical.

The following paragraphs review the results of four different particle arrays. Although these results are presented for arrays formed with touching, identical cylinders, the conclusions are identical for more complex cases which have been analyzed; viz., 2- and 3-D arrays of spheres which are presented elsewhere.[2]

LINEAR ARRAYS

The linear array, although not enclosing a pore, is instructive due to its simplicity and comparison to the closed arrays disscussed in the following section. Figure 59.1 illustrates a cross section of an infinite linear array of cylinders upon reaching its equilibrium configuration. For the case considered here, the centers of mass are allowed to approach one another (the case of mass transport by grain boundary, volume diffusion). The contact angle ϕ, that is, the angle between the tangents to the external surface at the grain boundary inter-section), is a convenient parameter to describe the geometrical changes as the cylinders "interpenetrate" one another. The angle $\phi = 0$ for initial contact (Fig. 59.1a), and increases as the contact area grows; as will be shown, $\phi = \phi_e$ when the array reaches its equilibrium configuration. The free energy per particle can be related simply to the initial particle radius r_0, γ_s, the external surface

(a)

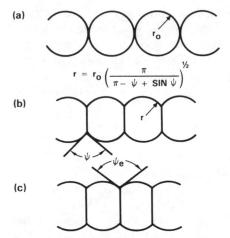

$$r = r_0 \left(\frac{\pi}{\pi - \psi + \text{SIN } \psi} \right)^{1/2}$$

(b)

(c)

Figure 59.1   Cross section of infinite linear arrays of identical cylinders (*a*) initial configuration, (*b*) after some mass transport that decreases center-to-center distance, and (*c*) configuration of lowest free energy.

energy per unit area, the contact angle ϕ, and the dihedral angle ϕ_e:

$$E = 2r_0\gamma_s \left(\frac{\pi}{\pi - \phi + \sin \phi} \right)^{1/2} \left(\pi - \phi + 2 \sin \frac{\phi}{2} \cos \frac{\phi_e}{2} \right) \qquad (4)$$

Figure 59.2 illustrates the energy per particle, normalized to its initial surface energy as a function of the contact angle for several different values of the dihedral angle. As shown, the equilibrium configuration exists when $\phi = \phi_e$;

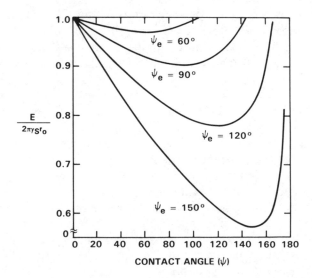

Figure 59.2   Free energy per cylinder (2-D particle), normalized to the particle's initial surface energy as a function of the contact angle formed where the grain boundary meets the surface. Different curves are for different values of dihedral angle.

any further penetration (shrinkage of the linear array) would increase the array's free energy. Similar calculations have been carried out for the case where the particle centers do not approach one another (mass transport via surface diffusion and/or evaporation–condensation). For values of $\phi_e < 90°$ the results are nearly identical to those shown in Fig. 59.2, suggesting that if surface diffusion and/or evaporation–condensation transport mechanisms were to dominate during early stages of sintering (e.g., during initial heating) to produce an equilibrium configuration, no further driving force would exist to produce shrinkage via grain boundary–volume diffusion once conditions (e.g., higher temperature) prevailed to favor these transport paths. For dihedral angles $> 100°$ transport paths that cause particle centers to approach always produce a further decrease in the array's free energy.

Figure 59.2 also shows that gradient of the free energy curves becomes more negative with increasing values of dihedral angle, that is, the driving force for sintering increases with dihedral angle. As detailed elsewhere, an expression for the effective compressive stress (i.e, the driving force for sintering) acting on the linear array can be obtained by relating the free energy per particle to the center to center distance (x) and by differentiating this expression with respect to x.

CLOSED ARRAYS

Closed arrays are defined by an arrangement of touching spheres (or cylinders) which form symmetrical rings or polyhedra. These closed arrays (i.e., particle clusters) are of greater interest for our discussion of sintering since each encloses a single pore which allows us to examine the conditions for pore closure. Similar to the linear array, the closed arrays will undergo mass transport to produce an equilibrium configuration. As will be shown for the case of a ring of cylinders, under certain conditions the pore within the array may disappear as the array approaches its equilibrium configuration; or, under other conditions, it may simply shrink to an equilibrium size. Conclusions for the more complex closed arrays (rings and polyhedra of identical spheres) are similar and are presented elsewhere.[2]

Figure 59.3 illustrates two configurations for the cross section of a ring of eight cylinders for the case where the centers of mass approach one another: (a) initial, and (b) the equilibrium configuration. Note that for the case illustrated here ($\phi_e < 135°$) the pore has decreased in size, but has not disappeared. For the case illustrated in Fig. 59.3, a simple relation exists between the coordination angle and the number (n) of particles coordinating the pore: $\theta = 2\pi/n$. At this point, once a contact area forms, the center of mass is no longer coincident with the center of curvature.

The free energy per particle can be derived in a similar manner for the linear array and results in identical plots for free energy versus contact angle as shown in Fig. 59.2 for conditions where the pore does not disappear (for conditions of

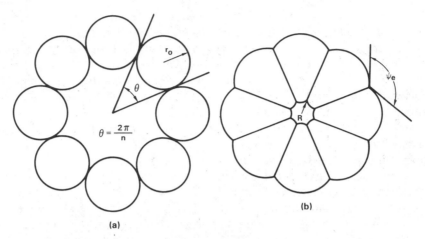

(a)

(b)

Figure 59.3 Cross section of cylinders arranged in a ring array (a) initial configuration, (b) after some mass transport, and (c) configuration of lowest free energy. (Dihedral angle is assumed to be less than 135° producing a stable pore when the coordination number is 8.)

pore closure, discussed below, the free energy gradient is less than that shown in Fig. 59.2 when $\phi > \phi_e$). Of greater interest is the relation between the free energy and the pore radius (R) shown in Fig. 59.3 as a circle that circumscribes the intersection of grain boundaries with the pore's surface. The free energy expression (per particle), rewritten in terms of the initial particle radius (r_0), the surface energy per unit area (γ_s), the dihedral angle (ϕ_e), the pore radius (R), and the coordination angle (θ) is plotted in Fig. 59.4, illustrating the case of $\phi_e = 150°$. The free energy per particle has been normalized with the particle's initial surface energy, and the pore radius is normalized with the particle's initial radius. As shown, for certain conditions the free energy function can exhibit a minimum; that is, the pore can have an equilibrium size, and for other conditions, the free energy function does not exhibit a minimum before the pore closes. The conditions for pore stability and closure can be obtained by examining the pore size function:

$$R = r_0 \left(\frac{\pi}{\pi - \phi + \sin \phi} \right)^{1/2} \left(\frac{\cos\left(\dfrac{\phi}{2} + \dfrac{\theta}{2} \right)}{\sin\left(\dfrac{\theta}{2} \right)} \right) \tag{5}$$

For the equilibrium configuration ($\phi = \phi_e$), the pore has a finite size ($R_e > 0$) when the sum of the dihedral angle and the coordination angle is $< \pi$ ($\phi_e + \theta < \pi$), whereas the pore closes when $\phi_e + \theta \geq \pi$. These conditions can be restated in terms of the pore's coordination number (n): pores will shrink to a equilibrium size when $n > n_c = 2\pi/(\pi - \phi_e)$ and pores will close when $n \leq n_c$. Thus the conditions of pore closure depend on the pore's coordination

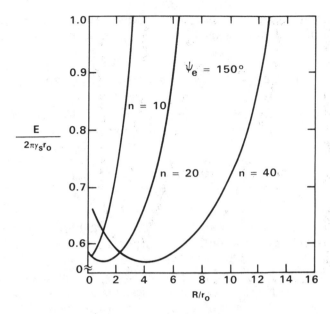

$$\frac{E}{2\pi\gamma_s r_0}$$

Figure 59.4 Free energy per cylinder (2-D particle), normalized to the particle's initial surface energy as a function of the pore radius for the case where the dihedral angle is 150°. Different curves as for ring arrays with different pore coordination numbers.

number and the ratio of the surface to grain boundary energies as expressed by the dihedral angle.

One can also view the ring of cylinders for the case where the center-to-center distances remain fixed (mass transportation via surface diffusion and/or evaporation–condensation). Once the contact area begins to grow, different surface curvatures develop on the outside of the ring relative to the inside (pore surface). Namely, the radius of curvature on the inner surface is larger than on the outer surface; when $n = n_c$ the pore's surface will become flat. If the inner and outer surfaces are connected (precluded for the case of the infinite ring of cylinders but not for closed arrays of spheres) then mass transport will take place from the surface of higher curvature (outer surface) to the surface of lower curvature (pore surface). Mass transport to the pore surface through surface diffusion and/or evaporation–condensation will cause a shift of each particle's center of mass toward the center of the array to produce array and pore shrinkage (total pore closure may be precluded if access between the inner and outer surfaces is closed prior to pore closure). Thus, array and pore shrinkage can take place for surface diffusion and/or evaporation–condensation mass transport paths—a result that was precluded for the linear array! This conclusion is only true for the individual particle clusters; and since mass has to be transported from some surface to another, it will not result in the shrinkage and/or densification of a powder compact. What this result implies is that surface diffusion and/or evaporation–condensation can cause some pores $(n < n_c)$ to shrink but, at the same time, can cause some pores $(n > n_c)$ to grow.

SHRINKAGE STRAIN AND STRAIN RATES

The strain (ε, defined as the change in center-to-center distance divided by the original distance) of the linear array can be shown to be only a function of the contact angle (ϕ):

$$\varepsilon = 1 - \left(\frac{\pi}{\pi - \phi + \sin \phi}\right)^{1/2} \cos\left(\frac{\phi}{2}\right) \tag{6}$$

The strain the linear array undergoes to reach its equilibrium configuration is determined simply by substituting $\phi = \phi_e$ into the above expression, which shows that the greater the dihedral angle, the greater the strain.

The strain of the closed array can be defined either with the circle (or sphere for the case of the polyhedra arrays) circumscribing the array or with the circle (sphere) intersecting the center of mass of each particle (remembering that the center of mass is not coincident with the center of curvature once a contact area is formed). Strain calculations based on both definitions have been performed, and conclusions have been similar. Results for the radial strain for the ring of cylinders based on the simpler definition (circle circumscribing the array) are given as

$$\varepsilon = 1 - \left(\frac{\pi}{\pi - \phi + \sin \phi}\right)^{1/2} \left(\frac{\cos(\phi/2) + \sin(\theta/2)}{1 + \sin(\theta/2)}\right) \tag{7}$$

The above expression shows that the strain is a function of contact angle *and* the coordination angle (or the pore's coordination number). Figure 59.5 illustrates strain versus contact angle plots for ring arrays with different pore coordination numbers. As shown, arrays with greater coordination number exhibit greater shrinkage strains. Figure 59.5 also shows the condition for pore closure ($\phi = \phi_e$). The result that the strain of the closed array depends on its coordination number is not unexpected; viz., lower density powder compacts will exhibit greater volumetric strains in achieving theoretical density than higher density compacts. The unexpected result is that the shrinkage strain rate will also be proportional to the pore's coordination number. This result is obtained either by choosing an expression from the literature derived for the center-to-center approach rate of two particles for a given mass transport mechanism and by using the expression with the geometical parameters used above; or, more simply, by recognizing that the time needed for an array to reach its equilibrium configuration will be the same as that for any other array. That is, diffusion path lengths either within the particle or at its grain boundary are independent of how the particle is packed with others in space. Thus, we must

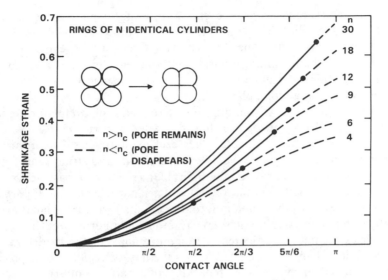

Figure 59.5 Shrinkage strain as a function of the contact angle for cylinders arranged in ring arrays. Different curves are for different pore coordination numbers. Pore stability/closure conditions are marked by the curve where the contact angle is equal to the dihedral angle required to cause pore closure of a given pore coordination number.

conclude that since higher coordinated arrays exhibit a greater shrinkage strain, they must also exhibit a greater strain rate.

The above results show that the shrinkage strain and strain rate depend on the pore coordination number of the closed array (or particle cluster). Clearly, one can see why a powder compact with lower initial bulk densities will exhibit a greater shrinkage strain and strain rate in achieving its equilibrium configuration relative to the same powder packed to a higher initial bulk density.[3] As discussed in the next section, the different strain and strain rates exhibited by different particle clusters result in forces between the different clusters and thus contribute to the phenomenon known as rearrangement.

REARRANGEMENT DUE TO DIFFERENTIAL STRAINS AND STRAIN RATES

Rearrangement is the term used to describe the nonuniform movement of particles during the initial stages of sintering. Exner's sintering studies (see Ref. 1) of random, planar arrays of identical spheres graphically illustrate this phenomenon, showing that some groups of particles shrink upon themselves and large, irregular voids open up between them. Mercury infiltration experiments on real ceramic powders also show this phenomenon,[4,5] viz., small equivalent capillaries close during the early stages of sintering, and concurrently large equivalent capillaries grow even larger. Exner explained his results by suggesting that the diffusion path to the neck region will depend on the angle three particles form with one another. His three-particle experiments did indicate a small change in angle during sintering, but this angular change did not appear significant relative to the much larger displacements exhibited by particles within the random, planar array.

To explain the forces arising during sintering causing nonuniform particle motion, one need only recognize that a powder compact is a collection of different particle clusters packed together, each exhibiting a different shrinkage strain and strain rate. That is, the differential strains and strain rates between adjacent clusters will give rise to nonuniform forces acting on particles common to the adjacent clusters and thus, on nonuniform particle motion. Figure 59.6a illustrates the effect of these nonuniform forces on an irregular planar array of partially sintered polystyrene, glass spheres; the nonuniform forces have caused the particles common to different adjoining planar arrays to deform their initial spherical shape. Figure 59.6b shows the large, irregular voids produced between ordered arrays in a 3-D packing of identical polystyrene spheres.

If the stresses produced by the differential strains and strain rates are analyzed from a continuum mechanics viewpoint, one concludes that any cluster with a pore coordination number greater than the average will be placed in tension, and clusters with a smaller coordination number will be placed in compression.

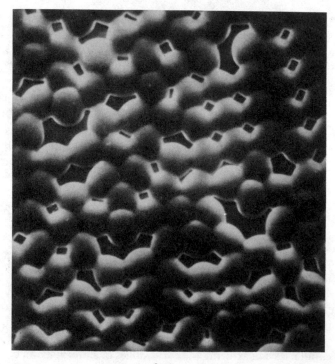

(a)

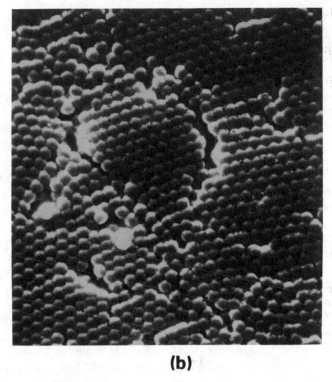

(b)

Figure 59.6 (*a*) Partial sintering of glassy, polystyrene identical (2 μm) spheres. Note shape distortion of particles connecting different arrays. (*b*) Partial sintering of a 3-D array. Note separations developed between ordered arrays.

As clusters separate and no longer contribute to the total shrinkage, the average strain and strain rate will decrease. This type of analysis suggests that pores with a smaller than average coordination number will disappear; and pores with a larger than average coordination number will increase their size to become coordinated by an even greater number of particles. This process is simply a mechanical transfer of void space from small coordinated pores to larger coordinated pores. One must conclude that the rearrangement process is most detrimental to densification since not only does it produce larger pores than initially present in the powder compact, but also pores with a higher coordination number which are more stable for conditions of short-range diffusion (interparticle diffusion).

GRAIN GROWTH (LONGER RANGE MASS TRANSPORT) IN PARTICLE ARRAYS

To this point, we have assumed that the particles under discussion were identical spheres. As pointed out above, if any one of these particles were slightly larger, the free energy of an array could be lowered as the larger particle grew at the expense of the others. As described below, the process of grain growth in a powder compact requires longer range mass transport relative to the shorter mass transport paths, viz., within the particle or along its grain boundary, to cause the contact area to grow. The driving force for contact area growth will also be shown initially to be much greater than that for grain growth. Thus, although both are expected to be concurrent, one might expect that contact area growth to dominate during the early stages of mass transport.

As detailed by Gupta[6] and others, grain (or particle) growth takes place through all stages of sintering. They have shown that the rate of grain growth is much lower at relative densities < 0.90 than the rapid growth rates at relative densities > 0.90. This observation suggests that the mechanism of grain growth is different in these two regimes.

Grain growth in dense materials is associated with boundary curvature; viz., due to the curvature, the chemical potential is greater on one side relative to the other, causing atoms or ions to hop across. Diffusion distances are very short in this process and boundary motion can be rapid. The driving force for grain growth in a powder compact can be different as depicted in Fig. 59.7, which shows two particles with different curvatures that have already developed an "equilibrium" configuration during contact area growth by short-range diffusion. The grain boundary between the two is curved (because it must meet the surface along the line defined by the bisector of the dihedral angle) and would like to move into the smaller grain, but doing so would increase its surface area and would increase the free energy of the two-particle system. Thus it is not the boundary curvature that will drive the growth of the larger grain, but instead, the differential external surface curvatures of the two grains themselves. As depicted, growth of the larger grain requires longer range transparticle mass transport; the boundary remains fixed in space as the larger particle grows at the expense of the smaller one. Figure 59.7 also shows that as the larger particle (grain) grows, a special geometrical relation occurs where the grain boundary can move spontaneously through the smaller grain without increasing its area to decrease the free energy of the system. As detailed elsewhere,[7] the size ratio of the two grains in which this special geometry occurs depends on the dihedral angle.

As depicted in Fig. 59.7, the consequence of this grain growth concept to the sintering process is very interesting. Figure 59.8 shows a linear array of two different size spheres. Although contact area growth by short-range inter-particle transport and grain growth by longer range transparticle transport are concurrent, the contact growth phenomenon is assumed dominant to produce an "equilibrium" configuration shown in Fig. 59.8b with the appropriate shrinkage strain from (a) to (b). We now "turn on" transparticle transport until the alternate, larger particles grow to contact one another as shown by (c); when this configuration is reached, the grain boundaries between the larger and smaller grains [shown in (c) by the broken lines] can move to reduce the free energy of the system and can form a single boundary between the larger grains.

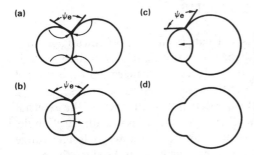

Figure 59.7 Two different size particles (*a*) "equilibrium" configuration due to short-range mass transport to form contact area, (*b*) growth of the larger particle by longer range, transparticle mass transport, (*c*) special geometry where grain boundary can move into smaller grain and decrease free energy, and (*d*) single grain before achieving equilibrium configuration.

As detailed elsewhere, the shrinkage strain of the linear array between (*b*) and (*c*) is relatively small for dihedral angles < 135° and is zero for larger dihedral angles. But as shown in (*c*), the contact angle formed between the newly formed larger grains is less than the dihedral angle, indicating that contact area growth via short-range interparticle transport will begin again finally to produce the equilibrium configuration shown in (*d*). As detailed elsewhere,[7] the shrinkage strain between (*c*) and (*d*) is large. If fewer larger particles were placed within the array (e.g., every fourth particle were larger) the phenomena depicted in Fig. 59.8 would be similar, but the transport path to reach sequentially the configuration shown by (*d*) would be larger and the time, longer.

One must conclude from the analysis presented in Fig. 59.8 that grain growth can lead to shrinkage. Extending this concept to closed arrays (cluster containing a single pore) the major conclusion is that grain growth within the cluster decreases the pore's coordination number and therefore overcomes the thermodynamic barrier to pore closure. Namely, grain growth within the powder compact is useful in achieving densification. (Abnormal grain growth that entraps pores during the last stage of sintering is still unwanted.)

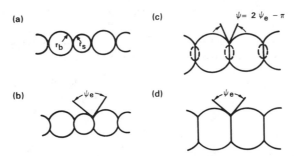

Figure 59.8 Linear array of two different size spheres (*a*) initial configuration, (*b*) "equilibrium" configuration due to short-range mass transport, (*c*) transparticle mass transport causing larger spheres to grow and disappearance of smaller particles due to grain boundary motion, (*d*) final equilibrium configuration developed by further contact area growth, all particles now have identical size and surface curvature.

CONCLUSIONS

Short-range (interparticle) mass transport can result in pore shrinkage to a stable size. This occurs for pores with particle coordination numbers greater than a critical value. Pores coordinated with fewer particles than the critical number disappear spontaneously. The critical coordination number separating the conditions of pore stability from closure depends on the ratio of the external surface energy to the grain boundary energy as defined by the dihedral angle. The thermodynamic barrier to pore closure can be overcome by longer range mass transport phenomena which produce grain growth, and reduce the pore's coordination number to less than the critical value. Grain growth is driven by differences in external surface curvature.

ACKNOWLEDGMENTS

This work has been supported by the Office of Naval Research. Dr. Lange's work was supported by contract number N00014-83-C-0469 and Mr. Kellett's graduate studies at UCLA were supported under contract number N00014-84-K-0286.

REFERENCES

1. H. E. Exner, Principles of Single Phase Sintering, *Rev. Powder Meta. Phys. Ceram.* **1** (1–4), 1–251 (1979).

2. B. Kellett and F. F. Lange (to be published).

3. F. F. Lange and B. I. Davis, "Relations Between Shrinkage Strain, Strain Rate, and Rearrangement During Sintering: Experimental Observations with Alumina, Zirconia, and Alumina/Zirconia Powder Compacts," Tech. Rpt. No. 4, Office of Naval Research Contract No. N00014-83-C-0469, July 1984.

4. O. J. Whittemore, Jr. and J. J. Sipe, Pore Growth During the Initial Stages of Sintering Ceramics, *Powder Technol.* **9**, 159 (1974).

5. F. F. Lange, Sinterability of Agglomerated Powders, *J. Am. Ceram. Soc.* **67**(2), 83 (1984).

6. T. K. Gupta, Possible Correlations Between Density and Grain Size During Sintering, *J. Am. Ceram. Soc.* **55**(5), 176 (1972).

7. F. F. Lange and B. Kellett (to be published).

PART 6

Quantum Chemistry

60

NEW DIRECTIONS IN QUANTUM CHEMICAL CALCULATIONS—PARTICULARLY AS TO NEW MATERIALS

PER-OLOV LÖWDIN

Departments of Chemistry and Physics
University of Florida
Gainesville, Florida

The objectives of this Keynote Lecture are (1) to review the key concepts underlying quantum chemical calculations, and (2) to comment on certain features and trends in the current research in this area. For this volume, only an abstract of the Keynote Lecture is given; the full test will be published elsewhere.

1.   Quantum Mechanics (Solution of the Schrödinger Equations)

The quantum theory of the electronic properties of matter is built either on quantum mechanics, or on quantum statistics. In quantum mechanics, one deals essentially with *pure states* characterized by a wave function ψ which satisfies the time-dependent Schrödinger equation.

$$-\frac{h}{2\pi i}\frac{\partial \psi}{\partial t} = H\psi \tag{1}$$

where H is the Hamiltonian of the system.

Solution of the Time-Dependent Schrödinger Equation; Volterra's Iterated Solution; Time-Dependent Perturbation Theory; Hilbert Space; Solution of the Time-Independent Schrödinger Equation.

2. Approximate Solution of the Time-Independent Schrödinger Equation; The Variation Principle; The Independent Particle Model and the Hartree–Fock Scheme; Slater Determinants; Fock–Dirac Density Matrix; Molecular Orbitals (MOs); Linear Combinations of Atomic Orbitals (LCAOs); Self-Consistent Field (SCF).

The SCF schemes occur today as standard computational programs, and as atomic orbitals $\Phi = \{\phi\mu\}$ one uses either Gaussians or Slater-type orbitals (STOs). Many valuable results are obtained in the applications to atomic, molecular, and solid-state systems, but it is also well known that the accuracy of the Hartree–Fock scheme is limited, since it neglects the *electronic correlation* associated with the interelectronic Coulomb repulsion e^2/r_{ij}.

Configuration-Interaction (CI) Methods; Gaussians.

Today there exists a fairly large number of molecular computer programs, which are built on the use of graphs and the unitary group approach (GUGA).

There is little question that the outcome of the molecular GUGA calculations depends to a large extent on the choice of basis, and that many figures are not as stable (and hence reliable) as we may desire. There is also the theoretical difficulty that a Gaussian basis can never describe the correct asymptotic behavior of the wave function away from the nuclei.

For various reasons, there has recently been a renewed interest in the Slater-type orbitals, in the expansion of the STOs on one atomic center around another by means of the so-called alpha functions, and in their utility in evaluating one-, two-, three-, or four-center integrals of the form.

$$\langle \phi_1 \phi_2 \left| \frac{e^2}{r_{12}} \right| \phi_3 \phi_4 \rangle \tag{2}$$

It is evident that the CI methods involve large-scale computations which would benefit from larger and larger "number crunchers" and particularly from the development from the scalar computers over the vector computers to the supercomputers. It is evident that the form of the matrix eigenvalue problem

$$\mathbf{HC} = \mathbf{EC} \tag{3}$$

should be particularly adapted to the use of vector computers and array processors. Scientists who do not yet have access to the supercomputers may still be able to beat them by using ingenuity and new concepts and ideas in the numerical analysis, but—in my personal opinion—the real future in molecular and solid-state calculations lies in the combination of deeper theoretical insight and new penetrating methods with the power of the supercomputers. The "computer graphics" has also become a new tool for improving our understanding of the numerical outcome of rather abstract calculations.

Semiempirical Methods.

Instead of going into the large-scale calculations associated with the *ab-initio* methods, one may feel inclined to "let experimental experience and insight help the theory"—and this is the basis for the semiempirical methods. They are often conceptually simple, and they are particularly useful in the planning of new experiments.

... In the semiempirical theories of molecular structure, one often starts from the Hartree–Fock scheme and then determines the key quantities occurring in theory from known experimental data and uses the results to predict other experimental data. In this way, the semiempirical theory is essentially a tool for correlating one set of experimental data with another set of experimental data. However, since the Hartree–Fock theory does not take the electronic correlation into proper account, one feels that it should not be able to reflect the true experimental connections.

There have been several attempts to include also the electronic correlation in the semiempirical theories. In 1953, Pariser, Pople, and Parr developed a modification of the semiempirical theory of the conjugated systems, which has played an essential role for the future development. This PPP model is partly based on the principle of the "neglect of differential overlap" (NDO). If $\Phi = \{\phi\mu\}$ is the set of given atomic orbitals, this principle may—in the opinion of the author—be partly justified if one introduces symmetrically orthonormalized atomic orbitals Φ through the relation:

$$\Phi = \Phi \langle \Phi | \Phi \rangle^{-1/2} \tag{4}$$

In fact, this idea has been studied and developed in greater detail. The PPP model has given rise to several different semiempirical schemes bases on the complete neglect of differential overlap (CNDO), the intermediate neglect of differential overlap (INDO), the modified intermediate neglect of differential overlap (MINDO), the neglect of diatomic differential overlap (NDDO), and other variations. We note also the connection between the PPP model for conjugated systems and the Hubbard–Hamiltonian in solid-state theory.

In concluding this subsection, we note that the semiempirical methods have been very successful in providing conceptual ideas and predict experimental results, but that they usually are very hard to extend beyond the area and purpose for which they were originally constructed. There is often the dilemma that the semiempirical parameter values determined from known experimental data do not agree with those calculated by *ab-initio* methods. This phenomenon depends on the fact that, in the original theory, one often has a very large number of parameters to be determined from only a few experimental data, and—in this case—one has to make a decision that certain parameters are going to be of importance, whereas others are negligible. The last part of this decision is usually wrong and leads to the above-mentioned discrepancy.

The semiempirical theories are often developed to such an extent that they give almost perfect agreement between theory and experiment, but they are then usually subject to the peculiar phenomenon that this agreement disappears if one tries to refine the theory. The obvious conclusion is that, if a simple semiempirical theory works well, one should not try to improve it.

In this connection, it is good to remember that a good agreement between theory and experiment is a necessary but *not sufficient* criterion for the correctness of a theory.

Special Approaches: The $X-\alpha$ Method and the Electron Density Functional Theory.

3.   More Recent Developments of Theoretical Methods for Solving the Time-Independent Schrödinger Equation; Resolvent Methods and Their Connection with Partitioning Technique; Weinstein Function; Many-Body Coupled-Cluster Theory (MBCCT); Connection with Perturbation Theory and Rational Approximations; Inner Projections.

4.   Quantum Statistics (Solution of the Liouville Equations):

In quantum mechanics, one deals only with pure states characterized by a wave function ψ, and this means that the results are essentially valid at absolute zero of temperature ($T = 0°\text{K}$). Anyone interested in the properties of materials and how they depend on the temperature T must use a more general form of the quantum theory—sometimes referred to as *quantum statistics*—in which the physical systems are described by system operators Γ having density matrices $\rho = \rho(X|X')$ as their kernels . . . The system operators Γ satisfy the time-dependent *Liouville equation:*

$$-\frac{h}{2\pi i}\frac{\partial \Gamma}{\partial t} = H\Gamma - \Gamma H$$

Liouvillian Superoperator; Superevolution Operator; Time-Independent Liouville Equation; Propagator Methods.

. . . [Propagator methods were] first developed in quantum electrodynamics and in nuclear physics, and [they were] introduced in molecular physics by Linderberg and Öhrn. Today, it is one of the most popular tools in quantum chemistry and solid-state theory.

5.   Some Properties of New Materials.

There are not too many *ab-initio* calculations of the properties of ceramics, glasses, and composites, and—as we have seen at this conference—most theoretical studies are of a semiempirical nature. One very fundamental property of amorphous material was found by Anderson, when he discovered the existence of dispersion-free wave packets, and important interpretation of this phenomenon was given by Mott. The technology of the practical use of such amorphous materials is still an open field of great importance for the future.

Since glasses are being used as shielding material for nuclear waste, the question of their properties and long-time behavior under radiation is of fundamental importance, and some of these problems have been theoretically studied by Hayns.

Other materials of great technological importance are, of course, liquid crystals and spin-glasses, and their properties are rather well theoretically understood.

An important new development deals with the properties of dispersion-free wave packets called "solitons," which occur in many parts of nonlinear dynamics. Of particular importance is the motion of the "kinks" in the long polyacethylene molecules, which consist of two conjugated systems which are put together so that the bond alternation is "out of phase" in one particular point—the kink. As a result, these polyacethylene molecules show interesting conductivity properties. One has great hopes of using organic molecules with such kinks as fundamental elements of molecular-size computers. In general, one can expect that the soliton theory is going to have many important technical applications.

There is little question that our understanding of the properties of new materials and particularly our ability to create "materials by design" are strongly going to depend on the development of new theoretical methods in quantum mechanics and quantum statistics. There are certainly going to be many surprises occurring in "quantum electronics," and many new developments would probably be started if one could establish a successful marriage between the special theory of relativity and quantum theory. It is certainly a sixty-year problem which would be well worth solving.

6.   Concluding Remarks.

The history of a theoretical scientific discovery is often fascinating: the original derivation may occur in a series of papers totaling perhaps 1500 pages of *Physical Review*, but, once the final result is known, the derivation may be shortened to 150 pages. This gives a new insight into the structure of the theory, which leads to a new derivation of only 15 pages, and—at this point—a brilliant young graduate student may appear and produce a derivation of only 1.5 pages. There are many examples of "information explosions" but also of condensations and simplifications which are essential for the "economy of thinking." Without these simplifications, it would be very hard and time consuming for the new graduate students to ever reach "research level."

In the physical sciences, there is an interesting interplay between experiments and theory which often leads to two different types of research—frontier research and axiomatic research. In the frontier research, one is interested in the explanation of new experimental data, and the agreement is so important that one sometimes forgets whether the theory has a true logical basis and is free from contradictions. When Bohr quantized the hydrogen atom in 1913, he was accused of using classical physics on Monday, Wednesday, and Friday and quantum theory on Tuesday, Thursday, and Saturday! The frontier research deals with the great new discoveries, and the derivations are often lengthy and complicated.

Once the new theory has been established, one is usually interested in giving it a true logical structure. In this "axiomatic research," one tries to find out exactly what fundamental assumptions or "axioms" are needed to produce the results desired, and usually this approach is full of surprises. Out of nothing, one can derive nothing—and one can never derive more than is hidden in the assumptions. The axiomatic research leads often to considerable simplifications, and it is hence of great importance for the "economy of thinking." Since the basic elements are often few, it may be easier to remember the derivations, and this is, of course pedagogically important. It is evident that the axiomatic research is of great importance for the teaching of the subject involved.

It is, of course, meaningless to try to separate "frontier research" and "axiomatic research"—they are both needed for the development of a particular field. In the same way, it is meaningless to try to separate "teaching" from "research"—the teaching of today is the basis for the new research to be carried out, and the research of today leads to new insight to be covered by the teaching of tomorrow. Hence, teaching and research go hand in hand and should never be separated.

In the same way, one should never try to separate basic research from applied research. It is exceedingly hard to carry out applications if one does not understand the underlying fundamental principles, and often the technological development goes very slowly, if it is not supported by a great deal of free-wheeling basic research. The most important point is perhaps that unrestricted basic research often leads to most unexpected and important consequences and applications.

INDEX

DATE	CHARGED TO:
8/4/92	W Rhodes